FUNDAMENTALS OF VIROID BIOLOGY

FUNDAMENTALS OF VIROID BIOLOGY

Edited by

CHARITH RAJ ADKAR-PURUSHOTHAMA
RNA Group, Department of Biochemistry and Functional Genomics, Faculty of Medicine and Health Sciences, Applied Cancer Research Pavilion, University of Sherbrooke, Sherbrooke, QC, Canada

TERUO SANO
Faculty of Agriculture and Life Science, Hirosaki University, Hirosaki, Japan

JEAN-PIERRE PERREAULT
RNA Group, Department of Biochemistry and Functional Genomics, Faculty of Medicine and Health Sciences, Applied Cancer Research Pavilion, University of Sherbrooke, Sherbrooke, QC, Canada

SREENIVASA MARIKUNTE YANJARAPPA
Department of Studies in Microbiology, University of Mysore, Mysuru, India

FRANCESCO DI SERIO
Consiglio Nazionale delle Ricerche (CNR), Istituto per la Protezione Sostenibile delle Piante (IPSP), Bari, Italy

JOSÉ-ANTONIO DARÒS
Instituto de Biología Molecular y Celular de Plantas (Consejo Superior de Investigaciones Científicas-Universitat Politècnica de València), Valencia, Spain

ACADEMIC PRESS
An imprint of Elsevier

ELSEVIER

Academic Press is an imprint of Elsevier
125 London Wall, London EC2Y 5AS, United Kingdom
525 B Street, Suite 1650, San Diego, CA 92101, United States
50 Hampshire Street, 5th Floor, Cambridge, MA 02139, United States
The Boulevard, Langford Lane, Kidlington, Oxford OX5 1GB, United Kingdom

Notices
Knowledge and best practice in this field are constantly changing. As new research and experience broaden our
understanding, changes in research methods, professional practices, or medical treatment may become
necessary.

Practitioners and researchers must always rely on their own experience and knowledge in evaluating and using
any information, methods, compounds, or experiments described herein. In using such information or methods
they should be mindful of their own safety and the safety of others, including parties for whom they have a
professional responsibility.

To the fullest extent of the law, neither the Publisher nor the authors, contributors, or editors, assume any liability
for any injury and/or damage to persons or property as a matter of products liability, negligence or otherwise, or
from any use or operation of any methods, products, instructions, or ideas contained in the material herein.

ISBN: 978-0-323-99688-4

For information on all Academic Press publications
visit our website at https://www.elsevier.com/books-and-journals

Publisher: Stacy Masucci
Acquisitions Editor: Kattie Washington
Editorial Project Manager: Teddy A. Lewis
Production Project Manager: Jayadivya Saiprasad
Cover Designer: Mark Rogers

Typeset by STRAIVE, India

Working together
to grow libraries in
developing countries

www.elsevier.com • www.bookaid.org

Contents

A

Introduction

1. Milestones in viroid research

Robert A. Owens

2. Viroid taxonomy

Francesco Di Serio and Michela Chiumenti

3. Structure of viroids

Jean-Pierre Perreault, François Bolduc, and
Charith Raj Adkar-Purushothama

4. Replication and movement of viroids in host plants

Beatriz Navarro, Gustavo Gómez, and Vicente Pallás

B

Prospecting for viroid

5. Viroids diseases and its distribution in Asia

G. Vadamalai, Charith Raj Adkar-Purushothama, S.S. Thanarajoo,
Y. Iftikhar, B. Shruthi, Sreenivasa Marikunte Yanjarappa,
and Teruo Sano

6. Viroid-associated plant diseases in Europe

Dijana Škorić

7. Naturally occurring viroid diseases of economically important plants in Africa

Amine Elleuch and Imen Hamdi

8. Finding the coconut cadang-cadang and tinangaja viroids, naturally occurring pathogens of tropical monocotyledons of Oceania

J.W. Randles, C.A. Cueto, G. Vadamalai, and D. Hanold

9. Viroids and their distribution in North America

Xianzhou Nie

10. Viroid-associated plant diseases in South America

Nicola Fiore and M. Francisca Beltrán

C
Viroid pathogenesis and viroid-host interaction

11. Viroid pathogenicity

Charith Raj Adkar-Purushothama, Francesco Di Serio,
Jean-Pierre Perreault, and Teruo Sano

12. Viroids and protein translation

Purificación Lisón, Francisco Vázquez-Prol, Irene Bardani,
Ismael Rodrigo, Nikoleta Kryovrysanaki, and Kriton Kalantidis

13. Viroid infection and host epigenetic alterations

Joan Marquez-Molins, German Martinez, Vicente Pallás, and Gustavo Gómez

14. Transcriptomic analyses provide insights into plant-viroid interactions

Jernej Jakše, Ying Wang, and Jaroslav Matoušek

15. Viroid-induced RNA silencing and its secondary effect on the host transcriptome

Charith Raj Adkar-Purushothama, Jean-Pierre Perreault, and Teruo Sano

16. Detection of viroids

Zhixiang Zhang and Shifang Li

17. Viroid disease control and strategies

Rosemarie W. Hammond

18. Policies, regulations, and production of viroid-free propagative plant materials for sustainable agriculture

Irene Lavagi-Craddock, Scott Harper, Robert Krueger, Paulina Quijia-
Lamiña, and Georgios Vidalakis

19. Bioinformatic approaches for the identification and discovery of viroid-like genomes

Maria José López-Galiano and Marcos de la Peña

20. Contributions of viroid research to methods for RNA purification, diagnostics, and secondary structure prediction

Gerhard Steger and Detlev Riesner

21. Future perspectives in viroid research

José-Antonio Daròs

Contributors

Charith Raj Adkar-Purushothama RNA Group, Department of Biochemistry and Functional Genomics, Faculty of Medicine and Health Sciences, Applied Cancer Research Pavilion, University of Sherbrooke, Sherbrooke, QC, Canada

Irene Bardani Department of Biology, University of Crete, Heraklion, Greece

M. Francisca Beltrán University of Chile, Faculty of Agricultural Sciences, Department of Plant Health, Santiago, Chile

François Bolduc RNA Group, Department of Biochemistry and Functional Genomics, Faculty of Medicine and Health Sciences, Applied Cancer Research Pavilion, University of Sherbrooke, Sherbrooke, QC, Canada

Michela Chiumenti Consiglio Nazionale delle Ricerche (CNR), Istituto per la Protezione Sostenibile delle Piante (IPSP), Bari, Italy

C.A. Cueto Philippine Coconut Authority-Albay Research Center, Guinobatan, Albay, Philippines

José-Antonio Daròs Instituto de Biología Molecular y Celular de Plantas (Consejo Superior de Investigaciones Científicas-Universitat Politècnica de València), Valencia, Spain

Marcos de la Peña Institute of Molecular and Cellular Biology of Plants, Politechnic University of Valencia-CSIC, Valencia, Spain

Francesco Di Serio Consiglio Nazionale delle Ricerche (CNR), Istituto per la Protezione Sostenibile delle Piante (IPSP), Bari, Italy

Amine Elleuch Laboratoire de Biotechnologie Végétale Appliquée à l'amélioration des plantes, Faculté des Sciences de Sfax, Université de Sfax, Sfax, Tunisia

Nicola Fiore University of Chile, Faculty of Agricultural Sciences, Department of Plant Health, Santiago, Chile

Gustavo Gómez Institute for Integrative Systems Biology (I2SysBio), Consejo Superior de Investigaciones Científicas (CSIC)—Universitat de València (UV), Paterna, Spain

Imen Hamdi Laboratoire de Protection des Végétaux, Institut National de la Recherche Agronomique de Tunis, Université de Carthage, Tunis, Tunisia

Rosemarie W. Hammond USDA ARS Molecular Plant Pathology Laboratory, Beltsville, MD, United States

D. Hanold School of Agriculture, Food and Wine, The University of Adelaide, Adelaide, SA, Australia

Scott Harper Clean Plant Center Northwest, Washington State University, Prosser, WA, United States

Y. Iftikhar Department of Plant Pathology, College of Agriculture, University of Sargodha, Sargodha, Pakistan

Jernej Jakše Department of Agronomy, Biotechnical Faculty, University of Ljubljana, Ljubljana, Slovenia

Kriton Kalantidis Department of Biology, University of Crete; Institute of Molecular Biology and Biotechnology, Foundation for Research and Technology, Heraklion, Greece

Robert Krueger USDA-ARS National Clonal Germplasm Repository for Citrus and Dates, Riverside, CA, United States

Nikoleta Kryovrysanaki   Institute of Molecular Biology and Biotechnology, Foundation for Research and Technology, Heraklion, Greece

Irene Lavagi-Craddock   Citrus Clonal Protection Program and the University of California, Riverside, CA, United States

Shifang Li   Institute of Plant Protection, Chinese Academy of Agricultural Sciences, Beijing, China

Purificación Lisón   Institute for Plant Molecular and Cellular Biology, Universitat Politècnica de València-Consejo Superior de Investigaciones Científicas, Valencia, Spain

Maria José López-Galiano   Institute of Molecular and Cellular Biology of Plants, Politechnic University of Valencia-CSIC; Department of Genetics, University of Valencia, Burjassot, Valencia, Spain

Joan Marquez-Molins   Institute for Integrative Systems Biology (I2SysBio), Consejo Superior de Investigaciones Científicas (CSIC)— Universitat de València (UV), Paterna, Spain; Department of Plant Biology, Uppsala BioCenter, Swedish University of Agricultural Sciences and Linnean Center for Plant Biology, Uppsala, Sweden

German Martinez   Department of Plant Biology, Uppsala BioCenter, Swedish University of Agricultural Sciences and Linnean Center for Plant Biology, Uppsala, Sweden

Jaroslav Matoušek   Biology Centre of the Czech Academy of Sciences, Department of Molecular Genetics, Institute of Plant Molecular Biology, České Budějovice, Czech Republic

Beatriz Navarro   Institute for Sustainable Plant Protection, National Research Council, Bari, Italy

Xianzhou Nie   Fredericton Research and Development Centre, Agriculture and Agri-Food Canada, Fredericton, NB, Canada

Robert A. Owens   Molecular Plant Pathology Laboratory, Beltsville Agricultural Research Center (retired), Beltsville, MD, United States

Vicente Pallás   Instituto de Biología Molecular y Celular de Plantas (IBMCP), Consejo Superior de Investigaciones Científicas (CSIC)— Universitat Politècnica de València (UPV), Valencia, Spain

Jean-Pierre Perreault   RNA Group, Department of Biochemistry and Functional Genomics, Faculty of Medicine and Health Sciences, Applied Cancer Research Pavilion, University of Sherbrooke, Sherbrooke, QC, Canada

Paulina Quijia-Lamiña   Citrus Clonal Protection Program and the University of California, Riverside, CA, United States

J.W. Randles   School of Agriculture, Food and Wine, The University of Adelaide, Adelaide, SA, Australia

Detlev Riesner   Institut für Physikalische Biologie, Heinrich Heine University Düsseldorf, Universitätsstraße 1, Düsseldorf, Germany

Ismael Rodrigo   Institute for Plant Molecular and Cellular Biology, Universitat Politècnica de València-Consejo Superior de Investigaciones Científicas, Valencia, Spain

Teruo Sano   Faculty of Agriculture and Life Science, Hirosaki University, Hirosaki, Japan

B. Shruthi   Department of Studies in Microbiology, University of Mysore, Mysuru, India

Dijana Škorić   University of Zagreb, Faculty of Science, Department of Biology, Zagreb, Croatia

Gerhard Steger   Institut für Physikalische Biologie, Heinrich Heine University Düsseldorf, Universitätsstraße 1, Düsseldorf, Germany

S.S. Thanarajoo   CABI South East Asia, Serdang, Selangor, Malaysia

G. Vadamalai   Department of Plant Protection, Faculty of Agriculture, Institute of Plantation Studies, Universiti Putra Malaysia, Serdang, Selangor, Malaysia

Francisco Vázquez-Prol   Institute for Plant Molecular and Cellular Biology,

Universitat Politècnica de València-Consejo Superior de Investigaciones Científicas, Valencia, Spain

Georgios Vidalakis Citrus Clonal Protection Program and the University of California, Riverside, CA, United States

Ying Wang Plant Pathology Department, University of Florida, Gainesville, FL, United States

Sreenivasa Marikunte Yanjarappa Department of Studies in Microbiology, University of Mysore, Mysuru, India

Zhixiang Zhang Institute of Plant Protection, Chinese Academy of Agricultural Sciences, Beijing, China

Foreword

An oft-cited quotation from physicist Erwin Schrodinger reminds scientists that "The task is not to see what has never been seen before, but to think what has never been thought before about what you see everyday." Diseases associated with potato spindle tuber and citrus exocortis viroids were long believed to be caused by conventional plant viruses, and it was only the creative application of then state-of-the-art technology that, in 1971, identified the true nature of their causal agents. The ability of these small, circular, noncoding RNA molecules to replicate autonomously in susceptible plant hosts immediately focused attention on their biological, biochemical, and evolutionary significance.

Beginning in 1979 with T.O. Diener's *Viroids and Viroid Diseases*, a series of monographs have documented advances in our understanding of how viroids replicate, move from cell to cell, and induce disease. This publication is the first book in the field designed to be used as both course material for students and a reference manual for researchers and the scientific community. Each chapter starts with a graphical summary of the chapter followed by an introduction and main content. To help researchers and students, every chapter contains an experimental design and study questions. The prospective and future implications sections in each chapter provide readers with an idea of the research questions that need to be addressed and the corresponding effects or consequences. All chapters end with a "Further reading" section that references sources the author has deemed valuable to readers seeking additional information or context about the research problem.

My own introduction to viroid research, in contrast, owed much to serendipity. As a graduate student in early 1972, I happened to pick up the issue of *Virology* in which T.O. Diener reported the discovery of potato spindle tuber viroid. The significance of this discovery was immediately apparent, and little more than 3 years later, I found myself in Beltsville working with Dr. Diener just as viroid research began to expand rapidly. New thoughts were seemingly everywhere in those early days of viroid research. For those now entering the field at a time when the knowledge base is so much larger and the experimental tools so much more powerful, this book will be particularly helpful in choosing the most promising questions to investigate.

As this volume was going to press, I received word that Dr. Diener had recently passed away at the age of 102. Ted was a truly remarkable scientist whose many achievements are widely recognized. Those not fortunate enough to have met him or heard him speak need only read his 1971 description of the discovery of potato spindle tuber viroid to appreciate the breadth of knowledge and creative vision that he brought to viroid research.

R.A. Owens
(*Retired*), Molecular Plant Pathology Laboratory, Beltsville Agricultural Research Center, Beltsville, MD, United States

Preface

Theodor O. Diener, in the year 1971, coined the term "viroid" when he identified the causal agent of potato spindle tuber disease, the smallest RNA replicon with links between "living" and "nonliving." Since the discovery of the first viroid, many researchers, including master's and PhD students, across the globe have contributed to the field of viroid biology, including viroid diseases, classification, pathogenicity, interaction with hosts, crop protection, biotechnology, and molecular evolution. However, most of this research has been and is being performed in a few laboratories, indicating viroid research has not yet achieved a worldwide reach. When we started to think more and more about this problem, we realized we had failed to attract young researchers to the field of viroids. Recent observations of the presence of long circular nonprotein-coding RNAs in mammalian cells and their role in modulating microRNAs, as well as the identification of viroid-like RNAs in fungi, may further widen the scope of the viroid host range. At this point, we thought of bringing out a student-friendly handbook that presents the fundamentals of viroid biology. Hence, we contacted the researchers who have made substantial contributions to the field of viroids and requested their contribution to their favorite topic. Therefore, as you read this book, you will understand that each chapter is written by authors who have dedicated themselves to viroid research for many years. We are wholeheartedly thankful to all the contributors for their patience, cooperation, and collaboration and for sharing their vast and excellent scientific expertise on viroids, especially in the midst of the uncertainty surrounding the SARS-CoV-2/COVID-19 pandemic. We are also indebted to the Elsevier staff for supporting the publication of this book. We hope this book serves as a "one-stop" resource for researchers, teachers, and students who want to know more about the biology of viroids and nonprotein-coding RNAs.

Charith Raj Adkar-Purushothama
Teruo Sano
Jean-Pierre Perreault
Sreenivasa Marikunte Yanjarappa
Francesco Di Serio
José-Antonio Darós

Introduction

A new entrant into the field of viroidology might be a scientist reading a research paper describes a new viroid. The host plant, infectivity, and pathogenicity of the viroid; its primary and secondary structure; and its classification based on a comparison of its nucleotide sequence with known viroids may be described. Each of these aspects is often a complex undertaking, and the authors of the chapters in this comprehensive book have been asked to explain their topic in sufficient detail for newcomers to understand the history, recent advances in this field, and the background to current thinking on the evolution of these autonomously replicating RNA molecules.

In his Foreword to *Viroids and Satellites* (2017), Theodor O. Diener stated that the discovery of viroids in 1971 ushered in the "third major extension in the history of the biosphere." His pioneering experiments shocked both virologists and molecular biologists. The science of virology had developed from the discovery in the 1890s that some diseases could be induced by agents smaller than bacteria. Viruses followed the "central dogma" of the flow of genetic information, competing with and parasitizing the host's transcription and translation processes for virus-directed replication. The ability of naked small RNA without a translatable genome to induce disease proved the existence of a subviral world where the host's RNA metabolism was parasitized for viroid replication. More than 30 viroids have been shown to cause plant diseases that previously were of unknown etiology. Although

the causal prions of animal subacute spongiform encephalopathies have also been described as subviral pathogens, there are no similarities in the pathologies or biologies of the viroids and prions.

Viroids concern both the agricultural and the biochemical sciences. Plant pathology originated from the need to control major epidemics of plant disease, even famine, due to any of a wide range of microscopic plant parasites. While pathologists take the holistic view that plant disease results from interactions between a pathogen, the host, the environment, and mode of spread, diagnosis of the cause is the first requirement for achieving control. For plant virus research, the rules of causation attributed to Henle, Koch, and Loeffler in the 19th century for human diseases have been adapted to include symptomatology, host range, and bioassays as the first steps in characterizing unknown viruses by various differential centrifugation techniques and electron microscopy. When the spindle tuber disease of the potato was subjected to standard virus purification procedures without any virions being found, it was thought in 1967 that a naked virus nucleic acid could be the cause. It was Diener's skills in basic plant virus and nucleic acid research that led to his first publication recognizing that the pathogen was a small unencapsidated autonomously replicating circular RNA, the potato spindle tuber viroid (PSTVd). His dedicated research uncovered a number of biological and morphological properties of this viroid. It was the first viroid to be sequenced by molecular biologists in Germany, opening up an era of structure-function analyses related to biology and viroid classification.

Diener's discovery of the cause of the economically important potato spindle tuber disease led rapidly to the identification of the causes of the economically damaging citrus exocortis, coconut cadang-cadang, hop stunt, and avocado sunblotch diseases. Viroid diseases are now known to affect orchards, plantations, greenhouses, and plant nurseries around the world, and advanced molecular diagnostic methods are used for the application of quarantine procedures to international plant germplasm movement. Little is known about the distribution of viroids in natural ecosystems, including those which have provided germplasm for commercial plantings, but the latest molecular diagnostic methods have the potential to find viroids and viroid-like molecules even where they do not cause overt disease. The signatures of viroids are now being found in some species without symptoms of disease. Relationships between nucleotide sequence variation and disease severity are being explored in attempts to discover how viroids interact with the host. Due to the high mutability of viroids and the accumulation of quasi-species during infection, populations of variants change during the infection cycle. High-throughput sequencing is the most recent tool for studying viroid population dynamics.

This volume emphasizes the biology of viroids, including their history, taxonomy, structure, replication, global distribution, host-viroid interactions, perceptions of their origins, and how they may have evolved. Future directions for viroid research are suggested, showing that a multitude of questions are yet to be addressed in the field and laboratory. Readers are introduced to a team of authors representing many of the laboratories that have collaborated in viroid research over the last five decades.

John W. Randles
School of Agriculture, Food and Wine,
The University of Adelaide, SA, Australia

Introduction

1

Milestones in viroid research

Robert A. Owens

Molecular Plant Pathology Laboratory, Beltsville Agricultural Research Center (retired), Beltsville, MD, United States

Graphical representation

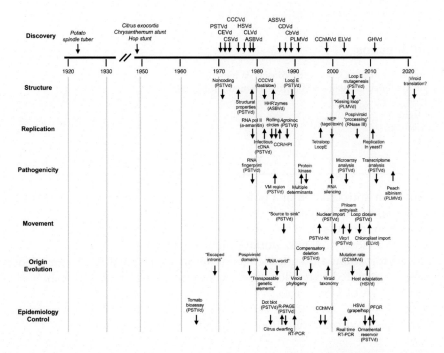

Viroid research milestones. For clarity, major advances in different areas of research (indicated by *vertical arrows*) have been arranged along separate timelines. Dates for the first reports of selected viroid diseases and identification of representative members of the eight currently recognized viroid genera are indicated at the top of the figure.

Abbreviations

HDV	hepatitis delta virus
HP I	secondary hairpin I
HP II	secondary hairpin II
PFOR	progressive filtering of overlapping small RNAs
RT-PCR	reverse transcription polymerase chain reaction
SHAPE	selective 2′-hydroxyl acylation analyzed by primer extension

Viroid species

ADFVd	apple dimple fruit viroid
ASBVd	avocado sunblotch viroid
ASSVd	apple scar skin viroid
CbVd	coleus blumei viroid
CCCVd	coconut cadang-cadang viroid
CChMVd	chrysanthemum chlorotic mottle viroid
CDVd	citrus dwarfing viroid
CEVd	citrus exocortis viroid
CLVd	columnea latent viroid
ELVd	eggplant latent viroid
GHVd	grapevine hammerhead viroid
HSVd	hop stunt viroid
PLMVd	peach latent mosaic viroid
PSTVd	potato spindle tuber viroid

Pospiviroid domains

C	central domain contains CCR (central conserved region)
P	pathogenicity domain contains VM (virulence-modulating) region
TL	terminal left
TR	terminal right
V	variable domain

Definitions

Agroinoculation: introduction of a potentially infectious viroid or viral cDNA into plant cells via the Ti plasmid of *A. tumefaciens*.

Cross protection: process in which exposure to a mild virus or viroid isolates induces resistance against a closely related, more virulent isolate.

Dot blot hybridization: a diagnostic technique in which total RNA extracted from the plant to be tested is spotted on a nylon or nitrocellulose membrane and then hybridized with a viroid-specific RNA or DNA probe.

Indicator host: plants specifically chosen for experimental studies because they either support high levels of viroid replication or exhibit strong symptoms.

RNA structure

Primary: linear sequence of ribonucleotides (A, C, G, and U) linked by phosphodiester bonds.

Secondary: RNA molecules contain both single- and double-stranded regions. Base pairing between complementary regions creates helices, bulged nucleotides, internal loops, and junctions.

Tertiary: three-dimensional arrangement of helical duplexes and other secondary structural components that is stabilized by long-range interactions.

Isosteric base pairs: base pairs in which the relative positions and distances between the two C1′ carbon atoms are very similar. Watson-Crick pairs are only one of 12 different families of base pairs containing at least two H bonds between atoms located on the three edges of nucleotide bases.

Loop E motif: type of internal loop first identified in 5S rRNA that contains several non-Watson-Crick base pairs.

Tetraloops: hairpin loops containing four unpaired nucleotides that cap many RNA hélices.

Ribozymes: RNA molecules able to catalyze specific biochemical reactions (e.g., self-cleavage) similar to the action of protein enzymes.

Rolling circle replication: the process of unidirectional nucleic acid replication that can rapidly synthesize multiple copies of a circular RNA or DNA molecule.

Temperature gradient gel electrophoresis: a specialized form of polyacrylamide gel electrophoresis used to study specific structural features of nucleic acids or nucleic acid-protein complexes.

Viroid chimera: viroid genomes portions of which are similar/identical to sequences in two or more other viroids.

Learning objectives

- Objective 1: Understand the fundamental differences between viroids and plant RNA viruses
- Objective 2: Understand how viroids have been used to examine the relationship between RNA structure and biological function(s)
- Objective 3: Understand the role of common agricultural practices such as vegetative propagation in the origin/spread of viroid diseases

Fundamental introduction

When announcing the discovery of the first viroid in 1971 T.O. Diener suggested that the "spindle tuber" disease caused by potato spindle tuber viroid (PSTVd) might reflect its ability

A. Introduction

to act as an "abnormal regulatory RNA." Fifty years later, much evidence has accumulated to support this hypothesis. This historical overview briefly describes the most significant findings, grouping them according to the research area. PSTVd was the first pathogen infecting a eukaryotic host to have its genome completely sequenced. Later advances include demonstrations of (i) the importance of viroid tertiary structure for replication, (ii) the role of RNA silencing in both pathogenicity and replication, and (iii) the role of specific sequence motifs in regulating intracellular viroid movement as well as movement across specific tissue boundaries. The possibility that viroids originated in the precellular "RNA world" remains a viable hypothesis, and new information about all aspects of viroid-host interaction continues to accumulate.

Viruses (especially bacteriophages) played a crucial role in the development of modern molecular biology. For early plant molecular biologists the small RNA genomes of tobacco mosaic virus (TMV), brome mosaic virus (BMV), and cowpea mosaic virus (CPMV) provided tools to probe many aspects of plant gene expression. The 1971 discovery of potato spindle tuber viroid (PSTVd) by T.O. Diener revealed a more extreme form of parasitism and, therefore, an even more powerful tool with which to examine the biochemical capacities of their host cells.

Initial characterization of PSTVd (Diener, 1971; Singh and Clark, 1971) and CEVd (Semancik and Weathers, 1972) revealed that their genomes were approximately 10-fold smaller than those of the smallest known RNA viruses. Furthermore, the lack of a protective capsid frees their genomes from the need to encode the corresponding structural protein(s). In announcing the discovery of viroids Diener noted that "The demonstration of a low molecular weight RNA that replicates in a variety of host species apparently uninfected by another agent entails a number of important implications for virology, molecular biology, and genetics." These implications include (i) the nature of the molecular signals allowing host enzymes to accept viroids as templates for replication, (ii) the possibility that such pathways might also be operative in uninfected cells, (iii) the mechanism(s) by which viroids induce disease in the infected host plant, (iv) the factors limiting viroids to higher plants, and (v) the question of viroid origin (Diener, 1987). Studies carried out over the past 50 years have addressed each of these questions.

Fig. 1 presents a timeline of viroid research milestones that begins with descriptions of the first two diseases later shown to be viroid-induced; namely, "potato spindle tuber" disease in the early 1920s (Schultz and Folsom, 1923) and "citrus exocortis" disease in the late 1940s (Fawcett and Klotz, 1948). At a time when sucrose density gradient centrifugation and polyacrylamide gel electrophoresis were the states of the art in plant virology neither potato nor citrus was particularly well-suited for molecular studies; thus, identification of more-amenable indicator hosts like tomato (PSTVd), Etrog citron, or *Gynura aurantiaca* (CEVd) in the 1960s played a key role in establishing the true nature of the causal agent. Shortly after the discoveries of PSTVd (Diener, 1971) and CEVd (Semancik and Weathers, 1972) several other diseases affecting vegetatively propagated crops like chrysanthemum and hops were also shown to be caused by viroids. More than 30 different species of viroids are now known (10th ICTV Report; https://talk.ictvonline.org/ictv-reports/ictv_online_report/subviral-agents/w/viroids), and a combination of modern high throughput RNA sequencing and

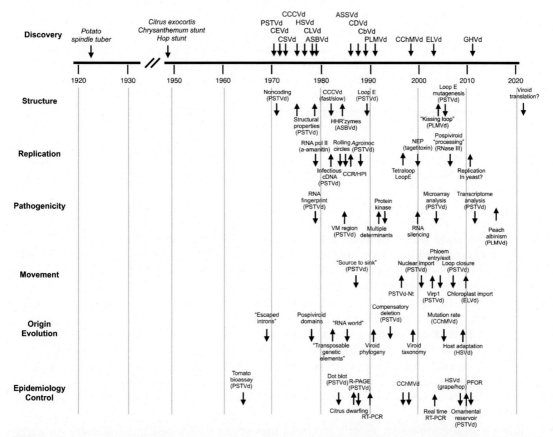

FIG. 1 Viroid research milestones. For clarity, major advances in different areas of research (indicated by *vertical arrows*) have been arranged along separate timelines. Dates for the first reports of selected viroid diseases and identification of representative members of the eight currently recognized viroid genera are indicated at the top of the figure.

bioinformatic analysis has revealed the presence of several novel viroids in asymptomatic plants (e.g., Wu et al., 2012).

By the early 1980s viroid genomes were known to be covalently closed, circular molecules (the first such RNAs to be described), the complete nucleotide sequences of PSTVd and several other viroids had been determined, and physical-chemical studies exploring the unusual structural properties of PSTVd were underway. Studies from several laboratories had shown that viroids replicate via a rolling circle mechanism, and demonstration of the ability of a greater-than-full-length cDNA copy of PSTVd to initiate infection in tomato seedlings (Cress et al., 1983) allowed recombinant DNA technology to be applied to studies of viroid replication, movement within the host, and disease induction.

As the molecular details of these processes became better understood, viroids were seen to form two natural groupings (or taxonomic families). PSTVd, CEVd, and other members of the more numerous group have an unbranched, rod-like structure with several conserved sequence motifs and replicate in the nucleus; members of the second group which includes avocado sunblotch viroid (ASBVd) are structurally more diverse and replicate in the chloroplast. Importantly, multimeric forms of ASBVd of both polarities cleave spontaneously due to the presence of novel "hammerhead" ribozymes. In 1989 Diener proposed that viroids and viroid-like satellite RNAs may represent "relics of precellular evolution" with origins in the RNA world. Two years later additional support for this hypothesis was presented by Elena et al. (1991); namely, the results of the phylogenetic analysis which indicated a monophyletic origin for viroids, viroid-like satellite RNAs, and the viroid-like domain of hepatitis delta virus (HDV) RNA.

Even before the publication of the first complete viroid nucleotide sequence (Gross et al., 1978), RNA fingerprinting of selected PSTVd isolates had shown that small changes in sequence could lead to dramatic differences in symptom expression. In 1985 Sänger's laboratory proposed the first molecular model for viroid pathogenicity. Nucleotides within the so-called "virulence modulating" region of PSTVd were proposed to modulate the binding- and hence the competition potential of the genomic RNA for an unknown host factor(s). Genetic analysis of a series of novel viroid chimeras (Sano et al., 1992) revealed the presence of multiple (i.e., three) pathogenicity determinants. Following several later reports that viroid infection induces RNA silencing (e.g., Itaya et al., 2001) there was a pronounced shift in the focus of studies of viroid pathogenicity. Rather than continuing to search for *direct interactions* between viroid genomic RNAs and host protein(s) that could explain symptom induction, several groups began to look for possible *indirect interactions* between the viroid and host genomes via viroid small RNA-mediated RNA silencing.

Interest in the molecular details of viroid movement developed more slowly. An early study (Palukaitis, 1987) used dot-blot hybridization to demonstrate that PSTVd (like most plant viruses) moves rapidly from inoculated leaf (source) to actively growing tissues (sink) via the phloem. More recently, site-directed mutagenesis identified a number of PSTVd variants that are able to replicate in single cells but unable to spread systemically in whole plants. In situ hybridization studies have shown that several naturally-occurring PSTVd variants are unable to cross specific cellular boundaries in the leaf. Other studies have identified structural motifs that allow viroids to leave the cytoplasm and enter either the nucleus (PSTVd) or chloroplast (ELVd) before replication.

When the first viroids were discovered in the early 1970s, bioassays using suitable indicator hosts provided the primary tool available to detect new viroids. The introduction of polyacrylamide gel electrophoresis and, later, dot-blot hybridization assays greatly facilitated efforts to control the increasing number of viroid diseases affecting crop species. RT-PCR analysis and high throughput RNA sequencing have now largely replaced these first-generation molecular techniques. In this process much has been learned about the epidemiology and, in some cases, the origin of the viroid diseases affecting economically important crop species (Fig. 2).

FIG. 2   Participants in a 1983 symposium on "Subviral pathogens of plants and animals: viroids and prions" held in Bellagio, Italy, and sponsored by the Rockefeller Foundation. From the left in the front row are HJ Gross, E Shikata, and HL Sänger. Beginning third from the left in the rear row are D Peters, RIB Francki, TO Diener, L Salazar, F Solymosy, G Boccardo, JW Randles, RA Owens, RH Symons, Tien Po, and A Branch. DC Gajdusek (second from left, rear row) and SB Prusiner (second from the right, front row) were awarded the Nobel Prize in Physiology and Medicine in 1976 and 1997 respectively for their studies of prion diseases.

Viroid conformation(s)

Fifty years ago, sucrose density gradient centrifugation and polyacrylamide gel electrophoresis were the methods most commonly used to fractionate RNAs extracted from diseased plants, and bioassays on a suitable indicator host were the most sensitive method available to detect new viroids. Diener (2003) vividly describes how these techniques were combined to define the essential properties of the first viroid (PSTVd) without directly visualizing its small RNA genome. Soon thereafter, the viroid-specific band on the polyacrylamide gel had been identified, and processing kilogram amounts of infected leaf tissue yielded the milligram amounts of purified viroid required for more detailed molecular studies. Thus, PSTVd was shown to lack mRNA activity in a cell-free wheat germ translation system (Davies et al., 1971), and electron microscopy of formaldehyde-denatured PSTVd revealed the unexpected presence of circular molecules (Sogo et al., 1973). Also, a comparison of their structural properties revealed that PSTVd, CEVd, and cucumber pale fruit viroid (CPFVd) were all single-stranded, covalently closed circular molecules with a highly base-paired, rod-like secondary structure (Sänger et al., 1976).

The next advance in viroid molecular biology was featured on the cover of the journal Nature. Determination of the complete nucleotide sequence and secondary structure of PSTVd by Gross et al. (1978) was a landmark event, the first genome of a eukaryotic pathogen to be characterized in such detail. DNA sequencing was still in its infancy in the mid-late 1970s, and PSTVd had to be *directly sequenced* using fragments of the genomic RNA that were produced by nuclease digestion in vitro and then radioactively-labeled. Shortly thereafter, Riesner et al. (1979) used a variety of biophysical techniques to show that all the base pairs in the rod-like conformation dissociate in one highly cooperative main transition, thereby forming three very stable "secondary hairpins". Two of these hairpins can also be formed by other pospiviroids, and the roles of HP I and HP II in replication are discussed in the next section.

Over the next several years complete nucleotide sequences were determined for a number of additional viroids. Comparison of seven different viroid genomes led Keese and Symons (1985) to propose that PSTVd and related viroids contain *five structural/functional domains*. These include (i) a central domain whose conserved central region (CCR) is capable of forming alternative structures that may regulate the viroid replication cycle, (ii) a domain associated with pathogenicity, (iii) a variable domain with high sequence variability, and (iv and v) two terminal domains that are interchangeable between viroids. These authors also point to the partial duplications that appear de novo in the genome of coconut cadang cadang viroid (CCCVd) during the infection process (Haseloff et al., 1982) as evidence for the importance of RNA rearrangements in viroid evolution. The possibility that similar rearrangements may have led to the repeated exchange of information between RNA pathogens and other RNA molecules in the host cell has attracted widespread attention (see the final section on "Origin/evolution"). ASBVd, the one viroid whose rod-like secondary structure lacks identifiable domains, was later shown to replicate in the chloroplasts rather than the nucleus of infected cells.

RNA folding programs (e.g., Zuker and Stiegler, 1981) have been widely used to investigate viroid structure/function relationships. In most cases, the lowest-free-energy secondary structure predicted in silico is reasonably consistent with the results of chemical or enzymatic probing in vitro. Probing results obtained using a technique known as SHAPE (selective 2'-hydroxyl acylation analyzed by primer extension) which interrogates the RNA backbone at single nucleotide resolution are of particular interest. For example, López-Carrasco and Flores (2017) report that certain nucleotides in the conserved central region of PSTVd (including several within its so-called "loop E motif") are more SHAPE-reactive in vitro than in vivo. Such localized differences could reflect interaction with proteins involved in replication (see next section). The overall similarity in SHAPE reactivities, however, supports the long-standing view that PSTVd accumulates *in planta* as a "naked" RNA rather than in a tight complex with host proteins.

First identified by Branch et al. (1985) on the basis of its susceptibility to UV crosslinking, several lines of evidence now indicate that the loop E motif plays a key role in regulating PSTVd replication (see "Replication" section). Many other RNAs including eukaryotic 5S rRNA and hepatitis delta virus (HDV) genomic RNA contain similar loop E motif(s), and their tertiary structures have been studied in great detail. The comprehensive mutational analysis of the PSTVd loop E motif carried out by Zhong et al. (2006) using isostericity matrix analysis of non-Watson-Crick base pairing to rationalize mutagenesis followed by systematic in vitro and in vivo functional assays of the resulting mutants illustrates the ability of such studies to elucidate complex structure-function relationships.

Unlike the rod-like secondary structure of PSTVd, the computer-predicted conformations of peach latent mosaic viroid (PLMVd) and chrysanthemum chlorotic mottle viroid (CChMVd) are branched. Co-variation analysis of naturally-occurring CChMVd variants indicates a very similar conformation in vivo. Furthermore, this pattern of natural variability also suggests that the branched conformation of CChMVd may be stabilized by a "kissing-loop" interaction similar to one previously proposed for PLMVd (Gago et al., 2005). Using a combination of site-directed mutagenesis, infectivity assays, and PAGE analysis under denaturing and nondenaturing conditions these authors showed that the kissing-loop interaction determines proper folding of CChMVd RNA in vitro. The presence of similar kissing-loop interactions in two hammerhead viroids exhibiting a low overall degree of sequence similarity suggests that this tertiary interaction facilitates the adoption and stabilization of a compact structure critical for viroid viability in vivo. Temperature gradient gel electrophoresis (Riesner et al., 1989) provides a powerful tool with which to detect just these sorts of structural changes in co-existing RNA structures.

Replication

All viroids replicate via a "rolling circle" mechanism, either an *asymmetric* mechanism for PSTVd and related viroids that replicate in the nucleus or a *symmetric* mechanism for ribozyme-containing viroids like PLMVd that replicate in the chloroplast. While an early study (Mühlbach and Sänger, 1979) revealed the involvement of α-amanitin sensitive RNA polymerase II in pospiviroid replication, the identity of the nuclear-encoded polymerase that catalyzes the replication of ASBVd and related viroids (Navarro et al., 2000) and the role of a type III RNase in the in vivo cleavage of multimeric (+)-strand pospiviroid RNAs (Gas et al., 2007) were established much more recently. Many important questions await future investigation.

In the early 1980s publications from several different laboratories described the characterization of replicative intermediates isolated from viroid-infected leaf tissue. The Robertson lab at Rockefeller University was the first to arrange various PSTVd-related RNAs into a rolling circle mechanism (Branch et al., 1981) and later proposed a symmetric mechanism to account for the replication of viroids and other small infectious RNAs (Branch and Robertson, 1984). The presence of comparatively large amounts of (+)-strand genomic RNA makes detection of low levels of a putative monomeric circular form of viroid (−)-strand RNA technically difficult (Hutchins et al., 1985), and fingerprint analysis of purified PSTVd replication intermediates failed to yield evidence for the presence of the monomeric circular (−)-strand RNA required by a symmetric mechanism (Branch et al., 1988).

Initial evidence that at least one viroid (ASBVd) replicates via a symmetric rolling circle mechanism was provided by Hutchins et al. (1986) who reported the ability of both (+) and (−)-ASBVd RNA transcripts synthesized from cloned dimeric cDNA templates to spontaneously self-cleave at two sites in each transcript, thereby giving rise to the corresponding monomeric (+)- and (−)-RNA. Circularization of the linear (−)-ASBVd monomer would then yield the second circular template required for replication via a symmetric (rather than asymmetric) mechanism. Many prokaryotic and eukaryotic genomes contain multiple copies of DNA sequences encoding similar "hammerhead ribozymes" (the name subsequently given to the RNA motif responsible for ASBVd self-cleavage).

Even as the mechanism(s) of viroid replication were being clarified, other studies sought to develop new tools with which to study viroid replication, which include the construction of the first infectious cloned PSTVd cDNA (Cress et al., 1983), development of an in vitro transcriptional system for in vitro synthesis of infectious viroid RNAs (Tabler and Sänger, 1985), and the use of *A. tumefaciens* to introduce infectious PSTVd cDNAs into potential host cells (Gardner et al., 1986). Similar "agroinoculation" strategies quickly became the method of choice to transmit a number of conventional plant viruses; for example, geminiviruses.

In the early-mid 1980s, the first generation of low copy number plasmid vectors each containing only a limited number of cloning sites was being replaced by high copy number plasmids whose more versatile multiple cloning sites were flanked by promoters for SP6 and T7 RNA polymerase. Completely by chance, the sequence of a portion of the PSTVd upper conserved central region (i.e., ...GGATCCCCGGG...) turns out to be identical to that of the overlapping BamHI/*Sma*I recognition sites present in several polylinkers. While cloned PSTVd cDNAs (or their corresponding RNA transcripts) which contain this 11-nt sequence duplication are highly infectious, those containing only a shorter 6-nt GGATCC duplication are essentially noninfectious (Tabler and Sänger, 1985). This single observation had several important consequences: First, an immediate focus on secondary hairpin I (or related structures) as the probable site of pospiviroid cleavage/ligation during normal replication (e.g., Diener, 1986); second, a dramatic increase in the ease of generating viroid variants containing precisely targeted mutations to be screened for their effects on infectivity or other biological properties. One particularly informative mutagenesis study compared the effect of systematically closing each of the predicted interior loops or bulges in the secondary structure of PSTVd genomic RNA on the viroid's ability to replicate in single cells (protoplasts) with the effect on its ability to spread systemically in whole *Nicotiana benthamiana* plants (Zhong et al., 2008). The results of this study provided the framework for later high-throughput studies of the role of RNA tertiary structures in regulating viroid replication and trafficking (see "Pathogenicity" and "Movement" sections).

Two different mechanisms have been proposed to explain the "processing" of pospivirioid replicative intermediates to release monomeric linear (+)-strand progeny. Both involve cleavage/ligation between nucleotides G_{95} and G_{96} in the upper portion of the conserved central region but differ in the type of nuclease involved. As described by Baumstark et al. (1997), efficient cleavage/ligation of a longer-than-unit-length PSTVd RNA transcript in a potato nuclear extract requires that the central conserved region folds into a multihelix junction containing at least one GNRA tetraloop hairpin. The first cleavage occurs within the stem of the GNRA tetraloop, and a local conformational change then converts the tetraloop into a loop E motif, thereby stabilizing its base-paired end (5′-hydroxyl). The second cleavage yields a unit-length linear intermediate whose base-paired 3′ end (2′, 3′-cyclic phosphate) can then be ligated to form mature circular progeny. In a separate series of experiments, Gas et al. (2007) examined the in vivo processing of dimeric CEVd RNA transcripts expressed in transgenic *A. thaliana*. In this case, the monomeric linear processing product was shown to contain 5′-Phosphate and 3′-Hydroxyl termini. Such termini are the hallmarks of a type III RNase activity, and the authors propose that the necessary dsRNA substrate is formed by a "kissing loop" interaction between two copies of secondary hairpin I.

Finally, baker's yeast (*S. cerevisiae*) provides a variety of powerful genetic and molecular tools that could find many applications in studies of viroid replication. Unlike higher plants, however, this particular species of budding yeast does not contain a functional RNA silencing pathway, and therefore, may lack either a suitable type III RNase or other protein(s) required for viroid replication. An initial report that ASBVd is able to replicate in yeast (Delan-Forino et al., 2011) has not been confirmed.

Pathogenicity

Interest in the molecular events responsible for viroid disease was greatly stimulated by two events, the determination of the complete nucleotide sequence of PSTVd-Intermediate strain (Gross et al., 1978) and RNA fingerprinting studies indicating only minor sequence differences between mild and severe strains of the same viroid (Dickson et al., 1979). At the time, disease induction was commonly assumed to involve *direct interaction* of either the viroid itself or some viroid-related RNA with host cell component(s). Thermodynamic calculations carried out with several naturally-occurring strains of PSTVd revealed an inverse correlation between isolate virulence and stability of a so-called "virulence modulating (VM) region" located within the pathogenicity domain (Schnölzer et al., 1985). The decreased base pairing was proposed to favor the interaction of the VM region with unidentified host cell component(s). One possible target might be a plant homolog of the mammalian interferon-induced, double-stranded RNA-activated protein kinase shown by Diener et al. (1993) to be differentially activated by PSTVd strains of varying severity. Unfortunately, intensive efforts to identify this putative kinase homolog were not successful (Hammond and Zhao, 2000).

Could portions of the viroid genome other than the pathogenicity domain play an important role in disease induction? Is it possible that the interaction between viroid and host is indirect rather than direct? To answer the first of these questions and assess the role of individual structural domains in viroid pathogenicity and replication Sano et al. (1992) compared the symptoms induced by a series of interspecific viroid chimeras constructed by exchanging the terminal left (TL) and/or pathogenicity (P) domains between CEVd and tomato apical stunt viroid (TASVd). Although their individual contributions to symptom induction could not be completely separated from that of viroid titer, the TL domain appeared to exert a greater effect upon symptom severity than did the P domain. Pathogenicity determinants uncovered by this genetic analysis may reside within three discrete regions of sequence and/or structural variability in the TL, P, V, and TR domains of these two viroids.

In 2001 two groups reported the presence of short (~21–26-nt) viroid-specific RNAs of both polarities in preparations of low molecular weight RNA isolated from PSTVd-infected plants (Papaefthimiou et al., 2001; Itaya et al., 2001). A hallmark of posttranscriptional gene silencing (PTGS), similar RNAs were soon detected in several other viroid-host combinations. Much effort has subsequently been devoted to (i) characterization of the distribution of small RNAs around the viroid genome and (ii) identification of host gene targeted by viroid-induced PTGS. Sometimes overlooked in all this activity is an earlier report by Wassenegger et al. (1994b) showing that PSTVd replication also induces de novo methylation of noninfectious

A. Introduction

PSTVd cDNA transgenes; that is, gene silencing at the transcriptional rather than posttranscriptional level.

The clearest evidence to date of a direct role for viroid-induced RNA silencing in symptom induction comes not from studies with PSTVd or other viroids that replicate in the nucleus but from studies of a phenomenon known as "peach calico" induced by PLMVd, a ribozyme-containing viroid that replicates in the chloroplast (Navarro et al., 2012). The chloroplast abnormalities characteristic of peach calico are strictly associated with PLMVd variants that contain a specific 12–14-nt hairpin insertion, and two PLMVd-sRNAs that contain this PC-associated insertion (PC-sRNA8a and PC-sRNA8b) were shown to target for cleavage an mRNA encoding the chloroplastic heat-shock protein 90 (cHSP90). This study also identified several other host genes potentially targeted by other PLMVd sRNAs. The Arabidopsis homolog of cHSP90 participates in chloroplast biogenesis and plastid-to-nucleus signal transduction.

The first genome-wide analysis of viroid pathogenesis (Itaya et al., 2002) was a microarray study comparing the effects of mild and severe strains of PSTVd on the expression of more than 1100 tomato genes with those of tobacco mosaic virus. The two PSTVd strains were shown to alter the expression of both common and unique tomato genes involved in a variety of metabolic pathways; for example, defense/stress response, cell wall structure, chloroplast function, and protein metabolism as well as other diverse functions. A subsequent, more comprehensive transcriptome analysis by Zheng et al. (2017) revealed that PSTVd infection is accompanied by a variety of regulatory changes such as genome-wide alteration in alternative splicing of host protein-coding genes, enhanced guided-cleavage activities of a host microRNA, and induction of the *trans*-acting function of phased secondary small interfering RNAs. Furthermore, PSTVd infection was shown to massively activate genes involved in plant immune responses, mainly those in the calcium-dependent protein kinase and mitogen-activated protein kinase cascades, as well as prominent genes involved in hypersensitive responses, cell wall fortification, and hormone signaling.

Movement

Systemic infection is a multistep process: initial entry of the viroid into a host cell is followed by movement to the site of replication; following replication, the resulting progeny exit from this site and invade neighboring cells; and finally, the viroid moves from the organ to organ to infect the entire plant. Early cell fractionation studies (e.g., Diener, 1971) indicated that PSTVd and CEVd accumulate in the nucleus, and these results were confirmed and extended by Riesner and colleagues who used a combination of biochemical (Schumacher et al., 1983b) and in situ hybridization (Harders et al., 1989) techniques to demonstrate the presence of both (+) and (−)-PSTVd RNAs in the nucleoli of highly purified nuclei. In situ hybridization analysis of ASBVd-infected avocado leaf tissue, in contrast, revealed this viroid to be localized in the chloroplast, mostly on thalykoid membranes (Bonfiglioli et al., 1994).

Returning again to the intracellular distribution of PSTVd Qi and Ding (2003) reported that (−)-strand PSTVd RNA was localized in the nucleoplasm, whereas the (+)-strand RNA was present in the nucleolus as well as in the nucleoplasm—each compartment exhibiting a distinct spatial pattern. These results supported a model in which (i) the synthesis of the (−)- and

(+)-strands of PSTVd RNAs occurs in the nucleoplasm, (ii) the (−)-strand RNA is anchored in the nucleoplasm, and (iii) the (+)-strand RNA is transported selectively into the nucleolus. Furthermore, the presence of (+)-PSTVd in the nucleolus resulted in the redistribution of a small nucleolar RNA.

Whole plant studies have used green fluorescent protein (GFP) as a reporter gene to examine viroid transport into either the nucleus (PSTVd) or chloroplast (ELVd). For example, an intron-containing subgenomic RNA encoding GFP and expressed in the cytoplasm from a potato virus X vector cannot produce functional GFP unless the RNA is imported into the nucleus, where the intron can be removed and the spliced mRNA returned to the cytoplasm for translation. The presence within the intron of either a full-length (Zhao et al., 2001) or a partial length PSTVd sequence containing the conserved hairpin I palindromic processing site (Abraitiene et al., 2008) was sufficient to restore GFP translation and the appearance of green fluorescence. A similar experimental strategy was used by Gómez and Pallás (2010) to demonstrate that replacement of the 5′UTR of GFP mRNA with ELVd allows this normally cytoplasm-limited mRNA to enter the chloroplast. Transport of numerous proteins encoded by nuclear genes into the chloroplast is mediated by specific transit-peptides. As the authors point out, the ability to use viroid (and presumably other noncoding) RNA sequences to redirect mRNAs to the chloroplast before translation promises to greatly facilitate the use of plant chloroplasts as bioreactors.

Limited progress has also been made in establishing the role of host proteins in mediating intracellular viroid movement. Thus, a search for RNA-binding proteins interacting specifically with PSTVd identified a bromodomain-containing tomato protein known as "viroid RNA-binding protein 1" (VIRP1) that binds strongly and specifically to monomeric and oligomeric PSTVd positive-strand RNA transcripts (Martínez de Alba et al., 2003). A member of a family of transcriptional regulators associated with chromatin remodeling, VIRP1 also contains a nuclear-localization signal, thereby pointing to its possible involvement in PSTVd transport to the nucleus.

Palukaitis (1987) used dot-blot hybridization to monitor PSTVd movement in tomatoes following mechanical inoculation. The movement pattern observed was consistent with the rapid systemic movement of PSTVd from the inoculated leaf to actively growing tissue via the phloem; that is, the same "source to sink" route utilized by most plant viruses. Phloem protein 2 (PP2), the most abundant component of phloem exudate, moves from cell to cell via plasmodesmata and then a long distance in the phloem. The ability of HSVd to interact with PP2 to form a stable ribonucleoprotein complex (Owens et al., 2001; Gómez and Pallás, 2001) suggests that this dimeric lectin may facilitate the systemic movement of viroids and, possibly, other RNAs in vivo. Subsequent analysis of its primary structure revealed the presence of a potential double-spaced-RNA-binding motif (Gómez and Pallás, 2004).

The plasmodesmata and phloem form a symplasmic network that mediates direct cell-cell communication and transport throughout a plant. Mutational analysis of PSTVd has identified a structural motif that facilitates the movement of this viroid from the bundle sheath into mesophyll in tobacco, a vital step in establishing systemic infection (Qi et al., 2004). Surprisingly, this motif is not required for trafficking in the reverse direction (i.e., from the mesophyll to bundle sheath) or between other cell types. An extension of earlier work by Wassenegger et al. (1996) this study illustrates the advantages of combining different experimental approaches to tease apart a complex phenomenon like viroid movement.

A. Introduction

Origin/evolution

As the first viroids were being discovered, several characteristics—their very limited information content, localization in the nucleus, and ability to induce severe disease symptoms—suggested that they might function as "abnormal regulatory RNAs." Subsequent research has amply verified this hypothesis. In announcing the discovery of PSTVd Diener (1971) speculated that viroids might represent the "missing link" between viruses and the host genome. As described below it now seems more likely that viroids are evolutionarily older than viruses—possibly originating in the precellular "RNA world."

When viroid nucleotide sequences began to accumulate in the early 1980s, software programs designed to align and compare multiple nucleotide or protein sequences (e.g., Wilbur and Lipman, 1983) were also coming into wide-spread use. Introns had only recently been discovered, and the similarities between transposable genetic elements (TGE) and certain retroviruses were also attracting increasing attention. Sequence similarities between viroids and the ends of both introns (Diener, 2003) and TGE (Kiefer et al., 1983) provided the first possible insight into the ancestors of present-day viroids. Several years later, however, Diener (1989) proposed that, rather than introns, viroids and certain viroid-like satellite RNAs are the more plausible candidates as "living fossils" from a precellular RNA world.

Their small size and circularity would have enhanced the probability of viroid survival under the primitive, error-prone conditions faced by self-replicating RNA systems and assured complete replication without the need for initiation or termination signals. All contemporary viroids possess an efficient mechanism for the precise cleavage of monomers from oligomeric replication intermediates, and those replicating in the chloroplast contain ribozymes far smaller and simpler than those derived from introns. Consistent with a possible origin in the precellular RNA world, subsequent phylogenetic analysis indicated a monophyletic origin for viroids, selected plant satellite RNAs, and the viroid-like domain of human HDV RNA (Elena et al., 1991).

In a recent theoretical study, Catalán et al. (2019) evaluated the likelihood of different steps along a parsimonious evolutionary pathway leading to the de novo emergence of viroid-like replicons. While *Avsunviroidae*-like structures were relatively easy to obtain through the evolution of a population of random RNA sequences of fixed length, the emergence of circular RNA replicons analogous to *Pospiviroidae* required seeding the process with minimal circular RNAs that grow through the gradual addition of nucleotides. Although the apparent absence of viroids from prokaryotic algae (ancestors of modern higher plants) remains to be explained, this RNA world hypothesis is currently widely accepted.

Prospects for future progress in this area are encouraging. Improvements in large-scale sequencing methodologies have allowed López-Carrasco et al. (2017) to compare the mutation rates for two viroids (ELVd and PSTVd) replicating in a common host (eggplant). Rates of spontaneous mutation for ELVd (replicating in the chloroplast) ranged from 1/1000 to 1/800, while those for PSTVd (replicating in the nucleus) ranged from 1/7000 to 1/3800, a rate more similar to those measured for conventional RNA viruses. These results confirmed a previous report (Gago et al., 2009) of an extremely high mutation rate for CChMVd, another ribozyme-containing viroid that replicates in the chloroplast.

Many of the diseases caused by viroid are believed to be the result of chance transfer from a symptomless reservoir host to susceptible crop species. To investigate the origin of hop stunt disease Kawaguchi-Ito et al. (2009) infected hops with naturally-occurring HSVd isolates derived from four host species (hop, grapevine, plum, and citrus), three of which represent possible sources of the ancestral viroid. Inoculated plants were maintained for 15 years under field conditions and periodically analyzed for the presence of novel HSVd variants. Upon prolonged infection, variants derived from the grapevine isolate were observed to undergo convergent evolution resulting in a limited number of adapted mutants. Some of these adapted mutants were identical in sequence to variants currently responsible for hop stunt epidemics in commercial hops.

Other, larger-scale changes in viroid sequence have also been observed in real-time. Examples include the spontaneous appearance of sequence duplications/insertions within the genomes of CCCVd (Haseloff et al., 1982), CEVd (Szychowski et al., 2005), and PLMVd (Malfitano et al., 2003) as the infection progresses. Furthermore, Wassenegger et al. (1994a) have described the recovery of a 341-nt infectious PSTVd RNA replicon from a transgenic tobacco plant whose genome contained a noninfectious 350 bp long PSTVd cDNA dimer. Details of the in vivo evolutionary process resulting in the deletion of nine additional nucleotides and restoration of the rod-like secondary structure remain to be determined. Finally, the likely role of RNA recombination in viroid evolution received additional support when Hammond et al. (1989) reported that the newly-sequenced columnea latent viroid (CLVd) appeared to be the product of in vivo recombination between two co-infecting viroids, one PSTVd-like and other HSVd-like.

Epidemiology/control

Identification of Rutgers tomato as a sensitive indicator host for the agent responsible for potato spindle tuber disease (Raymer and O'Brien, 1962) played a crucial role in the discovery of PSTVd. Rather than requiring an entire growing season like its natural host (potato), tomato bioassays carried out under favorable conditions in the greenhouse could be completed in only 3–4 weeks. The introduction of other widely used viroid indicators such as "Etrog" citron (CEVd and other citrus viroids) and "Suyo" cucumber (HSVd and other hostuviroids) yielded similar improvements. To detect strains of PSTVd that induce only mild symptoms on tomato Fernow (1967) exploited the "cross-protection" phenomenon often observed among closely-related viroids to develop a two-step bioassay protocol.

In the early 1980s molecular methods began to replace traditional bioassays for viroid detection. Two-step electrophoresis protocols in which samples of low molecular weight plant RNA are fractionated by polyacrylamide gel electrophoresis under, first, nondenaturing and, second, denaturing conditions proved to be particularly useful (e.g., Schumacher et al., 1983a,b; Rivera-Bustamante et al., 1986; Singh and Boucher, 1987). Taking advantage of the circular nature of viroids these PAGE methods further reduced the time required to detect their presence from several weeks to just 1–2 days. No sequence information was required, and sensitivity is often comparable to that of bioassays. For the first time, it became possible to rapidly screen dozens (or even hundreds) of germplasm accessions for their infection status. Before the wide-spread availability of large-scale RNA sequencing and bioinformatic strategies for identifying small circular RNAs (in the following section) PAGE was an essential part of every effort to identify and characterize new viroids.

Owens and Diener (1981) reported the development of a "dot-blot" test allowing rapid and sensitive diagnosis of potato spindle tuber viroid disease by nucleic acid hybridization. Nucleic acid sequencing was then still in its infancy, and this testing protocol relied upon the hybridization of a radioactively-labeled, cloned PSTV cDNA probe to RNA samples immobilized on a membrane. Although the times required to complete an assay are similar (i.e., 2–3 days), dot bot hybridization offers several advantages over PAGE protocols; for example, simplified sample preparation suitable for use in the field and the ability to distinguish between viroids of similar size. The introduction of nonradioactively labeled RNA probes greatly increased the ease of use for routine diagnostic screening.

Hadidi and Yang (1990) were the first to adopt RT-PCR methodology for routine viroid detection. The subsequent introduction of real-time RT-PCR assays (e.g., Boonham et al., 2004) eliminated the need to use PAGE to detect the expected amplification product; instead, product formation is constantly monitored over the course of the reaction. PCR-based assays are inherently more sensitive than other molecular methods, and improvements in both the instrumentation and reagents required have led to their wide-spread adoption for routine viroid screening. For example, Verhoeven et al. (2010) used this technology to gather epidemiological evidence implicating several vegetatively propagated ornamental species as the source of PSTVd inoculum leading to periodic disease outbreaks in tomato. Primer design does require at least partial knowledge of the target sequence, however.

The limitations are inherent in conventional strategies used to purify, clone, and sequence novel viroids are vividly illustrated by the difficulties encountered in efforts to characterize the causal agent of chrysanthemum chlorotic mottle disease (Navarro and Flores, 1997). Symptomatic leaf tissue contained only very low levels of CChMVd, and it's branched (rather than rod-like) structure rendered it insoluble in 2M LiCl. Purifying sufficient amounts of CChMVd to allow cloning of a partial length cDNA needed for the design of an RT-PCR primer set was a truly monumental task. More recently, Wu et al. (2012) have described a homology-independent approach for identifying viroids using data sets obtained by large-scale RNA sequence analysis. The unique computational algorithm used, progressive filtering of overlapping small RNAs (PFOR), progressively eliminates nonoverlapping small RNAs as well those that overlap but cannot be assembled into a direct repeat RNA, thereby identifying any small circular RNA(s) present in the original tissue sample. Application to a grapevine sRNA library identified a viroid-like circular RNA 375-nt long that exhibited no significant sequence similarity to known molecules and encoded active hammerhead ribozymes in RNAs of both plus and minus polarities. The availability of large-scale RNA sequencing and bioinformatic analysis on a fee-for-service basis is bringing this technology within the reach of more and more laboratories.

Questions for the reader

1. What are the fundamental differences between viroids and RNA plant viruses?
2. In what ways do their replication mechanisms differ?

A. Introduction

3. How do the molecular mechanisms responsible for the appearance of visible disease symptoms differ between viroid and virus infection?
4. What role do symptomless infections play in the origin/spread of viroid diseases?
5. Antibody-based techniques are widely used for routine virus detection/diagnosis. What techniques are used for viroid detection/diagnosis?
6. It has been proposed that viroids originated in the precellular "RNA world." Can you imagine an alternative scenario?

Further reading

Diener TO. 1971. Potato spindle tuber "virus" IV. A replicating, low molecular weight RNA. Virology 45, 411–428.

Diener TO. 1989. Circular RNAs: relics of precellular evolution? Proc Natl Acad Sci USA. 86 (23), 9370–9374. doi: https://doi.org/10.1073/pnas.86.23.9370.

Diener TO. 2003. Discovering viroids – a personal perspective. Nat Rev. Microbiol. 1(1):75–80. doi: https://doi.org/10.1038/nrmicro736.

10th ICTV Report (https://talk.ictvonline.org/ictv-reports/ictv_online_report/subviral-agents/w/viroids).

Acknowledgments

I thank the editors for this opportunity to revisit many memorable events from a career in viroid research. My first exposure to viroids came about by chance one evening when, as a graduate student at University of California, Davis, I picked up a recent issue of *Virology* in the departmental library. As I began to scan the article by T.O. Diener announcing his discovery of potato spindle tuber viroid, its significance was immediately obvious. Three years later I found myself at Beltsville working with Dr. Diener, and the ensuing 35 years saw a series of deeply satisfying collaborations with many other members of the viroid research community including Biao Ding, Luis Salazar, Teruo Sano, and Gerhard Steger.

References

Abraitiene, A., Zhao, Y., Hammond, R., 2008. Nuclear targeting by fragmentation of the potato spindle tuber viroid genome. Biochem. Biophys. Res. Commun. 368 (3), 470–475. https://doi.org/10.1016/j.bbrc.2008.01.043.

Baumstark, T., Schröder, A.R., Riesner, D., 1997. Viroid processing: switch from cleavage to ligation is driven by a change from a tetraloop to a loop E conformation. EMBO J. 16 (3), 599–610. https://doi.org/10.1093/emboj/16.3.599.

Bonfiglioli, R.G., McFadden, G.I., Symons, R.H., 1994. *In situ* hybridization localizes avocado sunblotch viroid on chloroplast thylakoid membranes and coconut cadang cadang viroid in the nucleus. Plant J. 6, 99–103. https://doi.org/10.1046/j.1365-313X.1994.6010099.x.

Boonham, N., Pérez, L.G., Mendez, M.S., Peralta, E.L., Blockley, A., Walsh, K., Barker, I., Mumford, R.A., 2004. Development of a real-time RT-PCR assay for the detection of potato spindle tuber viroid. J. Virol. Methods 116 (2), 139–146. https://doi.org/10.1016/j.jviromet.2003.11.005.

Branch, A.D., Robertson, H.D., 1984. A replication cycle for viroids and other small infectious RNA's. Science 223 (4635), 450–455.

Branch, A.D., Robertson, H.D., Dickson, E., 1981. Longer-than-unit-length viroid minus strands are present in RNA from infected plants. Proc. Natl. Acad. Sci. U. S. A. 78 (10), 6381–6385. https://doi.org/10.1073/pnas.78.10.6381.

Branch, A.D., Benenfeld, B.J., Robertson, H.D., 1985. Ultraviolet light-induced crosslinking reveals a unique region of local tertiary structure in potato spindle tuber viroid and HeLa 5S RNA. Proc. Natl. Acad. Sci. U. S. A. 82 (19), 6590–6594. https://doi.org/10.1073/pnas.82.19.6590.

Branch, A.D., Benenfeld, B.J., Robertson, H.D., 1988. Evidence for a single rolling circle in the replication of potato spindle tuber viroid. Proc. Natl. Acad. Sci. U. S. A. 85 (23), 9128–9132. https://doi.org/10.1073/pnas.85.23.9128.

Catalán, P., Elena, S.F., Cuesta, J.A., Manrubia, S., 2019. Parsimonious scenario for the emergence of viroid-like replicons de novo. Viruses 11 (5), 425. https://doi.org/10.3390/v11050425.

Cress, D.E., Kiefer, M.C., Owens, R.A., 1983. Construction of infectious potato spindle tuber viroid cDNA clones. Nucleic Acids Res. 11 (19), 6821–6835. https://doi.org/10.1093/nar/11.19.6821.

Davies, J.W., Kaesberg, P., Diener, T.O., 1971. Potato spindle tuber viroid. XII. An investigation of viroid RNA as a messenger for protein synthesis. Virology 61 (1), 281–286. https://doi.org/10.1016/0042-6822(74)90262-1.

Delan-Forino, C., Maurel, M.C., Torchet, C., 2011. Replication of avocado sunblotch viroid in the yeast Saccharomyces cerevisiae. J. Virol. 85 (7), 3229–3238. https://doi.org/10.1128/JVI.01320-10.

Dickson, E., Robertson, H.D., Niblett, C.L., Horst, R.K., Zaitlin, M., 1979. Minor differences between nucleotide sequences of mild and severe strains of potato spindle tuber viroid. Nature 277, 60–62.

Diener, T.O., 1971. Potato spindle tuber "virus" IV. A replicating, low molecular weight RNA. Virology 45, 411–428.

Diener, T.O., 1986. Viroid processing: a model involving the central conserved region and hairpin I. Proc. Natl. Acad. Sci. U. S. A. 83 (1), 58–62. https://doi.org/10.1073/pnas.83.1.58.

Diener, T.O., 1987. The viroids. In: Fraenkel-Conrat, H., Wagner, R.R. (Eds.), The Viruses. Plenum Press, New York. 344 pp.

Diener, T.O., 1989. Circular RNAs: relics of precellular evolution? Proc. Natl. Acad. Sci. U. S. A. 86 (23), 9370–9374. https://doi.org/10.1073/pnas.86.23.9370.

Diener, T.O., 2003. Discovering viroids – a personal perspective. Nat. Rev. Microbiol. 1 (1), 75–80. https://doi.org/10.1038/nrmicro736.

Diener, T.O., Hammond, R.W., Black, T., Katze, M.G., 1993. Mechanism of viroid pathogenesis: differential activation of the interferon-induced, double-stranded RNA-activated, M(r) 68,000 protein kinase by viroid strains of varying pathogenicity. Biochimie 75 (7), 533–538. https://doi.org/10.1016/0300-9084(93)90058-z.

Elena, S.F., Dopazo, J., Flores, R., Diener, T.O., Moya, A., 1991. Phylogeny of viroids, viroidlike satellite RNAs, and the viroidlike domain of hepatitis delta virus RNA. Proc. Natl. Acad. Sci. U. S. A. 88 (13), 5631–5634. https://doi.org/10.1073/pnas.88.13.5631.

Fawcett, H.S., Klotz, L.J., 1948. Exocortis on trifoliate orange. Citrus Leaves 28, 8.

Fernow, K.H., 1967. Tomato as a test plant for detecting mild strains of potato spindle tuber virus. Phytopathology 57, 1347–1352.

Gago, S., De la Peña, M., Flores, R., 2005. A kissing-loop interaction in a hammerhead viroid RNA critical for its in vitro folding and in vivo viability. RNA 11 (7), 1073–1083. https://doi.org/10.1261/rna.2230605.

Gago, S., Elena, S.F., Flores, R., Sanjuán, R., 2009. Extremely high mutation rate of a hammerhead viroid. Science 323 (5919), 1308. https://doi.org/10.1126/science.1169202.

Gardner, R.C., Chonoles, K.R., Owens, R.A., 1986. Potato spindle tuber viroid infections mediated by the Ti plasmid of Agrobacterium tumefaciens. Plant Mol. Biol. 6 (4), 221–228.

Gas, M.E., Hernández, C., Flores, R., Daròs, J.A., 2007. Processing of nuclear viroids in vivo: an interplay between RNA conformations. PLoS Pathog. 3 (11), e182. https://doi.org/10.1371/journal.ppat.0030182.

Gómez, G., Pallás, V., 2001. Identification of an in vitro ribonucleoprotein complex between a viroid RNA and a phloem protein from cucumber plants. Mol. Plant-Microbe Interact. 14 (7), 910–913. https://doi.org/10.1094/MPMI.2001.14.7.910.

Gómez, G., Pallás, V., 2004. A long-distance translocatable phloem protein from cucumber forms a ribonucleoprotein complex in vivo with hop stunt viroid RNA. J. Virol. 78 (18), 10104–10110. https://doi.org/10.1128/JVI.78.18.10104-10110.2004.

Gómez, G., Pallás, V., 2010. Noncoding RNA mediated traffic of foreign mRNA into chloroplasts reveals a novel signaling mechanism in plants. PLoS One 5 (8), e12269. https://doi.org/10.1371/journal.pone.0012269.

Gross, H.J., Domdey, H., Lossow, C., Jank, P., Raba, M., Alberty, H., Sänger, H.L., 1978. Nucleotide sequence and secondary structure of potato spindle tuber viroid. Nature 273 (5659), 203–208. https://doi.org/10.1038/273203a0.

A. Introduction

Hadidi, A., Yang, X., 1990. Detection of pome fruit viroids by enzymatic cDNA amplification. J. Virol. Methods 30 (3), 261–269. https://doi.org/10.1016/0166-0934(90)90068-q.

Hammond, R.W., Zhao, Y., 2000. Characterization of a tomato protein kinase gene induced by infection by potato spindle tuber viroid. Mol. Plant-Microbe Interact. 13 (9), 903–910. https://doi.org/10.1094/MPMI.2000.13.9.903.

Hammond, R., Smith, D.R., Diener, T.O., 1989. Nucleotide sequence and proposed secondary structure of columnea latent viroid: a natural mosaic of viroid sequences. Nucleic Acids Res. 17 (23), 10083–10094. https://doi.org/10.1093/nar/17.23.10083.

Harders, J., Lukács, N., Robert-Nicoud, M., Jovin, T.M., Riesner, D., 1989. Imaging of viroids in nuclei from tomato leaf tissue by in situ hybridization and confocal laser scanning microscopy. EMBO J. 8 (13), 3941–3949.

Haseloff, J., Mohamed, N., Symons, R., 1982. Viroid RNAs of cadang-cadang disease of coconuts. Nature 299, 316–321. https://doi.org/10.1038/299316a0.

Hutchins, C.J., Keese, P., Visvader, J.E., Rathjen, P.D., McInnes, J.L., Symons, R.H., 1985. Comparison of multimeric plus and minus forms of viroids and virusoids. Plant Mol. Biol. 4 (5), 293–304. https://doi.org/10.1007/BF02418248.

Hutchins, C.J., Rathjen, P.D., Forster, A.C., Symons, R.H., 1986. Self-cleavage of plus and minus RNA transcripts of avocado sunblotch viroid. Nucleic Acids Res. 14 (9), 3627–3640. https://doi.org/10.1093/nar/14.9.3627.

Itaya, A., Folimonov, A., Matsuda, Y., Nelson, R.S., Ding, B., 2001. Potato spindle tuber viroid as inducer of RNA silencing in infected tomato. Mol. Plant-Microbe Interact. 14 (11), 1332–1334.

Itaya, A., Matsuda, Y., Gonzales, R.A., Nelson, R.S., Ding, B., 2002. Potato spindle tuber viroid strains of different pathogenicity induces and suppresses expression of common and unique genes in infected tomato. Mol. Plant-Microbe Interact. 15 (10), 990–999. https://doi.org/10.1094/MPMI.2002.15.10.990.

Kawaguchi-Ito, Y., Li, S.F., Tagawa, M., Araki, H., Goshono, M., Yamamoto, S., Tanaka, M., Narita, M., Tanaka, K., Liu, S.X., Shikata, E., Sano, T., 2009. Cultivated grapevines represent a symptomless reservoir for the transmission of hop stunt viroid to hop crops: 15 years of evolutionary analysis. PLoS One 4 (12), e8386. https://doi.org/10.1371/journal.pone.0008386.

Keese, P., Symons, R.H., 1985. Domains in viroids: evidence of intermolecular RNA rearrangements and their contribution to viroid evolution. Proc. Natl. Acad. Sci. U. S. A. 82 (14), 4582–4586.

Kiefer, M.C., Owens, R.A., Diener, T.O., 1983. Structural similarities between viroids and transposable genetic elements. Proc. Natl. Acad. Sci. U. S. A. 80 (20), 6234–6238. https://doi.org/10.1073/pnas.80.20.6234.

López-Carrasco, A., Flores, R., 2017. Dissecting the secondary structure of the circular RNA of a nuclear viroid *in vivo*: a "naked" rod-like conformation similar but not identical to that observed *in vitro*. RNA Biol. 14 (8), 1046–1054. https://doi.org/10.1080/15476286.2016.1223005.

López-Carrasco, A., Ballesteros, C., Sentandreu, V., Delgado, S., Gago-Zachert, S., Flores, R., Sanjuán, R., 2017. Different rates of spontaneous mutation of chloroplastic and nuclear viroids as determined by high-fidelity ultra-deep sequencing. PLoS Pathog. 13 (9), e1006547. https://doi.org/10.1371/journal.ppat.1006547.

Malfitano, M., Di Serio, F., Covelli, L., Ragozzino, A., Hernández, C., Flores, R., 2003. Peach latent mosaic viroid variants inducing peach calico (extreme chlorosis) contain a characteristic insertion that is responsible for this symptomatology. Virology 313 (2), 492–501. https://doi.org/10.1016/s0042-6822(03)00315-5.

Martínez de Alba, A.E., Sägesser, R., Tabler, M., Tsagris, M., 2003. A bromodomain-containing protein from tomato specifically binds potato spindle tuber viroid RNA in *vitro* and in *vivo*. J. Virol. 77 (17), 9685–9694.

Mühlbach, H.P., Sänger, H.L., 1979. Viroid replication is inhibited by alpha-amanitin. Nature 278 (5700), 185–188. https://doi.org/10.1038/278185a0.

Navarro, B., Flores, R., 1997. Chrysanthemum chlorotic mottle viroid: unusual structural properties of a subgroup of self-cleaving viroids with hammerhead ribozymes. Proc. Natl. Acad. Sci. U. S. A. 94 (21), 11262–11267. https://doi.org/10.1073/pnas.94.21.11262.

Navarro, J.A., Vera, A., Flores, R., 2000. A chloroplastic RNA polymerase resistant to tagetitoxin is involved in replication of avocado sunblotch viroid. Virology 268 (1), 218–225. https://doi.org/10.1006/viro.1999.0161.

Navarro, B., Gisel, A., Rodio, M.E., Delgado, S., Flores, R., Di Serio, F., 2012. Small RNAs containing the pathogenic determinant of a chloroplast-replicating viroid guide the degradation of a host mRNA as predicted by RNA silencing. Plant J. 70 (6), 991–1003. https://doi.org/10.1111/j.1365-313X.2012.04940.x.

Owens, R.A., Diener, T.O., 1981. Sensitive and rapid diagnosis of potato spindle tuber viroid disease by nucleic acid hybridization. Science 213 (4508), 670–672. https://doi.org/10.1126/science.213.4508.670.

A. Introduction

Owens, R.A., Blackburn, M., Ding, B., 2001. Possible involvement of the phloem lectin in long-distance viroid move-ment. Mol. Plant-Microbe Interact. 14 (7), 905–909. https://doi.org/10.1094/MPMI.2001.14.7.905.

Palukaitis, P., 1987. Potato spindle tuber viroid: investigation of the long-distance, intra-plant transport route. Virol-ogy 158 (1), 239–241.

Papaefthimiou, I., Hamilton, A., Denti, M., Baulcombe, D., Tsagris, M., Tabler, M., 2001. Replicating potato spindle tuber viroid RNA is accompanied by short RNA fragments that are characteristic of post-transcriptional gene si-lencing. Nucleic Acids Res. 29 (11), 2395–2400. https://doi.org/10.1093/nar/29.11.2395.

Qi, Y., Ding, B., 2003. Differential subnuclear localization of RNA strands of opposite polarity derived from an au-tonomously replicating viroid. Plant Cell 15 (11), 2566–2577. https://doi.org/10.1105/tpc.016576.

Qi, Y., Pélissier, T., Itaya, A., Hunt, E., Wassenegger, M., Ding, B., 2004. Direct role of a viroid RNA motif in mediating directional RNA trafficking across a specific cellular boundary. Plant Cell 16 (7), 1741–1752. https://doi.org/10.1105/tpc.021980.

Raymer, W.B., O'Brien, M.J., 1962. Transmission of potato spindle tuber virus to tomato. Am. Potato J. 39, 401–408.

Riesner, D., Henco, K., Rokohl, U., Klotz, G., Kleinschmidt, A.K., Domdey, H., Jank, P., Gross, H.J., Sänger, H.L., 1979. Structure and structure formation of viroids. J. Mol. Biol. 133 (1), 85–115. https://doi.org/10.1016/0022-2836(79)90252-3.

Riesner, D., Steger, G., Zimmat, R., Owens, R.A., Wagenhöfer, M., Hillen, W., Vollbach, S., Henco, K., 1989. Temperature-gradient gel electrophoresis of nucleic acids: analysis of conformational transitions, sequence var-iations, and protein-nucleic acid interactions. Electrophoresis 10 (5–6), 377–389. https://doi.org/10.1002/elps.1150100516.

Rivera-Bustamante, R.F., Gin, R., Semancik, J.S., 1986. Enhanced resolution of circular and linear molecular forms of viroid and viroid-like RNA by electrophoresis in a discontinuous-pH system. Anal. Biochem. 156 (1), 91–95. https://doi.org/10.1016/0003-2697(86)90159-4.

Sänger, H.L., Klotz, G., Riesner, D., Gross, H.J., Kleinschmidt, A.K., 1976. Viroids are single-stranded covalently closed circular RNA molecules existing as highly base-paired rod-like structures. Proc. Natl. Acad. Sci. U. S. A. 73 (11), 3852–3856. https://doi.org/10.1073/pnas.73.11.3852.

Sano, T., Candresse, T., Hammond, R.W., Diener, T.O., Owens, R.A., 1992. Identification of multiple structural do-mains regulating viroid pathogenicity. Proc. Natl. Acad. Sci. U. S. A. 89 (21), 10104–10108.

Schnölzer, M., Haas, B., Raam, K., Hofmann, H., Sänger, H.L., 1985. Correlation between structure and pathogenicity of potato spindle tuber viroid (PSTV). EMBO J. 4 (9), 2181–2190.

Schultz, E.S., Folsom, D.J., 1923. Transmission, variation, and control of certain degeneration diseases of irish pota-toes. J. Agr. Res. 25 (2), 43.

Schumacher, J., Randles, J.W., Riesner, D., 1983a. A two-dimensional electrophoretic technique for the detection of circular viroids and virusoids. Anal. Biochem. 135 (2), 288–295. https://doi.org/10.1016/0003-2697(83)90685-1.

Schumacher, J., Sänger, H.L., Riesner, D., 1983b. Subcellular localization of viroids in highly purified nuclei from to-mato leaf tissue. EMBO J. 2 (9), 1549–1555.

Semancik, J.S., Weathers, L.G., 1972. Exocortis virus: an infectious free-nucleic acid plant virus with unusual prop-erties. Virology 47, 456–466.

Singh, R.P., Boucher, A., 1987. Electrophoretic separation of a severe from mild strains of potato spindle tuber viroid. Phytopathology 77, 1588–1591.

Singh, R.P., Clark, M.C., 1971. Infectious low-molecular weight ribonucleic acid from tomato. Biochem. Biophys. Res. Commun. 44 (5), 1077–1083.

Sogo, J.M., Koller, T., Diener, T.O., 1973. Potato spindle tuber viroid: X. Visualization and size determination by elec-tron microscopy. Virology 55, 70–80.

Szychowski, J.A., Vidalakis, G., Semancik, J.S., 2005. Host-directed processing of Citrus exocortis viroid. J. Gen. Virol. 86 (2), 473–477. https://doi.org/10.1099/vir.0.80699-0.

Tabler, M., Sänger, H.L., 1985. Infectivity studies on different potato spindle tuber viroid (PSTV) RNAs synthesized in vitro with the SP6 transcription system. EMBO J. 4 (9), 2191–2199.

Verhoeven, J.T.J., Jansen, C.C.C., Botermans, M., Roenhorst, J.W., 2010. Epidemiological evidence that vegetatively propagated, solanaceous plant species act as sources of potato spindle tuber viroid inoculum for tomato. Plant Pathol. 59, 3–12. https://doi.org/10.1111/j.1365-3059.2009.02173.x.

A. Introduction

Wassenegger, M., Heimes, S., Sänger, H.L., 1994a. An infectious viroid RNA replicon evolved from an *in vitro*-generated non-infectious viroid deletion mutant via a complementary deletion in vivo. EMBO J. 13 (24), 6172–6177.

Wassenegger, M., Heimes, S., Riedel, L., Sänger, H.L., 1994b. RNA-directed *de novo* methylation of genomic sequences in plants. Cell 76 (3), 567–576. https://doi.org/10.1016/0092-8674(94)90119-8.

Wassenegger, M., Spieker, R.L., Thalmeir, S., Gast, F.U., Riedel, L., Sänger, H.L., 1996. A single nucleotide substitution converts potato spindle tuber viroid (PSTVd) from a noninfectious to an infectious RNA for *Nicotiana tabacum*. Virology 226 (2), 191–197. https://doi.org/10.1006/viro.1996.0646.

Wilbur, W.J., Lipman, D.J., 1983. Rapid similarity searches of nucleic acid and protein data banks. Proc. Natl. Acad. Sci. U. S. A. 80 (3), 726–730. https://doi.org/10.1073/pnas.80.3.726.

Wu, Q., Wang, Y., Cao, M., Pantaleo, V., Burgyan, J., Li, W.X., Ding, S.W., 2012. Homology-independent discovery of replicating pathogenic circular RNAs by deep sequencing and a new computational algorithm. Proc. Natl. Acad. Sci. U. S. A. 109 (10), 3938–3943. https://doi.org/10.1073/pnas.1117815109.

Zhao, Y., Owens, R.A., Hammond, R.W., 2001. Use of a vector based on potato virus X in a whole plant assay to demonstrate nuclear targeting of potato spindle tuber viroid. J. Gen. Virol. 82 (6), 1491–1497. https://doi.org/10.1099/0022-1317-82-6-1491.

Zheng, Y., Wang, Y., Ding, B., Fei, Z., 2017. Comprehensive transcriptome analyses reveal that potato spindle tuber viroid triggers genome-wide changes in alternative splicing, inducible trans-acting activity of phased secondary small interfering RNAs, and immune responses. J. Virol. 91 (11), e00247-17. https://doi.org/10.1128/JVI.00247-17.

Zhong, X., Leontis, N., Qian, S., Itaya, A., Qi, Y., Boris-Lawrie, K., Ding, B., 2006. Tertiary structural and functional analyses of a viroid RNA motif by isostericity matrix and mutagenesis reveal its essential role in replication. J. Virol. 80 (17), 8566–8581. https://doi.org/10.1128/JVI.00837-06.

Zhong, X., Archual, A.J., Amin, A.A., Ding, B., 2008. A genomic map of viroid RNA motifs critical for replication and systemic trafficking. Plant Cell 20 (1), 35–47. https://doi.org/10.1105/tpc.107.056606.

Zuker, M., Stiegler, P., 1981. Optimal computer folding of large RNA sequences using thermodynamics and auxiliary information. Nucleic Acids Res. 9 (1), 133–148. https://doi.org/10.1093/nar/9.1.133.

A. Introduction

2

Viroid taxonomy

Francesco Di Serio and Michela Chiumenti

Consiglio Nazionale delle Ricerche (CNR), Istituto per la Protezione Sostenibile delle Piante (IPSP), Bari, Italy

Graphical representation

Family	Genus	Species
Pospiviroidae	Apscaviroid	Apple dimple fruit viroid
		Apple scar skin viroid
		Apscaviroid aclsvd
		Apscaviroid cvd-VII
		Apscaviroid dvd
		Apscaviroid glvd
		Apscaviroid lvd
		Apscaviroid plvd-I
		Apscaviroid pvd
		Apscaviroid pvd-2
		Australian grapevine viroid
		Citrus bent leaf viroid
		Citrus dwarfing viroid
		Citrus viroid V
		Citrus viroid VI
		Grapevine yellow speckle viroid 1
		Grapevine yellow speckle viroid 2
		Pear blister canker viroid
	Cocadviroid	Citrus bark craking viroid
		Coconut cadang cadang viroid
		Coconut tinangaja viroid
		Hop latent viroid
	Coleviroid	Coleus blumei viroid 1
		Coleus blumei viroid 2
		Coleus blumei viroid 3
		Coleviroid cbvd-5
		Coleviroid cbvd-6
	Hostuviroid	Dahlia latent viroid
		Hop stunt viroid
	Pospiviroid	Chrysanthemum stunt viroid
		Citrus exocortis viroid
		Columnea latent viroid
		Iresine viroid 1
		Pepper chat fruit viroid
		Pospiviroid plvd
		Potato spindle tuber viroid
		Tomato apical stunt viroid
		Tomato chlorotic dwarf viroid
		Tomato planta macho viroid
Avsunviroidae	Apscaviroid	Avocado sunblotch viroid
	Elaviroid	Eggplant latent viroid
	Pelamoviroid	Apple hammerhead viroid
		Chrysanthemum chlorotic mottle viroid
		Peach latent mosaic viroid

Current viroid classification.

Definitions

Viroid taxonomy: the discipline that classifies and names viroids.
Viroid variant: the complete RNA sequence of a viroid. A viroid variant differs slightly from other variants in the same species. Therefore, a viroid species includes several closely related variants that differ from each other at one or more positions. The number of variants in each species can be low or very high, depending on the intrinsic variability of each viroid. Viroid variants in the same species are more closely related to each other than to variants from other species.
Noncoding RNAs: RNAs unable to code for a protein.
Viroid structure: viroids are noncoding RNAs. Viroid nucleotide sequence (primary structure) assumes a specific conformation characterized by Watson-Crick base pairing between self-complementary regions (secondary structure), which is in turn arranged in a three-dimensional structure (tertiary structure).
Pairwise identity score (PWIS): the value of the percentage identity between two sequences over an alignment. Multiple alignments of sequences provide PWISs that can be arranged in a matrix where any element corresponds to the PWIS between two sequences of the alignment. The matrices calculated at the species level allow the identification of the minimum PWIS (mPWIS) between the variants of that species.
Threshold identity score (TIS) of a genus: the lowest mPWIS calculated for each species included in the analyzed genus (Table 1).

Chapter outline

The chapter will present an historical perspective of viroid taxonomy from early classification schemes to the last proposal for novel species demarcation criteria based on sequence identity matrices. It also provides general information on the procedure for the preliminary classification of new viroids.

Learning objectives

On completion of this chapter, you should be able to:

- Understand the scope of viroid classification.
- Recognize the criteria used to discriminate viroid species, genera and families.
- Perform the preliminary classification of a new viroid.

Fundamental introduction

Taxonomy is the discipline concerned with the classification and naming of entities. The systematic and consistent organization of entities according to a predetermined scheme occurs in all scientific disciplines, not just in biology, so that from a comprehensive

TABLE 1 Taxonomic ranks and species demarcation criteria established by the International Committee on Taxonomy of Viruses (ICTV) overtime.

Year	1991	1995	2000	2005	2012	2021	2022
ICTV report/release	5° (Randles and Rezaian, 1991)	6° (Flores, 1995)	7° (Flores et al., 2000)	8° (Flores et al., 2005)	9° (Owens et al., 2012)	10° (Di Serio et al., 2018 and Di Serio et al., 2021a,b)	Last release of viroid classification[a]
No. of species	19	20	27	28	32	34	44
No of genera	0	0	7	7	8	8	8
No. of families	0	0	2	2	2	2	2
Species demarcation criteria	Less than 90% sequence identity with the closest classified viroid	Less than 90% sequence identity with the closest classified viroid	Less than 90% sequence identity with the closest classified viroid	Less than 90% sequence identity over the entire genomes and distinct biological properties with respect to the closest classified viroid	Less than 90% sequence identity over the entire genomes and distinct biological properties with respect to the closest classified viroid	Less than 90% sequence identity over the entire genomes and distinct biological properties with respect to the closest classified viroid	Pairwise identity score with the closest related viroid must be below the threshold identity score of the genus in which the latter is classified. Distinct biological properties are required only in some critical cases

[a] https://talk.ictvonline.org/taxonomy/, accessed on 01/04/2022.

understanding of the similarities and differences of the classified elements, the information pertaining to a single entity can be more easily expanded on or compared with others classified in the same or in a different group. If the classification of biological entities is based upon phylogenetic and biological relationships among them, the result of such an exercise will be a general framework in which the classified units can be accommodated and connected to each other according to specific features that may have an evolutionary or biological meaning. The aims of this chapter are: (i) to provide a historical excursus of the milestones in viroid taxonomy and discuss the reasons behind an evolving classification scheme; (ii) to present the criteria currently used to classify viroids; (iii) to explain the application of these criteria in the classification of new viroids.

Viroids, with their small and circular RNA genomes unable to code for proteins, can infect plants. Likely due to the error prone host polymerases involved in their replication, viroids accumulate in the infected hosts as heterogeneous populations of closely related sequence variants differing slightly from each other. This means that a viroid species is expected not to be formed by identical sequence variants, but by variants that may differ from each other in a few nucleotides, a concept expressed by the term "quasispecies" that apply to viroids (Codoñer et al., 2006), as previously proposed for viruses (Biebricher and Eigen, 2006; Andino and Domingo, 2015). Sequence variability must be taken into consideration when classifying viroids.

Viroid classification is officially regulated by the International Committee on Taxonomy of Viruses (ICTV, https://talk.ictvonline.org/). Why would a committee focusing on virus classification expand its scope to viroids? Viroids differ from viruses in several key aspects; the structural and functional divergence between these two groups of infectious agents support their unlinked evolutionary origin. As detailed in Chapter 1, due to their inability to code for proteins, viroids can be regarded as parasites of the cellular transcriptional machinery. In contrast, the protein coding capability of viruses allows them to parasitize the cellular translational machinery. However, both viroids and viruses are endocellular parasites that are able to replicate and infect their hosts systemically, where they may evolve. In this respect, both viruses and viroids are mobile genetic elements (MGEs), as specified by the ICTV in the International Code of Virus Classification and Nomenclature (ICVCN) (https://ictv.global/about/code). The ICVCN states that the first objective of the ICTV is to develop "an internationally agreed taxonomy for viruses and other mobile genetic elements (MGEs) that are part of the virosphere." Therefore, despite its current name only referring to viruses, the ICTV regulates nomenclature and classification of viruses as well as those of all other replicons of the virosphere, which includes viroids. In fact, in the section on the scope of the classification the ICVCN clarifies that "members of the virosphere include selfish genetic elements, which are replicons that are subject to selective pressures mostly independent of other replicons and hence have distinct evolutionary histories but depend on cellular hosts for energy and chemical building blocks. The relationship between selfish genetic elements and hosts spans the spectrum from mutualism to aggressive parasitism. Typically, MGEs are selfish genetic elements that move between hosts and/or change their integration sites in host genomes. MGEs are distributed among viruses *sensu stricto* and the remaining replicator space of the virosphere (virus-like entities, such as satellite nucleic acids and viroids, and virus-derived elements, such as viriforms)" (https://ictv.global/about/code).

History of viroid classification

The discovery of the first two viroids, potato spindle tuber viroid (PSTVd) (Diener, 1971) and citrus exocortis viroid (Semancik and Weathers, 1972) in potato and citrus, respectively, was followed by the identification of several other viroids infecting other hosts in the following years, creating the need to classify these emerging infectious agents. Early classification attempts proposed to group viroids on the bases of similarities in sequence (Puchta et al., 1988) or structure such as the central conserved region (CCR) (Koltunow and Rezaian, 1989). The CCR is a motif conserved in most of the viroids that were known at the time (Keese and Symons, 1985) and plays a major role in their replication (Gas et al., 2007; Riesner et al., 1979). The ICTV 5th Report included the first official viroid classification scheme, which recognized 19 species, thus classifying viroids only at the lowest taxonomic rank, which corresponds to the species level (Randles and Rezaian, 1991). The first formal proposal of a viroid classification including higher taxonomic ranks, such as genera, was advanced based on a phylogenetic analysis by Elena et al. (1991). The proposed monophyletic origin from a common ancestor of all viroids and other small RNA replicons, such as viroid-like satellite RNAs, proposed by these authors was controversial (Jenkins et al., 2000; Elena et al., 2001). However, the proposal by Elena et al. (1991) had the advantage of taking into consideration the evolutionary history of viroids and established a nomenclature code to classify viroid species into a higher taxonomic rank than species (i.e., genera). Subsequently, following a proposal by Flores et al. (1998), the criteria to group viroid species in genera and families were established and a first classification framework, including two families, 7 genera and 20 viroid species was published in the 7th ICTV report (Flores et al., 2000). This original classification scheme has been essentially maintained until now, although the number of species and genera has increased overtime, mainly due to the discovery of new viroids (Table 1). The most recent ICTV report included two families, eight genera and 34 viroid species (Di Serio et al., 2018, 2021a). Very recently, the number of classified viroid species increased to 44, mainly due to the identification of new viroids through high-throughput sequencing (HTS) technology. As a result of HTS-based findings, in 2022, novel viroid species demarcation criteria have been adopted by the ICTV (see below). Table 1 summarizes the major historical phases of viroid taxonomy as inferred from the ICTV reports published since the adoption of the first viroid classification scheme about 20 years ago.

Demarcation criteria of viroid families

Family is the highest taxonomic rank of the current viroid classification. Two viroid families include members sharing relevant structural and biological features (Fig. 1). Viroids classified in the family *Pospiviroidae* have a genomic RNA that adopts a rod-like or quasi rod-like conformation in silico, in vitro and, as shown for some of them, also in vivo (Fig. 1A). All viroids belonging to this family possess two stretches of conserved nucleotide sequences located in the upper and lower strands of the rod-like structure, approximately in the middle, which form the central conserved region (CCR). The CCR upper strand is flanked by two imperfect repeats that, together with the CCR upper strand, form a transient structure (named hairpin I) implicated in the replication of the members of the family *Pospiviroidae*

A. Introduction

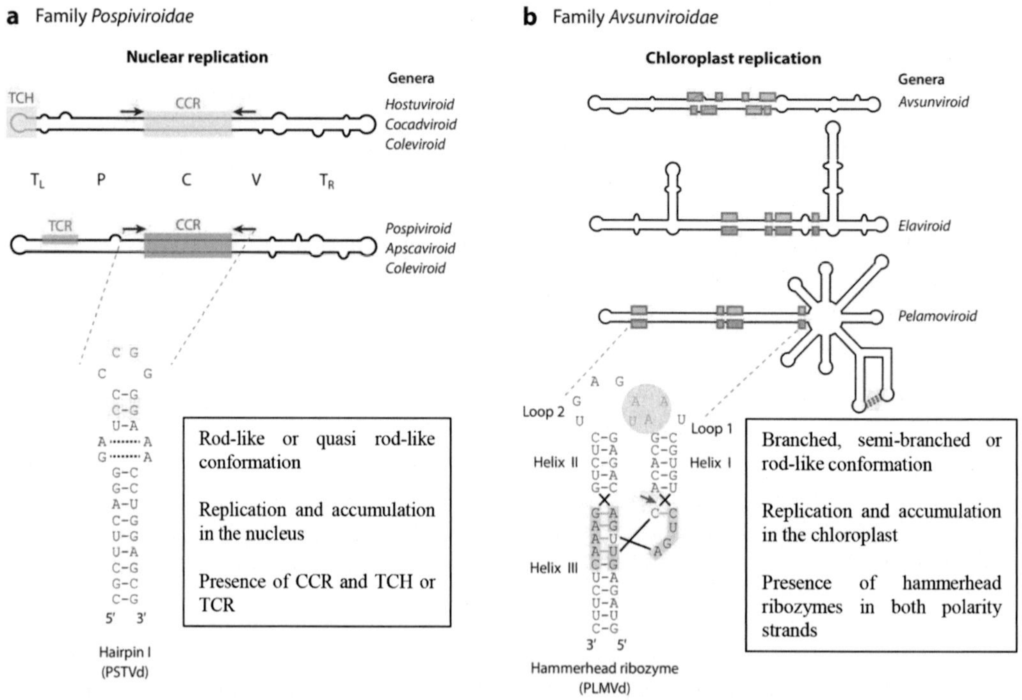

2. Viroid taxonomy

a Family *Pospiviroidae*

b Family *Avsunviroidae*

FIG. 1 Main features of viroids classified in the families *Pospiviroidae* and *Avsunviroidae*. (A) Viroids classified in the family *Pospiviroidae* replicate and accumulate in the nucleus, adopt a rod-like secondary structure containing a central conserved region (CCR) and the terminal conserved region (TCR) or the terminal conserved hairpin (TCH) (conserved domains are shaded with different colors). Members of the same genus share the same CCR. The upper CCR strand and inverted repeats (*red arrows*) flanking it form a metastable hairpin I involved in replication (*bottom left*). Nucleotides conserved in all members of the family *Pospiviroidae*, therefore corresponding to relevant positions, are marked in red. Inset: summary of major features of members of family *Pospiviroidae*. (B) Viroids classified in the family *Avsunviroidae* replicate and accumulate in the chloroplast, and may adopt rod-like, semibranched and branched conformation depending on the genus. Members of this family contain natural hammerhead structures in the plus and minus polarity strands (whose conserved domains are depicted by *pink* and *blue boxes*, respectively). The tertiary interaction between two loops identified in members of the genus *Pelamoviroid* is indicated with *dashed blue lines*. The hammerhead ribozyme structure (*bottom right*) is adopted only during replication and catalyzes the site-specific (*arrow*) self-cleavage of the replication intermediate viroid RNAs. *Continuous and dashed lines* represent Watson-Crick and noncanonical base pairs, respectively. Inset: summary of major features of members of family *Avsunviroidae*. *Modified from Navarro, B., Flores, R., Di Serio, F., 2021. Advances in viroid-host interactions. Annu. Rev. Virol. 8, 305–325.*

(see chapter viroid replication and movement) (Fig. 1A). In contrast, viroids classified in the family *Avsunviroidae* frequently assume branched or semibranched conformation (Fig. 1B) and lack the CCR. Instead, both polarity RNA strands of these viroids contain hammerhead ribozyme (HH) sequences (Fig. 1B), which are structural domains with catalytic activity. The term "ribozyme" is used to designate RNA motifs that catalyze chemical reactions, an activity generally provided by proteins (enzymes). In the case of viroids of the family *Avsunviroidae*, the site-specific cleavage of viroid RNAs during replication (a self-cleavage reaction) is mediated by the HH contained in the viroid RNA sequence itself, without the activity of any host enzyme. This is a relevant difference compared to the members of the family *Pospiviroidae*, for

A. Introduction

which the same activity is very likely provided by a host protein, which to date has not been identified. The subcellular site of replication and accumulation is also different for the members of the two families. Several representative members of the family *Pospiviroidae* have been shown to replicate and accumulate in the nucleus and, by extension, this organelle is considered the replication and accumulation site for all the members of the family. In contrast, members of the family *Avsunviroidae* replicate and accumulate in the chloroplast. Due to their different subcellular localization, the host enzymes involved in viroid transcription and ligation during replication are also different for the viroids of the two families, that is nuclear enzymes for *Pospiviroidae* viroids and chloroplastic enzymes for the members of the and *Avsunviroidae* family (reviewed in Navarro et al., 2021; see also the chapter "viroid replication and movement").

The divergent structural, biochemical and biological features of members belonging to the two families argue against their origin from a common ancestor viroid. In this respect, it has been recently proposed that viroids of families *Pospiviroidae* and *Avsunviroidae*, as well as all the other viroid-like RNA replicons, may have been originated from different protoviroids (small replicating RNAs with ribozymatic activities) populating the RNA world proposed to precede the appearance of life based on cells, DNA and proteins (reviewed by Flores et al., 2022).

Demarcation criteria of viroid genera

Structural features are taken into consideration to demarcate viroid genera. Members of the family *Pospiviroidae* are classified in a genus depending on: (i) the type of CCR, (ii) the presence or absence of two other conserved regions, the terminal conserved region (TCR) and the terminal conserved hairpin (TCH) (Flores et al., 1997; Koltunow and Rezaian, 1989; Puchta et al., 1988) and (iii) the clustering in phylogenetic trees based upon whole-genome sequences (Di Serio et al., 2021a). Based on these criteria, viroids of the family *Pospiviroidae* have been classified into the five genera *Apscaviroid*, *Avsunviroid*, *Coleviroid*, *Cocadviroid* and *Pospiviroid*. Members of each genus share a characteristic CCR, which differs from that of other genera, and contain the TCR or the TCH sequence. The functional roles of these two conserved motifs are unknown, but they have never been found simultaneously in the same viroid, thus appearing alternative to each other. Members of the genera *Pospiviroid* and *Apscaviroid* (also indicated with the vernacular name apscaviroids and pospiviroids) contain a TCR, while those of the genera *Hostuviroid*, *Cocadviroid* and *Coleviroid* (vernacular names hostuviroids, cocadviroids and coleviroids) generally contain a TCH sequence (Fig. 1A). Although these criteria provide the perfect framework for the classification of most *Pospiviroidae*, they are not completely applicable to some. Indeed, these motifs are assembled in a different combination in some viroids likely originated by recombination events. For example, coleus blumei viroid 1 (CbVd-1) and dahlia latent viroid (DLVd) contain the typical CCR of coleviroids and hostuviroids, but the TCH is replaced by a TCR. Columnea latent viroid (CLVd), a member of the genus *Pospiviroid*, contains the TCR as the other members of this genus, but its CCR is identical to that of hostuviroids. In these situations, the final classification was consistent with the results of the phylogenetic analyses and took into consideration other biological aspects such as the host range (Di Serio et al., 2021a).

A. Introduction

In the family *Avsunviroidae*, demarcation criteria of genera are based on the G/C content, the conformation of the genomic RNAs (rod-like, semibranched, branched), which is somehow related with the solubility in 2M LiCl, and with the morphology of the HHs (Fig. 1B). Based on these criteria, three genera have been created named *Avsunviroid*, *Pelamoviroid* and *Elaviroid* (Di Serio et al., 2018). Avocado sunblotch viroid, the unique member of the genus *Avsunviroid*, assumes a rod-like conformation in silico, in vitro and in vivo (Fig. 1B), is soluble in 2M LiCl and contains low G/C content and thermodynamically unstable HHs. These unstable HHs most likely became active when more stable double hammerhead structures are formed in oligomeric viroid RNAs during replication (Davies et al., 1991; Forster et al., 1988). In contrast, members of the genus *Pelamoviroid* have a branched genomic RNA stabilized by a tertiary structural element (kissing loop), are insoluble in 2M LiCl, have a high G/C content and contain thermodynamically stable HHs (Fig. 1B). Stable HH structures are also formed by variants of the unique species (*Eggplant latent viroid*) classified in the genus *Elaviroid*, which, however, have intermediate structural properties between the members of the other two genera (Fig. 1B).

Species demarcation criteria

Similar to other RNA replicons, viroids are quasispecies; they accumulate in a single host as a population of heterogeneous sequence variants (Codoñer et al., 2006). Likely due to the low fidelity of the DNA-dependent RNA polymerase involved in their replication (Navarro et al., 2021), the quasispecies nature of viroids poses serious challenges from a taxonomic point of view. In fact, each viroid species includes several sequence variants that differ slightly from each other in a few nucleotide positions. Therefore, it is difficult to establish how much two viroid variants must differ from one other to be considered members of different species. In the first classification scheme (Table 1), such a problem was addressed by establishing that members of two different species should have genomes sharing less than 90% sequence identity. Although this demarcation criterion was arbitrary, it was consistent with the different biological features of viroids known at that time. Starting from 2005, besides sequence identity lower that 90%, ICTV requested evidence of divergent biological features with respect the closest related viroid as an additional mandatory criterion for establishing a new viroid species.

Therefore, investigation on host range, pathogenesis, mode of transmission, or any other biological aspect was needed to finalize taxonomic proposals. However, these more stringent criteria were difficult to apply to several viroids that had a restricted host range or were latent in most hosts. In the case of viroids infecting woody hosts, for example, this kind of studies might be even more complex due to the time- and cost-demanding experiments needed. Moreover, the use of HTS for diagnostic purposes has allowed the discovery of many new viroids in the last few years (Di Serio et al., 2018). Consequently, the number of unclassified new viroids has increased significantly over time, generating a situation where about one-third of known viroids were unclassified as of 2021.

To find a solution to this inconvenient, ICTV recently adopted new species demarcation criteria for viroids. These criteria are based on pairwise identity matrices, adapting to viroids an approach also proposed for the classification of some virus taxa, including the families *Potyviridae* (Adams et al., 2005) and *Geminiviridae* (Muhire et al., 2013; Varsani et al., 2014;

Brown et al., 2015). In the case of viroids, it was shown that if pairwise identity scores (PWISs) are calculated for all the sequence variants of viroids already classified within each family, and if the resultant PWIS frequencies are plotted, those scores between variants classified in the same species had the highest values and clustered together in the distribution profiles (Fig. 2; Chiumenti et al., 2021).

Extending these analyses to the genera and species, the highest PWISs in the calculated matrices corresponded always to comparisons between variants of the same species that are generally grouped in a single cluster, separated from all the other pairwise values. Therefore, for each viroid species, it was possible to identify the lowest PWIS among the variants of that species (minimum PWIS, mPWIS) (Table 2). It was observed that (i) for most viroid species, the mPWIS was lower than the arbitrarily established value of 90% sequence identity used in the previous classifications schemes, and (ii) the mPWISs were quite different among viroid species (Table 2).

These findings made evident the need of adopting more appropriate sequence identity values for demarcating viroid species and highlighted the opportunity of not establishing a single threshold identity score (TIS) for all viroids, as previously done using a 90% sequence identity to classify viroids in new species (Chiumenti et al., 2021).

By the ratification of a proposal on new species demarcation criteria (Di Serio et al., 2021b), the ICTV decided to use a different TIS for each genus. The TIS of a genus corresponds to the lowest mPWIS calculated for the species classified in that genus (Table 2) rounded down to the first integer (i.e., for genus *Apscaviroid* mPWIS = 78.4%, the corresponding TIS will be 78%). Therefore, to establish that a new viroid should be classified in a new species, the relevant criteria to fill is that the PWIS with the closest viroid variants already classified is lower that the TIS of the pertinent genus (Fig. 3).

Based on this rule, eight new viroids reported in Fig. 3 have been recently classified in respective eight new species in the genus *Apscaviroid*. In fact, in all these cases, the maximum PWIS with the respective closest related variants (from 34.5% to 69.9%) was clearly lower than the TIS established for this genus (78%). In contrast, the two viroids, apple fruit crinkle viroid (AFCVd) and grapevine yellow speckle viroid 3 (GYSVd-3), had maximum PWISs higher than the TIS (99.4% and 89.4%, respectively, Fig. 3); thus, they do not fulfill the criterion to be classified as new species. In addition, for GYSVd-3, the maximum PWIS (99.4%) derived from a comparison with a variant of the species *Grapevine yellow speckle viroid 1*, and it was in the range of the PWISs calculated for the viroid variants already classified in this species (from 81.5% to 100%) (Table 1). It was concluded that GYSVd-3 should be considered as an additional viroid variant of the species *Grapevine yellow speckle viroid 1*. In agreement with this, both GYSVd-1 and GYSVd-3 infect grapevine. In the case of AFCVd, a viroid infecting apple, the maximum PWIS (89.2%) was derived from a comparison with a variant of the species *Australian grapevine viroid*. However, this value is below the PWIS range calculated for all the variants already classified in this species (from 93.8% to 100%, Table 1). In addition, variants of the *Australian grapevine viroid* are known to infect grapevine but not apple. In this case, the classification of AFCVd as an additional variant of the *Australian grapevine viroid* species could have the consequence of lowering the current minimum PWIS from 93.8% to 89.4%. In similar situations, ICTV requires provision of biological data to support the classification based on sequence comparison. In the absence of the required biological information, the classification of AFCVd remains pending. In the same proposal by Di Serio et al. (2021b), it is also

A. Introduction

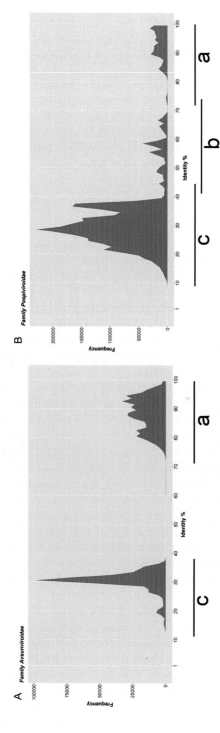

FIG. 2 Frequency distribution of PWISs calculated considering full-length viroid variants classified in the family *Avsunviroidae* (A) and *Pospiviroidae* (B). PWISs between variants classified in the same species are clustered together at high values (a) and separated from the PWISs corresponding to comparisons between variants classified in different species in the same genus (b) or in different genera (c). *Modified from Chiumenti, M., Navarro, B., Candresse, T., Flores, R., Di Serio, F., 2021. Reassessing species demarcation criteria in viroid taxonomy by pairwise identity matrices. Virus Evol. 7, veab001.*

TABLE 2 Minimum PWISs between the sequence variants of each viroid species recognized by the International Committee on Taxonomy of Viruses before the last ratification of new species in March 2022.

Family	Genus	Species	Abbreviation	Minimum PWIS*
Pospiviroidae	*Pospiviroid*	Potato spindle tuber viroid	PSTVd	88,4
		Citrus exocortis viroid	CEVd	**83,5**
		Chrysanthemum stunt viroid	CSVd	91,9
		Columnea latent viroid	CLVd	85,2
		Iresine viroid 1	IrVd1	96,5
		Pepper chat fruit viroid	PCFVd	92,6
		Tomato apical stunt viroid	TASVd	84,2
		Tomato chlorotic dwarf viroid	TCDVd	95,0
		Tomato planta macho viroid	TPMVd	90,9
	Hostuviroid	Hop stunt viroid	HSVd	**79,4**
		Dahlia latent viroid	DLVd	100
	Cocadviroid	Coconut cadang cadang viroid	CCCVd	97,6
		Coconut tinangaja viroid	CTVd	100
		Citrus bark craking viroid	CBCVd	**79,9**
		Hop latent viroid	HLVd	96,9
	Apscaviroid	Apple scar skin viroid	ASSVd	87,7
		Apple dimple fruit viroid	ADFVd	79,1
		Australian grapevine viroid	AGVd	93,8
		Citrus bent leaf viroid	CBLVd	**78,4**
		Citrus dwarfing viroid	CDVd	85,2
		Citrus viroid V	CVd-V	88,8
		Citrus viroid VI	CVd-VI	90,4
		Grapevine yellow speckle viroid 1	GYSVd-1	81,5
		Grapevine yellow speckle viroid 2	GYSVd-2	95,6
		Pear blister canker viroid	PBCVd	87,1
	Coleviroid	Coleus blumei viroid 1	CbVd-1	96,0
		Coleus blumei viroid 2	CbVd-2	98,7
		Coleus blumei viroid 3	CbVd-3	**91,8**
Avsunviroidae	*Avsunviroid*	Avocado sunblotch viroid	ASBVd	**92,4**
	Pelamoviroid	Peach latent mosaic viroid	PLMVd	**73,9**
		Chrysanthemum chlorotic mottle viroid	CCMVd	91,0
		Apple hammerhead viroid	AHVd	80,3
	Elaviroid	Eggplant latent viroid	ELVd	**83,9**

* The lowest PWISs among species of each genus are reported in red on a yellow background.

Modified from Chiumenti, M., Navarro, B., Candresse, T., Flores, R., Di Serio, F., 2021. Reassessing species demarcation criteria in viroid taxonomy by pairwise identity matrices. Virus Evol. 7, veab001.

A. Introduction

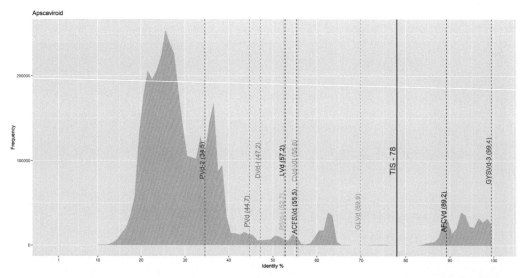

FIG. 3 Frequency distribution of PWISs between full-length sequence variants of viroids of the genus *Apscaviroid*. The TIS, used as a species demarcation criterion for each genus (in *red*), corresponds to the lowest PWISs among the minimum PWISs calculated for each species in the genus (see Table 2). According with the current species demarcation criteria, only the viroids that, in a PWIS matrix containing the comparisons with all the other viroid variants already classified in the genus, had a maximum PWIS (indicated on the bottom of the graphic for each viroid) lower than the TIS have been classified in new species. ACFSVd, AFCVd, CVd-VII, DVd, GLVd, GYSVd-3 LVd, PlVD-I PVd. PVd-2. *Modified from Chiumenti, M., Navarro, B., Candresse, T., Flores, R., Di Serio, F., 2021. Reassessing species demarcation criteria in viroid taxonomy by pairwise identity matrices. Virus Evol. 7, veab001.*

stated that the proposed novel species demarcation criterion based on pairwise identity matrices cannot solve other complex situations, such as those where the PWIS of the novel unclassified variant is very close to the threshold identified at the genus level. Likewise, in these cases, biological criteria should be taken into consideration to establish whether a new species should be created or not.

Another relevant point highlighted in the adopted ICTV proposal (Di Serio et al., 2021b) is that if the ability of these RNAs to systemically infect their hosts in the absence of coinfection of helper viruses has not been confirmed, the proposed method should not be considered sufficient for the classification of viroid-like RNAs containing hammerhead ribozymes in both polarity strands, which would be suggestive members of the *Avsunviroidae* family. Indeed, "akin to some viroids, several viroid-like satellite RNAs consist of small circular RNAs containing hammerhead ribozymes in both polarity strands. Therefore, biological data supporting the autonomous replication of the candidate viroid are considered as needed before proceeding to its classification, especially in the case of potential members of the family *Avsunviroidae*" (Di Serio et al., 2021b). There are several viroid-like RNAs containing hammerhead ribozymes in both polarity strands for which the autonomous replication capability is still pending; for these reasons, they have not yet been classified. Among them is grapevine hammerhead viroid-like RNA, which was identified in grapevine a decade ago (Wu et al., 2012).

The current species demarcation criteria as ratified by ICTV are reported here below (Di Serio et al., 2021b):

Family *Pospiviroidae*

Genus Apscaviroid

Viroids with rod-like or quasi rod-like conformation, with the TCR, with the CCR identical to that of members of the other species of the genus and with less than 78% pairwise sequence identity with respect to the members of the genus are classified in different species. For viroids with pairwise identity scores close to 78% evidence of distinct biological properties should be provided.

Genus Cocadviroid

Viroids with rod-like conformation, with the TCH, with the CCR identical to that of members of the species in the genus and with less than 79% sequence identity with respect to the other members of the genus, are classified in different species. For viroids with pairwise identity scores close to 79% evidence of distinct biological properties should be provided.

Genus Coleviroid

Viroids with a rod-like conformation, with the TCR, with the CCR identical to that of members of the other species of the genus and with less than 91% sequence identity with respect to the other members of the genus are classified in different species. Members of certain species of this genus, including the viroids with the smallest genome, may contain the TCH instead of the TCR. For viroids with pairwise identity scores close to 91% evidence of distinct biological properties should be provided.

Genus Hostuviroid

Viroids with a rod-like conformation, with the TCH (or with the TCR instead of the TCH), with the CCR identical to that of members of the other species of the genus and with less than 79% sequence identity with respect to the other members of the genus are classified in different species. For viroids with pairwise identity scores close to 79% evidence of distinct biological properties should be provided.

Genus Pospiviroid

Viroids with a rod-like conformation, with the TCH, with the CCR identical to that of members of the other species of the genus and with less than 83% sequence identity with respect to the other members of the genus, should be classified in different species. For viroids with pairwise identity scores close to 83% evidence of distinct biological properties should be provided.

Family *Avsunviroidae*

Genus Avsunviroid

Avocado sunblotch viroid is the only species of the genus *Avsunviroid* reported so far. Viroids with similar molecular features (low G+C content, a rod-like conformation and thermodynamically unstable hammerhead ribozymes), but with less than 92% sequence identity should be classified as different species. Evidence of infectivity in the absence of a helper virus should be provided.

A. Introduction

2. Viroid taxonomy

Genus **Pelamoviroid**

Viroids with similar molecular features [multibranched conformations stabilized by a kissing-loop interaction in the (+) strands, and thermodynamically stable hammerhead ribozymes] but with less than 73% sequence identity should be classified as different species. Evidence of infectivity in the absence of a helper virus should be provided.

Genus **Elaviroid**

Eggplant latent viroid is the only species of the genus *Elaviroid* reported. Viroids with similar molecular features (quasirod-like conformation and thermodynamically stable hammerhead ribozymes) but with less than 83% sequence should be classified as different species. Evidence of infectivity in the absence of a helper virus should be provided.

Current viroid classification

Based on these rules, the current classification of viroids has been recently updated as reported in Table 3.

TABLE 3 Current classification of viroids.

Family	Genus	Species
	Apscaviroid	Apple dimple fruit viroid
		Apple scar skin viroid
		Apscaviroid aclsvd
		Apscaviroid cvd-VII
		Apscaviroid dvd
Pospiviroidae		Apscaviroid glvd
		Apscaviroid lvd
		Apscaviroid plvd-I
		Apscaviroid pvd
		Apscaviroid pvd-2
		Australian grapevine viroid
		Citrus bent leaf viroid
		Citrus dwarfing viroid
		Citrus viroid V
		Citrus viroid VI
		Grapevine yellow speckle viroid 1
		Grapevine yellow speckle viroid 2
		Pear blister canker viroid

A. Introduction

TABLE 3 Current classification of viroids—cont'd

Family	Genus	Species
	Cocadviroid	*Citrus bark cracking viroid*
		Coconut cadang cadang viroid
		Coconut tinangaja viroid
		Hop latent viroid
	Coleviroid	*Coleus blumei viroid 1*
		Coleus blumei viroid 2
		Coleus blumei viroid 3
		Coleviroid cbvd-5
		Coleviroid cbvd-6
	Hostuviroid	*Dahlia latent viroid*
		Hop stunt viroid
	Pospiviroid	*Chrysanthemum stunt viroid*
		Citrus exocortis viroid
		Columnea latent viroid
		Iresine viroid 1
		Pepper chat fruit viroid
		Pospiviroid plvd
		Potato spindle tuber viroid
		Tomato apical stunt viroid
		Tomato chlorotic dwarf viroid
		Tomato planta macho viroid
Avsunviroidae	*Apscaviroid*	*Avocado sunblotch viroid*
	Elaviroid	*Eggplant latent viroid*
	Pelamoviroid	*Apple hammerhead viroid*
		Chrysanthemum chlorotic mottle viroid
		Peach latent mosaic viroid

Protocols/procedures/methods

As reported above, TISs were identified by generating complex pairwise identity matrices including all the sequence variants of each viroid species available in databases. This approach needs a certain level of expertise and time-consuming calculations. Although conclusive classification of novel viroids must be confirmed using all the available sequence

A. Introduction

variants, an alternative and simplified method based on matrices generated using only the best BLASTn-matching variants was tested and was deemed appropriate for preliminary classification efforts (Chiumenti et al., 2021). The main steps of this method are reported in Fig. 4. In this approach, a pairwise identity matrix between the sequence of the potential novel viroid and those of the first 100 best BLASTn-matching sequences is generated. This exercise allows to identify the genus in which the best BLASTn-matching sequences are classified and

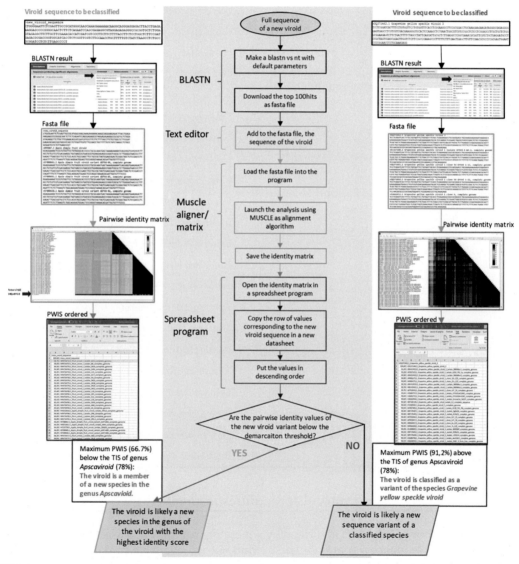

FIG. 4 Flowchart of the main steps of a fast method proposed to classify novel viroids based on pairwise identity scores (PWISs). A general diagram is shown in the middle *(green background)*, with examples of a viroid to be classified in a new or in an existing species reported on the left and on the right, respectively.

the maximum PWIS in the matrix. According with the novel criteria, these two values may be sufficient to establish how to proceed with the classification of the novel viroid: if the maximum PWIS in the matrix is lower than the TIS of the identified genus, the novel viroid could be proposed as a representative member of a novel species. In the opposite situation (where the maximum PWIS is higher than the TIS), the viroid should be classified in the same species of the most identical viroid sequence in the matrix. However, when PWIS and TIS are very close to each other, confirmation by generating more complex matrices, including as many as variants from the pertinent genus possible, is required.

Assigning a name to novel viroid species

To enter officially in the classification scheme, new viroid species need to be proposed, accepted and ratified by the ICTV. Proposals are usually done by the ICTV *Avsunviroidae* and *Pospiviroidae* study group, though anyone is welcome to prepare a taxonomy proposal. Proposals are generally reviewed and discussed by the ICTV study group before being taken into consideration by the executive committee of the ICTV. If considered acceptable, a proposal is ratified by a majority vote of the ICTV members.

ICTV recently adopted a binomial nomenclature for viruses, which was also extended to viroids. This means that the name of new viroid species is formed by two parts: the first one is identical to the name of the genus to which the species belongs, while the second part consists of the species epithet. The binomial nomenclature is quite different from the one used until 2021 and, possibly with additional adjustments, it will be eventually extended to all already classified viroid species.

Prospective and future implications

Although ICTV established taxonomic hierarchical levels from realm to species (ICTV code rule 3.2; https://ictv.global/about/code), the current viroid classification only includes species, genera and families. Whether higher ranks should be created for viroids is subject of much debate, given the unresolved issue of the phylogenetic relationships between the viroid members of the two families and between viroids and other circular RNAs sharing some structural feature with viroids, such as (a) the viroid-like satellite RNAs, which are small, circular RNAs unable to code for proteins and able to infect plants only in association with an helper virus (Navarro et al., 2017); (b) retroviroid-like RNAs, which are noninfectious viroid-like RNAs containing HHs in both polarity strands and associated with a DNA counterpart integrated in a viral DNA or in the host genome (Daròs and Flores, 1995); (c) retrozymes, which are circular RNAs containing a HH in one polarity strand and associated with a host DNA counterpart (Cervera et al., 2016); (d) hepatitis delta virus, a human pathogen consisting of a circular RNA containing ribozymes, coding for a protein and depending on hepatitis virus B for its infectivity (Flores et al., 2019); (e) other hepatitis delta virus-related RNAs recently identified in several animals (de la Peña et al., 2021); (f) several viroid-like RNAs containing HHs in both polarity strands the biological nature of which is still undetermined (Flores et al., 2022). Discovery of new viroids and viroid-like RNAs in the near future, mainly due to the application of HTS technologies, is expected to increase our knowledge on these biological entities and the phylogenetic relationships, thus providing new information to be integrated in a more complete and phylogenetically sound classification schemes.

Chapter summary

In this chapter, a historical overview of viroid taxonomy was presented highlighting the criteria currently used to demarcate viroid families, genera and species. The species demarcation criteria based on pairwise identity matrices, recently adopted by the International Committee on Taxonomy of Viruses, were discussed in more detail including the advantages and limits of these new criteria. Finally, a simplified and fast method for the preliminary classification of novel viroids was described.

Study question

Which are the viroid taxa and their demarcation criteria?
How can you establish whether a new viroid should be classified as a new viroid species?

References

Adams, M.J., Antoniw, J.F., Fauquet, C.M., 2005. Molecular criteria for genus and species discrimination within the family Potyviridae. Arch. Virol. 150, 459–479.

Andino, R., Domingo, E., 2015. Viral quasispecies. Virology 479, 46–51.

Biebricher, C.K., Eigen, M., 2006. What is a quasispecies? Curr. Top. Microbiol. Immunol. 299, 1–31.

Brown, J.K., Zerbini, F.M., Navas-Castillo, J., Moriones, E., Ramos-Sobrinho, R., Silva, J.C., Fiallo-Olivè, E., Briddon, R.W., Hernández-Zepeda, C., Idris, A., Malathi, V.G., Martin, D.P., Rivera-Bustamante, R., Ueda, S., Varsani, A., 2015. Revision of *Begomovirus* taxonomy based on pairwise sequence comparisons. Arch. Virol. 160, 1593–1619.

Cervera, A., Urbina, D., de la Peña, M., 2016. Retrozymes are a unique family of non-autonomous retrotransposons with hammerhead ribozymes that propagate in plants through circular RNAs. Genome Biol. 17, 135. https://doi.org/10.1186/s13059-016-1002-4.

Chiumenti, M., Navarro, B., Candresse, T., Flores, R., Di Serio, F., 2021. Reassessing species demarcation criteria in viroid taxonomy by pairwise identity matrices. Virus Evol. 7, veab001.

Codoñer, F.M., Darós, J.A., Solé, R.V., Elena, S.F., 2006. The fittest versus the flattest: experimental confirmation of the quasispecies effect with subviral pathogens. PLoS Pathog. 2, 1187–1193.

Daròs, J.A., Flores, R., 1995. Identification of a retroviroid-like element from plants. Proc. Natl. Acad. Sci. USA 92, 6856–6860.

Davies, C., Sheldon, C.C., Symons, R.H., 1991. Alternative hammerhead structures in the self-cleavage of avocado sunblotch viroid RNAs. Nucleic Acids Res. 19, 1893–1898.

de la Peña, M., Ceprián, R., Casey, J.L., Cervera, A., 2021. Hepatitis delta virus-like circular RNAs from diverse metazoans encode conserved hammerhead ribozymes. Virus Evol. 7 (1), veab016.

Di Serio, F., Li, S.F., Matoušek, J., Owens, R.A., Pallás, V., Randles, J.W., Sano, T., Verhoeven, J.T.J., Vodalakis, G., Flores, R., 2018. ICTV virus taxonomy profile: *Avsunviroidae*. J. Gen. Virol. 99, 611–612.

Di Serio, F., Owens, R.A., Li, S.F., Matoušek, J., Pallás, V., Randles, J.W., Sano, T., Verhoeven, J.T.J., Vidalakis, G., Flores, R., 2021a. ICTV virus taxonomy profile: *Pospiviroidae*. J. Gen. Virol. 102 (2), 001543.

Di Serio, F., Li, S.F., Matousek, J., Pallas, V., Randles, J.W., Sano, T., Verhoeven, J.T.J., Vidalakis, G., Owens, R.A., 2021b. Viroid Demarcation Criteria. https://ictv.global/files/proposals/approved?fid=4453#block-teamplus-page-title. (Accessed 9 July 2023).

Diener, T.O., 1971. Potato spindle tuber "virus". IV. A replicating, low molecular weight RNA. Virology 45, 411–428.

Elena, S.F., Dopazo, J., Flores, R., Diener, T.O., Moya, A., 1991. Phylogeny of viroids, viroid-like satellite RNAs, and the viroid-like domain of hepatitis delta virus. Proc. Natl. Acad. Sci. USA 88, 5631–5634.

Elena, S.F., Dopazo, J., De la Peña, M., Flores, R., Diener, T.O., Moya, A., 2001. Phylogenetic analysis of viroid and viroid-like satellite RNAs from plants: a reassessment. J. Mol. Evol. 53, 155–159.

Flores, R., 1995. Subviral agents: viroids. In: Murphy, F.A., Fauquet, C.M., Bishop, D.H.L., Said, A.G., Jarvis, A.W., Martelli, G.P., Mayo, M.A., Summers, M.D. (Eds.), Virus Taxonomy, Sixth Report of the International Committee on the Taxonomy of Viruses. Elsevier/Academic Press, San Diego, CA, pp. 1009–1024. 495–497.

Flores, R., Di Serio, F., Hernández, C., 1997. Viroids: the noncoding genomes. Semin. Virol. 8, 65–73.

Flores, R., Randles, J.W., Bar-Joseph, M., Diener, T.O., 1998. A proposed scheme for viroid classification and nomenclature. Arch. Virol. 143, 623–629.

Flores, R., Randles, J.W., Owens, R.A., Bar-Joseph, M., Diener, T.O., 2000. Subviral agents: viroids. In: Regenmortel, M.H.V., Fauquet, C.M., Bishop, D.H.L., Carsten, E.B., Estes, M.K., Lemon, S.M., Wickner, R.B. (Eds.), Virus Taxonomy, Seventh Report of the International Committee on the Taxonomy of Viruses. Elsevier/Academic Press, San Diego, CA, pp. 1009–1024.

Flores, R., Randles, J.W., Owens, R.A., Bar-Joseph, M., Diener, T.O., 2005. Subviral agents: viroids. In: Fauquet, C.M., Mayo, M.A., Maniloff, J., Desselberger, U., Ball, L.A. (Eds.), Virus Taxonomy, Eighth Report of the International Committee on the Taxonomy of Viruses. Elsevier/Academic Press, San Diego, CA, pp. 1147–1161.

Flores, R., Serra, P., Delgado, S., Navarro, B., Di Serio, F., 2019. Human hepatitis D virus and plant viroids: trans-kingdom similarities between small infectious circular RNAs. In: Rizzetto, M., Smadile, A. (Eds.), Hepatitis D: Virology, Management and Methodology. Il Pensiero Scientifico Editore, Rome, Italy, pp. 15–25.

Flores, R., Navarro, B., Serra, P., Di Serio, F., 2022. A scenario for the emergence of protoviroids in the RNA world and for their further evolution into viroids and viroid-like RNAs by modular recombinations and mutations. Virus Evol. 8, veab107.

Forster, A.C., Davies, C., Sheldon, C.C., Jeffries, A.C., Symons, R.H., 1988. Self-cleaving viroid and newt RNAs may only be active as dimers. Nature 334, 265–267.

Gas, M.E., Hernández, C., Flores, R., Darós, J.A., 2007. Processing of nuclear viroids in vivo: an interplay between RNA conformations. PLoS Pathog. 3, e182. https://doi.org/10.1371/journal.ppat.0030182.

Jenkins, G.M., Woelk, C.H., Rambaut, A., Holmes, E.C., 2000. Testing the extent of sequence similarity among viroids satellite RNAs and the hepatitis delta virus. J. Mol. Evol. 50, 98–102.

Keese, P., Symons, R.H., 1985. Domains in viroids: evidence of intermolecular RNA rearrangements and their contribution to viroid evolution. Proc. Natl. Acad. Sci. USA 82, 4582–4586.

Koltunow, A.M., Rezaian, M.A., 1989. A scheme for viroid classification. Intervirology 30, 194–201.

Muhire, B., Martin, D.P., Brown, J.K., Navas-Castillo, J., Moriones, E., Zerbini, F.M., Rivera-Bustamante, R., Malathi, V.G., Briddonn, R.W., Varsani, A., 2013. A genome-wide pairwise-identity-based proposal for the classification of viruses in the genus *Mastrevirus* (family *Geminiviridae*). Arch. Virol. 158, 1411–1424.

Navarro, B., Rubino, L., Di Serio, F., 2017. Small circular satellite RNAs. In: Hadidi, A., Flores, R., Randles, J.W., Palukaitis, P. (Eds.), Viroids and Satellites. Academic Press, Cambridge, UK, pp. 659–669.

Navarro, B., Flores, R., Di Serio, F., 2021. Advances in viroid-host interactions. Annu. Rev. Virol. 8, 305–325.

Owens, R.A., Flores, R., Di Serio, F., Li, S.F., Pallás, V., Randles, J.W., et al., 2012. Viroids. In: King, A.M.Q., Adams, M.J., Carstens, E.B., Lefkowitz, E.J. (Eds.), Virus Taxonomy, Ninth Report of the International Committee on Taxonomy of Viruses. Elsevier/Academic Press, London, UK, pp. 1221–1234.

Puchta, H., Ramm, K., Sänger, H.L., 1988. The molecular structure of hop latent viroid (HLVd), a new viroid occurring worldwide in hops. Nucleic Acids Res. 16, 4197–4216.

Randles, J.W., Rezaian, M.A., 1991. Viroids. In: Francki, R.I.B., Fauquet, C.M., Knudson, D.L., Brown, F. (Eds.), Classification and Nomenclature of Viruses, Fifth Report of the International Committee on the Taxonomy of Viruses. Springer-Verlag, New York, NY, pp. 403–405.

Riesner, D., Henco, K., Rokohl, U., Klotz, G., Kleinschmidt, A.K., Domdey, H., Jank, P., Gross, H.J., Sänger, H.L., 1979. Structure and structure formation of viroids. J. Mol. Biol. 133, 85115.

Semancik, J.S., Weathers, L.G., 1972. Exocortis virus: an infectious free-nucleic acid plant virus with unusual properties. Virology 47, 456–466.

A. Introduction

Varsani, A., Martin, D.P., Navas-Castillo, J., Moriones, E., Hernández-Zepeda, C., Idris, A., Murilo Zerbini, F.,
    Brown, J.K., 2014. Revisiting the classification of curtoviruses based on genome-wide pairwise identity. Arch.
    Virol. 159, 1873–1882.
Wu, Q.F., Wang, Y., Cao, M.J., Pantaleo, V., Burgyan, J., Li, W.X., Ding, S.W., 2012. Homology-independent discovery
    of replicating pathogenic circular RNAs by deep sequencing and a new computational algorithm. Proc. Natl.
    Acad. Sci. USA 109, 3938–3943.

A. Introduction

3

Structure of viroids

Jean-Pierre Perreault, François Bolduc,
and Charith Raj Adkar-Purushothama

RNA Group, Department of Biochemistry and Functional Genomics, Faculty of
Medicine and Health Sciences, Applied Cancer Research Pavilion, University of Sherbrooke,
Sherbrooke, QC, Canada

Graphical representation

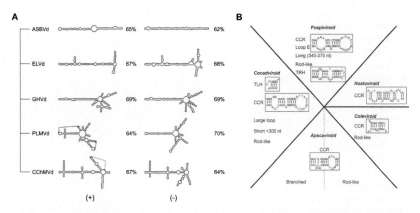

(A) A schematic representation of the most stable structures for both polarities of each species of the *Avsunviroidae* family. The structures are classified to illustrate the progression in complexity from the most rod-like to the most complex structure. The numbers shown represent the percentage of base pairs. *ASBVd*, avocado sunblotch viroid; *CChMVd*, chrysanthemum chlorotic mottle viroid; *ELVd*, eggplant latent viroid; *GHVd*, grapevine hammerhead viroid-like RNA; *PLMVd*, peach latent mosaic viroid (Source: Reproduced from doi:10.1111/mpp.12130 after minor modifications. License # 5566620773851). (B) The boxed structures are representative examples for each genus of the family *Pospiviroidae*. The color of the nucleotide represents the level of accessibility as determined by selective 2′-hydroxyl acylation analyzed by primer extension (SHAPE): namely the *black* nucleotides are of low reactivity (0–0.40), the *orange* nucleotides are of intermediate reactivity (0.40–0.85) and those in *red* are of high reactivity (>0.85). The *underlined* nucleotides are very reactive (>2.0). *CCR*, central conserved region; *TRH*, terminal right hairpin; *TLH*, terminal left hairpin (Source: Giguère, T., Perreault, J.-P., 2017. Classification of the *Pospiviroidae* based on their structural hallmarks. PLoS One 12(8), e0182536. Modified and used as per terms of CC BY 4.0).

Abbreviations

C	central domain
CCR	central conserved region
hSHAPE	high-throughput selective 2′-hydroxyl acylation analyzed by primer extension
P	pathogenic domain
T_L	terminal left domain
T_R	terminal right domain
V	variable domain
VMR	virulence modulating region

Viroid

ADFVd	apple dimple fruit viroid
AGVd	Australian grapevine viroid
AHVd	apple hammerhead viroid
ASBVd	avocado sunblotch viroid
ASSVd	apple scar skin viroid
CBCVd	citrus bark cracking viroid
CBLVd	citrus bend leaf viroid
CbVd-1-6	Coleus blumei viroid-1 to -6
CCCVd	coconut cadang-cadang viroid
CChMVd	chrysanthemum chlorotic mottle viroid
CDVd	citrus dwarfing viroid
CEVd	citrus exocortis viroid
CLVd	columnea latent viroid
CSVd	chrysanthemum stunt viroid
CTiVd	coconut tinangaja viroid
CVd-V	citrus viroid-V
CVd-VI	citrus viroid-VI (formerly CVd-OS)
DLVd	dahlia latent viroid
ELVd	eggplant latent viroid
GYSVd-1	grapevine yellow speckle viroid-1
GYSVd-2	grapevine yellow speckle viroid-2
HLVd	hop latent viroid
HSVd	hop stunt viroid
IrVd	Iresine viroid 1
MPVd	Mexican papita viroid
PBCVd	pear blister canker viroid
PCFVd	pepper chat fruit viroid
PLMVd	peach latent mosaic viroid
PSTVd	potato spindle tuber viroid
TASVd	tomato apical stunt viroid
TCDVd	tomato chlorotic dwarf viroid
TPMVd	tomato planta macho viroid

Definitions

Primary structure: The linear sequence of ribonucleotides (A, C, G, and U) linked by phosphodiester bonds constitutes the primary structure of an RNA molecule.

A. Introduction

Secondary structure: RNA secondary structure consists of nucleotides that are in one of two states, paired or unpaired, where pairing includes all base-base interactions. In general, most base pairings are adjacent and antiparallel with other base pairings to form secondary structure helices. The combination of one or more helical elements interspersed with unpaired, single-stranded nucleotides constitutes an RNA structure.

Tertiary structure: Tertiary structure refers to the locations of the nucleotides in three-dimensional space, taking into consideration geometrical and steric constraints. It is a higher order than the secondary structure, in which large-scale folding in a linear polymer occurs, and the entire chain is folded into a specific three-dimensional shape.

Quaternary structure: The quaternary structure refers to a higher-level of organization of RNA. More specifically, it refers to the interaction of nucleic acids with other molecules such as chemical ligands, proteins, and other RNA molecules.

Quasispecies: A quasispecies is a population structure of a given virus or a viroid which consists of a large number of sequence variants.

Chapter outline

Viroids are circular, single-stranded, infectious, noncoding RNAs that autonomously replicate in their host plants. Since viroids lack coding capability and also lack a capsid, they must rely on both their nucleotide sequence and structural motifs to utilize the host's systems to ensure that all of the different mechanisms that are required for their life cycle (e.g., replication, processing, transport, and pathogenesis) can take place. This chapter summarizes the established knowledge of the secondary and tertiary structural features of viroids that are critical for their existence and biological functioning in their hosts.

Learning objectives

- Objective 1: Secondary structure of viroids.
- Objective 2: Tertiary structure of viroids.
- Objective 3: Different methods used for viroid structure determination.
- Objective 4: Importance of structural features in the life cycle of a viroid.
- Objective 5: Viroid classification based on the structure.

Fundamental introduction

Viroids are plant infectious circular RNA molecules. Though recent research has reported the viroid infection of fungus, it is not conclusive. In terms of structure, viroids can be simply described by their characteristics. In brief, viroids consist of a covalently closed, circular, single-stranded, RNA molecule. They exhibit several significant differences when compared to viruses. Their genome sizes are significantly smaller than those of even the smallest known viruses (see Table 1). Unlike viruses, viroids completely lack any sort of protein layer around

A. Introduction

TABLE 1   Distinctive characteristics between viroids and viruses.

Characteristic	Virus	Viroid
Genetic material	• DNA or RNA • Linear • ssDNA or dsDNA or ssRNA or dsRNA • Size: >1.7 kb	• RNA • Circular • ssRNA • Size: between 246 and 401 nt
Translation	• Codes for proteins	• No
Capsid/protein coat	• Yes	• No
Replication	• Nonrolling circle • Viral polymerase	• Rolling circle • Host's DNA dependent RNA polymerase
Host	• Bacteria • Fungi • Plant • Animals	• Plants

their genetic material. Most importantly, to our current knowledge, the viroid RNA genome has not been demonstrated so far to code for peptide. This latter characteristic is fundamental to viroids and has as a consequence that viroids must rely on host components (i.e., enzymes and other host factors) to ensure their survival, replication, processing, transport, and pathogenesis. More precisely, their nucleotide sequence and structural motifs must trigger the host machinery to support their life cycle.

Viroids range in length from 246 to 401 nucleotides (nt), depending on the species and sequence variant. All known viroids have been classified into two families, the *Avsunviroidae* and the *Pospiviroidae* (Di Serio et al., 2014). Everything indicates that this classification, which was made possible by comparative sequence analyses and whether the viroid in question possessed a central conserved region (CCR), resulted from their cellular localizations and replication mechanisms. Viroids replicate via a rolling-circle mechanism using either an asymmetric or a symmetric mode (see Fig. 1 and Chapter 4). In either case, replication includes a processing step that converts multimeric replication intermediates to one-unit-length molecules. Multimeric strands of *Avsunviroidae* accomplish this step via a self-catalytic cleavage, while *Pospiviroidae* involves host proteinaceous nucleases. It has been proposed that this is most likely due to the specific cellular compartment in which their replication takes place (Bussière et al., 1999). Replication intermediates of *Avsunviroidae* are located in the chloroplast, while those of *Pospiviroidae* are found in the nucleus. The fact that the proteome of the nucleus is larger than that of the chloroplasts may be a factor in favor of the RNA self-cleavage activity demonstrated by the members of *Avsunviroidae*.

The elucidation of the structures adopted by viroids has always attracted a significant amount of attention from the research community. It is paramount to understand the different mechanisms implicated in their life cycles. In general, the secondary structure of viroid species has been predicted using computer software (Bussière et al., 1996; Fadda et al.,

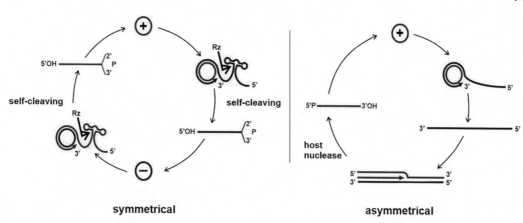

symmetrical   asymmetrical

FIG. 1 Rolling circle mechanism of viroid replication. *Avsunviroidae* uses the symmetrical mode of replication, while *Pospiviroidae* uses the asymmetrical mode. Rz, 5'OH and 2',3'P indicate the ribozyme (self-cleaving motif), the 5'-hydroxyl group, and the 2'-3'-phosphodiester groups, respectively.

2003; Navarro and Flores, 1997; Symons, 1981). The use of classical structural methods in solution for the determination of viroid structure was limited by the size of the RNA molecule involved. More recently, advances over the last decade have led to the elucidation of the structures in the solution of most known viroid species. All exhibit a high degree of base pairing throughout the molecule.

Avsunviroidae

The primary structure of the model peach latent mosaic viroid

Peach latent mosaic viroid (PLMVd) is the member of the family *Avsunviroidae* that has received the most attention over the years. The first report dated from the early 1990s and included both a primary structure (i.e., the nucleotide sequence) and a predicted secondary structure (Flores et al., 1990). Since, more than 300 distinct sequences of PLMVd have been reported after the cloning and small-scale sequencing of RT-PCR generated fragments that were amplified from total RNA isolated from either a single tree or from a group of trees (Ambrós et al., 1998; Fekih Hassen et al., 2007). These data were sufficient to demonstrate that PLMVd accumulates in the host plant as a complex mix of sequence variants, called quasispecies, with genome lengths varying from 336 to 339 residues, with one exception, specifically the peach calico variant. This particular variant (349 nucleotides), which is responsible for severe symptoms, includes an extra stem-loop structure that is located at the left-hand end of the structure (Fig. 2A, red left inset; Malfitano et al., 2003).

The development of high-throughput sequencing has permitted the deep investigation of PLMVd in terms of its nucleotide sequences. An original experiment including bioinformatic refinement of sequence libraries based on a systematic analysis of the self-cleavage activity, to remove technical sequencing errors that may significantly bias the data, yielded 3939 other distinct PLMVd sequence variants after 6 months of infection (Glouzon et al., 2014). Most

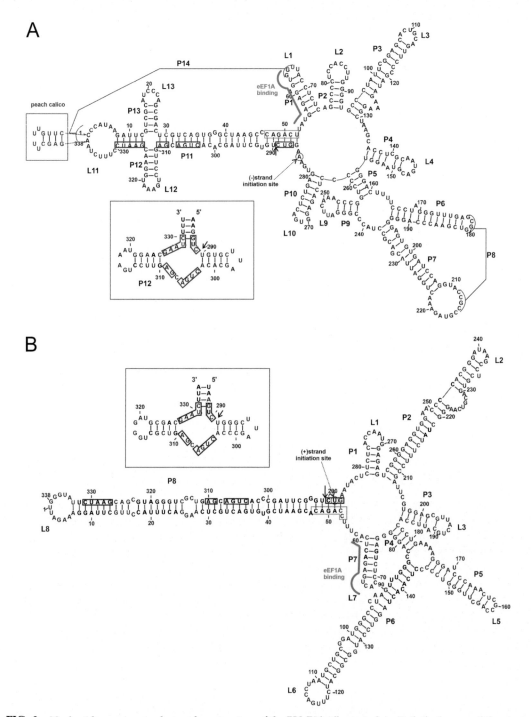

FIG. 2 Nucleotide sequence and secondary structure of the PLMVd Alberta isolate. Both the loops and the stems are indicated. (A) The (+) strand's structure. The *black arrowhead* indicates the hammerhead self-cleavage site, and the *red boxes* identify the core nucleotides of the hammerhead structure. The *hammerhead structure* is illustrated in the inset. The peach calico insertion consisting of a short stem-loop that is found in a few variants is shown in the *pink* inset. The *orange box* highlights a highly conserved CAGAC box that is reminiscent of the sequence found in the vicinity of the PLMVd initiation sites for strands of both polarities (Motard et al., 2008). The *black dotted-line arrow* pointing to the circled nucleotide indicates the universal initiation site. Finally, the *green lines* indicate the binding region of the elongation factor eEF1A. (B) The (−) strand's structure. The color code is as for panel A, except the hammerhead core nucleotide that is in *blue boxes*. P and L on the structure indicate paired and loop regions, respectively.

of the sequences obtained in this experiment contained an average of 4.6 to 6.4 mutations when compared to the initially inoculated PLMVd variant. However, variants with up to 20% mutated residues were also reported in the analysis. The use of a hierarchical clustering algorithm has permitted the subdivision of the pool of PLMVd variants into seven clusters. Further analysis revealed that each cluster exhibited key mutations that were the hallmark of that particular cluster. Even if, initially, it was assumed that most sequence positions would mutate, 50% remained perfectly conserved, including several small stretches as well as a small motif important for the self-cleaving structures and one motif reminiscent of a stable tetraloop. Together, these features revealed not only the quasispecies nature of the primary structure of PLMVd but also the fact that selective pressures, which remain to be determined, limited this variation.

The secondary structure of PLMVd

Over the years, the secondary structures of several PLMVd sequence variants have been elucidated in solution based on classical biochemical methods such as ribonuclease digestion and oligonucleotide binding assays (Bussière et al., 2000) and, more recently, by RNA selective 2′-hydroxyl acylation analyzed by primer extension (SHAPE) and high-throughput SHAPE (hSHAPE) (Dubé et al., 2011; Giguère et al., 2014a). In SHAPE, RNA structural information is determined by treatment with chemicals, such as *N*-methylisatoic anhydride and benzoyl cyanide that react selectively with flexible RNA nucleotides at the ribose's 2′-hydroxyl group, which discriminates between paired and unpaired nucleotides (Merino et al., 2005; Vasa et al., 2008). The 2′-*O*-adducts that are generated are then detected by their ability to inhibit primer extension by reverse transcriptase. This reactivity information is then used as input data in thermodynamic folding algorithms, thus generating a more accurate predicted secondary structure. hSHAPE is an evolution of SHAPE which uses fluorescent oligonucleotides (rather than radiolabeled oligonucleotides) and capillary electrophoresis (instead of denaturing polyacrylamide gels) for the resolution of the cDNAs generated by the reverse transcriptase. This evolution made it possible to resolve the structures of RNA species of more than 400-nt in length. In fact, the PLMVd species has been used as the model viroid for the development of hSHAPE (Steger and Perreault, 2016).

Globally, the secondary structures of the PLMVd variants contained 14 relatively well conserved stem regions, namely paired regions (P) in comparison to unpaired or loops (L) (P1 to P14; Fig. 2A; Giguère et al., 2014a). They can be described as a central loop surrounded by several stem-loop regions. Moreover, the PLMVd secondary structure can be divided in two domains: the left-hand domain, which is composed of the P11-L11 to P13-L13 and P1-L1 stem-loops, and, the right-hand domain which includes the P2-L2 to P10-L10 stem-loops. The left-hand domain is the more stable of the two domains in thermodynamic terms. Under mild denaturing conditions, electron microscopy experiments have permitted the visualization of the left-hand domain as being a stable rod-like structure, while the right-hand one remains unfolded, forming a large circle (Pelchat and Perreault, unpublished data).

The PLMVd structure illustrated in Fig. 2A is that obtained for the 338-nt long sequence variant "Alberta" isolated in Canada (NCBI GenBank accession number DQ680690). The left-hand domain folds into a complex, branched secondary structure that includes a cruciform motif formed by the P12-L12 and P13-L13 stem-loops, and a succession of short stems and internal loops that is located along the P11 and that is capped at the end by a relatively large loop. In this sequence variant, a P14 pseudoknot is also formed between residues from both the L1 and L11 loops. This is the most complex secondary structure for the left-handed domain of PLMVd reported to date. Some PLMVd sequence variants that included only a longer P11-L11 stem-loop that is deprived of both the cruciform motif and the P14 pseudoknot have been characterized (Dubé et al., 2011). Moreover, the presence of the cruciform motif and the pseudoknot were shown to be independent of each other as one can be detected in some sequence variants, but not in others. The role of the P14 remains to be determined. Conversely, the P12-L12 and the P13-L13 stem-loops are in fact part of the hammerhead self-cleaving motif (Fig. 2A, lower inset).

A hammerhead self-cleaving structure is folded during the synthesis of the nascent strand during replication. This is how the one-unit-length viroids are generated by the *Avsunviroidae* members. Within PLMVd, the lower strands of the P11 stem formed the hammerhead motifs of the (+) polarity strand, while the sequence of the upper strands form the hammerhead motif of the (−) polarity strand (Fig. 2A). These RNA motifs consist of three nonsequence specific helices that border a catalytic core of 14 conserved nucleotides which form a complex array of noncanonical interactions (Fig. 2A, see lower inset; Doudna, 1995; Dufour et al., 2009; de la Peña et al., 2017; Scott et al., 2013). The folding of the hammerhead structure in the presence of magnesium ions results in the self-cleavage of the RNA strand at a specific site (see the arrow in the Fig. 2A), creating both a $2'$-,$3'$-cyclic phosphodiester and a $5'$-hydroxyl termini.

The left- and right-hand domains of PLMVd

The left-hand domain has been proposed to include all of the features required by the replication mechanism of PLMVd (Fig. 2A). In addition to the hammerhead self-cleaving structures, several other features have been reported. Firstly, polymerization of (+)-strand polarity strand starts from U291 in (−)-strand (i.e. A50 in (+)-strand), and (−)-strand polarity strand starts from A284 in (+) strand (i.e. U55 in (−) strand; Delgado et al., 2005; Motard et al., 2008). It has been proposed that these positions of P11 represent the universal initiation sites for the strands of each polarity. Secondly, an in vitro selection experiment using a model system led to the suggestion that the conserved CAGACG sequence motif, which is reminiscent of the $_{47}$CAGACU$_{52}$ and $_{47}$CAGACC$_{52}$ motifs of both the (+) and the (−) P11 strands found in the vicinity of the initiation sites, may act like a promoter sequence (Motard et al., 2008). Thirdly, the peach elongation factor 1-alpha (eEF1A) was reported to bind to PLMVd in infected cells. In vitro experiments have shown that the binding of eEF1A occurs in the vicinity of the universal initiation sites, supporting a potential contribution to the initiation of the polymerization of the strands of both polarities, although physical evidence remains to be obtained in support of this hypothesis (Dubé et al., 2009). To date, this is the only established information concerning the quaternary structure of PLMVd.

The current knowledge of the right-hand domain is significantly more limited. In fact, as yet no biological contributions have been attributed to this domain. The presence of each stem received support from base-pair covariation data when several sequence variants were analyzed. Moreover, minor local structural variations resulting from sequence variation have been reported (Dubé et al., 2011). One peculiar structural feature of this domain is the tertiary structure P8 pseudoknot that is conserved in all PLMVd variants. This pseudoknot is formed by GC-base pairing residues from the loops capping both the P6 and P7 stems. The P8 pseudoknot is essential to be able to detect accumulating PLMVd one year postinoculation, although its precise contribution to the viroid life cycle remains unknown (Dubé et al., 2010).

The secondary structure of the (−) polarity PLMVd strand

The secondary structure of the (−) polarity PLMVd strand has been also studied using a classical biochemical approach as well as various versions of the SHAPE protocols (Dubé et al., 2010; Giguère et al., 2014a). In the cases of the Alberta variant and of several other sequence variants, only eight stems (P1 to P8) were identified, instead of the 14 that were identified for the (+) polarity strand (Fig. 2B). In addition, the P8 and P14 pseudoknots and the P12 and P13 stem-loops were absent. Moreover, several stems were longer (P2, P6, P8, and P11), which results in a more linear, or rod-like, structure. Most likely, this explains why the (−) polarity strand has been shown to exhibit slower electrophoretic mobility under native conditions as compared to that of the (+) polarity strand (Dubé et al., 2010). More importantly, the (+) polarity strand, which adopted a branched-like structure due to the presence of at least the P8 pseudoknot, was shown to be insoluble in the presence of lithium chloride, which is not the case for the (−) polarity strand (Dubé et al., 2010). It has been suggested that the presence of pseudoknot results in a more globular structure that is responsible for the differential biophysical properties observed.

Structure of the other *Avsunviroidae* members

The secondary structures of both the (+) and (−) polarity strands of most *Avsunviroidae* members have been elucidate using the hSHAPE method (Giguère et al., 2014a). When looking at the solved structures of the (+) polarity strands of the chrysanthemum chlorotic mottle viroid (CChMVd), the eggplant latent viroid (ELVd) and the avocado sunblotch viroid (ASBVd), in that order, they tend to become less complex (Fig. 3A–C). While CChMVd appears to be almost as complex as PLMVd, even including the presence of a pseudoknot, ELVd

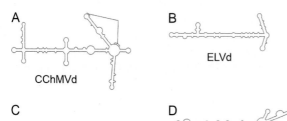

A CChMVd

B ELVd

C ASBVd

D GHV

FIG. 3 Structures of the viroids of the family. *Avsunviroidae*. Schematic representations of the elucidated (+) strands of CChMVd (A), ELVd (B), ASBVd (C) and the proposed *Avsunviroidae* GHV species (D).

A. Introduction

is notably more linear with only one extruding stem-loop located near the left-hand end, and a Y-shape located at the right-hand end. The structure of Apple hammerhead viroid (AHVd), which has been recently included as a member of genus Pelamoviroid by ICTV has not been elucidated so far. ASBVd adopts a simple rod-like structure. It is important to mention that when quite distinct sequence variants of ASBVd are used, the resulting shape data suggests the presence of a pseudoknot, but, as yet, no further physical support for this motif has been reported (Delan-Forino et al., 2014). Thus, the ASBVd structure appears to be more reminiscent of that of the *Pospiviroidae* (in the following section), although it does include a hammerhead self-cleaving sequence. Compared to the other *Avsunviroidae* members, for which the hammerhead sequence results from a contiguous sequence on the RNA strand, those of the ASBVd are divided into two domains, with one being located on the upper strand and the other on the lower strand. This latter fact was the origin of the first designed *trans*-acting cleaving ribozyme (Uhlenbeck, 1987).

Following the elucidation of the secondary structure of *Avsunviroidae* members, that of the grapevine hammerhead viroid-like RNA was then determined (Giguère et al., 2014a). This is an RNA species of 375 nucleotides in length that includes hammerhead self-cleaving sequences on both the (+) and (−) strands (Wu et al., 2012). However, to our knowledge, as yet no reported experiments are demonstrating that it satisfies Koch's postulate. In other words, it cannot be definitively classified as a viroid species. That said, the secondary structures of both strands have been elucidated and they look like that of PLMVd with the exclusion of the presence of the pseudoknot in the (+) polarity strand (Fig. 3D).

Pospiviroidae

The primary structure of the model potato spindle tuber viroid (PSTVd)

From the original discovery of viroids until today, PSTVd has been the most studied species. PSTVd is the type member of the family *Pospiviroidae*. Based on sequence similarities, Keese and Symons proposed that viroids of this family contain five structural domains: the central domain (C), the pathogenicity domain (P), the variable domain (V), and the terminal right (T_R) and left (T_L) domains (Keese and Symons, 1985). In the case of the P domain, it has been proposed that the local conformation, rather than the local secondary structure stability alone, may be crucial in determining pathogenicity (Hu et al., 1996). Based on sequence analysis, more than 20 different PSTVd isolates were known by the early 1990s. Already, it had been observed that a slight change in the nucleotide sequence of PSTVd could induce different disease symptoms in host plants. For instance, the PSTVd-intermediate strain (PSTVd-I) and PSTVd-RG1, both of which are 359-nt long but differ in three nucleotides, induce different symptoms in the sensitive tomato cultivar Rutgers. Specifically, PSTVd-I induces intermediate disease symptoms, while PSTVd-RG1 induces severe disease symptoms. Upon infection, viroids are known to produce a heterogenic population of sequence variants (i.e., to behave as a quasispecies), as has been observed for *Avsunviroidae*. This nature of viroids helps not only in viroid evolution but also in their adaptation to new host plants (see Chapter 2 for more details).

The secondary structure of PSTVd

The initially proposed secondary structures of PSTVd were based on both mapping data and computer prediction (Domdey et al., 1978; Gast et al., 1996; Gross et al., 1978). More recently, this subject has been revisited using hSHAPE for several sequence variants (Giguère et al., 2014b; Xu et al., 2012). Even if these experiments were not performed using a circular strand, all precaution was used to ensure that these most stable structures were accurate. All folded into a rod-like shape, and only minor local differences were observed. The structure of the PSTVd-M isolate is illustrated in Fig. 4. Clearly, this is a succession of short stems intercalated by small loops that are capped at both ends by terminal loops. The five domains (T_L, P, C, V, and T_R) are well established. This division into domains is based mainly on the presence of highly conserved sequences within the C domain and in part on the presence of sequence homologies located in the other domains. The boundaries between these domains are defined by sharp changes in the degree of sequence homology among the different viroids. Initially, these structural domains were proposed to have specific functional roles: for example, the P domain was associated with pathogenicity. However, experiments have shown that the situation is significantly more complex than that. The expression of disease symptoms is now thought to be controlled cooperatively or independently by discrete determinants, or structural motifs, located within the multiple domains, and not only in the P domain.

Left to right several features should be mentioned with respect to not only the PSTVd structures but, in many cases, that are of importance for other *Pospiviroidae* (Steger and Perreault, 2016). The T_L domain of PSTVd contains an imperfect repeat that could form either the rod-like or the Y-shaped structure (Fig. 4, left inset; Dingley et al., 2003). In PSTVd, the T_L domain has been reported to be important for the initiation of replication (Bojić et al., 2012; Kolonko et al., 2006). The P domain contains an oligopurine stretch located on the upper strand, and an oligopyrimide one located on the bottom strand for PSTVd as well as for most of the pospiviroids (genus *Pospivirus*) (Steger et al., 1984). Moreover, the presence of a single-stranded region rich in adenosine residues located on both strands, namely loop A, has been demonstrated. The resulting base-pairing of these sequences provides a structural region that has relatively low thermodynamic stability, hence the name "premelting" region has been attributed to it. It has been reported that sequence variations in this region have the potential to modulate pathogenicity. The C domain (which includes central conserved domain, CCR) is the most highly conserved sequence. In the solved structure of PSTVd it folds into a rod-like region. However, both strands can also form thermodynamically stable hairpins (see Fig. 4, insets) (Baumstark et al., 1997; Gas et al., 2007; Steger and Perreault, 2016). In the center of the C domain of PSTVd, there is a particular internal loop that shows homology to loop E of the eukaryotic 5S RNA (Wimberly et al., 1993). The residues of this loop adopt a network of noncanonical interactions that yield crosslinking upon UV irradiation. The loop E structural integrity was shown to be important for replication (Zhong et al., 2006). The V domain is the most variable in terms of sequence (Keese and Symons, 1985). Finally, the T_R of PSTVd comprises a relatively large internal loop with several consecutive uridines located on the bottom strand. Moreover, the RY motif (i.e. AGG/CCUUC) within the TR domain has been shown to enhance the recognition of TR by VirP1 (viroid RNA-binding protein 1) and could be involved in the systemic transport of viroids.

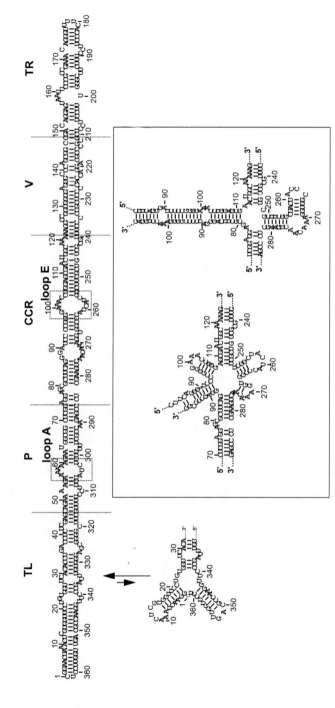

FIG. 4 Secondary structure of PSTVd. The nucleotides involved in the formation of the extra stable hairpins I and II (HPI, HPII) are *underlined*.

Structure of the other *Pospiviroidae* members

The secondary structure of at least one sequence variant of all *Pospiviroidae* members has been elucidated using hSHAPE (Giguère and Perreault, 2017; Giguère et al., 2014b). This work led to the proposal of structural hallmarks for the five genera of the *Pospiviroidae* (Table 2; see also Giguère and Perreault, 2017). These include the characteristics of the CCR for each genus, the specific accessibility for the reagent modification according to the hSHAPE data, the secondary structure (rod-like or not) and the presence of loops A and E, as well as other structural features.

Almost all members of the genus *Pospiviroid* (chrysanthemum stunt viroid [CSVd]; citrus exocortis [CEVd]; Iresine viroid [IrVd]; Mexican papita viroid [MPVd; Syn. TPMVd]; pepper chat fruit viroid [PCFVd]; tomato apical stunt viroid [TASVd]; tomato chlorotic dwarf viroid [TCDVd]; and tomato planta macho viroid [TPMVd]) fold into rod-like structures reminiscent of PSTVd (Table 2). The only exception is the columnea latent viroid (CLVd), which has a branched TL region. Importantly, they all possess a loop A located in domain P and

TABLE 2 Structural hallmarks of *Pospiviroidae* species.

Genus	Length (nt)	Examples of structural features
Pospiviroids	340–370	**PSTVd** — TL, P (loop A), CCR (loop E), V, TR
Hostuviroids	295–340	**HSVd** — TL, P (loop A), CCR, V, TR
Cocadviroids	Short <300	**CCCVd** — TL, P (loop A), CCR (loop E), V, TR
Apscaviroids	294–369	**ASSVd** — TL, P, CCR, V, TR
Coleviroids	250–360	**CbVd-1** — CCR

a loop E located in the C domain. Moreover, they all share the relatively large loop located in the T_R domain, a feature which is unique to this genus.

The two members of the *Hostuviroid* genus (hop stunt viroid [HSVd], and Dahlia latent viroid [DLVd]) folded into a rod-like structure (Table 2). The elucidation of the secondary structure of DLVd permitted its definitive classification as a member of *Hostuviroid*.

The four species classified in the genus *Cocaviroid* (coconut cadang-cadang viroid [CCCVd], citrus bark cracking viroid [CBCVd], coconut tinangaja viroid [CTiVd], and hop latent viroid [HLVd]) were reported to also adopt a rod-like structure (Table 2). They all possess a particularly large loop A located in domain P, as well as a loop E located in the C domain, in addition to sharing similar CCRs. Moreover, they all possess a one-nucleotide loop that caps the T_L domain.

In terms of size, the *Apscaviroid* genus demonstrates the most structural diversity (Table 2), a feature which is usually relatively conserved within the members of a genus. The smaller members of the genus, the citrus viroid V [CVd-V], and citrus viroid VI [CVd-VI] are 294 nucleotides in length, while the longer ones, the Australian grapevine viroid [AGVd], and the grapevine yellow speckle viroid 1 [GYSVd-1], are 369 and 367 nucleotides in length, respectively. All share a characteristic CCR. In terms of the secondary structure, five apscaviroids were reported to fold into a classical rod-like structure (citrus bent leaf viroid [CBLVd], citrus dwarfing viroid [CDVd], citrus viroid V [CVd-V], and grapevine yellow speckle viroid 1 and 2 [GYSVd-1 and GYSVd-2]). Conversely, apple scar skin viroid (ASSVd) and apple dimple fruit viroid (ADFVd) share a rod-like structure with the presence of an extra stem-loop in the T_L domain, resulting in a Y-shape. This structural feature has been the hallmark of the classification of ADFVd as a member of the genus *Apscaviroid* (Giguère and Perreault, 2017). The pear blister canker viroid (PBCVd) folds into a rod-like structure, but its T_R domain includes a capping loop that is unusually large. Finally, the Citrus viroid VI (CVd-VI; also reported as CVd-OS) adopts a rod-like structure that includes the presence of branched structures in both the T_L domain and at the end of the T_R domain. The latter is unique in terms of structure. The only motif common to all apscaviroids is their CCR.

All four *Coleviroid* members (Coleus blumei viroids-1, -2, -3, and -6 [CbVd-1, CbVd-2, CbVd-3 and CbVd-6]) fold into rod-like structures and share a common CCR with no presence of the other structural features discussed for the other genera (Table 2).

Finally, it is interesting to mention that some members of the family *Pospiviroidae* such as CbVd are known to have chimeric structures at various inter- or cross-species levels. Understanding how viroids could have evolved their molecular structures remains largely unstudied.

Perspective

Progress in the structural determination of viroids using hSHAPE has resulted in the elucidation of the secondary structures of at least one sequence variant per viroid species. Together, these secondary structures form a complete compendium of the viroids. Since the hSHAPE procedure is relatively simple and does not require special conditions, like the use of radioactivity, it should be used for the classification of any new viroid species reported in the future. These data provide structural information that contributes the proper classification of a viroid, as was illustrated earlier for both DLVd and CLVd. Moreover, performing hSHAPE is even more important when considering that data from in vivo SHAPE has validated that viroid folding is similar in both cells and test tubes (López-Carrasco and Flores, 2017a,b). This also suggests that most viroid RNAs accumulating in cells are not interacting with proteins.

Data from the tertiary structure determination of viroids have remained elusive. For a long time, the size of these RNA species limited their study by both NMR and crystallographic analysis. However, the recent direct visualization of the native structure of PLMVd at a single-molecule resolution using atomic force microscopy has been reported (Moreno et al., 2019). This study confirmed the stabilizing role of the tertiary structures.

The fact that today we have a compendium of viroid structures provides a great reference point for many future biological studies that have as their goal the assignment of biological contribution to the various structural motifs (Giguère and Perreault, 2017; Giguère et al., 2014a; Xu et al., 2012). Viroid sequence variation appears to be wider, in comparison to their secondary structures, which seem to be more conserved. Consequently, these might be a corner stone for a better understanding of viroid life cycles.

Questions for the reader

1. Describe the primary, secondary, and tertiary structure of viroids?
2. Give a brief account of two methods that are used for the elucidation of viroid secondary structure?
3. What are the distinctive structural features of *Pospiviroidae* and *Avsunviroidae*?
4. Name the five structural domains of *Pospiviroidae*?
5. Why establishing the structure of a viroid so important?

Acknowledgments

This work was supported by grants from the Natural Sciences and Engineering Research Council of Canada (NSERC, Grant number 155219-17) to J.-P.P. The RNA group was supported by grants from the Université de Sherbrooke. J.-P. P. holds the Research Chair at the Université de Sherbrooke in RNA Structure and Genomics and is a member of the Centre de Recherche du CHUS.

References

Ambrós, S., Hernández, C., Desvignes, J.C., Flores, R., 1998. Genomic structure of three phenotypically different isolates of peach latent mosaic viroid: implications of the existence of constraints limiting the heterogeneity of viroid quasispecies. J. Virol. 72, 7397–7406.

Baumstark, T., Schröder, A.R., Riesner, D., 1997. Viroid processing: switch from cleavage to ligation is driven by a change from a tetraloop to a loop E conformation. EMBO J. 16, 599–610.

Bojić, T., Beeharry, Y., Zhang, D.J., Pelchat, M., 2012. Tomato RNA polymerase II interacts with the rod-like conformation of the left terminal domain of the potato spindle tuber viroid positive RNA genome. J. Gen. Virol. 93, 1591–1600.

Bussière, F., Lafontaine, D., Perreault, J.P., 1996. Compilation and analysis of viroid and viroid-like RNA sequences. Nucleic Acids Res. 24, 1793–1798.

Bussière, F., Lehoux, J., Thompson, D.A., Skrzeczkowski, L.J., Perreault, J., 1999. Subcellular localization and rolling circle replication of peach latent mosaic viroid: hallmarks of group A viroids. J. Virol. 73, 6353–6360.

Bussière, F., Ouellet, J., Côté, F., Lévesque, D., Perreault, J.P., 2000. Mapping in solution shows the peach latent mosaic viroid to possess a new pseudoknot in a complex, branched secondary structure. J. Virol. 74, 2647–2654.

de la Peña, M., García-Robles, I., Cervera, A., 2017. The hammerhead ribozyme: a long history for a short RNA. Molecules 22, E78.

Delan-Forino, C., Deforges, J., Benard, L., Sargueil, B., Maurel, M.-C., Torchet, C., 2014. Structural analyses of avocado sunblotch viroid reveal differences in the folding of plus and minus RNA strands. Viruses 6, 489–506.

Delgado, S., Martínez de Alba, A.E., Hernández, C., Flores, R., 2005. A short double-stranded RNA motif of peach latent mosaic viroid contains the initiation and the self-cleavage sites of both polarity strands. J. Virol. 79, 12934–12943.

Di Serio, F., Flores, R., Verhoeven, J.T.J., Li, S.-F., Pallás, V., Randles, J.W., Sano, T., Vidalakis, G., Owens, R.A., 2014. Current status of viroid taxonomy. Arch. Virol. 159, 3467–3478.

Dingley, A.J., Steger, G., Esters, B., Riesner, D., Grzesiek, S., 2003. Structural characterization of the 69 nucleotide potato spindle tuber viroid left-terminal domain by NMR and thermodynamic analysis. J. Mol. Biol. 334, 751–767.

Domdey, H., Jank, P., Sänger, L., Gross, H.J., 1978. Studies on the primary and secondary structure of potato spindle tuber viroid: products of digestion with ribonuclease A and ribonuclease T1, and modification with bisulfite. Nucleic Acids Res. 5, 1221–1236.

Doudna, J.A., 1995. Hammerhead ribozyme structure: U-turn for RNA structural biology. Structure 3, 747–750.

Dubé, A., Bisaillon, M., Perreault, J.-P., 2009. Identification of proteins from *Prunus persica* that interact with peach latent mosaic viroid. J. Virol. 83, 12057–12067.

Dubé, A., Baumstark, T., Bisaillon, M., Perreault, J.-P., 2010. The RNA strands of the plus and minus polarities of peach latent mosaic viroid fold into different structures. RNA 16, 463–473.

Dubé, A., Bolduc, F., Bisaillon, M., Perreault, J.-P., 2011. Mapping studies of the peach latent mosaic viroid reveal novel structural features. Mol. Plant Pathol. 12, 688–701.

Dufour, D., de la Peña, M., Gago, S., Flores, R., Gallego, J., 2009. Structure-function analysis of the ribozymes of chrysanthemum chlorotic mottle viroid: a loop-loop interaction motif conserved in most natural hammerheads. Nucleic Acids Res. 37, 368–381.

Fadda, Z., Daròs, J.-A., Flores, R., Duran-Vila, N., 2003. Identification in eggplant of a variant of citrus exocortis viroid (CEVd) with a 96 nucleotide duplication in the right terminal region of the rod-like secondary structure. Virus Res. 97, 145–149.

Fekih Hassen, I., Massart, S., Motard, J., Roussel, S., Parisi, O., Kummert, J., Fakhfakh, H., Marrakchi, M., Perreault, J.-P., Jijakli, M.H., 2007. Molecular features of new peach latent mosaic viroid variants suggest that recombination may have contributed to the evolution of this infectious RNA. Virology 360, 50–57.

Flores, R., Hernández, C., Desvignes, J.C., Llácer, G., 1990. Some properties of the viroid inducing peach latent mosaic disease. Res. Virol. 141, 109–118.

Gas, M.-E., Hernández, C., Flores, R., Daròs, J.-A., 2007. Processing of nuclear viroids in vivo: an interplay between RNA conformations. PLoS Pathog. 3, e182.

Gast, F.U., Kempe, D., Spieker, R.L., Sänger, H.L., 1996. Secondary structure probing of potato spindle tuber viroid (PSTVd) and sequence comparison with other small pathogenic RNA replicons provides evidence for central non-canonical base-pairs, large A-rich loops, and a terminal branch. J. Mol. Biol. 262, 652–670.

Giguère, T., Perreault, J.-P., 2017. Classification of the Pospiviroidae based on their structural hallmarks. PLoS One 12, e0182536.

Giguère, T., Adkar-Purushothama, C.R., Bolduc, F., Perreault, J.-P., 2014a. Elucidation of the structures of all members of the Avsunviroidae family. Mol. Plant Pathol. 15, 767–779.

Giguère, T., Adkar-Purushothama, C.R., Perreault, J.-P., 2014b. Comprehensive secondary structure elucidation of four genera of the family Pospiviroidae. PLoS One 9, e98655.

Glouzon, J.-P.S., Bolduc, F., Wang, S., Najmanovich, R.J., Perreault, J.-P., 2014. Deep-sequencing of the peach latent mosaic viroid reveals new aspects of population heterogeneity. PLoS One 9, e87297.

Gross, H.J., Domdey, H., Lossow, C., Jank, P., Raba, M., Alberty, H., Sänger, H.L., 1978. Nucleotide sequence and secondary structure of potato spindle tuber viroid. Nature 273, 203–208.

Hu, Y., Feldstein, P.A., Bottino, P.J., Owens, R.A., 1996. Role of the variable domain in modulating potato spindle tuber viroid replication. Virology 219, 45–56.

A. Introduction

Keese, P., Symons, R.H., 1985. Domains in viroids: evidence of intermolecular RNA rearrangements and their contribution to viroid evolution. Proc. Natl. Acad. Sci. USA 82, 4582–4586.

Kolonko, N., Bannach, O., Aschermann, K., Hu, K.-H., Moors, M., Schmitz, M., Steger, G., Riesner, D., 2006. Transcription of potato spindle tuber viroid by RNA polymerase II starts in the left terminal loop. Virology 347, 392–404.

López-Carrasco, A., Flores, R., 2017a. Dissecting the secondary structure of the circular RNA of a nuclear viroid in vivo: a "naked" rod-like conformation similar but not identical to that observed in vitro. RNA Biol. 14, 1046–1054.

López-Carrasco, A., Flores, R., 2017b. The predominant circular form of avocado sunblotch viroid accumulates in planta as a free RNA adopting a rod-shaped secondary structure unprotected by tightly bound host proteins. J. Gen. Virol. 98, 1913–1922.

Malfitano, M., Di Serio, F., Covelli, L., Ragozzino, A., Hernández, C., Flores, R., 2003. Peach latent mosaic viroid variants inducing peach calico (extreme chlorosis) contain a characteristic insertion that is responsible for this symptomatology. Virology 313, 492–501.

Merino, E.J., Wilkinson, K.A., Coughlan, J.L., Weeks, K.M., 2005. RNA structure analysis at single nucleotide resolution by selective 2′-hydroxyl acylation and primer extension (SHAPE). J. Am. Chem. Soc. 127, 4223–4231.

Moreno, M., Vázquez, L., López-Carrasco, A., Martín-Gago, J.A., Flores, R., Briones, C., 2019. Direct visualization of the native structure of viroid RNAs at single-molecule resolution by atomic force microscopy. RNA Biol. 16, 295–308.

Motard, J., Bolduc, F., Thompson, D., Perreault, J.-P., 2008. The peach latent mosaic viroid replication initiation site is located at a universal position that appears to be defined by a conserved sequence. Virology 373, 362–375.

Navarro, B., Flores, R., 1997. Chrysanthemum chlorotic mottle viroid: unusual structural properties of a subgroup of self-cleaving viroids with hammerhead ribozymes. Proc. Natl. Acad. Sci. USA 94, 11262–11267.

Scott, W.G., Horan, L.H., Martick, M., 2013. The hammerhead ribozyme: structure, catalysis, and gene regulation. Prog. Mol. Biol. Transl. Sci. 120, 1–23.

Steger, G., Perreault, J.-P., 2016. Structure and associated biological functions of viroids. Adv. Virus Res. 94, 141–172.

Steger, G., Hofmann, H., Förtsch, J., Gross, H.J., Randles, J.W., Sänger, H.L., Riesner, D., 1984. Conformational transitions in viroids and virusoids: comparison of results from energy minimization algorithm and from experimental data. J. Biomol. Struct. Dyn. 2, 543–571.

Symons, R.H., 1981. Avocado sunblotch viroid: primary sequence and proposed secondary structure. Nucleic Acids Res. 9, 6527–6537.

Uhlenbeck, O.C., 1987. A small catalytic oligoribonucleotide. Nature 328, 596–600.

Vasa, S.M., Guex, N., Wilkinson, K.A., Weeks, K.M., Giddings, M.C., 2008. ShapeFinder: a software system for high-throughput quantitative analysis of nucleic acid reactivity information resolved by capillary electrophoresis. RNA 14, 1979–1990.

Wimberly, B., Varani, G., Tinoco, I., 1993. The conformation of loop E of eukaryotic 5S ribosomal RNA. Biochemistry 32, 1078–1087.

Wu, Q., Wang, Y., Cao, M., Pantaleo, V., Burgyan, J., Li, W.-X., Ding, S.-W., 2012. Homology-independent discovery of replicating pathogenic circular RNAs by deep sequencing and a new computational algorithm. Proc. Natl. Acad. Sci. USA 109, 3938–3943.

Xu, W., Bolduc, F., Hong, N., Perreault, J.-P., 2012. The use of a combination of computer-assisted structure prediction and SHAPE probing to elucidate the secondary structures of five viroids. Mol. Plant Pathol. 13, 666–676.

Zhong, X., Leontis, N., Qian, S., Itaya, A., Qi, Y., Boris-Lawrie, K., Ding, B., 2006. Tertiary structural and functional analyses of a viroid RNA motif by isostericity matrix and mutagenesis reveal its essential role in replication. J. Virol. 80, 8566–8581.

A. Introduction

4

Replication and movement of viroids in host plants

Beatriz Navarro[a], Gustavo Gómez[b], and Vicente Pallás[c]

[a]Institute for Sustainable Plant Protection, National Research Council, Bari, Italy
[b]Institute for Integrative Systems Biology (I2SysBio), Consejo Superior de Investigaciones Científicas (CSIC)—Universitat de València (UV), Paterna, Spain [c]Instituto de Biología Molecular y Celular de Plantas (IBMCP), Consejo Superior de Investigaciones Científicas (CSIC)—Universitat Politècnica de València (UPV), Valencia, Spain

Graphical representation

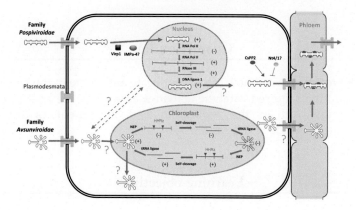

Definitions

Hammerhead self-cleaving motif: It is a type of RNA motif able to catalyze the self-cleavage of an RNA strand generating 5′OH and 2′,3′-cyclic phosphodiester (Hutchins et al., 1986). Hammerhead self-cleaving motif contained in both polarity strands of members of the family *Avsunviroidae* is involved in the self-cleavage of the viroid RNAs during replication.

Viroid RNA polarity strands: By convention, the plus polarity strand of a viroid RNA is the strand that accumulates at the higher level in the viroid infected tissues. The complementary and less concentrated strand is the minus polarity strand. The accumulation of both polarity strands may differ very much (i.e., potato spindle tuber viroid) or be quite low (i.e., peach latent mosaic viroid).

Central Conserved region (CCR): The CCR contains two stretches of conserved nucleotides, located in the upper and in the lower strand of the central region in the rod-like secondary structure proposed for certain viroids. The CCR is conserved in all members of the family *Pospiviroidae* and is involved in replication. Members of the same genus share the same CCR.

Plasmodesmata: Plasmodesmata are intercellular pores connecting adjacent plant cells allowing membrane and cytoplasmic continuity and are essential routes for intercellular trafficking, communication, and signaling in plant development and defense.

Agroinfiltration: It is a method used in plant biology to induce transient expression of genes in a plant.

Three-dimensional (3D) motifs: Three-dimensional (3D) motifs are patterns of local structure associated with function, typically based on residues in binding or catalytic sites. Viroids represent an excellent model to dissect the role of RNA three-dimensional (3D) structural motifs in regulating RNA movement.

Chapter outline

In this chapter we will present how viroids replicate and move in the host plant. The replication mechanism, the required catalytic activities as well as the host factors and viroid structural motifs involved in the replication and movement of the nuclear- and chloroplast-replicating viroids will be addressed.

Learning objectives

- Objective 1 to highlight the main characteristics of viroids and the principal steps of their life cycle.
- Objective 2 to show the mechanisms of viroid replication and the differences in this respect between nuclear and chloroplastic viroids.
- Objective 3 to learn which are the enzymatic activities involved in each step of the viroid replication mechanisms, and the role of the specific RNA structural motifs.

A. Introduction

- Objective 4 to learn how viroids subvert the plant-endogenous pathways that regulate the noncoding RNA compartmentalization into the nucleus and chloroplasts.
- Objective 5 to learn how viroids move from cell to cell.
- Objective 6 to learn how viroids reach, move within, and exit the phloem.

Fundamental introduction

Viroids are infectious agents of plants. They are the smallest pathogens known and represent the last step in the biological scale. Being exclusively composed of a circular RNA molecule of small dimensions (ranging between 246 and 414 nucleotides), viroids are approximately 20 times smaller than tobacco mosaic virus. One of the most peculiar characteristics of viroids, that distinguish them from viruses, is that they do not code for any protein. Moreover, viroids are not encapsidated and accumulate in vivo as "naked" nucleic acids. The genomic RNA of viroids folds in a compact conformation due to the high self-complementarity of its nucleotide sequence. This secondary structure, which can be rod-like or more or less branched, is formed by double-stranded regions separated by single-stranded loops. Despite their small size and simplicity, viroids are able to infect their host plants and, in several cases, induce diseases. The life cycle of viroids involves the following steps: after entering into the cell, viroids reach the intracellular organelle where they replicate to produce multiple copies of themselves; then viroids exit the organelle and move to neighboring cells; finally, viroids reach the phloem to move systemically to distal parts of the plant. Viroids replicate and move within the infected plant autonomously. Therefore, in the absence of viroid-encoded proteins, viroids contain in their RNA sequences and in the RNA structural motifs they can adopt, all the information necessary to redirect the plant host machinery for accomplishing their infectious life cycle (Navarro et al., 2021; Gómez and Pallás, 2013).

Based on the replication site and other structural characteristics, viroids have been classified in two families: *Pospiviroidae* and *Avsunviroidae* (see Chapter 2). Members of the family *Pospiviroidae*, with potato spindle tuber viroid (PSTVd) as type member, replicate in the nucleus and have a central conserved region (CCR), whereas those belonging to the family *Avsunviroidae*, with avocado sunblotch viroid (ASBVd) as type member, replicate in the chloroplast, do not have CCR and contain hammerhead ribozymes in both polarity strands (Di Serio et al., 2018, 2021).

Viroid replication occurs through a rolling circle mechanism

To accomplish the infectious process, viroids replicate generating multiple new copies of themselves. Viroid replication occurs exclusively through RNA intermediates without the involvement of DNA (Grill and Semancik, 1978), through a mechanism called rolling circle. As the name suggests, this mechanism is based on the reiterative transcription of the circular viroid RNA templates for the synthesis of oligomers of complementary strands. This model was proposed based on data showing the accumulation of viroid RNAs of opposite polarities in tomato tissues infected with citrus exocortis viroid (CEVd) (Grill and Semancik, 1978), and of multimeric RNA molecules with size longer than a genomic unit of the viroid in tissues infected with other viroids (Branch et al., 1981). Based on studies of the replicative intermediates accumulating in the infected tissues infected with viroids of both families, two variants of the replication model have been proposed: asymmetric for the members of the *Pospiviroidae* family and symmetric for members of the *Avsunviroidae* family, which will be described in detail in the next sections.

The accomplishment of the viroid replication through this mechanism requires three catalytic activities: an RNA polymerase for the RNA synthesis, an RNase for the cleavage of the RNA multimers, and an RNA ligase for the RNA circularization.

Replication of nuclear viroids

The asymmetric rolling circle mechanism

Viroids in the family *Pospiviroidae* replicate in the nucleus following the asymmetric variant of the rolling circle replication model. The infecting genomic circular RNA of plus polarity (by convention the polarity that accumulates at higher concentration in vivo) serves as a template for the generation, after several rounds of transcription, of multimers of the complementary minus strand which, in turn, are used as the template for the synthesis of RNA multimers of plus polarity strand. These concatemers of plus polarity are then cleaved in a specific site, to generate perfect linear monomers of plus polarity that in turn are circularized, by covalent ligation of their ends, to originate the mature monomeric circular forms of plus polarity, which are the final product of the replication process (Fig. 1). This variant of the model is called asymmetric because only one rolling circle is involved. It is characterized by the presence of circular forms of only plus polarity. In fact, circular RNAs of minus polarity were never detected in PSTVd-infected tissues (Branch et al., 1988) and in tissues infected by other nuclear replicating viroids. It has been shown that plus and minus polarity strands of PSTVd are distributed differently into the nucleus: while plus strands have been detected in the nucleolus and nucleoplasm, those of minus polarity accumulate only in the nucleoplasm, suggesting that after being synthesized only the plus strands are translocated to the nucleolus, likely to be cleaved at a specific site and generate the monomeric forms (Qi and Ding, 2003).

Although the asymmetric variant of the model is supported by experimental data performed principally with PSTVd, it is assumed that it is followed by all the members of the *Pospiviroidae* family.

Enzymatic activities involved in the replication of nuclear viroids

According to the asymmetric model, the accomplishment of the viroid replication requires three catalytic activities: an RNA polymerase for the RNA transcription, an RNase for the cleavage of the RNA multimers into monomers, and an RNA ligase for the RNA circularization. Since viroids do not code for proteins, these activities must derive from the host or be contained in the viroid RNA. In the case of nuclear viroids, the enzyme that catalyzes the synthesis of RNA multimers is the host DNA-dependent RNA polymerase II (RNA Pol II).

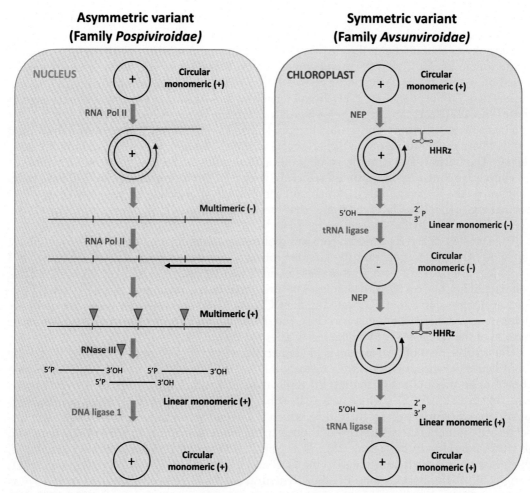

FIG. 1 Asymmetric (left panel) and symmetric (right panel) variants of the rolling circle mechanism proposed for replication of viroids belonging to the families *Pospiviroidae* and *Avsunviroidae*, respectively.

The transcription inhibition of representative nuclear viroids (PSTVd and CEVd) in the presence of a low concentration of α-amanitine, a fungal toxin that inhibits RNA Pol II (Mühlbach and Sänger, 1979; Flores and Semancik, 1982), and the immunoprecipitation of oligomers of CEVd and PSTVd with antibodies against RNA Pol II (Warrilow and Symons, 1999; Wang et al., 2016), support its involvement in the RNA synthesis of nuclear viroids. Interestingly, RNA Pol II transcribes cellular messenger RNAs using DNA as a template. Therefore, nuclear-replicating viroids are able to redirect RNA pol II to accept RNA as a template instead of the canonic DNA.

Due to the viroid circularity, the initiation of transcription could initiate at any site of the RNA molecule, but this is not the case. The transcription by RNA Pol II initiates in a specific position of the RNA molecule. In the circular plus polarity PSTVd templates, the starting transcription site has been determined in position C1 or U359 (Kolonko et al., 2006). However, the initiation site on the minus RNA multimers remains undetermined.

The enzyme that mediates the cleavage of RNA multimers of plus polarity into monomers is still unknown. However, the involvement of a cellular RNase type III has been suggested based on two features: (i) the linear PSTVd RNAs of plus polarity accumulating in infected tissues had 5′-Phosphate and 3′-Hydroxyl termini which are those generated by cleavage of these types of RNases; and, (ii) the cleavage site determined for several pospiviroids is located in a metastable structural motif called hairpin I, conserved in all nuclear viroids, and formed by the alternative folding of the upper strand of the CCR together with a flanking inverted repeat. The hairpin I in dimeric or oligomeric molecules could promote the formation of an alternative long double-stranded structure with a GC-rich central region of 10 base pairs containing the cleavage sites, which, in the double-stranded structure, leave two 3′-protruding nucleotides in each strand, that are the characteristic signature of RNase III enzymes (Gas et al., 2008).

The last step of the replication, consisting of the covalent ligation of the plus monomeric linear RNAs ends to generate the circular forms, is catalyzed by the plant DNA ligase I (Nohales et al., 2012a). Also, in this case, an enzyme that normally acts on DNA substrates is forced to operate on RNA molecules. The circularization is facilitated by an element of tertiary structure, called loop E, located in the CCR of PSTVd and related viroids, which likely determines the right positioning in close proximity and appropriate orientation of the two termini of the linear RNA to be ligated (Gas et al., 2007).

The involvement of host factors is likely necessary for redirecting the RNA Pol II and the DNA ligase I to accept viroid RNAs instead of DNA templates. Two host proteins have been shown to be modulators of the PSTVd replication: the transcription factor IIIA with seven zinc-fingers (TFIIIA-7ZF), which acts as an enhancer (Wang et al., 2016), and its splicing regulator, the ribosomal protein L5 (RPL5), which is a negative modulator. These proteins interact directly with the viroid. TFIIIA-7ZF binds to the left terminal region of PSTVd, where also the RNA Pol II binds (Bojić et al., 2012; Wang et al., 2016) and RPL5 interacts with the loop E, the tertiary structure motif located in the CCR and essential for replication (Eiras et al., 2011; Jiang et al., 2018). Besides loop E, other structural motifs have been identified in PSTVd as necessary for replication, such as several 3D-structured loops (Zhong et al., 2006, 2008; Wu et al., 2019) and GU pairs (Wu et al., 2020). Likely, these structural motifs are recognized by other still unknown host factors involved in viroid replication.

Replication of chloroplastic viroids

Symmetric rolling circle mechanism

Members of the family *Avsunviroidae* replicate in the chloroplast via a symmetric rolling circle model (Daròs et al., 1994). This model was proposed based on the identification of circular forms of both polarity strands in the tissues infected by chloroplastic viroids.

In the first step of the model, the infectious circular monomeric RNAs of plus polarity are reiteratively transcribed to generate linear concatemers of minus polarity. These longer-than-unit molecules are cleaved to generate monomers that in turn are circularized, forming the minus monomeric circular RNAs, which are the hallmark of the symmetric model. In fact, minus circular RNAs are not formed in the asymmetric variant of the model illustrated in the previous section. The circular RNA of the minus polarity of chloroplastic viroids serves as the template for a second round of synthesis of multimers of complementary polarity that are cleaved into monomers and ligated to generate the monomeric circular RNAs of plus polarity strand (Fig. 1).

Catalytic activities involved in the replication of chloroplastic viroids

The catalytic activities necessary for replication via a symmetric model are the same as in the asymmetric one. However, the enzymes involved are different because the replication of members of the *Avsunviroidae* family occurs in a different subcellular compartment, the chloroplast. The nuclear-encoded chloroplastic RNA polymerase (NEP) has been suggested as the enzyme that catalyzes the RNA synthesis for the viroids of this family. In the chloroplast, there is a second RNA polymerase (the plastid-encoded RNA polymerase, PEP) that also could catalyze the transcription of chloroplastic viroids. However, the involvement of PEP was excluded based on studies of the effects of the tagetitoxin, a PEP inhibitor, on the synthesis of ASBVd in purified chloroplasts (Navarro et al., 2000) and the high accumulation of the chloroplastic viroid peach latent mosaic viroid (PLMVd) in symptomatic tissues where PEP activity was almost absent (Rodio et al., 2007). The NEP is a DNA-dependent RNA polymerase in physiological conditions, but chloroplastic viroids are able to force it to use an RNA template. As in the case of nuclear viroids, transcription of members of the family *Avsunviroidae* starts at a defined position of the circular RNA molecule. The transcription initiation sites of ASBVd and PLMVd have been identified, for both polarity strands, at an AU-rich right terminal loop of the secondary structure in ASBVd (Navarro and Flores, 2000), and at a short double-stranded RNA motif containing also the self-cleaving site in PLMVd (Delgado et al., 2005). In the case of the chloroplastic viroid eggplant latent viroid, the starting transcription site for plus and minus RNA strands is located in different motifs of the respective RNA conformations (López-Carrasco et al., 2016).

For the members of the *Avsunviroidae* family, the cleavage of the RNA multimers into monomers is not mediated by a host protein, but it is a self-catalytic process mediated by the viroid RNA itself, through the hammerhead motifs (namely hammerhead ribozymes in this chapter; HHRzs). HHRzs adopt an active structure formed by a catalytic core of few conserved residues where the site of cleavage resides, surrounded by three double-stranded stems (helix), two of which are closed by loops in most natural hammerhead ribozymes. X-ray crystallography studies showed that actually, this ribozyme does not resemble a hammerhead (as initially proposed based on the predicted secondary structure), but it adopts a tertiary structure with a Y shape, stabilized by noncanonical tertiary interactions which are important for its catalytic activity (De la Peña et al., 2003; Khvorova et al., 2003) (Fig. 2). HHRzs motifs are present in both polarity strands of chloroplastic viroids and are active in vitro and in vivo. The activity in vivo and the role in viroid replication of HHRzs is supported by experimental data showing that the nucleotide termini of the linear monomeric RNAs accumulating in infected tissues were coincident with those predicted to be generated by a hammerhead self-cleavage (Marcos and Flores, 1993; Daròs et al., 1994;

FIG. 2 Structure of the hammerhead self-cleaving motif of peach latent mosaic viroid (plus polarity strand). Conserved nucleotides in most natural hammerhead structures are denoted with a *gray* background. The conserved core is surrounded by three helix, two of them closed by loops. Tertiary interactions between loops, indicated by a *green* background, stabilize the conformation and enhance the catalytic activity. This active conformation is adopted during viroid replication and determines the self-cleavage of the viroid RNA at a specific site (indicated by a scissor).

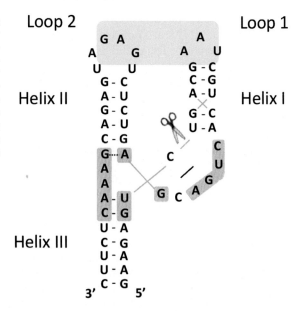

Navarro and Flores, 1997). To be active, the HHRzs must fold in a particular tertiary conformation that is adopted by the nascent RNAs during replication (Carbonell et al., 2006). The HHRzs are inactive in the most stable conformation adopted by the mature viroid where the HHRz conserved motifs form a very stable double-stranded region, called hammerhead arm that impedes the folding of the HHRzs in their active form, thus avoiding the cleavage of the genomic RNAs. Although the cleavage determined by HHRzs does not require the presence of proteins, it has been proposed that some host factors could facilitate the RNA cleavage by stabilizing the HHRz conformation. This is the case of PARB33, an RNA-binding protein that interacts in vivo with ASBVd and stimulates its self-cleavage catalytic activity in vitro (Daròs and Flores, 2002).

In line with the subcellular localization of the replicative process, a chloroplastic isoform of tRNA ligase has been identified as the enzyme responsible for the circularization of the monomeric linear RNAs for the members of the family *Avsunviroidae* (Nohales et al., 2012b).

How viroid RNAs reach the organelles where they replicate

Their small size and the fact that their biological functions must rely exclusively on their nucleotide sequence have made viroids an ideal system for studying RNA trafficking within and between plant cells (Ding et al., 2005; Steger and Perreault, 2016; Pallás and Gómez, 2017; Wang and Ding, 2010). From a practical viewpoint, systemic invasion of the plant is the most frequent scenario in which plant disease is revealed. As for plant viruses, the systemic invasion of a plant by a particular viroid requires the sequential concurrence of several phases: (1) intracellular movement, including subcellular compartmentalization for replication and subsequent exit; (2) export of the viroid progeny to neighboring cells; (3) entry to vascular tissue for long-distance trafficking to distant plant organs; and (4) systemic infection development (Ding and Wang, 2009; Navarro et al., 2019). As stated above, viroids can be classified into two main families depending, among other features, on the organelle where they replicate/accumulate, the nucleus for members of the *Pospiviroidae* family and chloroplasts for those belonging to the *Avsunviroidae* (Di Serio et al., 2018, 2021). Viroids must then subvert the plant-endogenous pathways that regulate the noncoding RNA compartmentalization into both cellular organelles.

How viroids are transported to the nucleus

The nuclear import of members of the family *Pospiviroidae* has been mainly studied for PSTVd. Woo et al. (1999) determined that PSTVd possesses a sequence and/or structural motif for nuclear import. Moreover, the nuclear import of this viroid is a cytoskeleton-independent process that is mediated by a specific and saturable receptor. By using a vector derived from the plant cytoplasmic virus potato virus X it was proposed that this nuclear import occurs through a precise process controlled by a sequence or structural motif mapping at the upper strand of the central conserved region (Zhao et al., 2001; Abraitiene et al., 2008).

In situ hybridization experiments revealed that (minus)-PSTVd-RNAs are localized in the nucleoplasm, but not in the nucleolus whereas (plus)-PSTVd-RNAs were detected in both the nucleolus and the nucleoplasm (Qi and Ding, 2003). Thus, it was suggested that PSTVd RNAs must contain a specific domain capable of recognizing the host factors commonly involved in the specific localization of endogenous RNAs in the plant-cell nucleolus. More recently, microinjected PSTVd in the abaxial leaf epidermal cells of *N. benthamiana* plants revealed that the time for half-maximal nuclear accumulation of the viroid was about 23 min and that despite the high level of cytosolic viroid injected, in some cells the nuclear import did not occur (Seo et al., 2020). Authors suggested that the nuclear import of PSTVd is not a simple concentration-dependent process but was probably under the regulation of diverse factors that may be missing from some cells.

A bromodomain-containing host protein (Viroid RNA-binding Protein 1, Virp1) with a nuclear localization signal showed in vitro and in vivo viroid-binding activity (Martínez de Alba et al., 2003) and it was proposed to mediate the nuclear compartmentalization of PSTVd-RNA in infected cells (Gozmanova et al., 2003). A direct correlation between this protein and infection by PSTVd and CEVd was demonstrated on protoplast assays using Virp1 downregulated cells (Kalantidis et al., 2007). Later on, by using a nuclear import assay system based on onion cell strips and observing the import of Alexa Fluor-594-labeled CEVd, Seo et al. (2021) demonstrated that Virp1 specifically bound to CEVd and promoted its nuclear import. More recently, Ma and Wang (2022) found that a particular RNA structure (C-loop) that is critical for PSTVd nuclear accumulation is recognized and bound by Virp1. Interestingly, Virp1 also binds to the cellular IMPORTIN ALPHA-4 (IMPa-4) which led to authors to propose a model that IMPa-4 transports the Virp1–PSTVd complex into the nucleus (Fig. 3, left upper part). Notably, nearly all nuclear-replicating viroids and viral satellite RNA contain a C-loop, suggesting that the C-loop is a conserved signal for RNA nuclear import.

The mechanism by which viroids are able to exit the nucleus is still unknown (Ding and Owens, 2003).

A. Introduction

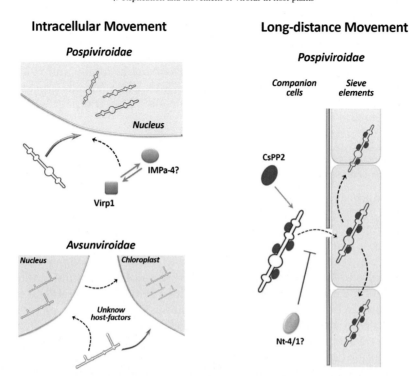

FIG. 3   Simplified picture representing the general overview of viroid movement in the infected host. The different aspects regarding intracellular (left part of the figure) and long distance (right part) trafficking are explained in the text. *CsPP2, Cucumis sativus* Phloem Protein 2; *IMPa-4*, IMPORTIN ALPHA-4; *Nt 4/1, Nicotiana tabacum* 4/1 protein; *Virp1*, bromodomain-containing viroid RNA-binding protein 1.

How viroids are transported to chloroplasts

Chloroplasts are specialized organelles, of cyanobacterial origin, that are of crucial importance for photosynthetic cells. For members of the family *Avsunviroidae*, chloroplast targeting is a critical step that permits access to the host machinery responsible for the replication of these viroids. Specific import of RNAs into chloroplasts has been suggested for both nuclear-encoded tRNAs (Bungard, 2004) and the mRNA that codes for translation factor eIF4E (Nicolai et al., 2007). However, members of the family *Avsunviroidae* are the only pathogenic RNAs known to selectively traffic into this organelle.

A partial-length RNA sequence derived from eggplant latent viroid (ELVd) acting as a 5'UTR end mediated the specific trafficking and accumulation of a functional foreign mRNA into *N. benthamiana* chloroplasts (Gómez and Pallas, 2010a). A confocal microscopy analysis in *N. benthamiana* leaves of the expression of the transcripts carrying partial deletions of the 5'UTR-end indicates that an internal 110 nucleotides length fragment is sufficient to mediate the traffic of functional GFP-mRNA into chloroplasts (Gómez and Pallas, 2010b). Later on, a potato virus X (PVX)-based in vivo assay that used a GFP-intron-ELVd construct as a reporter revealed that this viroid traffics from the cytoplasm into the nucleus, and subsequently from there into chloroplasts, which suggests a novel route to explain *Avsunviroidae* selective subcellular compartmentalization (Gómez and Pallás, 2012a,b). According to this pathway, once the ELVd invades the cell cytoplasm, it is imported into the nucleus. Next, the viroid uses this organelle as a subcellular port to be launched into the chloroplast, where replication occurs (see Fig. 3, left lower part). Consistently with these results, Baek et al. (2017) fused chrysanthemum chlorotic mottle viroid (CChMVd) to the leader sequence of a reporter gene (mRFP) and expressed it transiently in agroinfiltrated *N. benthamiana* showing that CChMVd can traffic into chloroplasts, thought to be the site of its replication. Isolated chloroplasts were shown by RT-PCR to contain the RNA sequences of both CChMVd and mRFP, if both were present, but not the mRFP sequence in the absence of the viroid sequences. The results suggest that RNA trafficking was probably due to an RNA structure, and not a particular sequence (Baek et al., 2017).

These results support the existence of a yet uncharacterized plant signaling mechanism mediated by noncoding RNAs that is able to regulate *in cis* the selective import of nuclear transcripts into chloroplasts.

Intercellular and vascular movement

To systemically invade the plant viroids must move from an infected cell to the neighboring cells to reach vascular tissue and through the phloem infect distal parts of the plant. Palukaitis (1987) showed that the movement of PSTVd in tomato plants is consistent with the pattern of movement of photosynthetic products in the phloem from fully expanded leaves to developing leaves in the upper portion of the plant and to the roots, a pattern mimicked by long distance virus transport (Pallás et al., 2011). Ten years later Ding et al. (1997) showed that viroid RNA moved rapidly from cell to cell via plasmodesmata (PD) when injected into symplasmically connected mesophyll cells and that PSTVd mediated PD transport of a fused nonmobile transcript.

The phloem is a plant-specific conduit that consists of enucleate conducting sieve elements (SEs) in close association with the nonconducting companion cells (CCs), the phloem parenchyma (PP), and the bundle sheath (BS) cells (Zhang and Turgeon, 2018). Phloem can be considered either as a communications integrator or as a checkpoint and surveillance post that limits the presence and movement of certain molecules based on their size and/or specific structural requirements (Navarro et al., 2019).

It soon became obvious that due to their lack of translational capacity and their small size, viroids could become extraordinary tools for studying the intercellular and vascular trafficking of RNAs (Zhu et al., 2001, 2002; Ding and Wang, 2009). The thermodynamically optimal, stable structure of most members of the family *Pospiviroidae* consists of an unbranched series of short helices and small loops with a high intramolecular base pairing that results in the formation of a highly structured double-stranded RNA-like rod-shaped molecule (Steger and Perreault, 2016; Wüsthoff and Steger, 2022). In the case of PSTVd its RNA genome folds into a well-established secondary structure that contains 26 base-paired stems and 27 loops (Ding, 2009; Ma and Wang, 2022). The potential involvement of these structural RNA motifs in the intercellular and vascular viroid movement has been studied in detail in recent years.

A. Introduction

After having moving from epidermis cells, through palisade and spongy mesophyll viroid long-distance movement involves the passage of viroid RNA through various cellular barriers including the bundle sheath, vascular parenchyma, and companion cells for viroid loading into sieve elements. Viroids are then transported through sieve elements to distant tissues. The invasion of distant cells requires viroid unloading from sieve elements into companion cells, followed by cell-to-cell movement into the bundle sheath and mesophyll cells. A large body of evidence has revealed that the uploading and unloading of viroid molecules are differentially regulated by these structural motifs.

How viroids enter the phloem

How viroid RNA traffics from epidermal to palisade mesophyll cells was recently revealed by using mechanical rub inoculation in *N. benthamiana* plants (Wu et al., 2019). Mutagenesis analysis of the PSTVd showed that loop 27 (5′-UUUUCA-3′) is essential for the spread of viroid RNA from the upper epidermis to underlying palisade mesophyll cells, but not for transit in the opposite direction (i.e., from palisade mesophyll to epidermal cells). Thus, requirements for RNA trafficking between epidermal cells and palisade mesophyll cells are unique and directional (Wu et al., 2019).

A loss-of-function genetic analysis identified an RNA motif (nucleotides U43/C318) forming a small loop (loop 7), which is required for PSTVd to traffic from nonvascular to vascular phloem tissue, specifically from the bundle sheath into the phloem regardless of the developmental stage of an infected leaf (Zhong et al., 2007). Takeda et al. (2011) provide evidence that loop 6 plays a decisive role in PSTVd trafficking from palisade mesophyll to spongy mesophyll cells of *N. benthamiana* leaves. This motif, called loop 6, contains the sequence 5′-CGA-3′… 5′-GAC-3′ flanked on both sides by *cis* Watson-Crick G/C and G/U wobble-base pairs. The authors propose that this motif could be recognized by distinct cellular factors, which are components of cell-specific RNA trafficking machinery. In situ hybridization demonstrated that the failure of loop 19 mutants to establish systemic infection is not due to deficiency in replication but rather to an inability to cross the cellular boundary between palisade mesophyll and spongy mesophyll in *N. benthamiana* leaves (Jiang et al., 2017). Later on, it was observed that two distinct structural motifs can replace loop 19 and retain the systemic trafficking capacity (Takeda et al., 2018). These results represent the first example where two RNA motifs regulate trafficking across the same cellular boundary.

A recent comprehensive mutational analysis demonstrated that nearly all G/U pairs are critical for the replication and/or systemic spread of PSTVd (Wu et al., 2020). Two selected G/U pairs were found to be required for PSTVd entry into, but not for exit from, the host vascular system. Thus, remarkably, PSTVd trafficking across the bundle sheath-phloem boundary can be unidirectional. This work adds further evidence for directional transit mechanisms, which would allow precise regulation of RNA transport and the establishment of distinct cellular boundaries.

How viroids move through phloem

It is commonly accepted that without the protection of specific RNA-binding proteins (RBPs), naked cellular RNA molecules quickly fall prey to degradative processes (Shyu et al., 2008). Thus, viroids, as naked RNAs most likely interact with some RBP to be transported through the phloem. The phloem of angiosperms is rich in a myriad of RBPs (Pallas and Gómez, 2013). Several groups launched in the first decade of this century to try to identify phloem proteins that could have a relevant role in the transport of these pathogenic RNAs and therefore of naked cellular RNAs. One of the most abundant RBPs of the phloem of cucurbits, phloem protein 2 (CsPP2), was shown, by two independent groups, to interact in vitro with hop stunt viroid (HSVd), strongly suggesting the involvement of this protein in viroid phloem transport (Fig. 3, right) (Owens et al., 2001; Gómez and Pallás, 2001). In vivo interaction was later demonstrated by using immunoprecipitation experiments and intergeneric graft assays revealed that both the CsPP2 and the viroid RNA were translocated to the scion (Gómez and Pallás, 2004). The translocated viroid was symptomatic in the nonhost scion, indicating that the translocated RNA was functional. A potential double-stranded-RNA binding motif in CsPP2, previously identified in a set of proteins that bind to highly structured RNAs, could explain its RNA-binding properties. Other phloem proteins with similar characteristics to CsPP2 have been identified in melon and shown to bind and translocate RNAs through intergeneric grafts (Gómez et al., 2005).

In addition to its critical involvement in the nuclear localization of PSTVd (mentioned previously), the tomato protein Virp1 also has been suggested to be involved in viroid RNA systemic spread (Maniataki et al., 2003). Authors proposed that the AGG/CCUUC motif bolsters recognition of the Terminal Right (TR) domain by Virp1 to achieve access of the viroid to pathways that propagate endogenous RNA systemic signals in plants.

More recently, an involvement of the *Arabidopsis thaliana* 4/1 (At-4/1) protein in viroid transport in the vasculature has been proposed (Morozov et al., 2014; Solovyev et al., 2013). Viroid accumulation and movement was altered in plants in which the *Nicotiana tabacum* orthologue Nt-4/1 expression was reduced by virus-induced gene silencing (Solovyev et al., 2013). The results suggest that the Nt-4/1 protein negatively regulates the long-distance transport of the viroid along the plant conducting system (Fig. 3, right).

How viroids exit from phloem

Although exit from the phloem can be considered one of the key control points in the phloem RNA transport, the information about this process is very limited (Lezzhov et al., 2021). Qi et al. (2004) analyzed PSTVd systemic trafficking in tobacco (*Nicotiana tabacum*) using a combination of cellular, genetic, and molecular approaches and identified a bipartite motif that is both necessary and sufficient to mediate the trafficking of viroid RNA from the bundle sheath to the mesophyll in young leaves. One part of the motif is located in the TR domain and the other in the pathogenicity domain. Remarkably, this motif was not necessary for trafficking in the reverse direction (i.e., from the mesophyll to bundle sheath). Authors also found that the requirement for this motif to mediate bundle sheath-to-mesophyll trafficking is dependent on leaf developmental stages. Remarkably, these findings revealed that movement across the bundle sheath–mesophyll boundary is a novel limiting step for an RNA that has exited the phloem to reach the mesophyll and that this limiting role diminishes at later stages, beyond the stage of sink–source transition, of leaf development (Qi et al., 2004).

A phloem-mobile in Etrog citron allowing the viroid transport from the scion to the root-stock has been suggested as the factor-enabling exit CEVd from vascular tissue to mesophyll cells (Bani-Hashemian et al., 2015).

Prospective and future implications

All the information necessary for viroid replication and movement in the host is condensate in a single RNA molecule of a few hundred nucleotides that does not code for any protein. Therefore, the life cycle of viroids relies on their capacity to exploit the enzymatic machinery and biological processes of the host, probably interacting with cellular factors by mimicking host RNA structural motifs. Thus, viroids are considered useful models to study RNA structure/function relationships.

Besides the identification of the host enzyme that mediates the cleavage of multimers of nuclear viroids, several other aspects of viroid replication remain still unresolved. One interesting question is how viroids are able to force host enzymes acting physiologically on a DNA template or substrate to accept an RNA template, such as the DNA-dependent RNA polymerases Pol II or the NEP, and the DNA ligase I. Recently, using a reconstituted in vitro transcription system, it has been demonstrated that the nuclear viroid PSTVd recruits a remodeled RNA Pol II complex, lacking several subunits with respect to the canonical RNA Pol II complex involved in the transcription of DNA templates. The RNA Pol II subunit Rpb9, which is responsible for the transcription fidelity of RNA Pol II, is also missing, thus giving an explanation for the high mutation rate of viroid RNAs. Moreover, the transcription complex on the viroid does not contain essential transcription elongation factors for DNA templates but only needs TFIIIA-7ZF for viroid transcription, which has been demonstrated to interact with PSTVd. This study shows that RNA Pol II complex acting on viroid replication has a different organization than that acting on DNA templates and provides a useful tool for future in vitro studies on viroid replication mechanism and machinery (Dissanayaka Mudiyanselage et al., 2022). The identification of host factors interacting with viroids and the structural motifs involved are expected to provide additional fundamental knowledge on viroid replication.

Studies on viroid movement go beyond plant virology and may shed light on how mammalian and plant cells select RNAs for transport to specific destinations (Blower, 2013). Although in recent years viroids have become uniquely simple and tractable models to elucidate the regulation of the cell-to-cell trafficking of RNAs (Ding and Wang, 2009) the precise requirements for the transport of viroids to their corresponding cellular organs where they replicate are still to be determined. We are far from having a complete vision of the possible relevant role of the RNA structure in its cellular compartmentalization. In addition, only a limited number of phloem proteins with RNA-binding capability have been identified and characterized. Thus, studies are needed to expand this list of central players in systemic viroid movement. Viroids can help to provide detailed insights into the emerging molecular mechanisms that might control the selective trafficking and delivery of phloem-mobile RNAs to target tissues and to understand how plants have evolved a systemic translocation system of RNA molecules to exert noncell-autonomous control over plant development. The nature of the possible ribonucleoprotein complex used by viroids, especially members of the family

Avsunviroidae, to move through the phloem is an issue that needs further attention. As has happened since their discovery, now a little over 50 years ago, the study of these fascinating RNA pathogens has served not only to try to solve the phytopathological problems inherent to their presence, but also to unravel essential functions and mechanisms of action of cellular RNAs.

Chapter summary

Despite the lack of protein-coding capacity, viroids are able to replicate and move autonomously in their hosts. Members of the family *Pospiviroidae* and *Avsunviroidae* replicate in the nucleus and in the chloroplast, respectively, through RNA intermediates, via a rolling circle mechanism consisting of three steps: (i) synthesis of RNA multimers, catalyzed by a host RNA polymerase; (ii) cleavage to monomers, that for chloroplast replicating viroids is mediated by hammerhead ribozymes; and, (iii) RNA circularization catalyzed by host ligases. Once viroids enter the cell, they must move intracellularly to reach the specific organelle for replication. Then, to achieve systemic infection, viroids exit from the organelle and invade the neighboring cells through plasmodesmata, and the distal part of the plant through the phloem using a complex trafficking pathway. Several structural RNA motifs and proteins involved in the replication of viroids and in the directional movement across different cell boundaries have been identified.

Study question

Question #1

"By Northern blot hybridization, it has been shown that circular RNA forms of both polarity strands of a 'viroid' accumulate in infected tissues. Which variant of the rolling circle mechanism do you propose for the replication of this viroid? In which subcellular compartment does this 'viroid' likely replicate?"

Question #2

"What are the main properties that a phloem protein should have to be a potentially necessary factor to transport viroid RNAs? Why?"

References

Abraitiene, A., Zhao, Y., Hammond, R., 2008. Nuclear targeting by fragmentation of the potato spindle tuber viroid genome. Biochem. Biophys. Res. Commun. 368, 470–475.
Baek, E., Park, M., Yoon, J.Y., Palukaitis, P., 2017. Chrysanthemum chlorotic mottle viroid-mediated trafficking of foreign mRNA into chloroplasts. Res. Plant Dis. 23 (3), 288–293. https://doi.org/10.5423/RPD.2017.23.3.288.
Bani-Hashemian, S.M., Pensabene-Bellavia, G., Duran-Vila, N., Serra, P., 2015. Phloem restriction of viroids in three citrus hosts is overcome by grafting with Etrog citron: potential involvement of a translocatable factor. J. Gen. Virol. 96, 2405–2410. https://doi.org/10.1099/vir.0.000154.
Blower, M.D., 2013. Molecular insights into intracellular RNA localization. Int. Rev. Cell Mol. Biol. 302, 1–39.

Bojić, T., Beeharry, Y., Zhang, D.J., Pelchat, M., 2012. Tomato RNA polymerase II interacts with the rod-like conformation of the left terminal domain of the potato spindle tuber viroid positive RNA genome. J. Gen. Virol. 93, 1591–1600.

Branch, A.D., Robertson, H.D., Dickson, E., 1981. Longer-than-unit-length viroid minus strands are present in RNA from infected plants. Proc. Natl. Acad. Sci. U. S. A. 78, 6381–6385.

Branch, A.D., Benenfeld, B.J., Robertson, H.D., 1988. Evidence for a single rolling circle in the replication of potato spindle tuber viroid. Proc. Natl. Acad. Sci. U. S. A. 85, 9128–9132.

Bungard, R.A., 2004. Photosynthetic evolution in parasitic plants: insight from the chloroplast genome. BioEssays 26, 235–247.

Carbonell, A., De la Peña, M., Flores, R., Gago, S., 2006. Effects of the trinucleotide preceding the self-cleavage site on eggplant latent viroid hammerheads: differences in co- and post-transcriptional self-cleavage may explain the lack of trinucleotide AUC in most natural hammerheads. Nucleic Acids Res. 34, 5613–5622.

Daròs, J.A., Flores, R., 2002. A chloroplast protein binds a viroid RNA in vivo and facilitates its hammerhead-mediated self-cleavage. EMBO J. 21, 749–759.

Daròs, J.A., Marcos, J.F., Hernández, C., Flores, R., 1994. Replication of avocado sunblotch viroid: evidence for a symmetric pathway with two rolling circles and hammerhead ribozyme processing. Proc. Natl. Acad. Sci. U. S. A. 91, 12813–12817.

De la Peña, M., Gago, S., Flores, R., 2003. Peripheral regions of natural hammerhead ribozymes greatly increase their self-cleavage activity. EMBO J. 22, 5561–5570.

Delgado, S., Martínez de Alba, E., Hernández, C., Flores, R., 2005. A short double-stranded RNA motif of peach latent mosaic viroid contains the initiation and the self-cleavage sites of both polarity strands. J. Virol. 79, 12934–12943.

Di Serio, F., Li, S.F., Matoušek, J., Owens, R.A., Pallás, V., et al., 2018. ICTV virus taxonomy profile: *Avsunviroidae*. J. Gen. Virol. 99, 611–612.

Di Serio, F., Owens, R.A., Li, S.F., Matoušek, J., Pallás, V., et al., 2021. ICTV virus taxonomy profile: *Pospiviroidae*. J. Gen. Virol. 102, 001543.

Ding, B., 2009. The biology of viroid-host interactions. Annu. Rev. Phytopathol. 47, 105–131.

Ding, B., Owens, R., 2003. Movement. In: Hadidi, A., Flores, R., Randles, J.W., Semancik, J.S. (Eds.), Viroids. CSIRO Publishing, Collingwood, Australia, pp. 49–54.

Ding, B., Wang, Y., 2009. Viroids: uniquely simple and tractable models to elucidate regulation of cell-to-cell trafficking of RNAs. DNA Cell Biol. 28, 51–56.

Ding, B., Kwon, M.O., Hammond, R., Owens, R., 1997. Cell-to-cell movement of potato spindle tuber viroid. Plant J. 12, 931–936.

Ding, B., Itaya, A., Zhong, X., 2005. Viroid trafficking: a small RNA makes a big move. Curr. Opin. Plant Biol. 8, 606–612.

Dissanayaka Mudiyanselage, S.D., Ma, J., Pechan, T., Pechanova, O., Liu, B., Wang, Y., 2022. A remodeled RNA polymerase II complex catalyzing viroid RNA-templated transcription. PLoS Pathog. 18, e1010850.

Eiras, M., Nohales, M.A., Katajima, E.W., Flores, R., Daros, J.A., 2011. Ribosomal protein L5 and transcription factor IIIA from Arabidopsis thaliana bind in vitro specifically Potato spindle tuber viroid RNA. Arch. Virol. 156, 529–533.

Flores, R., Semancik, J.S., 1982. Properties of a cell-free system for synthesis of citrus exocortis viroid. Proc. Natl. Acad. Sci. U. S. A. 79, 6285–6288.

Gas, M.E., Hernández, C., Flores, R., Daròs, J.A., 2007. Processing of nuclear viroids in vivo: an interplay between RNA conformations. PLoS Pathog. 3, 1813–1826.

Gas, M.E., Molina-Serrano, D., Hernández, C., Flores, R., Daròs, J.A., 2008. Monomeric linear RNA of *Citrus exocortis viroid* resulting from processing in vivo has 5′-phosphomonoester and 3′-hydroxyl termini: implications for the ribonuclease and RNA ligase involved in replication. J. Virol. 82, 10321–10325.

Gómez, G., Pallás, V., 2001. Identification of a vitro ribonucleoprotein complex between a viroid RNA and a phloem protein from cucumber. Mol. Plant-Microbe Interact. 14, 910–913.

Gómez, G., Pallás, V., 2004. A long distance translocatable phloem protein from cucumber forms a ribonucleoprotein complex in vivo with Hop stunt viroid RNA. J. Virol. 78, 10104–10110.

Gómez, G., Pallas, V., 2010a. Noncoding RNA mediated traffic of foreign mRNA into chloroplasts reveals a novel signaling mechanism in plants. PLoS One 5, e12269.

A. Introduction

Gómez, G., Pallas, V., 2010b. Can the import of mRNA into chloroplasts be mediated by a secondary structure of a small non-coding RNA? Plant Signal. Behav. 5, 1517–1519.

Gómez, G., Pallás, V., 2012a. Studies on subcellular compartmentalization of plant pathogenic non coding RNAs give new insights into the intracellular RNA-traffic mechanisms. Plant Physiol. 159, 558–564.

Gómez, G., Pallás, V., 2012b. A pathogenic non coding RNA that replicates and accumulates in chloroplasts traffics to this organelle through a nuclear-dependent step. Plant Signal. Behav. 7, 882–884.

Gómez, G., Pallás, V., 2013. Viroids: a light in the darkness of the lncRNA-directed regulatory netwoks in plants. New Phytol. 198, 10–15.

Gómez, G., Torres, H., Pallás, V., 2005. Identification of translocatable RNA-binding phloem proteins from melon, potential components of the long distance RNA transport system. Plant J. 41, 107–116.

Gozmanova, M., Denti, M.A., Minkov, I.N., Tsagris, M., Tabler, M., 2003. Characterization of the RNA motif responsible for the specific interaction of potato spindle tuber viroid RNA (PSTVd) and the tomato protein Virp1. Nucleic Acids Res. 31, 5534–5543.

Grill, L.K., Semancik, J.S., 1978. RNA sequences complementary to citrus exocortis viroid in nucleic acid preparations from infected *Gynura aurantiaca*. Proc. Natl. Acad. Sci. U. S. A. 75, 896–900.

Hutchins, C., Rathjen, P.D., Forster, A.C., Symons, R.H., 1986. Self-cleavage of plus and minus RNA transcripts of avocado sunblotch viroid. Nucleic Acids Res. 14, 3627–3640.

Jiang, D., Wang, M., Li, S., 2017. Functional analysis of a viroid RNA motif mediating cell-to-cell movement in Nicotiana benthamiana. J. Gen. Virol. 98, 121–125.

Jiang, J., Smith, H.N., Ren, D., Dissanayaka Mudiyanselage, S.D., Dawe, A.L., Wang, L., et al., 2018. Potato spindle tuber viroid modulates its replication through a direct interaction with a splicing regulator. J. Virol. 92, e01004-18.

Kalantidis, K., Denti, M., Tzortzakaki, S., Marinou, E., Tabler, M., Tsagris, M., 2007. Viroid binding protein 1 is necessary for the infection of potato spindle tuber viroid (PSTVd). J. Virol. 81, 12872–12880.

Khvorova, A., Lescoute, A., Westhof, E., Jayasena, S.D., 2003. Sequence elements outside the hammerhead ribozyme catalytic core enable intracellular activity. Nat. Struct. Biol. 10, 708–712.

Kolonko, N., Bannach, O., Aschermann, K., Hu, K.H., Moors, M., Schmitz, M., Steger, G., Riesner, D., 2006. Transcription of potato spindle tuber viroid by RNA polymerase II starts in the left terminal loop. Virology 347, 392–404.

Lezzhov, A.A., Morozov, S.Y., Solovyev, A.G., 2021. Phloem exit as a possible control point in selective systemic transport of RNA. Front. Plant Sci. 12, 739369. https://doi.org/10.3389/fpls.2021.739369.

López-Carrasco, A., Gago-Zachert, S., Mileti, G., Minoia, S., Flores, R., Delgado, S., 2016. The transcription initiation sites of eggplant latent viroid strands map within distinct motifs in their in vivo RNA conformations. RNA Biol. 13, 83–97. https://doi.org/10.1080/15476286.2015.1119365.

Ma, J., Wang, Y., 2022. Studies on viroid shed light on the role of RNA three-dimensional structural motifs in RNA trafficking in plants. Front. Plant Sci. 13, 836267. https://doi.org/10.3389/fpls.2022.836267.

Maniataki, E., Tabler, M., Tsagris, M., 2003. Viroid RNA systemic spread may depend on the interaction of a 71-nucleotide bulged hairpin with the host protein VirP1. RNA 9, 346–354.

Marcos, J.F., Flores, R., 1993. The 5' end generated in the in vitro self-cleavage reaction of avocado sunblotch viroid RNAs is present in naturally occurring linear viroid molecules. J. Gen. Virol. 74, 907–910.

Martínez de Alba, A.E., Sägesser, R., Tabler, M., Tsagris, M., 2003. A bromodomain-containing protein from tomato specifically binds potato spindle tuber viroid RNA in vitro and in vivo. J. Virol. 77, 9685–9694.

Morozov, S.Y., Makarova, S.S., Erokhina, T.N., Kopertekh, L., Schiemann, J., Owens, R.A., Solovyev, A.G., 2014. Plant 4/1 protein: potential player in intracellular, cell-to-cell and long-distance signaling. Front. Plant Sci. 5, 1–7.

Mühlbach, H.P., Sänger, H.L., 1979. Viroid replication is inhibited by α-amanitin. Nature 278, 185–188.

Navarro, B., Flores, R., 1997. Chrysanthemum chlorotic mottle viroid: unusual structural properties of a subgroup of viroids with hammerhead ribozymes. Proc. Natl. Acad. Sci. U. S. A. 94, 11262–11267.

Navarro, J.A., Flores, R., 2000. Characterization of the initiation sites of both polarity strands of a viroid RNA reveals a motif conserved in sequence and structure. EMBO J. 19, 2662–2670.

Navarro, J.A., Vera, A., Flores, R., 2000. A chloroplastic RNA polymerase resistant to tagetitoxin is involved in replication of avocado sunblotch viroid. Virology 268, 218–225.

Navarro, J.A., Sanchez-Navarro, J.A., Pallas, V., 2019. Key checkpoints in the movement of plant viruses through the host. Adv. Virus Res. 104, 1–64. https://doi.org/10.1016/bs.aivir.2019.05.001.

A. Introduction

Navarro, B., Flores, R., Di Serio, F., 2021. Advances in viroid-host interactions. Annu. Rev. Virol. 8, 305–325. https://doi.org/10.1146/annurev-virology-091919-092331.

Nicolai, M., Duprat, A., Sormani, R., Rodriguez, C., Roncato, M.A., Rolland, N., Robaglia, C., 2007. Higher plant chloroplasts import the mRNA coding for the eukaryotic translation initiation factor 4E. FEBS Lett. 581, 3921–3926.

Nohales, M.Á., Flores, R., Daròs, J.A., 2012a. Viroid RNA redirects host DNA ligase 1 to act as an RNA ligase. Proc. Natl. Acad. Sci. U. S. A. 109, 13805–13810.

Nohales, M.Á., Molina-Serrano, D., Flores, R., Daròs, J.A., 2012b. Involvement of the chloroplastic isoform of tRNA ligase in the replication of viroids belonging to the family Avsunviroidae. J. Virol. 86, 8269–8276.

Owens, R.A., Blackburn, M., Ding, B., 2001. Possible involvement of a phloem lectin in long distance viroid movement. Mol. Plant-Microbe Interact. 14, 905–909.

Pallás, V., Gómez, G., 2013. Phloem RNA-binding proteins as potential components of the long-distance RNA transport system. Front. Plant Sci. 4, 130. https://doi.org/10.3389/fpls.2013.00130.

Pallás, V., Gómez, G., 2017. Viroid Movement. In: Hadidi, A., Flores, R., Randles, J.W., Palukaitis, P. (Eds.), Viroids and Satellites. Academic Press, Elsevier, ISBN: 978-0-12-801498-1, pp. 83–91.

Pallás, V., Genoves, A., Sanchez-Pina, M.A., Navarro, J.A., 2011. Systemic movement of viruses via the plant phloem. In: Caranta, C., Aranda, M.A., Tepfer, M., Lopez Moya, J.J. (Eds.), Recent Advances in Plant Virology. Caister Academic Press, Norkfolk, pp. 75–102.

Palukaitis, P., 1987. Potato spindle tuber viroid: investigation of the long-distance, intra-plant transport route. Virology 158, 239–241.

Qi, Y., Ding, B., 2003. Differential subnuclear localization of RNA strands of opposite polarity derived from an autonomously replicating viroid. Plant Cell 15, 2566–2577. https://doi.org/10.1105/tpc.016576.

Qi, Y., Pélissier, T., Itaya, A., Hunt, E., Wassenegger, M., Ding, B., 2004. Direct role of a viroid RNA motif in mediating directional RNA trafficking across a specific cellular boundary. Plant Cell 16, 741–1752.

Rodio, M.E., Delgado, S., De Stradis, A., Gómez, M.D., Flores, R., Di Serio, F., 2007. A viroid RNA with a specific structural motif inhibits chloroplast development. Plant Cell 19, 3610–3626.

Seo, H., Wang, Y., Park, W.J., 2020. Time-resolved observation of the destination of microinjected potato spindle tuber viroid (PSTVd) in the abaxial leaf epidermal cells of Nicotiana benthamiana. Microorganisms 8, 2044. https://doi.org/10.3390/microorganisms8122044.

Seo, H., Kim, K., Park, W.J., 2021. Effect of VIRP1 protein on nuclear import of citrus exocortis viroid (CEVd). Biomolecules 11, 95.

Shyu, A.B., Wilkinson, M.F., van Hoof, A., 2008. Messenger RNA regulation: to translate or to degrade. EMBO J. 27, 471–481.

Solovyev, A.G., Makarova, S.S., Remizowa, M.V., Lim, H.S., Hammond, J., Owens, R.A., Kopertekh, L., Schiemann, J., Morozov, S.Y., 2013. Possible role of the Nt-4/1 protein in macromolecular transport in vascular tissue. Plant Signal. Behav. 8, e25784.

Steger, G., Perreault, J.P., 2016. Structure and associated biological functions of viroids. Adv. Virus Res. 94, 141–172. https://doi.org/10.1016/bs.aivir.2015.11.002.

Takeda, R., Petrov, A.I., Leontis, N.B., Ding, B., 2011. A three-dimensional RNA motif in potato spindle tuber viroid mediates trafficking from palisade mesophyll to spongy mesophyll in Nicotiana benthamiana. Plant Cell 23, 258–272.

Takeda, R., Zirbel, C.L., Leontis, N.B., Wang, Y., Ding, B., 2018. Allelic RNA motifs in regulating systemic trafficking of potato spindle tuber viroid. Viruses 10, 160. https://doi.org/10.3390/v10040160.

Wang, Y., Ding, B., 2010. Viroids: small probes for exploring the vast universe of RNA trafficking in plants. J. Integr. Plant Biol. 52, 17–21.

Wang, Y., Qu, J., Ji, S., Wallace, A.J., Wu, J., Li, Y., et al., 2016. A land plant-specific transcription factor directly enhances transcription of a pathogenic noncoding RNA template by DNA-dependent RNA polymerase II. Plant Cell 28, 1094–1107.

Warrilow, D., Symons, R.H., 1999. Citrus exocortis viroid RNA is associated with the largest subunit of RNA polymerase II in tomato in vivo. Arch. Virol. 144, 2367–2375.

Woo, Y.M., Itaya, A., Owens, R.A., Tang, L., Hammond, R.W., Chou, H.C., Lai, M.M.C., Ding, B., 1999. Characterization of nuclear import of potato spindle tuber viroid RNA in permeabilized protoplasts. Plant J. 17, 627–635.

Wu, J., Leontis, N.B., Zirbel, C.L., Bisaro, D.M., Ding, B., 2019. A three-dimensional RNA motif mediates directional trafficking of Potato spindle tuber viroid from epidermal to palisade mesophyll cells in Nicotiana benthamiana. PLoS Pathog. 15, e1008147. https://doi.org/10.1371/journal.ppat.1008147.

A. Introduction

Wu, J., Zhou, C., Li, J., Li, C., Tao, X., Leontis, N.B., Zirbel, C.L., Bisaro, D.M., Ding, B., 2020. Functional analysis reveals G/U pairs critical for replication and trafficking of an infectious non-coding viroid RNA. Nucleic Acids Res. 48, 3134–3155. https://doi.org/10.1093/nar/gkaa100.

Wüsthoff, K.P., Steger, G., 2022. Conserved motifs and domains in members of Pospiviroidae. Cells 11, 230. https://doi.org/10.3390/cells11020230.

Zhang, C., Turgeon, R., 2018. Mechanisms of phloem loading. Curr. Opin. Plant Biol. 43, 71–75.

Zhao, Y., Owens, R.A., Hammond, R.W., 2001. Use of a vector based on potato virus X in a whole plant assay to demonstrate nuclear targeting of potato spindle tuber viroid. J. Gen. Virol. 82, 1491–1497.

Zhong, X., Leontis, N., Qian, S., Itaya, A., Qi, Y., Boris-Lawrie, K., Ding, B., 2006. Tertiary structural and functional analyses of a viroid RNA motif by isostericity matrix and mutagenesis reveal its essential role in replication. J. Virol. 80, 8566–8581. https://doi.org/10.1128/JVI.00837-06.

Zhong, X., Tao, X., Stombaugh, J., Leontis, N., Ding, B., 2007. Tertiary structure and function of an RNA motif required for plant vascular entry to initiate systemic trafficking. EMBO J. 26, 3836–3846.

Zhong, X., Archual, A.J., Amin, A.A., Ding, B., 2008. A genomic map of viroid RNA motifs critical for replication and systemic trafficking. Plant Cell 20, 35–47. https://doi.org/10.1105/tpc.107.056606.

Zhu, Y., Green, L., Woo, Y.M., Owens, R., Ding, B., 2001. Cellular basis of potato spindle tuber viroid systemic movement. Virology 279, 69–77.

Zhu, Y., Qi, Y., Xun, Y., Owens, R., Ding, B., 2002. Movement of potato spindle tuber viroid reveals regulatory points of phloem-mediated RNA traffic. Plant Physiol. 130, 138–146.

A. Introduction

Prospecting for viroid

CHAPTER

5

Viroids diseases and its distribution in Asia

G. Vadamalai[a], Charith Raj Adkar-Purushothama[b], S.S. Thanarajoo[c], Y. Iftikhar[d], B. Shruthi[e], Sreenivasa Marikunte Yanjarappa[e], and Teruo Sano[f]

[a]Department of Plant Protection, Faculty of Agriculture, Institute of Plantation Studies, Universiti Putra Malaysia, Serdang, Selangor, Malaysia [b]RNA Group, Department of Biochemistry and Functional Genomics, Faculty of Medicine and Health Sciences, Applied Cancer Research Pavilion, University of Sherbrooke, Sherbrooke, QC, Canada [c]CABI South East Asia, Serdang, Selangor, Malaysia [d]Department of Plant Pathology, College of Agriculture, University of Sargodha, Sargodha, Pakistan [e]Department of Studies in Microbiology, University of Mysore, Mysuru, India [f]Faculty of Agriculture and Life Science, Hirosaki University, Hirosaki, Japan

Graphical representation

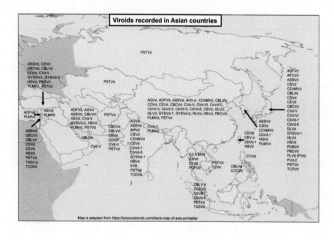

Definitions

- **Pospiviroids:** Members of the family *Pospiviroidae* are rod-like, circular RNA that replicate in the host plant's nucleus without self-cleaving activity and include most of the known viroids.
- *Apscaviroid*: The genus *Apscaviroid* belongs to the family *Pospiviroidae*. Members of this genus have a natural host range restricted to woody plants.
- **Leaf epinasty**: Downward bending and curling of the leaf.
- **Genus Citrus**: Plants of the genus *Citrus* produce fruits, including important crops such as oranges, lemons, grapefruits, pomelos, and limes.

Chapter outline

The chapter provides insights into the occurrence of different viroids in Asian countries, their natural plant hosts in the field and greenhouse, and their associated symptoms. It also specifies reliable sources for obtaining information on the viroids and their occurrence in Asia.

Learning objectives

On completion of this chapter, you should be able to understand:

- viroid diseases and their symptoms reported from the continent of Asia;
- the diversity of viroids infecting vegetables, fruits, ornamentals, and other plant species of Asia;
- the significance of viroid diseases in various crops in Asia; and
- importance of studying viroid diseases in Asian countries.

Fundamental introduction

To date, viroids are the smallest known infectious agents. Since viroids are noncoding RNA molecules, their life cycle depends on their genomic RNA structure and the interaction with certain host factors (Adkar-Purushothama and Perreault, 2019, 2020). Of the 44 viroid species recently endorsed by the International Committee on Taxonomy of Viruses (ICTV) (see Chapter 2, Viroid Taxonomy, Di Serio et al., 2017), 35 have been reported in Asia. Of these 35 viroid species, *Citrus viroid VI*, *Coconut cadang cadang viroid* and *Coconut tinangaja viroid* are only reported from Asian countries so far. On the other hand, viroids such as tomato planta macho viroid (TPMVd) and eggplant latent viroid (ELVd) have not been recorded in Asia. The palm-infecting viroids, i.e., CCCVd and CTiVd, caused the most important viroid diseases, cadang-cadang disease in the Philippines and tinangaja disease in Guam in history. Interestingly, they are the only two viroids known to cause disease in monocotyledons.

In this chapter, the viroid diseases reported from Asian countries as well as the scope for viroid research in Asian countries are discussed.

Viroid diseases of vegetable crops

Potato spindle tuber disease is instigated by potato spindle tuber viroid (PSTVd). The PSTVd RNA contains 359 nucleotides (nt) and its variants range from 356 to 363-nt (Lakshman and Tavantzis, 1993). Potato is considered the primary host due to acute symptoms and large-scale outbreaks. Other hosts recorded are tomato (*Solanum lycopersicum* L.) and pepper (*Capsicum annuum*) (Diermann et al., 2010; Hadjieva et al., 2021), which produce symptoms upon infection. Reports also noted symptomless infections in avocado (*Persea americana*), *Chrysanthemum* sp., *Dahlia* sp., *Datura* sp., and so on (Matsushita et al., 2021; Matoušek et al., 2014).

PSTVd has been reported in potato in China (Qiu et al., 2016), Iran (Arezou et al., 2008), Pakistan (Sial and Khan, 2018), Turkey (Güner et al., 2012), Afganistan, and Russia (Owens et al., 2009). PSTVd stimulates severe growth drop in potato. Leaves produced from infected plants are smaller than healthy ones. Tubers from the infected plants are smaller in size, elongated, misshapen and cleft. Their eyes might form knob-like bump that may even grow into small tubers (Fig. 1). Primary symptoms of PSTVd infection in tomato are growth reduction and chlorosis in the upper part of the plant, followed by stunting, and the chlorosis may become more severe, turning into reddening and/or purpling. In this stage, leaves may develop fragile (Gora et al., 1996).

Tomato is one of the most significant vegetables in the world. Asian countries are the center of hybrid tomato seed production. PSTVd epidemics in tomato was reported in China (Zhang et al., 2023), India (Shilpa et al., 2022), Japan (Matsushita et al., 2010), and Pakistan (Sial and

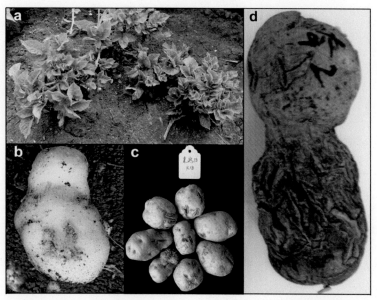

FIG. 1   Potato spindle tuber viroid (Spindle tuber of potato) symptoms in a field crop (A), and the potato tubers showing disease symptoms (B)–(D). *Original photographs are kindly provided by Prof. Zhixiang Zhang, State Key Laboratory of Biology of Plant Diseases and Insect Pests, Institute of Plant Protection, Chinese Academy of Agricultural Sciences, Beijing 100193, China. Source: https://doi.org/10.1016/S2095-3119(15)61175-3. Copyright license number: 5567080193505.*

Khan, 2019). Tomato apical stunt viroid (TASVd) infecting tomato plants (cv. *"Marmande"*) was reported in Israel. The viroid genome consists of 363 nt. The symptoms exhibited by the viroids are stunting, leaf deformation, yellowing, and brittleness were found in a few commercial plastic greenhouses at different locations in the coastal region of Israel during the spring and summer of 1999 and 2000. The shortened internodes resulted in a compact bunchy appearance of infected plants. The fruits from infected plants were considerably reduced in size and had a pale red discoloration (Antignus et al., 2002). The diseases are also reported in Asian countries like Indonesia, Malaysia, and India (Candresse et al., 1987). Pepper chat fruit viroid (PCFVd) and columnea latent viroid (CLVd) are two emerging and economically important pospiviroids infecting tomato. They are reported in the region of Thailand with the symptoms like plant stunting and leaf distortion observed on PCFVd or CLVd-infected tomato plants (Sombat et al., 2018: Reanwarakorn et al., 2011). Recently, both CLVd and PCFVd were identified in tomato and pepper plants in Vietnam (Choi et al., 2020). Tomato chlorotic dwarf viroid (TCDVd) was detected in India from symptomless *Vinca minor* (Singh and Dilworth, 2009). It occurred in greenhouse-grown tomato in Japan in 2006, and severe leaf chlorosis, yellowing, and dwarfing symptoms were observed (Matsuura et al., 2009), but the outbreak was temporary and eradicated (Tables 1 and 2).

TABLE 1   Geographical distribution of viroids of fruits in Asia.

Viroid	Host	Country
Citrus exocortis viroid (CEVd)	Citrus fruit	Widely distributed in citrus-growing regions of Asia: China[a], India[b], Iran[c], Japan[d], Israel[e], Turkey[f], Pakistan[g], Taiwan[h], Middle East Asian countries[i]
Citrus bent leaf viroid (CVd-I, CBLVd)	Citrus fruit	Widely distributed in citrus-growing regions of Asia: Japan[j], Pakistan[k], China[a], Malaysia[l], Turkey[m], Israel[n], UAE[o], Philippines[j], Iran[p]
Hop stunt viroid (CVd-II, HSVd)	Citrus fruit	Widely distributed in citrus-growing regions of Asia: Israel[q], Japan[s], China[a], India[b], Turkey[m], Iran[p]
Citrus dwarfing viroid (CVd-III, CDVd)	Citrus fruit	Turkey[m], China[a], Pakistan[k], Israel[r], Japan[t]
Citrus bark cracking viroid (CVd-IV, CBCVd)	Citrus fruit	China[u], Japan[v], Pakistan[w], Turkey[m], Israel[x]
Citrus viroid V (CVd-V)	Citrus fruit	China[a], Japan[y], Pakistan[z], Iran[aa], Turkey[ab], Oman[ac], Nepal[ac]
Apple dimple fruit viroid (ADFVd)	Apple	China[ad], Japan[ae], Lebanon[af], Iran[ag]
Apple fruit cirnkle viroid (AFCVd)	Apple	Japan[ah]
Apple scar skin viroid (ASSVd)	Apple	India[ai], Korea[aj], Japan[ak], China[al], Iran[am], Turkey[an]
	Apricot	China[ao]
	Pear, Japanese pear	Japan[ap], China[aq]

TABLE 1 Geographical distribution of viroids of fruits in Asia—cont'd

Viroid	Host	Country
Avocado sunblotch viroid (ASBVd)	Avocado	Israel[ar]
Hop stunt viroid (HSVd)	Sweet cherry	China[as]
Peach latent mosaic viroid (PLMVd)	Peach	China[at], Japan[au], Korea[av], Lebanon[aw], Turkey[ax], Syria[ay], Iran[az], Nepal[aaa]

[a] *Wang et al. (2008).*
[b] *Roy and Ramachandran (2003).*
[c] *Amiri Mazhar et al. (2014).*
[d] *Tanaka (1963).*
[e] *Ben-Shaul et al. (1994).*
[f] *Goral et al. (1993).*
[g] *Arif et al. (2005).*
[h] *Lin et al. (2015).*
[i] *Al-Harthi et al. (2013).*
[j] *Hataya et al. (1998).*
[k] *Cao et al. (2009).*
[l] *Khoo et al. (2017).*
[m] *Önelge (2010).*
[n] *Ashulin et al. (1991).*
[o] *Al-Shariqi et al. (2013).*
[p] *Amiri Mazhar et al. (2014).*
[q] *Puchta et al. (1989).*
[r] *Bar-Joseph (1993).*
[s] *Sano et al. (1986).*
[t] *Nakahara et al. (1998).*
[u] *Cao et al. (2010).*
[v] *Ito et al. (2002).*
[w] *Bakhtawar et al. (2022).*
[x] *Puchta et al. (1991).*
[y] *Ito and Ohta (2010).*
[z] *Ali et al. (2022).*
[aa] *Bani Hashemian et al. (2009).*
[ab] *Önelge and Yurtmen (2012).*
[ac] *Serra et al. (2008).*
[ad] *Ye et al. (2013).*
[ae] *He et al. (2010).*
[af] *Choueiri et al. (2007).*
[ag] *Roumi et al. (2017).*
[ah] *Koganezawa et al. (1989).*
[ai] *Walia et al. (2009).*
[aj] *Lee et al. (2001).*
[ak] *Koganezawa (1987).*
[al] *Wang et al. (2012).*
[am] *Yazarlou et al. (2014).*
[an] *Sipahioglu et al. (2009).*
[ao] *Zhao and Niu (2008).*
[ap] *Osaki et al. (1996).*
[aq] *Shamloul et al. (2004).*
[ar] *Vallejo-Pérez et al. (2015).*
[as] *Xu et al. (2019).*
[at] *Xu et al. (2008).*
[au] *Osaki et al. (1999).*
[av] *Jo et al. (2015).*
[aw] *Abou Ghanem-Sabanadzovic and Chouriri (2003).*
[ax] *Gazel et al. (2008).*
[ay] *Ismaeil et al. (2002).*
[az] *Yazarlou et al. (2012).*
[aaa] *Hadidi et al. (1997).*

TABLE 2 Geographical distribution of viroids of ornamentals in Asia.

Sl. No	Viroid	Host	Country
1	Chrysanthemum stunt viroid (CSVd)	Chrysanthemum	India[a], Japan[b], Korea[c], Thailand[d], China[e]
2	Chrysanthemum stunt viroid (CSVd)	Dahlia spp.	Japan[f]
3	Chrysanthemum chlorotic mottle viroid	Chrysanthemum plants	India[g], Thailand[d], Korea[h], China[e], Japan[i]
4	Coleus blumei viroid 1-6 (CBVd 1-6)	Coleus blumei	India[j], Malaysia[k], China[l], Korea[m], Japan[n]
5	Dahlia latent viroid (DLVd)	Dahlia	Japan[o], China[p]
6	Potato spindle tuber viroid (PSTVd)	Dahlia	Japan[q]

[a] *Adkar-Purushothama et al. (2017).*
[b] *Matsushita (2013).*
[c] *Chung et al. (2001).*
[d] *Supakitthanakorn et al. (2022).*
[e] *Zhang et al. (2011).*
[f] *Nakashima et al. (2007).*
[g] *Adkar-Purushothama et al. (2015).*
[h] *Chung et al. (2006).*
[i] *Yamamoto and Sano (2005).*
[j] *Adkar-Purushothama et al. (2013)*
[k] *Roslan et al. (2017).*
[l] *Li et al. (2006).*
[m] *Chung and Choi (2008).*
[n] *Ishiguro et al. (1996).*
[o] *Tsushima et al. (2015).*
[p] *Yan et al. (2019).*
[q] *Tsushima et al. (2011).*

Viroids infecting ornamental plants

Many ornamental plants have been found infected with viroids, and consecutively, numerous viroids were termed after their original ornamental hosts (e.g., chrysanthemum stunt viroid (CSVd) infects chrysanthemum). With the exception of the Chrysanthemum chlorotic mottle viroid (CChMVd), they all belong to the family *Pospiviroidae*.

Chrysanthemum

Chrysanthemum stunt disease is caused by CSVd. The CSVd RNA comprises 356 nt with 70% base pairing or 354 nt with 68% base pairing (Gross et al., 1982). Most chrysanthemum cultivation extents are susceptible to CSVd infection. The natural host of CSVd is chrysanthemum (*Dendranthema morifolium*) and susceptibility differs between cultivars. CSVd is reported in China, India (Adkar-Purushothama et al., 2017), Thailand (Supakitthanakorn et al., 2022), Japan (Matsushita et al., 2007), Korea (Chung et al., 2001), and Russia

(Matsushita et al., 2021). Distinctive symptoms which can be detected include: abridged plant size (one-half to two-third of the average plant size), leaves and flowers, and poor root development. Infected floras also yield flowers of reduced fresh weight by 65% (Matsushita et al., 2021). Also, the leaves of the diseased plants are often thinner and paler green compared to healthy ones. Other CSVd infection characteristics include the formation of small upper leaves on flowering stems, lower leaves that curl upward sharply at the margins, round pale diffuse spots near the margin of the leaves and the measles pattern in the "Mistletoe" varieties. CSVd outbreak caused an approximate $3 million worth of damage to the Australian chrysanthemum industry in 1987 (Hill et al., 1996). CSVd often coexist with chrysanthemum viruses, so it is difficult to determine figures on yield loss.

Chrysanthemum chlorotic mottle viroid (CChMVd, genus *Pelamoviroid*, family *Avsunviroidae*) has been reported to infect chrysanthemum in China, India, Japan, Thailand, and Korea. In India, 10% of 80 chrysanthemum samples examined were CChMVd-positive, and some of the isolates in India had a UUUC sequence at nucleotide position 82 to 85 in the tetraloop of the molecule, which is often associated with induction of symptoms in susceptible chrysanthemum cultivars. However, majority of chrysanthemum cultivars were symptomless (Adkar-Purushothama et al., 2015). Similar results have been reported in Japan (Yamamoto and Sano, 2006), so the occurrence of the chrysanthemum chlorotic mottle disease has been overlooked in the fields due to the tolerant nature of cultivated chrysanthemums to CChMVd in Asia.

Coleus

Members of the genus *Coleviroid*; *coleus blumei* viroid-1, -2, -3, -5, -6 (CbVd-1, -2, -3, -5, -6), the family *Pospiviroidae* have been reported in coleus (*Plectranthus scutellarioides*, synonym. *Coleus blumei*) in China (Li et al., 2006), India (Adkar-Purushothama et al., 2013), Indonesia (Jiang et al., 2013), Japan (Ishiguro et al., 1996), Korea (Chung and Choi, 2008), and Malaysia (Roslan et al., 2017). Although most of CbVd infections are asymptomatic, CbVd-1 isolates detected from symptomless coleus cultivator "Kong Scarlet" in Korea, comprised of 249 nt with 100% identical to CbVd-1 reported from China, was shown to cause discoloration and growth retardation in several cultivars but is symptomless in others (Chung and Choi, 2008).

Dahlia

A novel PSTVd variant was recorded for the first time from asymptomatic dahlia grown in Japan. The PSTVd dahlia isolates formed a quasispecies and a major sequence variant consisting of 361 nt in length, including five substitutions, three insertions, and one deletion compared to the intermediate strain from potato (Tsushima et al., 2015). Although the dahlia isolate showed mild symptoms in tomato, potato cultivars experimentally infected developed severe disease symptoms, indicating that the isolate has a potential risk of devastating damage to potato production (Fujiwara et al., 2013). Dahlia latent viroid (DLVd), potentially asymptomatic to dahlia, was also recorded in Japan and China (Yan et al., 2019).

B. Prospecting for viroid

Viroids infecting pome fruits

Apple

Apple (*Mallus domestica*) is one of the most significant commercial fruit crops cultivated worldwide. Apple is reported to be infected with viroids like apple scar skin viroid (ASSVd), apple dimple fruit viroid (ADFVd), apple hammerhead viroid (AHVd), and apple fruit crinkle viroid (AFCVd).

Apple dimple fruit disease is instigated by ADFVd. The ADFVd RNA consists of 300 to 311 nt single-stranded, circular RNA. Natural infections are reported from apples, figs (Italy), and pomegranates (Spain), but a broader host range could be present as ADFVd has been experimentally transmitted to pear (Di Serio et al., 1996). ADFVd is reported in Lebanon (Choueiri et al., 2007), Japan (He et al., 2010), China (Ye et al., 2013), and Iran (Roumi et al., 2017). The common evident symptom induced by ADFVd is depressed yellow-green spots of 3–4 mm in diameter on the fruits. These spots may ultimately merge into large discolored areas. ADFVd-infected fruits of apple cultivar "Starking Delicious" displayed wide necrotic areas of the flesh underlying skin depressions (Di Serio et al., 1996), and tarnished skin has been observed on apple cultivar "Golden Delicious" (Di Serio et al., 1996). The isolates from China were similar to those in Italy in size (306–307 nt), nucleotide sequence, and the symptoms on apple cultivars such as "Fuji" and "Gala." Meanwhile, the isolates in Japan were smaller in size (300–303 nt) and shared 85.5%–87.3% nucleotide sequence similarity with the Italian and Chinese variants. Graft-inoculation experiments showed that the symptoms were variable depending on the cultivars, and symptoms on some cultivars were similar in part to those by AFCVd or ASSVd, indicating that symptoms on any specific cultivar cannot discriminate them (Kasai et al., 2017).

ASSVd infection is a major limitation to apple fruit quality and roots huge economic losses. ASSVd is the type of species of the genus *Apscaviroid*, family *Pospiviroidae*. Apple scar skin disease was first reported in China (Ohtsuka, 1935), and its viroid etiology was reported in 1980s in Japan (Koganezawa et al., 1982). The viroid is composed of 330 nt (Hashimoto and Koganezawa, 1987). It also infects pear (*Pyrus communis* L) species and cultivars and causes fruit dimple disease in Japanese pear (Osaki et al., 1996) but is generally symptomless (e.g., *P. amygdalyformis, P. bretschneideri, P. communis, P. pyrifolia, P. ussuriensis*). It is reported in China (Wang et al., 2012), India (Walia et al., 2009), Iran (Yazarlou et al., 2014), Japan (1987), South Korea (Lee et al., 2001), and Turkey (Sipahioglu et al., 2009). Visible symptoms produced by ASSVd in apples vary by cultivars, can be observed at the calyx end of the pome, and include scar skin (reddish brown areas with brownish scar-like tissue), cracking, or dapple (spotting) indications (Fig. 2). The fruits are often small, rigid, do not mature sufficiently, and produce unpleasant flavor. At the time of harvest, the dapple symptoms develop on the red-skinned cultivars such as "Jonathan," "Red Gold," and "Fuji." Scar skin is very common on "Ralls Janet" and "Indo," and both types of symptoms can occur on apple cultivars "Starking Delicious" and "Red Delicious." As in "Indo," they are more prone to scars covering more than 50% of the fruit surface. Some apple cultivars also develop leaf roll or leaf epinasty like symptoms (Hadidi et al., 2017). The symptoms persuaded by ASSVd infections are limited to fruits and include color dappling, cracking, scarring, and distortion depending upon the cultivar (Walia et al., 2009). ASSVd causes significant reductions in fruit size and quality in the susceptible cultivars which causes the produce to be not marketable.

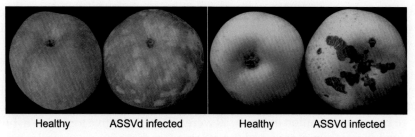

Healthy   ASSVd infected   Healthy   ASSVd infected

FIG. 2 Apple scar skin viroid (ASSVd)-induced symptoms on apples. *Source: Photographs are kindly provided by Prof. Zhixiang Zhang, State Key Laboratory of Biology of Plant Diseases and Insect Pests, Institute of Plant Protection, Chinese Academy of Agricultural Sciences, Beijing 100193, China.*

AFCVd, a tentative member of *Apscaviroid*, was first reported in Japan from apple (Koganezawa et al., 1989), then from hops (Sano et al., 2004) and Japanese persimmon (Nakaune and Nakano, 2008). The apple isolate causes apple fruit crinkle disease and the hop isolate causes hop stunt disease (see later), but the persimmon isolate is asymptomatic in Japanese persimmon. Apple fruit crinkle disease has been recorded only in Japan, and the outbreaks are sporadic.

The existence of AHVd in apples was first reported in China in 2014 (Zhang et al., 2014). Later AHVd was also reported in United States, Japan, Italy, Spain, and New Zealand (Szostek et al., 2018). In 2020, AHVd infecting four apple cultivars was reported from India, where apples showed mosaic and ringspot symptoms in infected plants. However, the role of AHVd is unclear because these trees were also infected with other viruses (Nabi and Baranwal, 2020).

Pear

Pear blister canker disease is mainly caused by pear blister canker viroid (PBCVd). The disease can occur in several cultivars of pear. Most pear cultivars are though tolerant and remain symptomless. PBCVd is reported in pear, quince, wild pear, nashi, etc. The PBCVd RNA is 315 nt in length (Hernandez et al., 1992); other variants range in size from 312 to 316 nt in length (Ambros et al., 1995); PBCVd is reported in many countries globally, including Japan (Sano et al., 1997), China, and Turkey (Yesilcollou et al., 2010). Petiole and leaf necrosis, back pustules, bark scaling, back splitting, and tree death can be observed on the pear indicator host, cv. "A20" and proposed replacement indicator pear indicator hosts, cv. "Fieudiére" (Fieud 37 and Fieud 110), and some susceptible pear varieties. Most commercial pear cultivars are tolerant to PBCVd and do not express bark symptoms (Desvignes et al., 1999). No known significant economic impact in Asia.

Peach

The peach latent mosaic disease is caused by peach latent mosaic viroid (PLMVd). The PLMVd genome consists of RNA of 335 to 338 nt. The natural host of PLMVd is peach, apricot, and cherry (St-Pierre et al., 2009; Hadidi et al., 1997). PLMVd is considered a quarantine pathogen by the European Plant Protection Organization (EPPO) and in many other countries. PLMVd is reported to be present in China (Turturo et al., 1998), Iran (Yazarlou et al.,

2012), Japan (Osaki et al., 1999), Lebanon (Abou Ghanem-Sabanadzovic and Choueiri, 2003), South Korea (Jo et al., 2015), Syria (Ismaeil et al., 2002), and Turkey (Gazel et al., 2008). Vegetative growth is affected due to the bud death which leads to reduction of peach fruit production. Other symptoms which can occur include yellow-creamy mosaic, blotch, and necrosis of leaves. The young plants also show discoloration of fruits. As a result of surveillance of several local orchards in Japan, PLMVd was detected in 33 out of 35 peach samples (94.3%); thus, infection rate is very high. Meanwhile, among 43 peach cultivars preserved in the Fruit Tree Experiment Station in Japan, 30 were RT-PCR positive for PLMVd, but only one cultivar showed mosaic symptoms, and the others were virtually symptomless (Osaki et al., 1999). A variant of PLMVd which has an insertion of 12 to 14 nt that can fold into a hairpin with a U-rich loop, causes, peach calico (PC) in susceptible peach cultivars. PC is characterized by white pattern on leaves and fruits. Peach calico is reported from China (Zhou et al., 2019) and Korea (Cho et al., 2018).

Viroids infecting avocado

Avocado sunblotch viroid (ASBVd) is recorded in Israel (Spiegel et al., 1984). This economically significant pathogen decreases the yield and quality standards of infected avocado trees. Viroid symptoms on fruits appear like depressed spots and yellow lines, undeveloped shoots and branches display yellow longitudinal bands, and leaves develop yellowish-white spots and deformations. Macroscopic symptoms caused by the ASBVd in fruits are the consequence of structural and chemical changes in the structure of the exocarpic and mesocarpic cells, which are characterized by marked cellular disorganization, accretion of phenolic compounds in the cytoplasm and cell walls, and decrease of cytoplasmic content resulting in cell collapse and death (Spiegel et al., 1984).

Viroids infecting citrus

Citrus from the *Rutaceae* family is a health fruit that is traded globally. Citrus viroids are known to constitute a risk to citrus production. Eight viroids have been reported including a tentative one (i.e., CVd-VII) so far, of them, seven were detected in Asia, i.e., citrus exocortis viroid (CEVd), citrus bent leaf viroid (CBLVd), hop stunt viroid (HSVd), citrus dwarfing viroid (CDVd), citrus bark cracking viroid (CBCVd), citrus viroid V (CVd-V), and CVd-VI (synonym, CVd-OS, Ito et al., 2001).

Citrus exocortis disease is caused by CEVd. All *Citrus* species are susceptible to infection by CEVd, but most remain symptomless. Leaf blotching and bark splitting or cracking are observed in some limes and lemons. Citrange (*Citroncirus webberi*), citrumelo (*Citrus reticulata × Poncirus trifoliata*), *P. trifoliata* and *Rangpur lime* (*Citrus limonia*) are sensitive to CEVd and express bark scaling and shelling and a decrease in tree vigor. Grafting a CEVd-infected scion variety onto susceptible rootstocks causes economic loss. CEVd is reported in China (Wang et al., 2009), India (Roy and Ramachandran, 2006), Iran (Amiri Mazhar et al., 2014), Israel (Ben-Shaul et al., 1994), Japan (Tanaka, 1963), Middle East Asia (Oman, Lebanon, Syria, Egypt, and Jordon) (Al-Harthi et al., 2013), Pakistan (Arif et al., 2005), Taiwan (Lin et al.,

2015), Turkey (Goral et al., 1993), and Middle East (El-Dougdoug et al., 2017). Citrus spp. on *P. trifoliata*, citrange (*Citroncirus webberi*) or citrumelo (*C. reticulata* × *P. trifoliata*) rootstock: bark shelling and longitudinal cracking occur about 1–2 years after budding. As time passes, the tree growth become stunted which eventually reduces the yield (Fig. 3). *C. limon*: longitudinal bark shelling and cracking occur; leaves may have a blotchy mottle. *Citrus medica* (Etrog citron): severe leaf epinasty and rugosity, cracking and browning of the veins and leaf tip underside, and stunting. *C. limonia* (Rangpur lime): yellow blotches on the older twigs. *Gynura aurantiaca* (velvet plant)—the principal herbaceous indicator plant. Symptoms of leaf epinasty and rugosity develop 10–30 days after inoculation.

CBLVd is an emerging and widely distributed viroid along with its variants in citrus growing areas of the world, including China, Israel, Japan, Pakistan, Philippines, and UAE. This disease was reported in some citrus cultivars of mandarin and sweet orange in the Pakistani regions. Chlorophyll contents were pointedly lower in the diseased leaves samples of all the citrus cultivars as related to healthy ones, whereas total soluble phenolics (TSP) were found in higher concentration in the CBLVd-infected samples of citrus cultivars. Similarly, the activities of polyphenol oxidase (PPO) and phenylalanine ammonia lyase (PAL) were amplified significantly in leaves of citrus cultivars diseased with CBLVd as compared to healthy plants. Asymptomatic citrus traits cause decrease in the size of fruit, including low yield (Bakhtawar et al., 2022). A 328-nt variant of CBLVd was characterized from 14 citrus varieties in Malaysia, showing leaf bending, stunting, and midvein necrosis. CBLVd was detected by RT-PCR assay using CBLVd specific primers in 12 out of 90 samples, collected 16 from six different areas in Malaysia. The viroid was present in species of citrus, namely *Citrofortunella microcarpa*, *Citrus aurantifolia*, and *C. sinensis* (Iftikhar et al., 2019, 2022; Khoo et al., 2017).

The first case of an HSVd isolate (HSVd-RL) from citrus rootstocks showing bark scaling and splitting symptoms in Central India was reported with an uncommon symptom of reduced size and yellowing of newly emerging leaves (Ramachandran et al., 2005). A viroid was isolated and purified from Kagzi lime (*C. aurantifolia*) leaves affected by yellow corky vein disease and consisted of 295 nt. In BLAST analysis, the sequence aligned with different HSVd variants showing nearly 100% sequence identity with six citrus cachexia isolates of HSVd. The viroid was tentatively named yellow corky vein variant of HSVd-ycv (Roy and Ramachandran, 2003).

FIG. 3   (A) Bark scaling appeared on trifoliata orange rootstock infected with citrus exocortis viroid *(arrow)*. (B) Diseased citrus tree infected with citrus exocortis viroid.

B. Prospecting for viroid

Although the adverse effects of individual citrus viroids other than CEVd on citrus are limited, it is known that mixed infection with multiple citrus viroids can cause disease symptoms similar to those of exocortis, indicating that certain exocortis-like diseases in Japan were caused by some combination of citrus viroids complex (Ito et al., 2002).

Viroid infecting coconut palm

The coconut palm (*Cocos nucifera*) grows all over the wet tropics. CCCVd endures to impend the coconut industry and is a newly recognized factor in the global oil palm (*Elaeis guineensis* Jacq.) industry (Rodriguez et al., 2017).

CCCVd belongs to genus *Cocadviroid*, the family *Pospiviroidae* causes the lethal cadang-cadang disease of coconut palms in the Philippines. Total losses exceed 40 million palms (Randles and Rodriguez, 2003). The basic form of CCCVd is a 246 nt RNA species, the smallest viroid size known and the smallest known infectious pathogen. There are also numerous variants of the basic 246 nt sequence (Haseloff et al., 1982; Imperial et al., 1981; Mohamed et al., 1982). Coconut palms with lamina depletion and further rapid lethal decay of coconut palms known as the "brooming" phenotype contain single mutations compared to the basic 246 nt (Randles and Rodriguez, 2003). All variants are infectious. CCCVd spread naturally, but the mode of spread is still unknown and several aspects of its epidemiology still require investigation. There are no known control measures for CCCVd, with exclusion being the only method considered effective in controlling the spread of this viroid. In addition, CCCVd is also associated with an orange spotting leaf disorder formerly described as "genetic" orange spotting (GOS) in commercially grown African oil palm in South East Asia and the South Pacific region (Hanold and Randles, 1991; Rodriguez et al., 2017). CCCVd is a quarantinable viroid, and variants of CCCVd have been characterized from oil palms in Malaysia (Vadamalai et al., 2006; Wu et al., 2013). Furthermore, several palm species and monocotyledons in the Pacific region and South East Asia (Indonesia, Malaysia, Thailand, Papua New Guinea, the Philippines) have been revealed by molecular hybridization to comprise CCCVd-like RNAs (see Chapter 8).

Coconut tinangaja or "yellow mottle decline" disease is caused by CTiVd. The molecular structure of CTiVd is similar to that of CCCVd, but it has only about 64% sequence identity cadang-cadang and tinangaja diseases differ in symptomatology. Tinangaja disease is distinguished by small elongated nuts without a kernel. It progresses through a reduction in the number of fronds and the size and number of fruits, shrivelling and deformation of fruit, cessation of fruit production, failure to produce inflorescences, tapering of the distal end of the trunk, persistence of stipules, stippling of leaflets (fine chlorotic spots), and brittleness of leaflets and fronds (Vadamalai et al., 2017). CTiVd has only been characterized from coconut palm (Fig. 4).

Viroids infecting grapevine

Grapevine (*Vitis vinifera* L.) belonging to family *Vitaceae* is a commercially significant fruit worldwide. Six distinct viroids (i.e., Australian grapevine viroid (AGVd), grapevine yellow speckle viroid 1 (GYSVd-1), grapevine yellow speckle viroid 2 (GYSVd-2), CEVd, HSVd, and

FIG. 4 Symptoms of OS on oil palm. (A) A six-year-old oil palm exhibiting OS symptoms; (B) OS translucent under the sunlight; (C) bright orange spotting on leaflets of an OS palm; (D) healthy oil palm.

grapevine latent viroid (GLVd)) and a tentative member named Japanese grapevine viroid (JGVd) have been identified in grapevines worldwide. All these six viroids infecting grapevines in Asian countries are recorded such as India (Adkar-Purushothama et al., 2014), China (Jiang et al., 2009), Iran (Zaki-Aghl et al., 2013), Tunisia (Elleuch et al., 2002), Korea (Heo et al., 2022), Japan (Chiaki and Ito, 2020), and Turkey (Buzkan et al., 2018). However, none of these viroids are known to induce severe disease symptoms in grapevines.

Viroids of hop

Hop (*Humulus lupulus*) is a "niche" crop. This plant of the family Cannabaceae can be successfully grown only in zones of the northern hemisphere with temperate climatic regions and sufficient humidity. Four viroids were known to infect hops; i.e., HSVd, hop latent viroid (HLVd), AFCVd, and CBCVd, three of them (HSVd, HLVd, and AFCVd) were reported in Asia.

Hop stunt disease first emerged in Japan in the 1940s–1950s and caused significant economic damage to the hop production in Japan during the 1960s–80s (Yamamoto et al., 1970). The causative agent was identified as a viroid, named HSVd (Sasaki and Shikata, 1977). The disease was later reported in South Korea (Lee et al., 1988) and more recently in China (Guo et al., 2008). The typical symptoms include stunting, leaf curling, reduction of α-acids content in the cones, and decrease in cone yields which result in about 1/2–1/3 of that of the healthy.

Hop stunt disease was also incited by AFCVd, a tentative member of *Apscaviroid* (Sano et al., 2004). Symptoms are similar to those caused by HSVd and virtually indistinguishable in the fields.

B. Prospecting for viroid

HLVd-infecting hops were reported in China (Liu et al., 2008) and Japan (Hataya et al., 1992). HLVd does not show any obvious symptoms in hop plants but can cause a considerable decrease in the content of α-bitter acid, an essential ingredient to give a bitter taste to beer, in the cones of sensitive cultivars.

Viroids infecting other fruit trees and weeds

Other than hops, HSVd infects a wide range of host species, including grapevines, citrus, prunus species, pear mulberry, and so on, where it causes specific disorders such as citrus cachexia in some cases but infects asymptomatic in many cases (Hataya et al., 2017). HSVd is distributed widely in the world, including Asia, i.e., India (Ramachandran et al., 2005), Iran (Maddahian et al., 2019), Israel (Puchta et al., 1989), Jordan (Al Rwahnih et al., 2001), Lebanon (Abou Ghanem-Sabanadzovic and Choueiri, 2003), South Korea (Cho et al., 2011), Syria (Elbeaino et al., 2012), Taiwan (Lin et al., 2015), and Turkey (Gazel et al., 2008). Except for hop stunt and citrus cachexia diseases, they are virtually asymptomatic. However, some susceptible cultivars of plum such as "Oishiwase-sumomo" and peach such as "Xiangshanhong" cultivated in China (Zhou et al., 2006), Japan (Sano et al., 1989), and South Korea (Cho et al., 2011) were reported to develop dapple fruit symptoms.

A novel viroid-like RNA was reported in four contigs in the latest Transcriptome Shotgun Assembly (TSA) from lychee fruit (*Litchi chinensis*) in China. Portions of this sequence are closely associated with the central conserved region (CCR) and terminal conserved region (TCR) present in members of the genus *Apscaviroid*, sequences that are important criteria for viroid classification. RT-PCR with two divergent adjacent primers amplified the full 304-nt sequence of this viroid-like RNA, which can be folded into a rod-like secondary structure, a typical feature of viroids in the family *Pospiviroidae* (Jiang et al., 2017). Although the detection rate was high (~76%) from samples collected in China, any specific diseases have not been observed.

A distinct variant of AFCVd and a new species named persimmon latent viroid (PLVd; synonym. PVd) were reported from Japanese persimmon (*Diospyrus kaki*), and persimmon viroid 2 (PVd-2) from American persimmon (*Diospyros virginiana*), both reported in Japan. They belong to the genus *Apscaviroid*, and any damage to the host has not been reported (Ito et al., 2013; Nakaune and Nakano, 2008).

A variant of Iresine viroid (IrVd) has been reported in India from a perennial weed herb, red joy weed (*Alternanthera sessilis*) (Singh et al., 2006).

Prospective

Asia is the largest region on earth, comprised of nearly 50 countries, home to about 4.4 billion people, accounting for about 60% of the world's population. It is geographically diverse, divided into East, Southeast, North, South, Central, and West, each standing on its own distinctive civilizations and cultures. From the east to the south, a lot of rainfall and fertile land spreads, and various agricultural products are produced. As described in this chapter, reflecting such regional diversity, viroids occurring in Asia are also diverse, and not only those most of the currently known species but even those not yet found in the other regions of the world have been detected. For example, coconut "cadang-cadang" and "tinangaja" diseases are those found only in Asia and hop stunt and apple scar skin diseases are among those first reported in Asia. All of them have caused devastating damages to the local crop production. Along with economic development, research on viroids in Asian countries is progressing rapidly, and since the beginning of the 21st century, the amount of information on viroids has been steadily increasing. On the other hand, information is still scarce in some regions in Asia. Therefore, it is still insufficient to grasp the overall picture of viroid ecology and epidemiology of viroid diseases in all Asian countries. It is necessary to steadily collect information related to viroids and viroid diseases, from basic to applied fields.

Future implications

Before 2000, several viroids were reported only from Japan. However, access to molecular biology techniques in the last two decades have increased detection of viroid diseases in Asian countries. This increase in the number of viroid is also a result of introduction of global plant species to new geographical areas. Further application of the deep-sequencing method led to the discovery of new viroid species such as PVd-2, GLVd, and AHVd, as well as possible members of grapevine and lychee viroid-like RNAs in Asian countries, some of them are now reported to have been detected from other parts of the world as well. This indicates a possibility of the presence of more diverse viroid species in Asian countries than what is reported so far. On the other hand, this also suggested the requirements for information exchanges and robust quarantine regulations based on scientific knowledge, as well as screening for viroids and viroid-like RNAs in the production fields.

Chapter summary

This chapter concentrated much on the viroids associated with different plants and their distribution in the Asia continent. This chapter also helps to understand the diversity of viroids associated with vegetables, fruits, ornamentals, special crops like hop and coconut in Asia, and different viroids reported in the regions of the Asia continent and their symptoms.

Study questions

1. Name one viroid species from each family that infects the chrysanthemum.
2. Name a viroid species that infects both ornamental and horticultural plants.
3. Name four viroids of pome fruits and their associated symptoms.
4. What is the size of hop stunt viroid? Mention at least three host plants of the viroid.
5. Name two pospiviroids that infect only palms.

Further reading

Parakh, D.B., Zhu, S, Sano, T. 2017. Chapter 47: Geographical distribution of viroids in south, southeast, and east Asia. In: Hadidi, A., et al. (Eds.), Viroids and Satellites. Academic Press, pp. 507–518. https://doi.org/10.1016/B978-0-12-801498-1.00047-4.

Verhoeven, J., Hammond, R., Stancanelli, G. 2017. Chapter 3: Economic significance of viroids in ornamental crops. In: Hadidi et al., (Eds.), Viroids and Satellites. pp. 27–38. https://doi.org/10.1016/B978-0-12-801498-1.00003-6.

References

Abou Ghanem-Sabanadzovic, N., Choueiri, E., 2003. Presence of peach latent mosaic and hop stunt viroid in Lebanon. In: Myrta, A., Di Terlizzi, B., Savino, V. (Eds.), Virus and Virus-like Diseases of Stone Fruits, with Particular Reference to the Mediterranean Region. Vol. 2003. CIHEAM, Bari, pp. 139–141. http://om.ciheam.org/article.php?IDPDF=3001787.

Adkar-Purushothama, C.J., Perreault, J-P., 2019. Suppression of RNA-dependent RNA polymerase 6 favors the accumulation of potato spindle tuber viroid in Nicotiana benthamiana. Viruses 11, 345. https://doi.org/10.3390/v11040345.

Adkar-Purushothama, C.R., Perreault, J.P., 2020. Current overview on viroid–host interactions. RNA 11 (2), 1–22. https://doi.org/10.1002/wrna.1570.

Adkar-Purushothama, C.R., Nagaraja, H., Sreenivasa, M.Y., Sano, T., 2013. First report of coleus blumei viroid infecting coleus in India. Plant Dis. 97 (1), 149. https://doi.org/10.1094/PDIS-08-12-0715-PDN.

Adkar-Purushothama, C.R., Kanchepalli, P.R., Yanjarappa, S.M., Zhang, Z.X., Sano, T., 2014. Detection, distribution, and genetic diversity of Australian grapevine viroid in grapevines in India. Virus Genes 49 (2), 304–311. https://doi.org/10.1007/s11262-014-1085-5.

Adkar-Purushothama, C.R., Chennappa, G., Poornachandra Rao, K., Sreenivasa, M.Y., Nagendra Prasad, M.N., Maheshwar, P.K., Sano, T., 2015. Molecular identification of chrysanthemum chlorotic mottle viroid infecting chrysanthemum in Karnataka, India. Plant Dis. 99 (12), 1868. https://doi.org/10.1094/PDIS-04-15-0428-PDN.

Adkar-Purushothama, C.R., Chennappa, G., Poornachandra Rao, K., Sreenivasa, M.Y., Maheshwar, P.K., Nagendra Prasad, M.N., Sano, T., 2017. Molecular diversity among viroids infecting chrysanthemum in India. Virus Genes 53 (4), 636–642. https://doi.org/10.1007/s11262-017-1468-5.

Al Rwahnih, M., Myrta, A., Abou-Ghanem, N., di Terlizzi, B., Savino, V., 2001. Viruses and viroids of stone fruits in Jordan. EPPO Bull. 31 (1), 95–98. https://doi.org/10.1111/j.1365-2338.2001.tb00973.x.

Al-Harthi, S.A., Al-Sadi, A.M., Al-Saady, A.A., 2013. Potential of citrus budlings originating in the Middle East as sources of citrus viroids. Crop Prot. 48, 13–15. https://doi.org/10.1016/j.cropro.2013.02.006.

Al-Shariqi, R.M., Al-Hammadi, M.S., Al-Sadi, A.M., 2013. First report of citrus bent leaf viroid in the United Arab Emirates. J. Plant Pathol. 96, S4.71.

Ali, A., Umar, U., Akhtar, S., Shakeel, M.T., Rehman, A.u., Tahir, M.N., Atta, S., Moustafa, M., Ölmez, F., Parveen, R., 2022. Identification and primary distribution of Citrus viroid V in citrus in Punjab, Pakistan. Mol. Biol. Rep. 49, 11433–11441.

Ambros, S., Desvignes, J.C., Lldcer, G., Flores, R., 1995. Pear blister canker viroid: sequence variability and causal role in pear blister canker disease. J. Gen. Virol. 76 (1), 2625–2629.

Amiri Mazhar, M., Bagherian, S.A.A., Ardakani, A.S., Izadpanah, K., 2014. Nucleotide sequence and structural features of hop stunt viroid and citrus bent leaf viroid variants from blighted citrus plants in Kohgiluyeh-Boyerahmad province of Iran. J. Agric. Sci. Technol. 16 (1), 657–665.

Antignus, Y., Lachman, O., Pearlsman, M., Gofman, R., Bar-Joseph, M., 2002. A new disease of greenhouse tomatoes in Israel caused by a distinct strain of tomato apical stunt viroid (TASVd). Phytoparasitica 30 (5), 502–510. http://www.ncbi.nlm.nih.gov/gorf/wblast.cgi.

Arezou, Y., Jafarpour, B., Rastegar, M.F., Javadmanesh, A., 2008. Molecular detection of potato spindle tuber viroid in Razavi and Northern Khorasan provinces. Pak. J. Biol. Sci. 11 (12), 1642–1645. https://doi.org/10.3923/pjbs.2008.1642.1645.

Arif, M., Ahmad, A., Ibrahim, M., Hassan, S., 2005. Occurence and distribution of virus and virus-like diseases of citrus in North-East frontier province in Pakistan. Pak. J. Bot. 37, 407–421.

Ashulin, L., Lachman, O., Hadas, R., Bar-Joseph, M., 1991. Nucleotide sequence of a new viroid species, citrus bent leaf viroid (CBLVd) isolated from grapefruit in Israel. Nucleic Acids Res. 19, 4767.

Bakhtawar, F., Wang, X., Manan, A., Iftikhar, Y., Atta, S., Bashir, M.A., Mubeen, M., Sajid, A., Hannan, A., Hashem, M., Alamri, S., 2022. Biochemical characterization of citrus bent leaf viroid infecting citrus cultivars. J. King Saud Univ. Sci. 34 (1), 101733. https://doi.org/10.1016/j.jksus.2021.101733.

Bani Hashemian, S.M., Taheri, H., Duran-Vila, N., Serra, P., 2009. First report of Citrus viroid V in Moro blood sweet orange in Iran. Plant Dis. 94 (1), 129. APS Publisher http://doi.org/10.1094/PDIS-94-1-0129A.

Bar-Joseph, M., 1993. Citrus viroids and citrus dwarfing in Israes. Acta Hortic. 349, 271–276.

Ben-Shaul, A., Guang, Y., Mogliner, N., Hadar, R., Mawassi, M., Gafny, R., Joseph, M., 1994. Genomic diversity among populations of two citrus viroids from different graft-transmissible dwarfing complexes in Israel. Mol. Plant Pathol. 85 (3), 359–364.

Buzkan, N., Kılıç, D., Balsak, S.C., 2018. Distribution and population diversity of Australian grapevine viroid (AGVd) in Turkish autochthonous grapevine varieties. Phytoparasitica 46 (3), 295–300. https://doi.org/10.1007/s12600-018-0668-4.

Candresse, T., Smith, D., Diener, T.O., 1987. Nucleotide sequence of a full-length infectious clone of the Indonesian strain of tomato apical stunt viroid (TASV). Nucleic Acids Res. 15 (24), 10597. https://doi.org/10.1093/nar/15.24.10597.

Cao, M.J., Atta, S., Liu, Y.Q., Wang, X.F., Zhou, C.Y., Mustafa, A., Iftikhar, Y., 2009. First report of citrus bent leaf viroid and citrus dwarfing viroid from citrus in Punjab, Pakistan. Plant Dis. 93 (8), 840. https://doi.org/10.1094/PDIS-93-8-0840C.

Cao, M., Liu, Y.Q., Wang, X., Yang, F., 2010. First report of Citrus bark cracking viroid and Citrus viroid V infecting Citrus in China. Plant Dis. 94. 922–922.

Chiaki, Y., Ito, T., 2020. Characterization of a distinct variant of hop stunt viroid and a new apscaviroid detected in grapevines. Virus Genes 56 (2), 260–265. https://doi.org/10.1007/s11262-019-01728-1.

Cho, I.-S., Chung, B.-N., Cho, J.-D., Choi, S.-K., Choi, G.-S., Kim, J.-S., 2011. Hop stunt viroid (HSVd) sequence variants from dapple fruits of plum (*Prunus salicina* L.) in Korea. Res. Plant Dis. 17 (3), 358–363. https://doi.org/10.5423/rpd.2011.17.3.358.

Cho, I.S., Kim, S.J., Kwon, S.J., Yoon, J.Y., Chung, B.N., Hammond, J., Ju, H.K., Lim, H.S., 2018. First report of a typical calico-associated isolate of Peach latent mosaic viroid from calico disease-affected peach trees in Korea. Plant Dis. 102 (5), 1044–1045. http://apsjournals.apsnet.org/loi/pdis.

Choi, H., Jo, Y., Cho, W.K., Yu, J., Tran, P.T., Salaipeth, L., Kwak, H.R., Choi, H.S., Kim, K.H., 2020. Identification of viruses and viroids infecting tomato and pepper plants in Vietnam by metatranscriptomics. Int. J. Mol. Sci. 21 (20), 1–16. https://doi.org/10.3390/ijms21207565.

Choueiri, E., Zammar, S.E., Jreijiri, F., Hobeika, C., Myrta, A., Di Serio, F., 2007. First report of apple dimple fruit viroid in Lebanon. J. Plant Pathol. 89 (1), 301–304.

Chung, B-N, Choi, G-S., 2008. Incidence of Coleus blumei viroid 1 in seeds of commercial coleus in Korea. Plant Pathol. J. 24, 305–308.

Chung, B.N., Choi, G.S., Kim, H.R., Kim, J.S., 2001. Chrysanthemum stunt viroid in Dendranthema grandiflorum. Plant Pathol. J. 17 (4), 194–200.

Chung, B., Kim, D., Kim, J., Cho, J., 2006. Occurrence of Chrysanthemum chlorotic mottle viroid in Chrysanthemum (Dendranthema grandiflorum) in Korea. Plant Pathol. J. 22, 334–338.

Desvignes, J.C., Cornaggia, D., Ambrós, S., Flores, R., 1999. Pear blister canker viroid: host range and improved bioassay with two new pear indicators, Fieud 37 and Fieud 110. Plant Dis. 83 (5), 419–422.

Di Serio, F., Aparicio, F., Alioto, D., Ragozzino, A., Flores, R., 1996. Identification and molecular properties of a 306 nucleotide viroid associated with apple dimple fruit disease. J. Gen. Virol. 77 (11), 2833–2837. https://doi.org/10.1099/0022-1317-77-11-2833.

Di Serio, F., Li, S.F., Pallás, V., Owens, R.A., Randles, J.W., Sano, T., Verhoeven, J.T.J., Vidalakis, G., Flores, R., 2017. Viroid taxonomy. In: Viroids and Satellites. Elsevier Inc, pp. 135–146.

Diermann, N., Matoušek, J., Junge, M., Riesner, D., Steger, G., 2010. Characterization of plant miRNAs and small RNAs derived from potato spindle tuber viroid (PSTVd) in infected tomato. Biol. Chem. 391 (12), 1379–1390. https://doi.org/10.1515/BC.2010.148.

B. Prospecting for viroid

El-Dougdoug, K.A., Çağlayan, K., Elleuch, A., Al-Tuwariqi, H.Z., Gyamera, E.A., Hadidi, A., 2017. Geographical distribution of Viroids in Africa and the Middle East. In: Viroids and Satellites. Elsevier Inc, pp. 485–496, https://doi.org/10.1016/B978-0-12-801498-1.00045-0.

Elbeaino, T., Kubaa, R.A., Ismaeil, F., Mando, J., Digiaro, M., 2012. Viruses and hop stunt viroid of fig trees in Syria. J. Plant Pathol. 94 (3), 687–691. https://www.jstor.org/stable/45156298.

Elleuch, A., Fakhfakh, H., Pelchat, M., Landry, P., Marrakchi, M., Perreault, J.-P., 2002. Sequencing of Australian grapevine viroid and yellow speckle viroid isolated from a Tunisian grapevine without passage in an indicator plant. Eur. J. Plant Pathol. 108 (1), 815–820.

Fujiwara, Y., Nomura, Y., Hiwatashi, S., Shiki, Y., Itto, T., Hamanaka, D., Saito, N., 2013. Pathogenicity of potato of potato spindle tuber viroid isolated from dahlia and its transmissibility in dahlia. Res. Bull. Plant Prot. Jpn. 49 (1), 41–46. http://www.pps.go.jp/.

Gazel, M., Serce, C.U., Caglayan, K., Faggioli, F., 2008. Sequence variability of hop stunt viroid isolates from stone fruits in Turkey. J. Plant Pathol. 9 (1), 23–28. https://www.jstor.org/stable/41998454.

Gora, A., Candresse, T., Zagrski, W., 1996. Use of intramolecular chimeras to map molecular determinants of symptom severity of potato spindle tuber viroid (PSTVd). Arch. Virol. 141, 2045–2055. https://link.springer.com/article/10.1007/BF01718214.

Goral, T., Tasdemir, H.A., Tavarci, T., Mermer, S., Goral, S., Kelten, M., Tasmedir, T., Gunes, S., Gocmen, M., 1993. The citrus variety improvement program in Turkey. Int. Org. Citrus Virol. Conf. Proc. 12 (12). https://doi.org/10.5070/c505w5v6qj.

Gross, H.J., Krupp, G., Domdey, H., Raba, M., Jank, P., Lossow, C., Alberty, H., Sanger, H.I., Ramm, K., 1982. Nucleotide sequence and secondary structure of citrus cxocortis and chrysanthemum stunt viroid. Eur. J. Biochem. 121 (2), 249–257. https://doi.org/10.1111/j.1432-1033.1982.tb05779.x.

Güner, Ü., Sipahioğlu, H.M., Usta, M., 2012. Incidence and genetic stability of potato spindle tuber pospiviroid in potato in Turkey. Turk. J. Agric. For. 36 (3), 353–363. https://doi.org/10.3906/tar-1103-54.

Guo, L.H., Liu, S.X., Wu, Z.J., Mu, L.X., Xiang, B.C., Li, S.F., 2008. Hop stunt viroid (HSVd) newly reported from hop in Xinjiang, China. Plant Pathol. 57 (4), 764. https://doi.org/10.1111/j.1365-3059.2008.01875.x.

Hadidi, A., Shamloul, A.M., Amer, M.A., 1997. Occurrence of peach latent mosaic viroid in stone fruits and its transmission with contaminated blades. Plant Dis. 81 (2), 154–158.

Hadidi, A., Barba, M., Hong, N., Hallan, V., 2017. Apple scar skin viroid. In: Viroids and Satellite. Elsevier Inc, pp. 217–228, https://doi.org/10.1016/B978-0-12-801498-1.00021-8.

Hadjieva, N., Apostolova, E., Baev, V., Yahubyan, G., Gozmanova, M., 2021. Transcriptome analysis reveals dynamic cultivar-dependent patterns of gene expression in potato spindle tuber viroid-infected pepper. Plan. Theory 10 (12), 2687. https://doi.org/10.3390/plants10122687.

Hanold, D., Randles, J.W., 1991. Detection of coconut cadang-cadang viroid-like sequences in oil and coconut palm and other monocotyledons in the south-west Pacific. Ann. Appl. Biol. 118, 139–151.

Haseloff, J., Mohamed, N.A., Symons, R.H., 1982. Viroid RNAs of cadang-cadang disease of coconuts. Nature 299, 316–321.

Hashimoto, J., Koganezawa, H., 1987. Nucleotide sequence and secondary structure of apple scar skin viroid. Nucl. Acids Res. 15, 7045–7052.

Hataya, T, Hikage, K, Suda, N, Nagata, T, Li, S, Itoga, Y, Tanikoshi, T, Shikata, E., 1992. Detection of hop latent viroid (HLVd) using reverse transcription and polymerase chain reaction (RT-PCR). Ann. Phytopathol. Soc. Jpn. 58, 677–684.

Hataya, T., Nakahara, K., Ohara, T., Ieki, H., Kano, T., 1998. Citrus viroid Ia is a derivative of citrus bent leaf viroid (CVd-Ib) by partial sequence duplications in the right terminal region. Arch. Virol. 143, 971–980.

Hataya, T., Tsushima, T., Sano, T., 2017. Hop stunt viroid. In: Viroids and Satellites. Elsevier Inc, pp. 199–210, https://doi.org/10.1016/B978-0-12-801498-1.00019-X.

He, Y.H., Isono, S., Kawaguchi-Ito, Y., Taneda, A., Kondo, K., Iijima, A., Tanaka, K., Sano, T., 2010. Characterization of a new apple dimple fruit viroid variant that causes yellow dimple fruit formation in "Fuji" apple trees. J. Gen. Plant Pathol. 76 (1), 324–330. https://doi.org/10.1007/s10327-010-0258-x.

Heo, J.Y., Lee, D.H., Lee, C.H., 2022. First report of Australian grapevine viroid infecting grapevines in Korea. J. Plant Pathol. 104 (1), 1569. https://doi.org/10.1007/s42161-022-01199-8.

Hernandez, C., Elena, S.F., Moya, A., Flores, R., 1992. Pear blister canker viroid is a member of the apple scar skin subgroup (apscaviroids) and also has sequence homology with viroids from other subgroups. J. Gen. Virol. 73 (10), 2503–2507. https://doi.org/10.1099/0022-1317-73-10-2503.

B. Prospecting for viroid

Hill, M.F., Giles, R.J., Moran, J.R., Hepworth, G., 1996. The incidence of chrysanthemum stunt viroid, chrysanthemum B carlavirus, tomato aspermy cucumovirus and tomato spotted wilt tospovirus in Australian chrysanthemum crops. Australas. Plant Pathol. 25 (1), 174–178. https://link.springer.com/article/10.1071/AP96030.

Iftikhar, Y., Khoo, Y.W., Murugan, T., Roslin, N.A., Adawiyah, R., Kong, L.L., Vadamalai, G., 2019. Charecterization and sap transmission of citrus bent leaf viroid in Malaysia. Ann. Phytopathol. Soc. Jpn. https://doi.org/10.1101/751560.

Iftikhar, Y, Khoo, Y.W., Murugan, T., Roslin, N.A., Adawiyah, R., Kong, L.L., Vadamalai, G., 2022. Molecular and biological characterization of citrus bent leaf viroid from Malaysia. Mol. Biol. Rep. 49 (2), 1581–1586.

Imperial, J.S., Rodriguez, M.J.B., Randles, J.W., 1981. Variation in the viroid-like RNA associated with cadang-cadang disease: evidence for an increase in molecular weight with disease progress. J. Gen. Virol. 56, 77–85.

Ishiguro, A., Sano, T., Harada, Y., 1996. Nucleotide sequence and host range of coleus viroid isolated from coleus (*Coleus blumei* Benth.) in Japan. Ann. Phytopathol. Soc. Jpn. 62 (1), 84–86. https://doi.org/10.3186/jjphytopath.62.84.

Ismaeil, F., Myrta, A., Ghanem-Sabanadzovic, N.A., Al Chaabi, S., Savino, V., 2002. Viruses and viroids of stone fruits in Syria. EPPO Bull. 32 (3), 485–488. https://doi.org/10.1046/j.1365-2338.2002.00594.x.

Ito, T., Ohta, S., 2010. First report of citrus viroid V in Japan. J. Gen. Plant Pathol. 76 (5), 348–350. https://doi.org/10.1007/s10327-010-0254-1.

Ito, T., Ieki, H., Ozaki, K., Ito, T., 2001. Characterization of a new citrus viroid species tentatively termed citrus viroid OS. Arch. Virol. 146 (1), 975–982. http://springer.com/article/10.1007/s007050170129.

Ito, T., Ieki, H., Ozaki, K., Iwanami, T., Nakahara, K., Hataya, T., Ito, T., Isaka, M., Kano, T., 2002. Virology multiple citrus viroids in citrus from Japan and their ability to produce exocortis-like symptoms in citron. Virology 92 (5), 542–547.

Ito, T., Suzaki, K., Nakano, M., Sato, A., 2013. Characterization of a new apscaviroid from American persimmon. Arch. Virol. 158, 2629–2631.

Jiang, D.M., Peng, S., Wu, Z.J., Cheng, Z.M., Li, S.F., 2009. Genetic diversity and phylogenetic analysis of Australian grapevine viroid (AGVd) isolated from different grapevines in China. Virus Genes 38 (1), 178–183. https://doi.org/10.1007/s11262-008-0306-1.

Jiang, D.M., Li, S.F., Fu, F.H., Wu, Z.J., Xie, L.H., 2013. First report of *Coleus blumei* viroid 5 from *Coleus blumei* in India and Indonesia. Plant Dis. 97 (4), 561. https://doi.org/10.1094/PDIS-09-12-0815-PDN.

Jiang, J.H., Zhang, Z.X., Hu, B., Hu, G.B., Wang, H.Q., Faure, C., Marais, A., Candresse, T., Li, S.F., 2017. Identification of a viroid-like RNA in a lychee transcriptome shotgun assembly. Virus Res. 240 (1), 1–7. https://doi.org/10.1016/j.virusres.2017.07.012.

Jo, Y., Yoo, S.H., Chu, H., Cho, J.K., Choi, H., Yoon, J.Y., Choi, S.K., Cho, W.K., 2015. Complete genome sequences of peach latent mosaic viroid from a single peach cultivar. Genome Announc. 3 (5), 1–2. https://doi.org/10.1128/genomeA.01098-15.

Kasai, H., Ito, T., Sano, T., 2017. Symptoms and molecular characterization of apple dimple fruit viroid isolates from apples in Japan. J. Gen. Plant Pathol. 83 (4), 268–272. https://doi.org/10.1007/s10327-017-0718-7.

Khoo, Y.W., Iftikhar, Y., Murugan, T., Roslin, N.A., Adawiyah, R., Kong, L.L., Vadamalai, G., 2017. First report of Citrus bent leaf viroid in Malaysia. J. Plant Pathol. 99 (1), 293.

Koganezawa, H., Yanase, H., Sakuma, T., 1982. Viroid-like RNA associated with apple scar skin (or dapple apple) disease. Acta Hortic. 130, 193–197.

Koganezawa, H., Ohnuma, Y., Sakuma, H., Yanse, H., 1989. 'Apple fruit crinkle', a new graft-transmissible fruit disorder of apple. Bull. Fruit Tree Res. Stn. MAFF Ser. C. (Morioka) 16, 57–62.

Lakshman, D.K., Tavantzis, S.M., 1993. Primary and secondary structure of a 360-nucleotide isolate of potato spindle tuber viroid. Arch. Virol. 128 (1), 128 (319–331).

Lee, J.Y., Puchta, H., Sänger, H.L., 1988. Nucleotide sequence of Korean isolate of hop stunt viroid (HSVd). Nucl. Acids Res. 16, 8708.

Lee, J.H., Park, J.K., Lee, D.H., Uhm, J.Y., Ghim, S.Y., Lee, J.Y., 2001. Occurrence of apple scar skin viroid-Korean strain (ASSVd-K) in apples cultivated in Korea. Plant Pathol. J. 17, 300–304.

Li, S.F., Su, Q.F., Guo, R., Tsuji, M., Sano, T., 2006. First report of *Coleus blumei* viroid from coleus in China. Plant Pathol. 55 (4), 565. https://doi.org/10.1111/j.1365-3059.2006.01382.x.

Lin, C.Y., Wu, M.L., Shen, T.L., Yeh, H.H., Hung, T.H., 2015. Multiplex detection, distribution, and genetic diversity of hop stunt viroid and citrus exocortis viroid infecting citrus in Taiwan. Virol. J. 12 (1), 1–11. https://doi.org/10.1186/s12985-015-0247-y.

B. Prospecting for viroid

Liu, S.X., Li, S.F., Zhu, J., Xiang, B.C., Cao, L., 2008. First report of hop latent viroid (HLVd) in China. Plant Pathol. 57 (2), 400. https://doi.org/10.1111/j.1365-3059.2007.01794.x.

Maddahian, M., Massumi, H., Heydarnejad, J., Hosseinipour, A., Khezri, A., Sano, T., 2019. Biological and molecular characterization of hop stunt viroid variants from pistachio trees in Iran. J. Phytopathol. 167 (3), 163–173. https://doi.org/10.1111/jph.12783.

Matoušek, J., Piernikarczyk, R.J.J., Dĕdič, P., Mertelík, J., Uhlířová, K., Duraisamy, G.S., Orctová, L., Kloudová, K., Ptáček, J., Steger, G., 2014. Characterization of potato spindle tuber viroid (PSTVd) incidence and new variants from ornamentals. Eur. J. Plant Pathol. 138 (1), 93–101. https://doi.org/10.1007/s10658-013-0304-6.

Matsushita, Y., 2013. Chrysanthemum stunt viroid. Jpn. Agric. Res. Q 47 (3), 237–242. http://www.jircas.affrc.go.jp.

Matsushita, Y., Tsukiboshi, T., Ito, Y., Chikuo, Y., 2007. Nucleotide sequences and distribution of chrysanthemum stunt viroid in Japan. Jpn. Soc. Hortic. Sci. 76 (4), 333–337.

Matsushita, Y., Usugi, T., Tsuda, S., 2010. Development of a multiplex RT-PCR detection and identification system for potato spindle tuber viroid and tomato chlorotic dwarf viroid. Eur. J. Plant Pathol. 128 (2), 165–170. https://doi.org/10.1007/s10658-010-9672-3.

Matsushita, Y., Yanagisawa, H., Khiutti, A., Mironenko, N., Ohto, Y., Afanasenko, O., 2021. Genetic diversity and pathogenicity of potato spindle tuber viroid and chrysanthemum stunt viroid isolates in Russia. Eur. J. Plant Pathol. 161 (3), 529–542. https://doi.org/10.1007/s10658-021-02339-z.

Matsuura, S., Matsushita, Y., Kozuka, R., Shimizu, S., Tsuda, S., 2009. Transmission of Tomato chlorotic dwarf viroid by bumblebees (Bombus ignitus) in tomato plants. Eur. J. Plant Pathol. 126, 111–115.

Mohamed, N.A., Haseloff, J., Imperial, J.S., Symons, R.H., 1982. Characterisation of the different electrophoretic forms of the cadang-cadang viroid. J. Gen. Virol. 63, 181–188.

Nabi, S.U., Baranwal, V.K., 2020. First report of apple hammerhead viroid infecting apple cultivars in India. Plant Dis. 104 (11), 3086. https://doi.org/10.1094/PDIS-12-19-2731-PDN.

Nakahara, K., Hataya, T., Uyeda, I., Ieki, H., 1998. An improved procedure for extracting nucleic acids from citrus tissues for diagnosis of citrus viroids. Ann. Phytopathol. Soc. Jpn. 64, 532–538.

Nakashima, A., Hosokawa, M., Maeda, S., Yazawa, S., 2007. Natural infection of Chrysanthemum stunt viroid in dahlia plants. J. Gen. Plant Pathol. 73, 225–227.

Nakaune, R., Nakano, M., 2008. Identification of a new Apscaviroid from Japanese persimmon. Arch. Virol. 153, 969–972.

Ohtsuka, Y., 1935. A new disease of apple, on the abnormality of fruit. J. Japan Soc. Hort. Sci. 6, 44–53 (in Japanese).

Önelge, N., 2010. Citrus viroids in Turkey. Proceedings, 17th Conference IOCV, 2010 – Short Communications.

Önelge, N., Yurtmen, M., 2012. First report of citrus viroid V in Turkey. J. Plant Pathol. 94 (4, Suppl), S4.85–S4.105.

Osaki, H., Kudo, A., Ohtsu, Y., 1996. Japanese pear fruit dimple disease caused by apple scar skin viroid (ASSVd). Ann. Phytopathol. Soc. Jpn. 62, 379–385.

Osaki, H., Yamaguchi, M., Sato, Y., Tomita, Y., Kawai, Y., Miyamoto, Y., Ohtsu, Y., 1999. Peach latent mosaic viroid isolated from stone fruits in Japan. Ann. Phytopathol. Soc. Jpn. 65 (1), 3–8.

Owens, R.A., Girsova, N.V., Kromina, K.A., Lee, I.M., Mozhaeva, K.A., Kastalyeva, T., 2009. Russian isolates of potato spindle tuber viroid exhibit low sequence diversity. Plant Dis. 93 (7), 752–759. https://doi.org/10.1094/PDIS-93-7-0752.

Puchta, H., Ramm, K., Hadast, R., Bar-Joseph, M., Luckinger, R., Freimiiller, K., Sanger, H.L., 1989. Nucleotide sequence of a hop stunt viroid (HSVd) isolate from grapefruit in Israel. Nucleic Acids Res. 17 (3), 1247. https://doi.org/10.1093/nar/17.3.1247.

Puchta, H., Ramm, K., Luckinger, R., Hadas, R., Bar-Joseph, M., Sanger, H.L., 1991. Primary and secondary structure of citrus viroid IV (CVd IV), a new chimeric viroid present in dwarfed grapefruit in Israel. Nucleic Acids Res. 19 (23), 6640.

Qiu, C.L., Zhang, Z.X., Li, S.F., Bai, Y.J., Liu, S.W., Fan, G.Q., Gao, Y.L., Zhang, W., Zhang, S., Lü, W.H., Lü, D.Q., 2016. Occurrence and molecular characterization of potato spindle tuber viroid (PSTVd) isolates from potato plants in North China. J. Integr. Agric. 15 (2), 349–363. https://doi.org/10.1016/S2095-3119(15)61175-3.

Ramachandran, P., Agarwal, J., Roy, A., Ghosh, D.K., Das, D.R., Ahlawat, Y.S., 2005. First record of a hop stunt viroid variant on Nagpur mandarin and Mosambi sweet orange trees on rough lemon and Rangpur lime rootstocks. Plant Pathol. 54 (4), 571. https://doi.org/10.1111/j.1365-3059.2005.01194.x.

Randles, J.W., Rodriguez, M.J.B., 2003. In: Hadidi, A., Flores, R., Randles, J.W., Semancik, J.S. (Eds.), Coconut Cadang-Cadang Viroid. CSIRO Publishing, Collingwood, Australia, pp. 233–241.

Reanwarakorn, K., Klinkong, S., Porsoongnurn, J., 2011. First report of natural infection of pepper chat fruit viroid in tomato plants in Thailand. N. Dis. Rep. 24 (1), 6. https://doi.org/10.5197/j.2044-0588.2011.024.006.

Rodriguez, M.J.B., Vadamalai, G., Randles, J.W., 2017. Economic significance of palm tree viroids. In: Viroids and Satellites. Elsevier Inc, pp. 39–49, https://doi.org/10.1016/B978-0-12-801498-1.00004-8.

Roslan, N., Thanarajoo, S.S., Vadamalai, G., 2017. First report of coleus blumei viroid in Malaysia. J. Plant. Pathol. 99, 800.

Roumi, V., Gazel, M., Caglayan, K., 2017. First report of apple dimple fruit viroid in apple trees in Iran. N. Dis. Rep. 35 (1), 3. https://doi.org/10.5197/j.2044-0588.2017.035.003.

Roy, A., Ramachandran, P., 2003. Occurrence of a hop stunt viroid (HSVd) variant in yellow corky vein disease of citrus in India. Curr. Sci. 85 (11), 1608–1612. https://www.jstor.org/stable/24110026.

Roy, A., Ramachandran, P., 2006. Characterization of a citrus exocortis viroid variant in yellow corky vein disease of citrus in India. Curr. Sci. 91 (6), 798–803. https://www.jstor.org/stable/24093911.

Sano, T., Hataya, T., Sasaki, A., Shikata, E., 1986. Etrog ctron is latently infected with hop stunt viroid-like RNA. Proc. Jpn. Acad. Ser. B 62, 325–328.

Sano, T., Hataya, T., Terai, Y., Shikata, E., 1989. Hop stunt viroid strains from dapple fruit disease of plum and peach in Japan. J. Gen. Virol. 70 (6), 1311. https://doi.org/10.1099/0022-1317-70-6-1311.

Sano, T., Ogata, T., Ochiai, M., Suzuki, C., Ohnuma, S., Shikata, E., 1997. Pear blister canker viroid isolated from European pear in Japan. Ann. Phytopathol. Soc. Jpn. 63 (1), 89–94.

Sano, T., Yoshida, H., Goshono, M., Monma, T., Kawasaki, H., Ishizaki, K., 2004. Characterization of a new viroid strain from hops: evidence for viroid speciation by isolation in different host species. J. Gen. Plant Pathol. 70, 181–187.

Sasaki, M., Shikata, E., 1977. On some properties oh hop stunt disease agent, a viroid. Proc. Jpn. Acad. 53 (Ser. B), 109–112.

Serra, P., Eiras, M., Bani-Hashemian, S.M., Murcia, N., Kitajima, E.W., Daròs, J.A., Flores, R., Duran-Vila, N., 2008. Citrus viroid V: Occurrence, host range, diagnosis, and identification of new variants. Phytopathology 98, 1199–1204.

Shamloul, A.M., Han, L., Hadidi, A., 2004. Characterization of a new variant of Apple scar skin viroid associated with pear fruit crinkle disease [Pyrus communis L.; China]. J. Plant Pathol. 86, 249–256.

Shilpa, N., Dhir, S., Janardhana, G.R., 2022. Molecular detection and characterization of potato spindle tuber viroid (PSTVd) infecting tomato (*Solanum lycopersicum* L.) in Karnataka state of India. Virus Dis. 33 (3), 261–269. https://doi.org/10.1007/s13337-022-00782-y.

Sial, Z.K., Khan, F., 2018. First report on potato spindle tuber viroid (PSTVd) from field grown infected potato plants (*Solanum tuberosum*) in Pakistan. Biologia (Pakistan) 64 (2), 241–246. https://biolspk.com/download/12-2-8/.

Sial, Z.K., Khan, F., 2019. Identification of potato spindle tuber viroid (PSTVd) from diseased plants of tomato (*Lycopersicon esculentum* MILL.) in Pakistan. Int. J. Biol. Biotechnol. 16 (3), 697–702.

Singh, R.P., Dilworth, A.D., 2009. Tomato chlorotic dwarf viroid in the ornamental plant Vinca minor and its transmission through tomato seed. Eur. J. Plant Pathol. 123, 111–116.

Singh, R.P., Dilworth, A.D., Baranwal, V.K., Gupta, K.N., 2006. Detection of citrus exocortis viroid, Iresine viroid, and tomato chlorotic dwarf viroid in new ornamental host plants in India. Plant Dis. 90, 1457.

Sipahioglu, H.M., Usta, M., Ocak, M., 2009. Development of a rapid enzymatic cDNA amplification test for the detection of apple scar skin viroid (ASSVd) in apple trees from Eastern Anatolia, Turkey. Arch. Phytopathol. Plant Protect. 42 (4), 352–360. https://doi.org/10.1080/03235400601070496.

Sombat, S., Reanwarakorn, K., Ling, K.S., 2018. Developing a multiplex real-time RT-PCR for simultaneous detection of pepper chat fruit viroid and Columnea latent viroid. Australas. Plant Pathol. 47 (6), 615–621. https://doi.org/10.1007/s13313-018-0597-1.

Spiegel, S., Alper, M., Allen, R.N., 1984. Evaluation of biochemical methods for the diagnosis of the avocado sunblotch viroid in Israel. Phytoparasitica 12 (1), 37–43. https://link.springer.com/article/10.1007/BF02980796.

St-Pierre, P., Hassen, I.F., Thompson, D., Perreault, J.P., 2009. Characterization of the siRNAs associated with peach latent mosaic viroid infection. Virology 383 (2), 178–182. https://doi.org/10.1016/j.virol.2008.11.008.

Supakitthanakorn, S., Vichittragoontavorn, K., Kunasakdakul, K., Ruangwong, O.U., 2022. Phylogenetic analysis and molecular characterization of chrysanthemum chlorotic mottle viroid and chrysanthemum stunt viroid from chrysanthemum in Thailand. J. Phytopathol. 170 (10), 700–710. https://doi.org/10.1111/jph.13134.

Szostek, S.A., Wright, A.A., Harper, S.J., 2018. First report of apple hammerhead viroid in the United States, Japan, Italy, Spain, and New Zealand. Plant Dis. 102 (12), 2670. https://doi.org/10.1094/PDIS-04-18-0557-PDN.

Tanaka, S., 1963. Citrus exocortis disease. Ann. Phytopathol. Soc. Jpn. 28, 88 (in Japanese).

Tsushima, T., Murakami, S., Ito, H., He, Y.-H., Adkar-Purushothama, C.R., Sano, T., 2011. Molecular characterization of Potato spindle tuber viroid in dahlia. J. Gen. Plant Pathol. 77, 253–256.

B. Prospecting for viroid

Tsushima, T., Matsushita, Y., Fuji, S., Sano, T., 2015. First report of dahlia latent viroid and potato spindle tuber viroid mixed-infection in commercial ornamental dahlia in Japan. N. Dis. Rep. 31 (1), 11. https://doi.org/10.5197/j.2044-0588.2015.031.011.

Turturo, C., Minafra, A., Ni, H., Wang, G., di Terlizzi, B., Savino, V., 1998. Occurrence of peach latent mosaic viroid in China and development of an improved detection method. J. Plant Pathol. 80 (2), 165–169. https://about.jstor.org/terms.

Vadamalai, G., Hanold, D., Rezaian, M.A., Randles, J.W., 2006. Variants of coconut cadang-cadang viroid isolated from an African oil palm (Elaies guineensis Jacq.) in Malaysia. Arch. Virol. 151 (7), 1447–1456. https://doi.org/10.1007/s00705-005-0710-y.

Vadamalai, G., Thanarajoo, S.S., Joseph, H., Kong, L.L., Randles, J.W., 2017. Coconut Cadang-Cadang Viroid and Coconut Tinangaja Viroid. In: Viroids and Satellites. Elsevier Inc, pp. 263–273, https://doi.org/10.1016/B978-0-12-801498-1.00025-5.

Vallejo-Pérez, M.R., Téliz-Ortiz, D., Colinas-León, M.T., de La Torre-Almaraz, R., Valdovinos-Ponce, G., Nieto-Ángel, D., Ochoa-Martínez, D.L., 2015. Alterations induced by avocado sunblotch viroid in the postharvest physiology and quality of avocado 'Hass' fruit. Phytoparasitica 43 (3), 355–364. https://doi.org/10.1007/s12600-015-0469-y.

Walia, Y., Kumar, Y., Rana, T., Bhardwaj, P., Ram, R., Thakur, P.D., Sharma, U., Hallan, V., Zaidi, A.A., 2009. Molecular characterization and variability analysis of apple scar skin viroid in India. J. Gen. Plant Pathol. 75 (4), 307–311. https://doi.org/10.1007/s10327-009-0168-y.

Wang, X.F., Zhou, C.Y., Tang, K.Z., Li, Z.A., 2008. Occurrence of four citrus viroids in Chongqing, China. Plant Dis. 92 (6), 978. https://doi.org/10.1094/PDIS-92-6-0978B.

Wang, X.F., Zhou, C.Y., Tang, K.Z., Zhou, Y., Li, Z.A., 2009. A rapid one-step multiplex RT-PCR assay for the simultaneous detection of five citrus viroids in China. Eur. J. Plant Pathol. 124 (1), 175–180. https://doi.org/10.1007/s10658-008-9386-y.

Wang, Y., Zhao, Y., Niu, J., 2012. Molecular identification and sequence analysis of the apple scar skin viroid (ASSVd) isolated from four kinds of fruit trees in Xinjiang province of China. Mol. Pathogens 3, 12–18.

Wu, Y.H., Cheong, L.C., Meon, S., Lau, W.H., Kong, L.L, Joseph, H., Vadamalai, G., 2013. Characterization of Coconut cadang-cadang viroid variants from oil palm affected by orange spotting disease in Malaysia. Arch. Virol. 158, 1407–1410.

Xu, W., Hong, N., Wang, G., Fan, X., 2008. Population structure and genetic diversity within peach latent mosaic viroid field Isolates from peach showing three symptoms. J. Phytopathol. 156, 565–572.

Xu, L., Wang, J.W., Chen, X., Zhu, D.Z., Wei, H.R., Hammond, R.W., Liu, Q.Z., 2019. Molecular characterization and phylogenetic analysis of hop stunt viroid isolates from sweet cherry in China. Eur. J. Plant Pathol. 154 (3), 705–713. https://doi.org/10.1007/s10658-019-01693-3.

Yamamoto, H., Sano, T., 2005. Occurrence of chrysanthemum chlorotic mottle viroid in Japan. J. Gen. Plant Pathol. 71, 156–157.

Yamamoto, H., Sano, T., 2006. An epidemiological survey of Chrysanthemum chlorotic mottle viroid in Akita Prefecture as a model region in Japan. J. Gen. Plant Pathol. 72, 387–390.

Yamamoto, H., Kagami, Y., Kurokawa, M., Nishimura, S., Kubo, S., Inoue, M., Murayama, D., 1970. Studies on hop stunt disease I. Memoirs Fac. Agric. Hokkaido Univ. 7, 491–512. (in Japanese).

Yan, C., Yin, H., Xia, F., Deng, C., Li, Y., Zhang, Y., 2019. First report of dahlia latent viroid infecting dahlia in China. Plant Dis. APS Publisher https://doi.org/10.1094/PDIS-08-19-1774-PDN.

Yazarlou, A., Jafarpour, B., Habili, N., Randles, J., 2012. First detection and molecular characterization of new Apple scar skin viroid variants from apple and pear in Iran. Australas. Plant Dis. Notes 7, 99–102.

Yazarlou, A., Jafarpour, B., Tarighi, S., Habili, N., Randles, J.W., 2012. New Iranian and Australian peach latent mosaic viroid variants and evidence for rapid sequence evolution. Arch. Virol. 157 (2), 343–347. https://doi.org/10.1007/s00705-011-1156-z.

Yazarlou, A., Jafarpour, B., Koohi Habibi, M., Tarighi, S., Randles, J.W., Sohrabi, S., Rashed Mohassel, M.H., Nassiri Mahalati, M., 2014. First characterization of apple scar skin viroid from apple and pear cultivars in Iran and determination of genetic diversity study the phenology of lesser Celandine (*Ranunculus ficaria*) and effect of planting depth on sprouting of its tuberous-roots. J. Plant Prot. 28 (3), 33.

Ye, T., Chen, S.Y., Wang, R., Hao, L., Chen, H., Wang, N., Guo, L.Y., Fan, Z.F., Li, S.F., Zhou, T., 2013. Identification and molecular characterization of apple dimple fruit viroid in China. J. Plant Pathol. 95 (3), 637–641. https://www.jstor.org/stable/23721589.

B. Prospecting for viroid

Yesilcollou, S., Minoia, S., Torchetti, E.M., Kaymak, S., Gumus, M., Myrta, A., Navarro, B., Di Serio, F., 2010. Molecular characterization of Turkish isolates of pear blister canker viroid and assessment of the sequence variability of this viroid. J. Plant Pathol. 92 (3), 813–819. https://www.jstor.org/stable/41998877.

Zaki-Aghl, M., Izadpanah, K., Niazi, A., Behjatnia, S.A.A., Afsharifar, A.R., 2013. Molecular and biological characterization of the Iranian isolate of the Australian grapevine viroid. J. Agric. Sci. Technol. 15 (1), 855–865.

Zhang, Z., Ge, B., Pan, S., Zhao, Z., Wang, H., Li, S., 2011. Molecular detection and sequences analysis of Chrysanthemum stunt viroid. Acta Hortic. Sin. 38, 2349–2356.

Zhang, Z.X., Qi, S.S., Tang, N., Zhang, X.X., Chen, S.S., Zhu, P.F., Ma, L., Cheng, J.P., Xu, Y., Lu, M.G., Wang, H.Q., Ding, S.W., Li, S.F., Wu, Q.F., 2014. Discovery of replicating circular RNAs by RNA-Seq and computational algorithms. PLoS Pathog. 10 (12), e1004553. https://doi.org/10.1371/journal.ppat.1004553.

Zhang, Y.H., Li, Z.X., Du, Y.J., Li, S.F., Zhang, Z.X., 2023. A universal probe for simultaneous detection of six pospiviroids and natural infection of potato spindle tuber viroid (PSTVd) in tomato in China. J. Integr. Agric. 22 (3), 2–10. https://doi.org/10.1016/j.jia.2022.08.119.

Zhao, Y., Niu, J.X., 2008. Apricot is a new host of apple scar skin viroid. Aust Plant Dis Notes 3 (1), 98. https://doi.org/10.1071/dn08039.

Zhou, Y., Guo, R., Cheng, Z., Sano, T., Li, S.F., 2006. First report of hop stunt viroid from peach (Prunus persica) with dapple fruit symptoms in China. New Dis. Rep. 12, 43.

Zhou, J., Hou, W., Zhang, Z., Wang, C., Jie, L., Wang, H., Li, S., 2019. First report of peach calico isolate of peach latent mosaic viroid from peach trees in China. Plant Dis. APS Publisher https://doi.org/10.1094/PDIS-04-19-0763-PDN.

Viroid-associated plant diseases in Europe

Dijana Škorić

University of Zagreb, Faculty of Science, Department of Biology, Zagreb, Croatia

Graphical representation

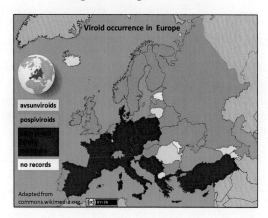

Definitions

Avsunviroids: Members of the viroid family *Avsunviroidae* are few and using ribozyme activity for replication in the chloroplast.

Biological indicator: A plant experimentally infected by a viroid (or other pathogen) as a secondary host displaying characteristic symptoms that can be specifically linked to that pathogenic agent and sometimes even to the variants of that agent.

Epinasty: Downward bending and curling of the leaf.

Plant germplasm: Living material (seeds, tissue, plants) maintained as genetic resources for plant breeding, research, conservation or other uses.

Pome fruits: Apple-like fruits or, botanically, the type of fruits produced by the flowering plants of the family *Rosaceae*, subtribe *Malinae*.

Pospiviroids: Members of the family *Pospiviroidae* replicate in the plant nucleus without ribozyme activity and include most of the known viroids.

Stone fruits: Fruits with a large "stone" inside, also called a pit, like in peach or plum. Botanical term for this type of fruit is a drupe and may be produced by flowering plants of different families.

sp.: Abbreviation used when a species in a genus cannot be specified.

spp.: Abbreviation used for indicating several species of the same genus.

Chapter outline

The chapter provides data on the occurrence of different types of viroids in European countries, their natural plant hosts in the field or greenhouse conditions and importance for plant production, biotechnology and biology. It also specifies reliable sources for obtaining information on the viroids and their occurrence in Europe and wider.

Learning objectives

On completion of this chapter, you should be able to:

- understand the distribution of different viroids in European countries
- recognize the influence of viroid diseases in various crops
- discuss the possible outcomes of uncontrolled spread of viroids on quarantine lists
- identify the sources of information for viroid occurrence in Europe and other countries
- describe some symptoms of viroid diseases.

Fundamental introduction

Viroids have been researched extensively in Europe since the beginning of the viroid discovery mainly due to the plant diseases they may cause, or contribute to, in different crops (Pallás et al., 2003). Some viroids are able to impact plant production so severely that they are classified as quarantine plant pests by the European Union (EU). This means they are surveyed on the country borders to stop their entering into the EU or, if they are already present, they must be eliminated from the plant propagation material, germplasm and their spread stopped. These measures are proposed on the basis of scientific data provided to the EU agency called the European Food Safety Authority (EFSA) by a group of scientists and experts included in the Panel for Plant Health. The European and Mediterranean Plant Protection Organization (EPPO) thus currently lists four viroids either as very harmful pests absent from Europe within its A1 list (coconut cadang cadang viroid, CCCVd) or as locally present but mandatory controlled pests within A2 list (chrysanthemum stunt viroid, CSVd; citrus bark cracking viroid, CBCVd; and potato spindle tuber viroid, PSTVd). EPPO also proposes the Alert list with emerging pests and diseases that may become a threat, and some viroids were on it in the past. Essentially, viroids are considered here as invasive alien species. The lists are updated yearly and can be found at the EPPO website along with the documents regarding pest risk analyses that served for their composition (https://www.eppo.int/ACTIVITIES/quarantine_activities).

National legislation in most European countries usually reflects EPPO regulations along with the procedures for the production of viroid-free planting material and viroid spread control in the fields and greenhouses. Enforcing these regulatory measures is ensured by each of the countries and take up considerable financial and scientific resources. Some countries have even more restrictive measures in their national phytosanitary regulations for some of the viroids than EPPO, depending on the local situation. An example is tomato planta macho viroid, TPMVd, added as a quarantine pest in the UK even though it is not listed as such for the entire EPPO space. That information is also available in the EPPO Global Database (https://gd.eppo.int) categorization section for this case (https://gd.eppo.int/taxon/TPMVD0/categorization) and similar ones in other European countries.

Despite the rules and regulations governing plant health, diverse viroids (Tables 1 and 2) have entered Europe with planting material (plants, tubers, seeds) from other parts of the world in the past (Hadidi et al., 2017). This will likely happen on occasion in future. The application of high-throughput sequencing (HTS), metagenomic or transcriptomic studies, faster publication rate together with the intense international trade of the plant propagative material in the 21st century are conducive to quick discovery of a viroid introduction into a new country. In addition, viroids previously unknown to science are continuously discovered locally, and their pathogenic potential in different plants needs yet to be assessed. Apple chlorotic fruit spot viroid (ACFSVd) is probably the most recent example of a new viroid discovered in Europe (Leichtfried et al., 2019). However, for ACFSVd all scientific criteria still need to be fulfilled to be officially recognized as a viroid species by the International Committee on Taxonomy of Viruses (ICTV). This organization encompasses experts for viruses and viroids who are in charge of their classification and naming (Di Serio et al., 2018a; Chiumenti et al., 2021; Table 3).

With a growing number of viroid first reports in new countries, it is quite possible that new records will be available between the finishing of this textbook and its publication. The reader is therefore advised to take the information given here with this limitation in mind. Also, it is worth noting that the viroid presence in a certain plant host and a country does not always entail the development of a plant disease. The discovery of eggplant latent viroid (ELVd) happened accidentally almost three decades ago (Fagoaga et al., 1994) while screening symptomless eggplants and other vegetables for the presence of PSTVd. Ever since, neither a host of ELVd other than eggplant has been found nor the evidence of symptoms. Some of the latest possible additions to the two viroid families (Table 3) that were either found in asymptomatic plants (e.g., portulaca latent viroid, PLVd) or could not be linked to any specific cluster of symptoms because of the mixed infections with viruses and/or other viroids (e.g., grapevine latent viroid, GLVd) may prove to be as harmless as ELVd. Nevertheless, they should be investigated further for biological reasons or to prevent the risk to the plant production or biodiversity.

TABLE 1   Viroid occurrence in Europe—members of the family *Avsunviroidae*.

Genus	Species	Acronym	Country (present)	Natural host plant	Country (intercepted or eradicated)
Avsunviroid	*Avocado sunblotch viroid*	ASBVd	Greece (Crete)	Avocado	Spain
Pelamoviroid	*Peach latent mosaic viroid*	PLMVd	Albania Austria Croatia Cyprus Czechia Republic Bosnia and Herzegovina France Greece Italy Montenegro Poland Romania Serbia Spain Turkey	Peach & nectarine Apricot European plum Japanese plum Mume Pear Sweet cherry Quince Wild pear	
	Chrysanthemum chlorotic mottle viroid	CChMVd	Denmark France	Chrysanthemum	Netherlands
	Apple hammerhead viroid	AHVd	Belgium Germany Italy Spain	Apple L oquat	
Elaviroid	*Eggplant latent viroid*	ELVd	Spain	Eggplant	

Blank table fields indicate there are no records in Europe.

Viroids of herbaceous plants

Vegetables

PSTVd is the longest known viroid and the most studied pospiviroid. Judging by its continuous or occasional presence in the vast majority of European countries (Table 2), it is a continuing risk to the production of herbaceous crops. In potato (*Solanum tuberosum*), its original host, PSTVd is present in all parts of the plant, including tubers and true seeds, and may be transmitted by their propagation. It is highly stable infectious RNA molecule reaching high concentrations in the plant. It is transmissible by contact between the plants or with sap on contaminated hands, surfaces, tools and clothes, maybe even via irrigation water (Hadidi et al., 2017). The PSTVd is on the EPPO A2 list, and control measures have been strong enough to minimize losses to the potato tuber production in Europe. Nevertheless, recent detection of quarantine pathogens PSTVd and CSVd in Russian seed potatoes, that should be viroid-free, and in *Solanum nigrum* (blackberry nightshade) is a cause for concern (Matsushita et al., 2021).

TABLE 2 Viroid occurrence in Europe—members of the family *Pospiviroidae*.

Genus	Species	Acronym	Country (present)	Natural host plant	Country (intercepted or eradicated
Pospiviroid	*Potato spindle tuber viroid*	PSTVd	Austria Azerbaijan Belarus Belgium Croatia Czechia Georgia Germany Greece Hungary Italy Malta Montenegro Netherlands Russia Slovenia Spain Switzerland Turkey UK Ukraine	Pepper Pepino Potato *Physalis peruviana* Tomato Solanaceous ornamentals (*Brugmansia* spp., *Datura* spp., *Cestrum* sp., *Lycianthes rantonetii*, *Petunia* sp. *Solanum* *jasminoides*), Chrysanthemum *Weed* *Solanum nigrum*	Bulgaria Denmark Estonia Finland France Ireland Portugal Slovakia
	Citrus exocortis viroid	CEVd	Austria Azerbaijan Belgium Bosnia and Herzegovina Croatia Cyprus Czechia France Germany Greece Italy Montenegro Netherlands Portugal Russia Slovenia Spain Turkey Ukraine	*Citrus* spp. Broad bean Carrot Eggplant Grapevine Tomato Turnip Ornamentals (*Cestrum* sp., *S. jasminoides*, *Verbena* sp.)	
	Chrysanthemum stunt viroid	CSVd	Belgium Czechia Germany Italy Netherlands	Chrysanthemum *S. jasminoides, Petunia* sp. Potato Invasive plants (*Cardamine bonariensis*,	Austria Denmark Finland France Hungary

Continued

B. Prospecting for viroid

TABLE 2 Viroid occurrence in Europe—members of the family *Pospiviroidae*—cont'd

Genus	Species	Acronym	Country (present)	Natural host plant	Country (intercepted or eradicated
			Norway Poland Russia Sweden Turkey UK	*Oxalis latifolia*) Weed *S. nigrum*	Latvia Slovenia Spain
	Columnea latent viroid	CLVd	France Netherlands Italy	Pepper Tomato Ornamentals (*Brunfelsia* sp., *Gloxinia* spp.)	Belgium Denmark Germany UK
	Iresine viroid 1	IrVd1	Italy Netherlands Slovenia	Ornamentals (*Celosia argentea* var. *cristata, C. plumosa, Iresine herbstii, Portulaca* sp.)	
	Pepper chat fruit viroid	PCFVd	Netherlands	Pepper *Solanum sisymbriifolium*	Netherlands
	Tomato apical stunt viroid	TASVd	Belgium Croatia Germany Italy Netherlands Poland Slovenia	Tomato Ornamentals (*Brugmansia* spp., *Cestrum* sp., *Lycianthes rantonetii, S. jasminoides, Solanum pseudocapsicum, Verbena* sp.)	Austria Estonia Finland France
	Tomato chlorotic dwarf viroid	TCDVd	Czechia France Slovenia	Eggplant Tomato Ornamentals (*Brugmansia* sp., *Calibrachoa* sp., *Petunia* sp., *Pitosporum tobira, Verbena* sp.)	Netherlands Belgium Finland Norway Spain UK
	Tomato planta macho viroid	TPMVd			
Hostuviroid	*Hop stunt viroid*	HSVd	Austria Azerbaijan Bosnia and Herzegovina Hungary Croatia Cyprus Czechia France Germany Greece Italy	Almond Apple Apricot *Citrus* spp. Cucumber Grapevine *Hibiscus rosa-sinensis* Hop Peach Pear Pistachio	Finland

B. Prospecting for viroid

TABLE 2 Viroid occurrence in Europe—members of the family *Pospiviroidae*—cont'd

Genus	Species	Acronym	Country (present)	Natural host plant	Country (intercepted or eradicated
			Portugal Serbia Slovenia Spain Turkey	Pomegranate Sweet cherry White mulberry Wild apple	
	Dahlia latent viroid	DLVd	Netherlands Turkey UK	*Dahlia* spp.	
Cocadviroid	*Coconut cadang cagang viroid*	CCCVd			
	Coconut tinangaja viroid	CTVd			
	Citrus bark cracking viroid (former CVd-IV)	CBCVd	Cyprus Greece Italy Slovenia Spain Turkey	*Citrus* spp. Hop	Germany
	Hop latent viroid	HLVd	Belgium Czechia France Germany Hungary Poland Portugal Russia Serbia Slovenia Spain UK	Hop Weed stinging nettle (*Urtica dioica*)	
Apscaviroid	*Apple scar skin viroid*	ASSVd	Greece France Germany Italy Spain UK	Apple Pear Sweet cherry Wild pear	
	Apple dimple fruit viroid	ADFVd	Italy	Apple Fig	
	Australian grapevine viroid	AGVd	Greece Italy Russia	Grapevine	
		CBLVd		*Citrus* spp.	

Continued

B. Prospecting for viroid

TABLE 2 Viroid occurrence in Europe—members of the family *Pospiviroidae*—cont'd

Genus	Species	Acronym	Country (present)	Natural host plant	Country (intercepted or eradicated
	Citrus bent leaf viroid (former CVd-Ib)		France Italy Spain probably widespread		
	Citrus dwarfing viroid (former CVd-III)	CDVd	Croatia Italy Spain Turkey probably widespread	*Citrus* spp.	
	Citrus viroid V	CVd-V	Spain Turkey	*Citrus* spp. and relatives *Atalantia citroides*	
	Citrus viroid VI (former CVd-OS)	CVd-VI			
	Grapevine yellow speckle viroid 1	GYSVd-1	Albania Austria Bosnia and Herzegovina Bulgaria Croatia Cyprus Czechia France Germany Greece Hungary Italy Russia Slovakia Spain Turkey	Grapevine	
	Grapevine yellow speckle viroid 2	GYSVd-2	Croatia Germany Greece Italy Spain Turkey	Grapevine	
	Pear blister canker viroid	PBCVd	Albania Bosnia and Herzegovina France Greece Italy	Apple Pear Quince Wild apple Wild pear	

B. Prospecting for viroid

TABLE 2 Viroid occurrence in Europe—members of the family *Pospiviroidae*—cont'd

Genus	Species	Acronym	Country (present)	Natural host plant	Country (intercepted or eradicated
			Malta Spain Turkey UK		
Coleviroid	*Coleus blumei viroid 1*	CbVd-1	Croatia Germany probably widespread	Coleus	
	C. blumei viroid 2	CbVd-2	Germany	Coleus	
	C. blumei viroid 3	CbVd-3	Croatia Germany	Coleus	

Blank table fields indicate there are no records in Europe.

TABLE 3 Viroid occurrence in Europe—officially unrecognized viroids of both families.

Family/ Genus	Proposes species	Acronym	European country (present)	Host in Europe	Country of origin/host
Avsunviroidae					
Pelamoviroid	*Grapevine latent viroid*	GLVd	France Greece (Crete) Italy	Grapevine	Grapevine
Pospiviroidae					
Apscaviroid	*Apple chlorotic fruit spot viroid*	ACFSVd	Austria	Apple *Viscum album* (mistletoe)	Austria/ Apple
	Apple fruit crinkle viroid	AFCVd			Japan/ Apple, Hop
	Citrus viroid VII	CVd-VII			Australia/ Citrus
	Dendrobium viroid	DVd			China/ Dendrobium orchid
	Grapevine latent viroid	GLVd	Italy	Grapevine	China/ Grapevine
	Lychee viroid	LVd			China/ lychee

Continued

B. Prospecting for viroid

TABLE 3   Viroid occurrence in Europe—officially unrecognized viroids of both families—cont'd

Family/ Genus	Proposes species	Acronym	European country (present)	Host in Europe	Country of origin/host
	Persimmon viroid	PVd			Japan/ Japanese persimmon (*Diospyros kaki*)
	Persimmon viroid 2	PVd-2			Japan/ American persimmon (*D. virginiana*)
	Plum viroid 1	PlVd-1			South Africa/ Japanese plum
Coleviroid	*Coleus blumei viroid 5*	CbVd-5			China/ Coleus
	C. blumei viroid 6	CbVd-6			China/ Coleus
	C. blumei viroid 7	CbVd-7			Canada/ Coleus
Pospiviroid	*Portulaca latent viroid*	PLVd	Netherlands	Portulaca	Netherlands/ *Portulaca* sp.

Blank table fields indicate there are no records in Europe.

S. nigrum is a common weed that may act as a reservoir for these two viroids, and if PSTVd and CSVd, seed transmissible in potato, can also be transmitted by seeds of this weed, it could enable viroids introduction to a new potato crop in the following season.

The production of other vegetable crops of the family *Solanaceae*, especially tomato (*Solanum lycopersicum*), may also be severely affected by PSTVd. After inoculation with severe PSTVd strains, tomato plants develop deformed bunchy tops with shortened internodes, purplish and yellowing twisted and curled leaves (Fig. 1). The whole plant is stunted, flowers may be degenerated and, consequently, the fruit production can be halved or reduced even more. Pepper (*Capsicum annuum*) and pepino (*S. muricatum*) may be afflicted too, but also botanically unrelated species like sweet potato (*Ipomea batatas*) from the bindweed or morning glory family (*Convolvulaceae*) which is increasingly cultivated in Europe. Besides plant cultivar and species, the severity of PSTVd disease symptoms depends on the viroid variant, but growing conditions with higher light intensity and temperatures exacerbate the symptoms. It is also the case in most other, viroid infections. Hence, the greenhouse conditions are particularly suitable for developing and disseminating viroid diseases.

It is no wander that greenhouse grown tomato and pepino plants were found PSTVd infected in many European countries with extensive greenhouse vegetable growing. It is not easy to trace the origin of these infections but, at least for some in the Netherlands and UK, the seeds imported from countries overseas (e.g., Oceania for tomato seeds) may have been what started them. Tomato seems to be particularly "hospitable" to viroids. Besides PSTVd, citrus exocortis viroid (CEVd), columnea latent viroid (CLVd), tomato apical stunt viroid (TASVd), tomato chlorotic dwarf viroid (TCDVd) and tomato planta macho viroid (TPMVd) have been isolated from it. Only TPMVd has not been reported from Europe (Table 2). All these

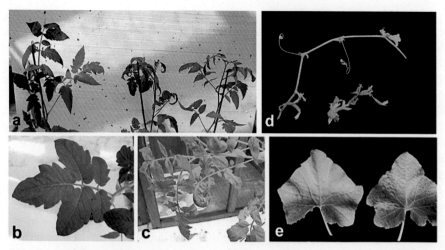

FIG. 1 Effects of potato spindle tuber viroid (PSTVd) on tomato (*Solanum lycopersicum*) "Rutgers"; (A) *left*—uninfected, *middle and right*—PSTVd-infected stunted tomato plants with downward leaf curling (epinasty) and shortened internodes, (B) close-up of uninfected tomato leaves and (C) PSTVd-infected chlorotic, mottled leaves with epinasty. Cucumber (*Cucumis sativus*) "Suyo" stems with flowers and fruits; (D) uninfected *(up)* and infected *(down)* with hop stunt viroid (HSVd) displaying shortened internodes and small fruits, (E) uninfected *(left)* and HSVd-infected *(right)* cucumber leaves.

viroids are harmful to tomato, even CLVd whose name can be misleading. It was isolated from healthy-looking ornamental plants of *Columnea erythrophae* (lipstick vine) but when inoculated to tomato, or potato, it may cause a similarly severe disease as PSTVd. Apart from the usual viroid transmission modes (mechanical inoculation, propagative material), TASVd and TCDVd can be transmitted by bumblebees often used in the greenhouses as pollinators.

Peppers host no such big variety of pospiviroids as tomatoes. PSTVd and CLVd can infect pepper, but pepper chat fruit viroid (PCFVd) is the only one substantially reducing the fruit production. The spread of this viroid is fortunately limited to a small number of plants in the greenhouses and only in the Netherlands where it is well controlled.

Other vegetables too have been found infected by viroids in Europe. CEVd host range expands in the field to carrot (symptomatic), broad bean and turnip (Fagoaga et al., 1996), while the very wide host range of hop stunt viroid (HSVd) includes cucumber as well (Fig. 1). Its production in the greenhouse may be threatened by HSVd-elicited cucumber pale fruit disease (Hadidi et al., 2017). Stunted cucumber plants with shortened internodes, curled leaves and deformed small flowers produce small, pale, unmarketable fruits as seen in the Netherlands and Finland.

Viroids of ornamentals

Over the decades of research, many ornamental plants have been found infected with viroids and, in turn, many viroids were named after their original ornamental hosts. Interestingly, all but one of these viroids are members of the family *Pospiviroidae*. PSTVd was detected in asymptomatic infections of many solanaceous ornamentals, and some of them are perennial woody plants (*Brugmansia* sp., *Cestrum* sp., *Lycianthes rantonettii*) that can serve as reservoir hosts for viroids because they are asymptomatically infected. If the vegetable species sensitive to PSTVd (e.g., tomato) are grown in the vicinity of these ornamentals, viroids may be transmitted to them if there is no spacing or strict sanitation rules are not applied for tools, machine implements and personnel involved. Similarly, CEVd wide host range (Figs. 2 and 4) includes solanaceous woody ornamentals but also *Verbena* sp. as a herbaceous one. TASVd and TCDVd likewise infect some of the aforementioned species (Table 2) in Europe, plus TCDVd can be hosted by *Pitosporum tobira* (*Pitosporaceae*), evergreen shrub growing as a garden plant in warmer climates and used for cut foliage throughout Europe. Iresine viroid 1 (IrVd1) seems to be a little peculiar in terms of the host range. Isolated from *Iresine herbstii* (*Amaranthaceae*), also called beefsteak plant, it has a different array of symptomless hosts among tender herbaceous plants of the families *Amaranthaceae* (*Celosia* spp.) and *Portulacaceae* (*Portulaca* sp.).

Ornamental plants that are probably the only ones significantly affected by viroid diseases so far are chrysanthemums. The florist chrysanthemum (*Chrysanthemum indicum*) is economically extremely important crop worldwide, and Netherlands is at the forefront of the world production. CSVd can cause a spectrum of diverse symptoms, depending on the viroid and the host variety (Fig. 2). Stunting is the most pronounced, as reflected in its name. It can stunt the plants by 50%, or even 75% in extreme cases, and reduce fresh weight and quality of the flower shoots (Hadidi et al., 2017). It occurs worldwide, including many European countries growing chrysanthemums, but it can affect many other types of ornamentals as well. Chrysanthemum chlorotic mottle (CChMVd) causes mild mottling and chlorosis of leaves accompanied by dwarfing of the whole plant. This viroid has a host range restricted to various chrysanthemum types and has only been recorded in three European countries. Thus, it can be considered of minor importance.

Coleus blumei viroids (CbVd) is a group of agents interesting not so much as plant pathogens, because most of the infections are asymptomatic, but for their biology. The host range is

FIG. 2 (A) Garden mums (*Chrysanthemum* spp.) uninfected plant tips with flower buds *(left)* and infected with Chrysanthemum stunt viroid (CSVd) devoid of flower buds, showing leaf chlorosis and stunting (right); (B) *Gynura aurantiaca* grown in vitro healthy *(right)* and infected with citrus exocortis viroid (CEVd) showing leaf epinasty, crinkling and stunting *(left)*.

restricted mainly to an ornamental plant *C. blumei*, now renamed *Coleus scutellarioides*, and only several other species (basil, mint) of the family *Lamiaceae*. The viroid genus (*Coleviroid*) includes three ICTV-recognized viroids (CbVd-1, -2 and -3). CbVd-1 is recorded in three European countries, but it is assumed to be widespread. CbVd-2 was detected only in Germany and CbVd-3 in Germany and Croatia. New *Coleviroid* members could be CbVd-5, CbVd-6 and very recently discovered CbVd-7 (Smith et al., 2021). The latter is a recombinant between CbVd-1 and CbVd-5, and some other coleviroids seem to have resulted from recombination events as well. The evolutionary plasticity of this viroid group makes them very interesting for small RNA molecular evolution studies. Dahlia latent viroid (DLVd) is a similar example. It was discovered in the Netherlands in the inspection of symptomless dahlias. Its molecule is a combination of PCFVd and hop stunt viroid (HSVd) probably resulted from a recombination event in its evolutionary past (Hadidi et al., 2017).

Viroids of specialized crops

Hop (*Humulus lupulus*) is sort of a "niche" crop. This plant of the family *Cannabaceae* can be successfully grown only in zones of the northern hemisphere with temperate climatic regions and sufficient humidity. Effectively, this is around the 48° of the northern latitude. The climatic conditions of Germany, Czechia, Poland, Slovenia, UK/England, and to a smaller degree Spain and France, enable hop cultivation. These countries are among top eleven most important producers, although hop is cultivated in smaller gardens in some other European countries. Hop is a climbing herbaceous perennial having impressive tall hardy flexible stems called bines. Its fruit cones are used in the production of beer and many pharmaceuticals, as it is known for its medicinal benefits. Hop is a host of four viroids, hop latent viroid (HLVd), HSVd, apple fruit crinkle viroid (AFCVd) and citrus bark cracking viroid (CBCVd). AFCVd and HSVd have not been reported in Europe, but other two viroids do occur in Europe and present a big concern for the hop industry. HLVd is latent only in the leaves but devastating for cones. It causes significant reduction in the cone yield and a big change in the cone secondary metabolite profile making them inadequate for beer production. All hop-producing countries in Europe have reported its occurrence at some point (Table 2). It seems that hop may not be the exclusive host of HLVd because stinging nettle (*Urtica dioica, Urticaceae*) was found infected in Germany and *Cannabis sativa*, another member of the *Cannabaceae* family (Bektaş et al., 2019). The latter had cannabis stunting disease, and it was reported in California. The cannabis disease causes plant stunting and changes in the flowers and trichomes (glandular hairs of the plant containing secondary metabolites). Even though this disease is not recorded in Europe, industrial and medicinal cannabis production is growing in Europe, and it should be kept in mind due to the ability of HLVd to change the hop secondary metabolite levels, also crucial in the *C. sativa* production.

The emerging and possibly the most important viroid disease of hop now in Europe is the one caused by CBCVd (Fig. 3). However, the co-infection of hop with HLVd, commonly seen in Europe, makes the disease even more severe due to the synergistic effect of the two viroids (Štajner et al., 2019). CBCVd has been known for a long time to occur in European citrus with limited spread. It used to be called citrus viroid IV (CVd-IV) but renamed CBCVd as an agent of mild disorder of bark cracking of the *Poncirus trifoliata* (trifoliate orange) citrus rootstock

FIG. 3 Severe hop stunt disease caused by Citrus bark cracking viroid (CBCVd) on susceptible hop (*Humulus lupulus*) cultivar Celeia; (A) stunting, epinasty and yellowing of leaves in two plants *(front right)* before flowering, (B) stunted plants before harvesting *(right)* and uninfected plant *(left)*, (C) severe cracking of primary bines. *Courtesy of Dr. Sebastjan Radišek, Slovenian Institute of Hop Research and Brewing.*

FIG. 4 (A) Peach (*Prunus persica*) infected with severe strain of peach latent mosaic viroid (PLMVd) showing calico *(front)* and *yellow* mottling symptoms of the leaves, (B) cracked sutures of the fruit; (C) small and deformed fruit of lemon reminiscent of acorn (*Citrus limon*, type Lisbon) infected with CEVd and (D) bark shelling of its rootstock *Poncirus trifoliata.*

B. Prospecting for viroid

(Hadidi et al., 2017). A devastating disease of hop was reported from Slovenia in 2007, and by 2013 it spread epidemically. Now even hop growers in southern Germany reported the same problem, and the disease is being eradicated. As in many viroid diseases, leaves showed mild yellowing, downward rolling (epinasty), leaf malformation but more severe symptoms ensued (Fig. 3). Severe leaf malformations, stunting, bine cracking, abnormal fruit cone development and severe yield loss prompted research efforts to identify the causal agent. It turned out to be CBCVd (Jakše et al., 2014), a previously known mild citrus pathogen. The zone of the hop production is not the place where citrus can be cultivated. How a citrus viroid spilled over to hop and wreaked such havoc? The contact of hop with improperly processed organic waste containing citrus material infected with CBCVd is assumed, and the ease of viroid mechanical transmission may have done the rest in the disease spread. Spillover events are characteristic of emerging pathogens, and CBCVd surely can be viewed as such according to its EPPO A2 listing.

Viroids of woody plants

Most of the fruit species grown in Europe are trees, or dicotyledonous woody perennial plant species. Depending on the botanical characteristics of their fruits, they can be further divided in agronomical categories such as pome fruits (e.g., apple, pear, quince), stone fruits (e.g., apricot, cherry, peach, plum, etc.), citrus, grapevine and some other minor groups. The types of fruit trees grown in a country are dictated by the soil and climate conditions, and as a result of a warmer climate, the European countries in the Mediterranean basin grow a wide variety of all these fruit tree species. In the countries with temperate or colder climate, the diversity of grown fruit tree species diminishes with lowering north geographical latitudes. Expectedly, the northernmost European countries have no conditions for growing citrus, pomegranate or grapevine in the open field.

The tradition of growing some of these species (e.g., grapevine in warm and temperate climate) dates back a couple of millennia giving ample opportunities for viroids to be distributed inadvertently among countries and wide regions. Importing viroid-infected planting material, or cutting and pruning with contaminated tools are well-known viroid transmission routes and fruit trees are handled by man extensively during their long lifetime (Hadidi et al., 2017). Actually, most of those plants are scions of a cultivated species with desirable fruits grafted onto vigorous and hardy rootstock of a different but compatible plant species (e.g., pear grafted on quince) presenting an important pathway for an unchecked viroid to enter planting material from a wider range of hosts. Thus, many types of viroids have been recorded in many types and species of fruit trees all over Europe (Tables 1–3). Their variety in this group of plants is remarkable, and some serious diseases are linked to their occurrence.

Pome and stone fruit trees

Pome fruits can be infected by eight viroids out of which two, apple chlorotic fruit spot viroid (ACFSVd) and apple fruit crinkle viroid (AFCVd), are as yet unrecognized viroid species (Table 3). While AFCVd is not reported in Europe, ACFSVd was discovered only few years ago in the Austrian region southern Burgenland where chlorotic spots and bumps were observed in 2016 on the fruit of apple (*Malus domestica*) cultivar Ilzer Rose (Leichtfried et al., 2019). The fruits were effectively unmarketable. ACFSVd is transmissible by grafting, budding (bud grafting), apple seeds and supposedly by mistletoe (*Viscum album* subsp. *album*), a hemiparasitic plants whose vegetative parts and seeds apparently contained this viroid (Leichtfried et al., 2020). Theoretically, a mistletoe growing on a viroid infected apple could take up the viroid in its phloem by haustoria (sap sucking organs), and in turn to seeds, and the mistletoe seeds could be distributed by birds to new apple hosts. Nonetheless, this interesting assumption on the potentially novel pathway of viroid transmission requires direct proof. ACFSVd could not be transmitted from apple to pear and quince by grafting (Leichtfried et al., 2020). The investigations of the ACFSVd have only started, and it remains to be seen if it may negatively influence the apple production in Europe. Apple hammerhead viroid (AHVd) is also discovered in the last decade by HTS in China but already recognized by ICTV as a true viroid (Serra et al., 2018). It was revealed in Italian apples with apple scar skin disease, trunk splitting, shoot decline and dieback. Nevertheless, the association of this viroid to a specific disease, or its contribution to the diseases caused by major apple viruses with which it has been found in mixed infections, remains to be determined. Meanwhile, more evidence from Germany (Zikeli et al., 2021) and Belgium (Fontdevila Pareta et al., 2022) accumulate suggesting AHVd may be present for a long time in Europe as this assumption is based on the results from old apple cultivars. Loquat (*Eriobotrya japonica*) is also a pome fruit tree whose commercial cultivation is not so important in Europe, but it is often grown as a garden tree in Euro-Mediterranean countries. Spain is the primary European commercial producer where AHVd has been reported in loquat very recently. It is only the second known host of this viroid (Canales et al., 2021), but the data on its pathogenicity, if any, is still lacking.

The other five viroids infecting pome fruit trees are: apple scar skin viroid (ASSVd), apple dimple fruit viroid (ADFVd), HSVd, pear blister canker viroid (PBCVd) and peach latent mosaic viroid (PLMVd). ASSVd causes brownish scar-like marks or dapples (greenish small round spots that may fuse into bigger ones) in apple fruits lowering the fruit quality. It has been reported in some European apple growing countries (Table 2), but it seems to be much more of a problem in apple production in China and Japan (Hadidi et al., 2017). ASSVd infects latently cultivated pear (*Pyrus communis*), wild pear (*P. amygdaliformis*) and sweet cherry in Greece, but only some pear cultivars seem to be more susceptible and develop blemished unmarketable fruits. However, as PLMVd was found in some of those pears in mixed infections with ASSVd, the latter still needs to be linked to the specific pear symptoms (Di Serio et al., 2018b). ADFVd has been reported in apple in southern Italy, but it does not seem to present a big concern in the Euro-Mediterranean basin as it has very limited distribution. HSVd was recorded in the European apple, wild apple and pear but, unlike in stone fruits where it has been linked to plum and peach dapple fruit disease in Italy, it does not seem to cause major problems on its own in pome fruits. The PBCVd presence in Europe is known since 1991 in pear and has been linked to a severe pear bark disorder only in a couple of pear cultivars used as biological indicators. Dark pustules and cracks appear after a couple of years on the outer layer of the bark turning into cankers and scaly bark leading to the death of the tree (Di Serio et al., 2018b). Commercial pear cultivars, wild pear, quince, apple and wild apple seem to be latently infected.

Regardless of the word latent in the PLMVd name, this viroid can cause a serious disease in peach and nectarine. The disease is described as latent because the symptoms are delayed for about two years after the infection. The most common symptoms, besides reduced tree vigor, are late blooming, leaf production and fruit ripening. Less common symptoms may include fruit and leaf blotches, deformed fruits with cracked sutures, enlarged pits and faster tree aging. In addition, the existence of severe PLMVd strain called peach calico provoking severe leaf chlorosis (albinism) in peach may obviously represent a big problem for the fruit growers (Fig. 4). The producers of peach propagation material trading internationally face additional challenges because PLMVd must be absent from certified material and some European countries still have it on their A1 lists. Plum (European and Japanese), mume, apricot, sweet cherry, apple, quince and pear varieties, even wild pear, in many parts of Europe may be infected but with no serious symptoms or economic damage reported.

Stone fruit, much like pome fruit, trees in Europe seem to be viroid infected quite often. Nevertheless, the diversity of viroid species in them is much smaller. Apart from ASSVd with negligible influence to stone fruit production and PLMVd as a serious problem in peach growing, only HSVd occurs. Larger-scale testing from many European countries demonstrated PLMVd and HSVd are frequently found in mixed infections, although HSVd alone can cause certain problems. Fortunately, most of the European stone fruit cultivars belonging to the genus *Prunus* (almond, apricot, cherry, peach, plum, etc.) do not develop fruit symptoms in HSVd infections. A few exceptions like the cases of plum dapple disease in some Italian cultivars and apricots with rugose and insipid fruits in Spain (Hadidi et al., 2017) call for caution regarding this viroid, especially because it has a very wide distribution and host range including many plants from different families.

Other fruit trees and shrubs

Besides loquat already mentioned above as a minor pome fruit tree species in Europe and its infection with AHVd, a number of other species have been found to harbor viroids. One of them is pomegranate (*Punica granatum*, family *Lythraceae*, subfamily *Punicoideae*), a deciduous shrub cultivated in the Euro-Mediterranean zone as a fruit tree. It can stand lower temperatures, probably down to −10°C, and its dwarfed cultivars (*P. granatum* var. *nana*) can be grown in Europe farther north as ornamental plants. HSVd was found in pomegranates in Turkey (Önelge, 2000) and Spain (Astruc et al., 1996). Fig is another example of fruit tree grown in this warmer part of Europe harboring viroids. ADFVd, a viroid thought to be limited to cultivated apple in southern Italy, was reported from symptomatic figs also hosting some viruses. No information is available yet on ADFVd-related symptoms or potential adverse impact to the fig production (Chiumenti et al., 2014). White mulberry (*Morus alba*) is, aside from fig, the second known HSVd host in the family *Moraceae*, and the trees tested in Italy showed no symptoms that could be linked to the viroid infection (Elbeaino et al., 2012). Pistacchio (*Pistacia vera*), a stone fruit in the cashew family (*Anacardiaceae*), is also a symptomless HSVd host in the Mediterranean part of Turkey (Balsak et al., 2017).

Perhaps the most exotic example for Europe in this group of cultivated trees is avocado (*Persea americana*, family *Lauraceae*) whose fruits are nowadays enjoyed in parts of the world far away from its mid-American center of origin. With over 90,000 tons (http://www.fao.org/faostat/) of avocado produced Spain is listed in the last quarter of the top 20 producing countries in the world. Turkey, Greece, France and Cyprus can be found after the 40th place on this list (https://en.wikipedia.org/wiki/List_of_countries_by_avocado_production). Avocado sunblotch viroid (ASBVd) can cause severe losses to the fruit production. There are records of ASBVd presence in the Spanish plantations from the late 1980s (Pallás et al., 2003), and the viroid has been under control ever since. In recent times, a single tree with fruits showing typical sunblotch disease symptoms (greenish, creamy-white depressions in the skin) was found in the Greek province of Chania (Crete). ASBVd presence was confirmed (Lotos et al., 2018) in this case that seems to be an isolated one so far.

Citrus

Citrus is grown commercially in the Euro-Mediterranean countries, and the Croatian commercial groves of Satsuma mandarin (*Citrus unshiu*) above the 43° of the northern latitude (43.5406 N) are probably the northernmost sites where it can be commercially cultivated. Citrus is known for hosting many viroids. So far, seven recognized and one pending viroid species have been described. Several of them often infect a plant simultaneously, and new ones like Citrus viroid V (CVd-V), CVd-VI and CVd-VII were discovered in the 21st century (Tables 2 and 3). Previously known viroids have been renamed which can be confusing when searching for information. For this reason, the former citrus viroid names are retained in the table showing their occurrence (Table 2).

Citrus exocortis viroid (CEVd) is the causal agent of the exocortis disease in oranges, lemons, limes and other sensitive citrus species grafted on *P. trifoliata* rootstock (Fig. 4). The name comes from the bark disorder of this rootstock (bark cracking and shelling). However, the whole plant may be affected in sensitive scion/rootstock combinations. Tree dwarfing, damage to the roots, leaf epinasty, and serious yield losses with deformed or acorn shaped fruits (Fig. 4) in some varieties (citron, lemon) are well-described symptoms documented even in antique mosaic tile floors of old Israeli synagogues (Bar-Joseph, 2003). A variant of HSVd may cause cachexia/xyloporosis disease in citrus with symptoms of stunting, gummy deposits in the phloem and yield reduction in mandarins, tangerines and clementine trees. CEVd and HSVd are widespread in the citrus-growing countries all over the world nonetheless, not all citrus types are equally sensitive to them (e.g., citrus grafted on sour orange rootstocks are tolerant). The selection of propagative material, both scions and rootstocks, has contributed to maintaining the situation in Europe under control.

Out of other five citrus viroids, the pending viroid species CVd-VII recently discovered in Australia (Chambers et al., 2018) and CVd-VI have not been reported in Europe. CVd-V is rare. Only Spain, where it had been originally identified, and Turkey reported it so far. Citrus bent leaf viroid (CBLVd), citrus dwarfing viroid (CDVd) and CBCVd are regularly found. They can induce different types of milder symptoms (tree stunting, epinasty, vein necrosis, bark disorder, depending on the viroid), but their cumulative negative impact on citrus production is not big. It seems that out of all citrus viroids HSVd and CDVd are encountered most frequently, and often in combination, CBLVd less so and CBCVd has been found only in Italy, France and Spain. CBCVd is also rarely found in the European citrus groves (Hadidi et al., 2017). Nonetheless, CBCVd has become a major problem in the hop production (Jakše et al., 2014) which was already discussed above.

The group of citrus viroids is the only one containing variants approved as biotechnological tools. For example, variants of CDVd (formerly CVd-III) are in use in California for over 20 years in obtaining dwarfed trees. Navel oranges (*C. sinensis*) on *P. trifoliata* rootstock have canopies that are 50% in size compared to the plants of the same age uninoculated by CDVd variant used for dwarfing (Lavagi-Craddock et al., 2020). CDVd has proved to be phenotypically stable in those long-term field experiments. The advantages of dwarfed trees are higher density plantings, enabling the more efficient use of land without losing yield, reduction in human labor and plant management costs (irrigation water, pesticides, etc.). Citrus viroids have no insect vectors, and uncontrolled dissemination of such a small regulatory RNA is not a problem making this approach both appealing for the growers and environmentally friendly.

Grapevine

Grapevine has been known to host a plethora of viruses. At least 86 have been isolated so far (Fuchs, 2020), and many are economically important. The number of viroids infecting grapevine is about ten times lower. Only seven are known to date; however, the number is growing. Viroids, or viroid like RNAs with ribozyme activity isolated from grapevine still awaiting to be recognized as viroids (Table 3), are grapevine latent viroid (GLVd) and grapevine hammerhead viroid (GHVd). In addition, there are five pospiviroids found in grapevine. CEVd and HSVd, encountered above as pathogens of fruit trees, have very wide host ranges that encompass grapevine, but they elicit no major diseases in this host. Three out of these five pospiviroids are linked exclusively to the grapevine as a host: Australian grapevine viroid (AGVd), GYSVd-1 and GYSVd-2. AGVd has been detected in Europe long ago (Gambino et al., 2014) with additional records only for Greece and Russia, so far (Table 2). GYSVd-1 is present all over Europe (Hadidi et al., 2017). The distribution of GYSVd-2 is somewhat more restricted for the time being (Table 2), but that picture may change as more studies are done as in the recent example from Greece (Sassalou et al., 2020). GYSVd-1 and GYSVd-2 can independently elicit tiny yellow spots (speckles) along the veins of the grapevine leaves of no real economic impact. Generally speaking, grapevine viroids are not considered economically significant. One exception is the synergism of GYSVd-2 and grapevine fanleaf virus (GFLV) linked to the serious vein banding disease, normally not induced by either of these agents alone (Hadidi et al., 2017).

Procedures

The data on the occurrence of viroids in Europe have been gathered from multiple sources. Several books have been published on viroids over the years (Hadidi et al., 2003, 2017). All contain data on this subject in various chapters, and all have been consulted. The data for each viroid species was gathered from the original research papers and compared with the data publicly available in the EPPO Global Database (https://gd.eppo.int), EFSA reports and occasionally at the CABI Invasive Species Compendium (https://www.cabi.org/isc) where datasheets, including distribution tables, maps and references, exist for some of the viroids considered as "invasive species" (e.g., PSTVd). The data found in these online resources has been crosschecked and updated with the latest findings from the original research papers. The findings not relevant for Europe were filtered out as well as the reports from experimental viroid hosts (plants that can be infected with a certain viroid in experimental conditions but have not been recorded as natural hosts in the field or greenhouse growing conditions). Only the findings where viroids were confirmed with two independent methods (biological indexing + laboratory, two different laboratory methods) were considered as reliable.

Prospective

Viroid diseases of plants have become a problem in the modern times. The practice of monoculture along with the intense handling of some cultivated plants (e.g., pruning) by man over the plant lifespan, especially those long living like fruit trees and grapevine, probably facilitated viroid distribution over large space, time periods and in a large spectrum of hosts. Europe does not only seem to be in the center of viroid research but also in the emergence center of many viroids with new ones discovered locally. Some viroids are controlled due to their pathogenic potential, at least in some combinations of plant hosts and environmental conditions. Nevertheless, should we do it for more viroid species that we do not perceive as a problem now? The case of CBCVd, a viroid previously not listed as quarantine pest in citrus, spilling over to hop and causing a devastating disease is an instructive example. It is still unclear how viroids influence metabolic pathways of plants used for the production of bioactive compounds (cannabis, coleus, hop) and herbs. We have seen some negative examples so far, but it would be interesting to see if there are positive ones.

Future implications

There are only three viroids of monocotyledonous plants known so far. Two have been identified in coconut and oil palms (Table 2) and one in the orchid of the genus *Dendrobium* (Table 3). None have been recorded in Europe. Monocotyledonous plants are important with cereals and corn being the main sources of staple food for humans and animal feed in the Old Continent. It will be interesting to see whether new viroids will be discovered from transcriptome studies of most prominent monocots grown in the temperate climate zones. Ornamentals that mostly have been found latently infected with many different viroids repeatedly are particularly interesting because of their potential to serve as reservoirs of infection to vegetables. The role of weeds in viroid diseases of vegetables as well as viroid infections of wild plants and their impact on biodiversity on the borders of agro-ecosystems, or wider, is completely unknown. European landscapes are often urbanized and enable close contacts of cultivated species and autochthonous flora. The generic capacity of viroids to act as regulatory molecules changing signaling pathways in plants makes them very interesting models for research of basic biochemical processes and potential biotechnological applications. New comprehensive research studies and application of new methodologies, as usual, hold the keys to answering those questions.

Chapter summary

European countries have established a number of rules and regulations to stop the spread of the viroids causing economical damage. All members of the family *Avsunviroidae* and most of the viroids of the family *Pospiviroidae* were recorded in 38 European countries infecting naturally at least 60 different plant host species. Members of the family *Avsunviroidae* have been recorded in one country alone, *Pospiviroidae* in 20 countries alone, and together in 17 countries. Viroid species, the country of occurrence and host plants are listed in appropriate tables. The impact of viroids on the plant production varies, depending on the influence of a certain viroid to a plant host and the control of its spread in specific country which is discussed for different types of crop plants. The newly discovered viroid-like pathogens are included as some of them are likely to be recognized soon as viroid species and may be of interest not only as plant pathogens emerging in Europe but as biological agents important in evolutionary, molecular or biotechnological studies.

Study questions

1. What does a status of EPPO quarantine pest means in practice for the control of viroids?
2. What type of ornamental plant is most affected by viroid diseases in Europe?
3. What is the emerging viroid disease of hop in Europe?
4. What viroids of ornamentals are known for evolutionary plasticity?
5. Is it true that CLVd does not cause symptoms in tomato?
6. What is economically the most important viroid of stone fruit trees in Europe?
7. Is it true that avsunviroids are present in Europe?

Further reading

EFSA Plant Health Panel, Bragard, C., Dehnen-Schmutz, K., Gonthier, P., Jacques, M.-A., Jaques Miret, J.A., Justesen, A.F., MacLeod, A., Magnusson, C.S., Milonas, P., Navas-Cortes, J.A., Parnell, S., Potting, R., Reignault, P.L., Thulke, H.-H., Van der Werf, W., Vicent Civera, A., Yuen, J., Zappala, L., Candresse, T., Chatzivassiliou, E., Finelli, F., Winter, S., Chiumenti, M., Di Serio, F., Kaluski, T., Minafra, A., Rubino, L. 2019. Scientific opinion on the pest categorisation of non-EU viruses and viroids of *Cydonia* Mill., *Malus* Mill. and *Pyrus* L. EFSA J., 17 (9), 5590, 81 pp. doi: 10.2903/j.efsa.2019.5590.
Hadidi, A., Flores, R., Randles, J.W., Palukaitis, P., (Eds.) Viroids and Satellites. Academic Press, Elsevier, London, San Diego, Cambridge (for specific viroids and their biology, occurrence and economic significance).
https://www.cabi.org/isc
https://www.eppo.int
https://gd.eppo.int

References

Astruc, N., Marcos, J.F., Macquaire, G., 1996. Studies on the diagnosis of hop stunt viroid in fruit trees: identification of new hosts and application of a nucleic acid extraction procedure based on non-organic solvents. Eur. J. Plant Pathol. 102, 837–846. https://doi.org/10.1007/BF01877053.

Balsak, S.C., Buzkan, N., Ay, M.Z., Gürbüz, M., 2017. Occurrence of Hop stunt viroid (HSVd) in Turkish pistachio trees. Phytopathol. Mediterr. 56, 376.

Bar-Joseph, M., 2003. Natural history of viroids – horticultural aspects. In: Hadidi, A., Flores, R., Randles, J.W., Semancik, J.S. (Eds.), Viroids. CSIRO Publishing, Collingwood, pp. 246–251.

Bektaş, A., Hardwick, K.M., Waterman, K., Kristof, J., 2019. Occurrence of hop latent viroid in *Cannabis sativa* with symptoms of cannabis stunting disease in California. Plant Dis. 103. https://doi.org/10.1094/PDIS-03-19-0459-PDN.

Canales, C., Moran, F., Olmos, A., Ruiz-Garcia, A.B., 2021. First detection and molecular characterization of apple stem grooving virus, apple chlorotic leaf spot virus, and apple hammerhead viroid in loquat in Spain. Plan. Theory 10, 2293. https://doi.org/10.3390/plants10112293.

Chambers, G.A., Donovan, N.J., Bodaghi, S., Jelinek, S.M., Vidalakis, G., 2018. A novel citrus viroid found in Australia, tentatively named citrus viroid VII. Arch. Virol. 163, 215–218. https://doi.org/10.1007/s00705-017-3591-y.

Chiumenti, M., Torchetti, E.M., Di Serio, F., Minafra, A., 2014. Identification and characterization of a viroid resembling apple dimple fruit viroid in fig (*Ficus carica* L.) by next generation sequencing of small RNAs. Virus Res. 188, 54–59. https://doi.org/10.1016/j.virusres.2014.03.026.

Chiumenti, M., Navarro, B., Candresse, T., Flores, R., Di Serio, F., 2021. Reassessing species demarcation criteria in viroid taxonomy by pairwise identity matrices. Virus Evol. 7 (1), veab001. https://doi.org/10.1093/ve/veab001.

Di Serio, F., Li, S.-F., Matoušek, J., Owens, R.A., Pallas, V., Randles, J.W., Sano, T., Verhoeven, J.T.J., Vidalakis, G., Flores, R., ICTV Report Consortium, 2018a. ICTV virus taxonomy profile: *Avsunviroidae*. J. Gen. Virol. 99, 611–612. https://doi.org/10.1099/jgv.0.001045.

Di Serio, F., Ambrós, S., Sano, T., Flores, R., Navarro, B., 2018b. Viroid diseases in pome and stone fruit trees and Koch's postulates: a critical reassessment. Viruses 10, 612. https://doi.org/10.3390/v10110612.

Elbeaino, T., Abou Kubaa, R., Choueiri, E., Digiaro, M., Navarro, B., 2012. Occurrence of Hop stunt viroid in mulberry (*Morus alba*) in Lebanon and Italy. J. Phytopathol. 160, 48–51. https://doi.org/10.1111/j.1439-0434.2011.01855.x.

Fagoaga, C., Pina, J.A., Duran-Vila, N., 1994. Occurrence of small RNAs in severely diseased vegetable crops. Plant Dis. 78, 749–753.

Fagoaga, C., Pina, J.A., Duran-Vila, N., 1996. Naturally occurring variants of citrus exocortis viroid in vegetable crops. Plant Pathol. 45, 45–53.

Fontdevila Pareta, N., Lateur, M., Steyer, S., Blouin, A.G., Massart, S., 2022. First reports of apple luteovirus 1, apple rubodvirus 1 and apple hammerhead viroid infecting apples in Belgium. New Dis. Rep. 45, e12076. https://doi.org/10.1002/ndr2.12076.

Fuchs, M., 2020. Grapevine viruses: a multitude of diverse species with simple but overall poorly adopted management solutions in the vineyard. J. Plant Pathol. 102, 643–653. https://doi.org/10.1007/s42161-020-00579-2.

Gambino, G., Navarro, B., Torchetti, E.M., La Notte, P., Schneider, A., Mannini, F., Di Serio, F., 2014. Survey on viroids infecting grapevine in Italy: identification and characterization of Australian grapevine viroid and grapevine yellow speckle viroid 2 Eur. J. Plant Pathol. 140, 199–205. https://doi.org/10.1007/s10658-014-0458-x.

Hadidi, A., Flores, R., Randles, J.W., Semancik, J.S.S. (Eds.), 2003. Viroids. CSIRO Publishing, Collingwood.

Hadidi, A., Flores, R., Randles, J.W., Palukaitis, P. (Eds.), 2017. Viroids and Satellites. Academic Press, Elsevier, London, San Diego, Cambridge.

Jakše, J., Radišek, S., Pokorn, T., Matoušek, J., Javornik, B., 2014. Deep-sequencing revealed a CBCVd viroid as a new and highly aggressive pathogen on hop. Plant Pathol. 64, 831–842. https://doi.org/10.1111/ppa.12325.

Lavagi-Craddock, I., Campos, R., Pagliaccia, D., Kapaun, T., Lovatt, C., Vidalakis, G., 2020. Citrus dwarfing viroid reduces canopy volume by affecting shoot apical growth of navel orange trees grown on trifoliate orange rootstock. J. Cit. Pathol., 1–6. https://doi.org/10.5070/C471045369.

Leichtfried, T., Dobrovolny, S., Reisenzein, H., Steinkellner, S., Gottsberger, R.A., 2019. Apple chlorotic fruit spot viroid: a putative new pathogenic viroid on apple characterized by next-generation sequencing. Arch. Virol. 164, 3137–3140. https://doi.org/10.1007/s00705-019-04420-9.

Leichtfried, T., Reisenzein, H., Steinkellner, S., Gottsberger, R.A., 2020. Transmission studies of the newly described apple chlorotic fruit spot viroid using a combined RT-qPCR and droplet digital PCR approach. Arch. Virol. 165, 2665–2671. https://doi.org/10.1007/s00705-020-04704-5.

Lotos, L., Kavroulakis, N., Navarro, B., Di Serio, F., Olmos, A., Ruiz-Garcia, A.B., Katis, N.I., Maliogka, V.I., 2018. First report of avocado sunblotch viroid (ASBVd) naturally infecting avocado (*Persea americana*) in Greece. Plant Dis. 102, 1470. https://doi.org/10.1094/PDIS-12-17-1980-PDN.

Matsushita, Y., Yanagisawa, H., Khiutti, A., Mironenko, N., Ohto, Y., Afanasenko, O., 2021. Genetic diversity and pathogenicity of potato spindle tuber viroid and chrysanthemum stunt viroid isolates in Russia. Eur. J. Plant Patol. 161, 529–542. https://doi.org/10.1007/s10658-021-02339-z.

Önelge, N., 2000. Occurrence of hop stunt viroid (HSVd) on pomegranate (*Punica granatum*) trees in Turkey. J. Turk. Phytopathol. 29, 49–52.

Pallás, V., Gómez, G., Duran-Vila, N., 2003. Viroids in Europe. In: Hadidi, A., Flores, R., Randles, J.W., Semancik, J.S. (Eds.), Viroids. CSIRO Publishing, Colingwood, pp. 268–278.

Sassalou, C.-L., Katsarou, K., Lotos, L., Orifanidou, C.G., Maglioka, V.I., Kalantidis, K., Katis, N.I., Pappi, P.G., 2020. First report of grapevine yellow speckle viroid-2 in grapevine in Greece. Plant Dis. 104, 1879. https://doi.org/10.1094/PDIS-12-19-2540-PDN.

Serra, P., Messmer, A., Sanderson, D., James, D., Flores, R., 2018. Apple hammerhead viroid-like RNA is a bona fide viroid: autonomous replication and structural features support its inclusion as a new member in the genus *Pelamoviroid*. Virus Res. 249, 8–15. https://doi.org/10.1016/j.virusres.2018.03.001.

Smith, R.L., Shukla, M., Creelman, A., Lawrence, J., Singh, M., Xu, H., Li, X., Gardner, K., Nie, X., 2021. *Coleus blumei* viroid 7: a novel viroid resulting from genome recombination between *Coleus blumei* viroids 1 and 5. Arch. Virol. 166, 3157–3163. https://doi.org/10.1007/s00705-021-05201-z.

Štajner, N., Radišek, S., Mishra, A.K., Nath, V.S., Matoušek, J., Jakše, J., 2019. Evaluation of disease severity and global transcriptome response induced by Citrus bark cracking viroid, hop latent viroid, and their co-infection in hop (*Humulus lupulus* L.). Int. J. Mol. Sci. 20, 3154. https://doi.org/10.3390/ijms20133154.

Zikeli, K., Berwarth, C., Faus, S., Jelkmann, W., 2021. Viroide in Apfel in Deutschland. In: 62. Deutsche Pflanzenschutztagung, 21. – 23. September 2021, Julius-Kühn-Archiv. 467, pp. 572–573 (in German).

B. Prospecting for viroid

Naturally occurring viroid diseases of economically important plants in Africa

Amine Elleuch[a] and Imen Hamdi[b]

[a]Laboratoire de Biotechnologie Végétale Appliquée à l'amélioration des plantes, Faculté des Sciences de Sfax, Université de Sfax, Sfax, Tunisia [b]Laboratoire de Protection des Végétaux, Institut National de la Recherche Agronomique de Tunis, Université de Carthage, Tunis, Tunisia

Graphical representation

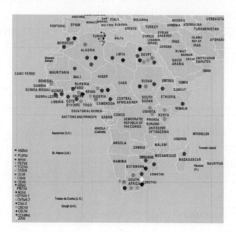

Distribution of viroids occurring and reported in Africa.

Definitions

Stone fruits: Stone fruits are deciduous tree species originating from the temperate zone of the northern hemisphere. Plum, peach, nectarine, sweet and sour (tart) cherry, apricot, and cornelian cherry belong to the group of stone fruits.

Pospiviroidae: Members of the family *Pospiviroidae* have a genome composed of a single-stranded circular RNA molecule that adopts a rod-like conformation containing a central conserved region (CCR) involved in their replication.

Avsunviroidae: Members of the family *Avsunviroidae* have a single-stranded circular RNA genome that adopts a rod-like or branched conformation and can form, in the strands of either polarity, hammerhead self-cleaving motifs involved in their replication.

EPPO organization: European and Mediterranean Plant Protection Organization (EPPO) is an intergovernmental organization responsible for cooperation and harmonization in plant protection within the European and Mediterranean regions.

Chapter outline

This chapter updates the occurrence and distribution of viroids in African countries. New viroids and diseases are occurring in the continent, and new hosts are reported.

Learning objectives

This chapter aims to:

- Provide insight on the occurrence and distribution of different viroids in African countries.
- Identify the impact of viroid diseases on different crops in Africa.
- Illustrate some symptoms of viroid diseases.
- Pointing out new emergent viroid diseases and threats on agriculture and trade in the African continent.

Introduction

The hot or warm climate of the African continent can enhance the expression of viroid disease symptoms. Viroids and viroid diseases occur widely in African countries. Viroids from *Avsunviroidae* and *Pospiviroidae* families have been reported from different regions of the continent.

Occurrence of *Avsunviroidae* members in fruits trees in Africa

Avocado sunblotch viroid

Avocado sunblotch viroid (ASBVd) is the causal agent of Avocado sunblotch disease, which is one of the important diseases of avocado that affects yield and quality worldwide. Typical symptoms are found on the tree's leaves, fruit, and bark (Fig. 1).

FIG. 1 Symptoms of avocado sunblotch disease. *Yellowish* sunken areas on fruits (A); discolored and necrotic depressions on infected twigs (B); distortion and variegation on leaves (C); cracked bark ("Alligator skin") appearance on some mature branches (D); fruits with *reddish color* areas (E); necrosis on severely affected fruits (F); multiple yellowish sunken areas in fruits (G). *Source: Saucedo Carabez, J.R., Teliz Ortiz, D., Ochoaascencio, S., Ochoa-Martinez, D., Vallejopérez, M.R., Beltrán Peña, H., 2014. Effect of Avocado sunblotch viroid (ASBVd) on avocado yield in Michoacan. Mexico. Eur. J. Plant Pathol. 138, 799–805.*

Avocado sunblotch disease, now known to be caused by avocado sunblotch viroid, was recorded in South Africa in 1954 (Da Graca and van Vuuren, 2003). Infection by ASBVd reduced fruit yield by up to 82% for symptomless trees but by only about 14% in trees with symptoms. The distribution of ASBVd in avocado orchards in two provinces (i.e., Limpopo and Mpumalanga) was investigated. In this survey, 11.2% of the trees sampled tested positive for ASBVd (Jooste and Zwane, 2018). This study showed that from a tree bearing asymptomatic fruit, all the fruit tested positive for the presence of ASBVd.

The incidence of ASBVd among Ghanaian accessions was investigated (Acheampong et al., 2008). Of the 185 symptomatic and symptomless avocado trees tested, one accession tested positive (0.01%). The incidence of the disease in Ghana was found to be very low.

Peach latent mosaic viroid

Peach latent mosaic viroid (PLMVd) is the causal agent of peach latent mosaic disease and ranges in size from 335 to 351 nucleotides. It mainly infects peach trees. The incidence of

PLMVd infection in Egypt on different tree species evaluated by RT-PCR was 45% in peach, 35% in plum, 23% in pear, 18% in mango, 5% in apple, and 2% in apricot (El-Dougdoug et al., 2012a). In the most recent study in Tunisia, the incidence of PLMVd in peach trees was estimated at 59.5% and mixed infection with hop stunt viroids (HSVd) was 2.5% (Mahfoudhi et al., 2010).

PLMVd was also reported in Algeria (Rouag et al., 2008; Meziani et al., 2010) and Morocco (EPPO, 1993). In Algeria, TPH (tissue printing hybridization) assays disclosed that 28 of 531 samples (5.2%) were positive for PLMVd and HSVd, both of which were present in peach (14% infection) and apricot (5.8% infection), but not in cherry, almond, plum, and myrobalan (Rouag et al., 2008).

Apple hammerhead viroid

To our knowledge, in Africa, apple hammerhead viroid (AHVd) was reported only in Tunisia (Hamdi et al., 2022), with an infection rate of 46%. AHVd was found mostly in samples showing dieback symptoms. Three selected AHVd isolates (amplified by the two primer pairs) were cloned and sequenced, confirming AHVd infection. The genome sequences obtained in this study showed that the three AHVd isolates shared 81%–93% of sequence identity with reference sequences from China (GenBank Acc. No.: KR605506) (Zhang et al., 2014). The closest variant was SD17-142 (GenBank Acc. No.: MK188694) from Canada, showing identities ranging from 89% to 96%.

Occurrence of *Pospiviroidae* members in fruits trees in Africa

Citrus viroid complex

Citrus viroids have been classified into distinct groups based on their biological and physical properties. Seven viroids reported to infect Citrus spp. (orange, mandarin, tangerine, clementine, grapefruit, pomelo, lemon, and lime) belong to four genera of the family of Pospiviridae (Duran-Vila et al., 1988): citrus exocortis viroid (CEVd), citrus bent leaf viroid (CBLVd), hop stunt viroid (HSVd), citrus dwarfing viroid (CDVd), citrus bark cracking viroid (CBCVd), citrus viroid V (CVd-V), and citrus viroid VI (CVd-VI) which has only been reported in Japan (Ito et al., 2003).

In the case of Egypt, citrus budlings tested contain citrus viroid complexes consisting of CBCVd (79%), CDVd (68%), HSVd (54%), CEVd (29%), and CBLVd (28%) (Al-Harthi et al., 2013). These findings illustrate the high rate of vegetative transmission of citrus viroids by budlings in Africa (Al-Harthi et al., 2013). Elleuch et al. (2006) have characterized a mixed infection by CEVd, HSVd, and CDVd from a tree nursery in Tunisia. In a more recent study, CEVd, HSVd, and CDVd were widespread and accounted for 70.3%, 72.3%, and 78.2% of the tested trees, respectively. CBLVd and CBCVd were found in 28.2% and 3.0% of trees, respectively. The most frequent viroid combinations found were CEVd-HSVd-CDVd (34.7%) and HSVd-CDVd (22.3%). Other combinations, such as CBLVd–HSVd-CDVd (12.9%), CEVd–CBLVd-HSVd (11.9%), and CEVd-CDVd (10.8%), were less frequent. Selected CEVd sequences had 99% similarity with each other and shared 99% similarity with those from Iran,

Greece, and Syria. CDVd isolates were 100% similar and shared more than 96% similarity with other isolates from Brazil, Cyprus, Greece, Uruguay, and Spain (Najar et al., 2017).

In Sudan, all described citrus viroid seem to be present and widely spread throughout a surveyed area except CVd-V (Mohamed et al., 2009).

Citrus exocortis viroid

Citrus exocortis viroid (CEVd) was reported in different African countries, i.e., South Africa (Da Graca and van Vuuren, 2003), Ghana (Opoku, 1972), Sudan (Mohamed et al., 2009), Morocco (Bibi et al., 2020), Algeria, Tunisia, Libya, Egypt, (Hadidi et al., 2003). In Egypt, the infection rate of CEVd was about 15% in sweet oranges, 4%–12% in grapevine, and 1% in mango (El-Dougdoug et al., 2012a; Nasr-Eldin et al., 2012). In Tunisia, CEVd is the prevalent viroid in citrus (Najar and Duran-Vila, 2004), and its variants belong to class B (mild in tomato) (Elleuch et al., 2006). CEVd was also found to infect figs in Tunisia (Yakoubi et al., 2007). The presence of CEVd in citrus viroid complex were also studied (refer section "Citrus viroid complex"). The most frequent symptoms were: (i) Bark gumming, (ii) Symptoms of stunting and epinasty, and (iii) Mid-vein necrosis (Najar et al., 2017) (Fig. 2).

Recently in Morocco, an extensive survey was performed in 100 commercial citrus groves of different varieties (Bibi et al., 2020). Symptoms of exocortis and cachexia were inspected, and 5390 samples were collected for imprint hybridization analysis. The results were confirmed by reverse transcription polymerase chain reaction (RT-PCR) assays using specific primers. The incidence of infection by viroids was about 29.0%, regardless of citrus species and location, although a slightly higher number of infected trees were located in the Gharb

A   B   C

FIG. 2 Characteristic citrus viroid symptoms. (A) Bark gumming on Cassar clementine. (B) Symptoms of stunting and epinasty. (C) Mid-vein necrosis on Citron Etrog. *Source: Najar, A., Hamdi, I., Varsani, A., Duran-Vila, N., 2017. Citrus viroids in Tunisia, prevalence and molecular characterization. J. Plant Pathol. 99, 781–786. https://doi.org/10.4454/jpp. v99i3.3989.*

region (41.5%). CEVd was detected in 18.2% of the tested samples, while HSVd was detected in up to 10.8% of the samples tested (Bibi et al., 2020).

According to the EPPO reports, CEVd also occurs in many other African countries such as Cameroon, Ethiopia, Mozambique, Nigeria, Sierra Leone, Somalia (EPPO, 2006), Madagascar, and Mauritius (EPPO, 2007).

HSVd-associated diseases of citrus

The history of the discovery and spread of citrus cachexia (xyloporosis) disease in the Middle East has been reviewed previously (Hadidi et al., 2003; Bar-Joseph, 2015). The disease, which is caused by variants of HSVd (Reanwarakorn and Semancik, 1999), was reported in Egypt, Libya, Tunisia, Algeria, Morocco, and Sudan (Hadidi et al., 2003). In Tunisia, HSVd was widespread and accounted for 72.3% of the citrus-tested trees (Elleuch et al., 2006; Najar et al., 2017). HSVd was also reported in South Africa (EPPO, 1972).

The history and properties of citrus gummy bark (CGB) disease were described by Önelge and Semancik (2004). An HSVd etiology for CGB disease of sweet orange is supported by the similarity of symptom expression to cachexia disease of mandarins and tangelos caused by HSVd, as well as by the detection of variants thereof in CGB-infected Washington navel and the Turkish orange cultivar "Dortyol" (Önelge et al., 2004). In Egypt, CGB was shown to be induced by a variant of HSVd characterized at the molecular level (Sofy and El-Dougdoug, 2014; Sofy et al., 2010, 2012b). In addition, the association of an HSVd variant with gumming and stem pitting on *Citrus volkameriana* rootstock was described, and it shared 100% identity with HSVd variants CVd-IIc (or Ca-905; Sofy and El-Dougdoug, 2014). CGB disease was also reported in Africa, Sudan, and Egypt (Sofy et al., 2012a).

HSVd in other fruit trees and grapevine

In Tunisia, several studies showed that HSVd occurs in fruit trees. The incidence of HSVd in apricot and peach trees was 25.7% and 5%, respectively, while none was found in almond, plum, and cherry (Mahfoudhi et al., 2010). The infection rate of HSVd in peach and pear samples in Egypt was about 93% (El-Dougdoug et al., 2010). In a subsequent study, HSVd infection in several fruit species was about 65% in peach, 40% in pear, 25% in sweet orange, 16% in mango, 10% in mandarin, 7% in apricot, 6% in plum, and 2% in apple (El-Dougdoug et al., 2012a). In grapevine, the HSVd infection rate was 8%–12% (El-Dougdoug et al., 2012a; Nasr-Eldin et al., 2012).

The first report of HSVd in fig and pistachio trees was from Tunisia as a new host species for HSVd (Yakoubi et al., 2007; Elleuch et al., 2013). HSVd was detected in symptomless pomegranate trees from Tunisia with an infection rate of 6% (Gorsane et al., 2010). A recent study on HSVd infecting pistachio showed that HSVd was detected in 50 of the 130 samples (38.4%). The sequences of six HSVd samples showed 95%–100% nucleotide identity with each other and clustered in two of the five distinct groups that compose this species (i.e., plum-hop/citrus and hop groups; (Chouk, 2021)).

HSVd was reported for the first time on the grapevine in Nigeria, with an infection rate reaching 34%. Complete genomes of HSVd derived from Nigeria shared 96% to 100% identity with several HSVd GenBank isolates. No discernible symptoms were associated with HSVd in the source grapevines (Zongoma et al., 2018). In Algeria, HSVd was detected for the first time by high-throughput small RNA sequencing (HTS) on a symptomless grapevine (Eichmeier et al., 2020).

Apscaviroid infecting pome fruit

Pear blister canker viroid (PBCVd), apple dimple fruit viroid (ADFVd), and apple scar skin viroid (ASSVd) are species that infect pome trees. Pear blister canker disease is a bark disorder induced by PBCVd, a member of the genus Apscaviroid within the family *Pospiviroidae* (Fig. 3I). PBCVd was detected only in Tunisia in the Sahel Region in 2.7% of pear trees (Fekih-Hassen et al., 2004). A recent study performed in the Cape Bon region of Tunisia showed an infection rate with PBCVd reaching 25% (unpublished result by Imen Hamdi).

ADFVd was the causal agent of apple dimple fruit disease (Di Serio et al., 2001). Symptoms are roundish and depressed green spots, 3–4mm in diameter, scattered on the fruit skin (Malfitano et al., 2004). ADFVd, reported first in Italy, was later identified in Lebanon, China, and Japan, always in apple trees affected by dimple fruit disease (Di Serio et al., 2017).

ASSVd is the type species of the genus Apscaviroid (family Pospiviroidae). This viroid is the etiological agent of two apple diseases: apple scar skin and dapple apple (Di Serio et al., 2018). ASSVd causes serious yield losses in apples, and affected fruits are significantly downgraded or unmarketable (Di Serio et al., 2017).

Considering the limited information on the presence and incidence of pome fruit viroid diseases in Africa, a preliminary assessment was taken first in Morocco to explore the existence of pome fruit viroid. A total of 168 apples and 81 pears were sampled and tested for pome fruit viroids ASSVd, ADFVd, and PBCVd by tissue printing hybridization. No viroid was detected from Morocco (Afechtal et al., 2010).

Recently, a study was conducted in Tunisia and the preliminary results show that ASSVd and ADFVd were present in Red Delicious cultivars in the Kasserine region (central region) (Fig. 3II) (unpublished data from Imen Hamdi), and more in-depth studies are in progress.

Apscaviroid infecting grapevine

Australian grapevine viroid (AGVd) and grapevine yellow speckle viroid 1 (GYSVd-1) were found in Tunisia, and their genomic nucleotide sequences were determined (Elleuch et al., 2002, 2003). Sequence variability of AGVd was not clustered in any specific domain or region of the genome (Elleuch et al., 2003). Recently, in Algeria, GYSVd-1 was reported for the first time on grapevine using HTS (Eichmeier et al., 2020). In Nigeria, in 2016, surveys were conducted in the Northern Guinea Savannah agroecological zone of Nigeria to document the occurrence of grapevine viruses and viroids. A total of 318 leaf tissue samples belonging to 5 cultivars were collected during the survey across 28 vineyards and analyzed in the laboratory (Zongoma et al., 2018). GYSVd-1, GYSVd-2, and HSVd were detected based on various molecular analyses (Zongoma et al., 2018). Incidence was estimated at 34%, 16%, and 6% for HSVd, GYSVd-1, and GYSVd-2, respectively.

FIG. 3   I: (A) Bark symptoms observed in the field on the local pear variety "Meski Ahrach" grafted on quince Tunisia (Hamdi, I; unpublished). II: (a) Apple orchard in the Kasserine region of Tunisia. (b) Apple orchard showing symptoms of dieback. (c) Swelling and radial cracking of branches. (d) Symptoms caused by ASSVd. *Hamdi., I; unpublished data.*

A recent study on 229 Vitis accessions from the field-maintained South African Vitis germ-plasm collection vineyard showed that five of the seven known grapevine-infecting viroids (HSVd, GYSVd-1, GYSVd-2, AGVd, CEVd, GLVd, and JGVd) were very commonly found (Morgan et al., 2022). HSVd was the most abundant viroid, present in 170 samples (70.4%), followed by GYSVd-1, present in 106 samples (46.3%), then GYSVd-2, present in 74 samples (32.3%), AGVd, present in 19 samples (8.3%), and JGVd, in one sample (0.29%) (this is the first report of JGVd outside Japan) (Morgan et al., 2022). No significant sequence diversity was found for AGVd and GYSVd-2 as previously described, and the four major variants of GYSVd-1 occurring worldwide were identified in this vineyard (Morgan et al., 2022).

Apscaviroid infecting citrus

An extensive study on CDVd from Israel revealed that point mutation and RNA recombination contribute to viroid sequence diversity (Owens et al., 2000). The natural variability of CDVd from Tunisia was also reported (Elleuch et al., 2006). A study on the incidence of CBLVd and CDVd in Tunisia shows that they were only found in 32.7% and 2.3%, respectively, of the sources (Najar and Duran-Vila, 2004); mixed infection with CEVd+CBLVd +CVd-III+CDVd was found in a single source (Najar and Duran-Vila, 2004). CBLVd and CDVd were also detected in citrus budlings originating in Egypt (Al-Harthi et al., 2013).

Citrus viroid V (CVd-V) is among the six citrus viroids belonging to the genus Apscaviroid (family *Pospiviroidae*) which induces symptoms of mild necrotic lesions and cracks (Ali et al., 2022). CVd-V has been detected only in Tunisia in most citrus varieties (Hamdi et al., 2015). The Tunisian variants showed 80%–91% nucleotide identity with other known variants.

Cocadviroid infecting citrus

Citrus bark cracking viroids (CBCVd) was reported in South Africa (Cook et al., 2012). Nucleotide sequences of the South African isolates were identical to reference sequences from the Middle East. CBCVd was reported in Tunisia on citrus (Najar and Duran-Vila, 2004) and recently on pistachio (Chouk, 2021). In Tunisia, CBCVd was detected in pistachio in 50 (38.4%) of the 130 samples. PCR amplicons from three CBCVd-pis infected accessions showed 100% nucleotide sequence identity with US isolate W11 (Chouk, 2021). CBCVd was found only in citrus in Sudan (Mohamed et al., 2009) and Egypt (Al-Harthi et al., 2013).

Occurrence of viroids in herbaceous plants

Vegetables and ornamental plants

Potato spindle tuber viroids (PSTVd) was detected in potato in Egypt (Abdel-Aziz, 2009), Nigeria (Ladipo, 1977), in tomato in Ghana (Batuman et al., 2013, 2019) (Fig. 4I-A), and ornamental plants in Tunisia from a plant nursery (A. Elleuch, unpublished data). PSTVd was also detected for the first time in the seed of uncultivated *Solanum anguivi* (Fig. 4I-B), *Solanum coagulans*, and *S. dasyphyllum* collected from Ghana, Kenya, and Uganda (Skelton et al., 2019). These results confirm the potentials for *Pospiviroids* to be distributed through noncommercial seed.

Tomato chlorotic dwarf viroid (TCDVd) is seed-transmitted in tomato plants (Singh and Dilworth, 2009). The nucleotide sequence has a high identity with variants from Canada and Netherlands (A. Elleuch, unpublished data). In Tunisia, the viroid was detected in 2008 from the ornamental plant, *Pittosporum tobira* showing stunting and dwarfing.

Tomato apical stunt viroid (TASVd) is a severe pathogen of tomato plants. Pathways for introduction include tomato seedlings, tomato seeds, and ornamentals. If spread to tomato, considerable losses could result. No symptoms appear on infected ornamental solanaceous plants, but these plants can act as a reservoir for spreading viroids in tomato production, especially in greenhouse conditions. TASVd outbreaks in tomato are rare, although it has occurred in several

FIG. 4 (I) Symptoms of PSTVd infection. (I-A) On potato tubers of the cv. "Nicola" (spindle tuber of potato). (I-B) On *Solanum anguivi* acknowledge "Plant Genetic Resources Research Institute (PGRRI), Ghana." (II) Symptoms of rasta disease in commercial tomato fields in Ghana and Mali. (II-A) Epinasty, short internodes, and crumpling of leaves (Ghana). (II-B) Stunting and a bushy or rosette appearance and crumpling and down curling of leaves (Ghana). (II-C) Severe stunting, distorted growth, and a bunchy or rosette appearance; chlorosis and leaves curling (Ghana). (II-D) Extreme stunting, chlorosis, and necrosis of a plant, with the production of a few small-sized fruits (Mali). *(I-A): Source: The Netherlands ©National Plant Protection Organization (NPPO)/The Netherlands. (I-B): Source: Skelton, A., Buxton-Kirk, A., Fowkes, A., Harju, V., Forde, S., Ward, R., Fox, A., 2019. Potato spindle tuber viroid detected in seed of uncultivated Solanum anguivi, S. coagulans and S. dasyphyllum collected from Ghana, Kenya and Uganda, New. Dis. Rep. 39, 23–23. https://doi.org/10.5197/j.2044-0588.2019.039.023. (IIA-D): Source: Batuman, O., Çiftçi, Ö.C., Osei, M.K., Miller, S.A., Rojas, M.R., Gilbertson, R.L., 2019. Rasta disease of tomato in Ghana is caused by the Pospiviroids potato spindle tuber viroid and tomato apical stunt viroid. Plant Dis.103, 1525–1535. https://doi.org/10.1094/PDIS-10-18-1751-RE.*

countries in Asia, Africa, and Europe. After severe symptoms and losses in greenhouse-tomato (Antignus et al., 2002, 2007), EPPO listed this viroid on their "EPPO Alert List" (EPPO, 2011). TASVd in tomato plants was reported in Tunisia (Verhoeven et al., 2006), Senegal (Candresse et al., 2007), Ivory Coast (EPPO, 1999), and Ghana (Batuman et al., 2013).

Rasta disease is a virus-like disease with unknown etiology affecting the tomato plant (*Solanum lycopersicum*) in Ghana. Symptoms include stunting, epinasty, crumpling, chlorosis of leaves, and necrosis of leaf vein, petioles stems (see Fig. 4II). The symptomatic plants showed a double infection with PSTVd and TASVd in a host range. PSTVd and TASVd isolates from

Ghana induced rasta symptoms in the highly susceptible tomato cultivar. The PSTVd and TASVd isolates from Ghana had high levels of identity (94%–100%) with previously characterized isolates from other geographical regions and were placed in phylogenetic clades with some of these isolates. This is suggestive of long-distance dissemination. Most likely in tomato seed. The results of the host range experiment confirmed that PSTVd and TASVd isolates from Ghana cause rasta disease in the highly susceptible tomato "Early Pak 7." This finding supports the hypothesis that these viroids can cause this disease in Ghana (Batuman et al., 2019).

Columnea latent viroid (CLVd) was reported in Africa only in tomatoes in Mali (Batuman and Gilbertson, 2012). In host range studies, the CLVd isolate from Mali induced symptoms in all 48 mechanically inoculated tomato plants, whereas no symptom developed (up to 90 days after inoculation) in different indicator plants; thus, it is possible that CLVd was introduced to Mali in association with seed.

Chrysanthemum stunt viroid (CSVd) is the causal agent of a major disease in cultivated chrysanthemum and is distributed worldwide. CSVd isolates show considerable variation in sequence within and between countries of isolation and produce a variety of symptoms on chrysanthemum plants, determined largely by the cultivar and the environment. CSVd in chrysanthemum was also reported in South Africa (Watermeyer, 1984) and Egypt (El-Dougdoug et al., 2012b) from the naturally infected plant in the greenhouse.

Chrysanthemum chlorotic mottle viroid (CChMVd) is the causal agent of chrysanthemum chlorotic mottle (CChM) disease (Navarro and Flores, 1997) inducing leaf mottling and chlorosis, delay in flowering, and dwarfing (Flores et al., 2017). Chrysanthemum is the only natural and experimental host for CChMVd (Flores et al., 2017). CChMVd was recently reported in South Africa for the first time on *Chrysanthemum* × *morifolium* cultivars (Read et al., 2022). Samples were collected from plants showing vein clearing, mosaic, and chlorosis, in the Western Cape province (Read et al., 2022). In South Africa, chrysanthemum represents one of the most important cut-flower exports in the country (Read et al., 2022). New CChMVd isolates from South Africa shared 99% nucleotide sequence similarity with isolate CMNS35 (AJ247121) (Read et al., 2022).

HSVd infects a broad range of natural hosts. It has been reported to be the causal agent of five different diseases (citrus cachexia, cucumber pale fruit, peach and plum dapple fruit, and hop stunt). HSVd is transmitted mechanically and by seed materials such as scions or cuttings. It is distributed worldwide.

Procedure

In Tunisia, the procedure to follow for citrus viroids detection is as follows:

Field surveys were carried out in the spring. Leaf samples were collected from around the canopy of each tree and stored at 4°C until assayed for viruses or viroid.

Biological indexing was performed using Etrog citron Arizona 861-S1 (*Citrus medica* L.) grafted on rough lemon (*Citrus jambhiri* Lush.) as the biological indicator species (Roistacher et al., 1977). Each sample was graft-inoculated and maintained in a greenhouse under controlled temperature (28–32°C) and symptoms were observed 2–6 months after inoculation. CEVd induces "Bark scaling" and bumps. HSVd induces stunting and epinasty.

B. Prospecting for viroid

To determine if citrus viroid RNAs were present in the tested trees, total nucleic acid (TNA) preparations from the inoculated citron were obtained following the silica extraction (Foissac et al., 2005). The purified nucleic acids were analyzed by sequential polyacrylamide gel electrophoresis (sPAGE) (Duran-Vila et al., 1993). To confirm the identity of the viroid bands observed by sPAGE analysis, the nucleic acid preparations were further subjected to dot-blot or northern-blot hybridization using viroid-specific probes for CEVd, CBLVd, HSVd, CDVd, CBCVd, and CVd-V.

After confirming the presence of viroid by hybridization methods, two-step RT-PCR was performed using specific primers. For example, in the case of CEVd and HSVd, we used specific primers described by Semancik et al. (1993) and Koflavi et al. (1997), respectively: CEVdS (5′ CCGGGGATCCCTGAAGAAC 3′)/CEVdAS (5′ GGAAACCTGGAGGAAGTCG3′); VP-19 (5′   GCCCCGGGGCTCCTTTCTCAGGTAAG-3′)/VP-20   (5′-CGCCCGGGGCAACTCTTCTCAGAATCC-3′). The RT-PCR products were determined by 2% agarose gel electrophoresis. Samples representing the size ~371bp for CEVd and ~300bp for HSVd were removed from the gel, ligated to pGEM-T Easy vector (Promega, USA), and cloned into *E. coli*. After that, the positive clones were sequenced.

Prospective

More research is needed on viroids in Africa in the coming years, given the number of countries that have not been prospected until now. Africa seems to have great potential in relation to the presence of viruses and viroids. Surveys must be carried out first on the continent's strategic crops, then on ornamental plants, which can constitute a reservoir for several viroids species.

Future implications

The improvement of scientific research on viroids in Africa will have a very beneficial impact on agricultural production in the African continent, and the research on viroids will make it possible to avoid an epidemic. Epidemics like observed in the Philippines where every year one million coconut palms are killed by CCCVd and over 30 million coconut palms have been killed since Cadang-cadang was discovered. Surveys and fundamental research on viroid (Recombination, evolution, new strain) can help the government to establish new rules to preserve strategic crops and set up new certification schemes for fruit trees as examples, and control the transfer of plant material between different countries.

Chapter summary

This chapter describes major viroids and viroid diseases occurring and reported in the African continent to the best of our knowledge. Twenty-two African countries are affected by at least one viroid species. All members of the family *Avsunviroidae* except ELVd and CChMVd have been recorded in Africa in six countries (Egypt, Tunisia, Morocco, Algeria, South Africa, and Ghana). Pospiviroidae were recorded in 22 countries. It is important to note that Tunisia has the largest number of viroids ever reported in Africa, with 14 species of viroids, two belonging to *Avsunviroidae* (AHVd and PLMVd) and 12 belonging to the *Pospiviroidae* family. Viroids are known to infect up to 20 plant species in African countries naturally. The summary of occurrence country and host plants is listed in Table 1. The spread and the impact of certain viroid in some countries are discussed. Unfortunately, few African countries have established rules and regulations to stop the spread of viroids causing economic damage like European countries. Only Tunisia, Morocco, and Algeria are members of the EPPO organization.

This chapter shows that viroids are widespread in almost all the regions of Africa, North, South, West, and East. As this continent is known for hot temperatures in certain regions and these conditions are optimal for viroids multiplication and propagation. It should be interesting to make a more intense study in this area with the aim to find new viroids species and new hosts. In Tunisia, the work on viroids seems well advanced; several viroids are characterized and introduced into the certification schemes of the majority of fruit trees.

It is important to point out that central Africa seems to be the region least inspected against viroids. For this purpose, it is preferable to establish a network between the African countries to follow the presence and the evolution of viroids in Africa.

Study questions

1. In Africa, how are quarantine organisms controlled?
2. What type of ornamental plant is most affected by viroid diseases in Africa?

TABLE 1 Summary of several viroids reported in Africa.

Viroid	Countries	Host
Apple hammerhead viroid (AHVd)	Tunisia	Apple
Apple dimple fruit viroid (ADFVd)	Tunisia	Apple
Apple scar skin viroid (ASSVd)	Tunisia	Apple
Australian grapevine viroid (AGVd)	Tunisia; South Africa	Grapevine
Avocado sunblotch viroid (ASBVd)	Ghana, South Africa	Avocado
Chrysanthemum stunt viroid (CSVd)	Egypt, South Africa	Chrysanthemum

Continued

B. Prospecting for viroid

TABLE 1   Summary of several viroids reported in Africa—cont'd

Viroid	Countries	Host
Citrus bark cracking viroid (CBCVd)	Sudan, South Africa, Tunisia, Egypt	Citrus, Pistachio
Citrus bent leaf viroid (CBLVd)	Egypt, Tunisia	Citrus
Citrus dwarfing viroid (CDVd)	Egypt, Tunisia	Citrus
Citrus exocortis viroid (CEVd)	Algeria, Egypt, Ghana, Libya, Morocco, South Africa, Sudan, Tunisia, Cameroon, Ethiopia, Madagascar, Mozambique Nigeria, Réunion, Sierra Leone, Somalia	Citrus, tomato, fig
Citrus viroid V (CVd-V)	Tunisia	Citrus
Columnea latent viroid (CLVd)	Mali	Columnea
Grapevine yellow speckle viroid 1 (GYSVd-1)	Tunisia; Algeria; Nigeria; South Africa	Grapevine
Grapevine yellow speckle viroid 2 (GYSVd-2)	Ghana; Nigeria; South Africa	Grapevine
Hop stunt viroid (HSVd)	Algeria, Egypt, Libya, Morocco, Sudan, Tunisia; Nigeria; South Africa	Citrus, several other fruit trees, grapevine, and ornamental plants
Peach latent mosaic viroid (PLMVd)	Egypt, Tunisia, Algeria, Morocco	Peach, several other fruit trees (Peach, Apricot)
Pear blister canker viroid (PBCVd)	Tunisia	Pear, Quince
Potato spindle tuber viroid (PSTVd)	Egypt, Ghana, Nigeria, Tunisia, Uganda, Kenya	Potato, tomato, ornamental plants
Tomato apical stunt viroid (TASVd)	Ghana, the Ivory Coast, Senegal, Tunisia	Tomato, ornamental plants
Tomato chlorotic dwarf viroid (TCDVd)	Tunisia	Tomato, ornamental plants
Japanese grapevine viroid (JGVd)	South Africa	Grapevine
Chrysanthemum chlorotic mottle viroid (CChMVd)	South Africa	Chrysanthemum

B. Prospecting for viroid

3. What are the main emerging viroid diseases occurring in Africa?
4. What is economically the most important viroid of fruit trees in Africa?
5. Is it true that ELVd and CChMVd are present in Africa?

Further reading

EFSA Plant Health Panel, Bragard, C., Dehnen-Schmutz, K., Gonthier, P., Jacques, M.-A., JaquesMiret, J.A., Justesen, A.F., MacLeod, A., Magnusson, C.S., Milonas, P., Navas-Cortes, J.A., Parnell, S., Potting, R., Reignault, P.L., Thulke, H.-H., Van derWerf, W., VicentCivera, A., Yuen, J., Zappala' L., Candresse, T., Chatzivassiliou, E., Finelli, F., Winter, S., Chiumenti, M., Di Serio, F., Kaluski, T., Minafra, A., Rubino, L. 2019. Scientific opinion on the pest categorisation of non-EU viruses and viroids of *Cydonia* Mill., *Malus* Mill. and *Pyrus* L. EFSA J., 17(9), 5590, 81 pp. https://doi.org/10.2903/j.efsa.2019 5590.
Adkar-Purushothama CR, Perreault J-P. Current overview on viroid–host interactions. WIREs RNA. 2020;11:e1570. https://doi.org/10.1002/wrna.1570
Khaled A. El-Dougdoug; Kadriye Çağlayan; Amine Elleuch; Hani Z. Al-Tuwariqi; Ebenezer A. Gyamera; Ahmed Hadidi. Chapter 45: Geographical Distribution of Viroids in Africa and the Middle East. https://www.sciencedirect.com/science/article/pii/B9780128014981000450.

Acknowledgment

A special acknowledgment to "Plant Genetic Resources Research Institute (PGRRI), Ghana" for providing photos corresponding to symptoms of PSTVd infection on Solanum anguivi. Special thanks to Dr. Jose Ramon Saucedo Carabez (Departamento de Investigación Aplicada-Driscolls, Libramiento, Jacona, Michoacán, Mexico) for providing photos of ASBVd symptoms on Avocado. We thank all the scientists and editors for their permission to use photos.

References

Abdel-Aziz, S.H., 2009. Molecular analysis of genetic variation of potato cultivars infected with viroid. Aus. J. B. A. Sci. 3, 4293–4301.
Acheampong, A.K., Akromah, R., Ofori, F.A., Takrama, J.F., Zeidan, M., 2008. Is there avocado sunblotchviroid in Ghana? Afr. J. Biotechnol. 7, 3540–3545.
Afechtal, M., Djelouah, K., D' Onghia, A.M., 2010. The first survey of pome fruit viruses in Morocco. In: 21st International Conference on Virus and Other Graft Transmissible Diseases of Fruit Crops. Julius-Ktihn-Archiv. 427, pp. 253–256.
Al-Harthi, A., Al-Sadi, A.M., Al-Saady, A.A., 2013. Potential of citrus budlings originating in the Middle East as sources of citrus viroids. Crop Prot. 48, 13–15. https://doi.org/10.1016/j.cropro.2013.02.006.
Ali, A., Umar, U.U., Akthar, S., Shakeel, M.T., Tahir, M.N., Atta, S., Ölmez, F., Parveen, R., 2022. Prevalence, symptomology, detection and molecular characterization of citrus viroid V infecting new citrus cultivars in Pakistan. Res. Square. https://doi.org/10.21203/rs.3.rs-1310164/v1.
Antignus, Y., Lachman, O., Pearlsman, M., Gofman, R., Bar-Joseph, M., 2002. A new disease of greenhouse tomatoes in Israel caused by a distinct strain of tomato apical stunt viroid (TASVd). Phytoparasitica 30, 502–510.
Antignus, Y., Lachman, O., Pearlsman, M., 2007. Spread of tomato apical stunt viroid (TASVd) in greenhouse tomato crops is associated with seed transmission and bumble bee activity. Plant Dis. 91, 47–50. https://doi.org/10.1094/PD-91-0047.

Bar-Joseph, M., 2015. Xyloporosis: A history of the emergence and eradication of a citrus viroid disease. IOCV J. Cit. Pathol. 2, 27202. https://doi.org/10.5070/C421027202.

Batuman, O., Gilbertson, R.L., 2012. First report of columnea latent viroid (CLVd) in tomato in Mali. Plant Dis. 97, 692. https://doi.org/10.1094/PDIS-10-12-0920-PDN.

Batuman, O., Osei, M.K., Mochiah, M.B., Lamptey, J.N., Miller, S., Gilbertson, R.L., 2013. The first report of tomato apical stunt viroid (TASVd) and potato spindle tuber viroid (PSTVd) in tomatoes in Ghana. Phytopathology 103 (Suppl 2), 12. http://hdl.handle.net/123456789/1216.

Batuman, O., Çiftçi, O.C., Osei, M.K., Miller, S.A., Rojas, M.R., Gilbertson, R.L., 2019. Rasta disease of tomato in Ghana is caused by the Pospiviroids potato spindle tuber viroid and tomato apical stunt viroid. Plant Dis. 103, 1525–1535. https://doi.org/10.1094/PDIS-10-18-1751-RE.

Bibi, I., Kharmach, E.Z., Chafik, Z., Ben Yazid, J., Afechtal, M., 2020. Incidence of Citrus exocortis viroid and hop stund viroid in commercial citrus groves from Morocco. Moroccan J. Agri. Sci. 1, 142–145.

Candresse, T., Marais, A., Ollivier, F., Verdin, E., Blancard, D., 2007. First report of the presence of tomato apical stunt viroid on tomato in Senegal. Plant Dis. 91, 330. https://doi.org/10.1094/PDIS-91-3-0330C.

Chouk, G., 2021. The presence of virus and virus-like diseases of pistachio (*Pistacia vera*) in a varietal collection plot (Tunisia). Master of science: Precision Integrated Pest Management (IPM) for fruit and vegetable crops http://www.secheresse.info/spip.php?article116054.

Cook, G., van Vuuren, S.P., Breytenbach, J.H.J., Manicom, B.Q., 2012. Citrus viroid IV detected in *Citrus sinensis* and *C. reticulata* in South Africa. Plant Dis. 96, 772. https://doi.org/10.1094/PDIS-11-11-0951-PDN.

Da Graca, J.V., van Vuuren, S.P., 2003. Viroids in Africa. In: Hadidi, A., Flores, R., Randles, J.W., Semancik, J.S. (Eds.), Viroids. CSIRO Publishing, Collingwood, VIC, Australia, pp. 290–292.

Di Serio, F., Malfitano, M., Ragozzino, A., Desvignes, J.C., Flores, R., 2001. Apple dimple fruit viroi: fulfillment of Koch postulates and symptom characteristics. Plant Dis. 85, 179–182. https://doi.org/10.1094/PDIS.2001.85.2.179.

Di Serio, F., Torchetti, E.M., Flores, R., Sano, T., 2017. Chapter 22: Other Apscaviroids infecting pome fruit trees. In: Viroids and Satellites. Academic Press, pp. 229–241. ISBN 9780128014981 https://doi.org/10.1016/B978-0-12-801498-1.00022-X.

Di Serio, F., Ambrós, S., Sano, T., Flores, R., Navarro, B., 2018. Viroid diseases in pome and stone fruit trees and Koch's postulates: a critical assessment. Viruses 10, 612. https://doi.org/10.3390/v10110612.

Duran-Vila, N., Roistacher, C.N., Rivera-Bustamante, R., Semancik, J.S., 1988. A definition of citrus viroid groups and their relationship to the exocortis disease. J. Gen. Virol. 69, 3069–3080.

Duran-Vila, N., Pina, J.A., Navarro, L., 1993. Improved indexing of citrus viroids. In: Moreno, P., da Graca, J.V., Timmer, L.W. (Eds.), Proceedings 12th Conference of the International Organization of Citrus Virologist (IOCV). Riverside, CA, USA, pp. 202–221.

Eichmeier, A., Peňázová, E., Čechová, J., Berraf-Tebbal, A., 2020. Survey and diversity of grapevine pinot gris virus in Algeria and comprehensive high-throughput small RNA sequencing analysis of two isolates from *Vitis vinifera* cv. Sabel revealing high viral diversity. Genes 11, 1110. https://doi.org/10.3390/genes11091110.

El-Dougdoug, K.A., Osman, M.E., Hayam, S.A., Rehab, D.A., Elbaz, R.M., 2010. Biological and molecular detection of HSVd-infecting peach and pear trees in Egypt. Austr. J. Basic Appl. Sci. 4, 19–26.

El-Dougdoug, K.A., Rehab, D.A., Rezk, A.A., Sofy, A.R., 2012a. Incidence of fruit trees viroid diseases by tissue print hybridization in Egypt. Int. J. Virol. 8, 114–120. https://doi.org/10.3923/ijv.2012.114.120.

El-Dougdoug, K.A., Rezk, A.A., Rehab, D.A., Sofy, A.R., 2012b. Partialnucleotide sequence and secondary structure of chrysanthemum stunt viroid Egyptian isolate from infected-chrysanthemum plants. Int. J. Virol. 8, 191–202. https://doi.org/10.3923/ijv.2012.191.202.

Elleuch, A., Fakhfakh, H., Pelchat, M., Landry, P., Marrakchi, M., Perreault, J.-P., 2002. Sequencing of Australian grapevine viroid and yellow speckle viroid I isolated from a Tunisian grapevine without passage in an indicator plant. Eur. J. Plant Pathol. 108, 815–820.

Elleuch, A., Perreault, J.-P., Fakhfakh, H., 2003. First report of Australian grapevine viroid from the Mediterranean region. J. Plant Pathol. 85, 53–57. https://www.jstor.org/stable/41998119.

Elleuch, A., Djilani-Khouaja, F.I., Bsais, N., Perreault, J.P., Marrakchi, M., Fakhfakh, H., 2006. Sequence analysis of three citrus viroids infecting a singleTunisian citrus tree (Citrusreticulata, Clementine). Gen. Mol. Biol. 29, 705–710.

Elleuch, A., Hamdi, I., Ellouze, O., Ghrab, M., Fakhfakh, H., Drira, N., 2013. Pistachio (*Pistacia vera* L.) is a new natural host of hop stunt viroid. Virus Genes 47, 330–337. https://doi.org/10.1007/s11262-013-0929-8.

EPPO, 1972. PQR Database. European and Mediterranean Plant Protection Organization, Paris, France. http://www.eppo.int/DATABASES/pqr/pqr.htm.

B. Prospecting for viroid

EPPO, 1993. PQR database. European and Mediterranean Plant Protection Organization, Paris, France. http://www.eppo.int/DATABASES/pqr/pqr.htm.

EPPO, 1999. PQR Database. European and Mediterranean Plant Protection Organization, Paris, France. http://www.eppo.int/DATABASES/pqr/pqr.htm.

EPPO, 2006. PQR database. European and Mediterranean Plant Protection Organization, Paris, France. http://www.eppo.int/DATABASES/pqr/pqr.htm.

EPPO, 2007. PQR database. European and Mediterranean Plant Protection Organization, Paris, France. http://www.eppo.int/DATABASES/pqr/pqr.htm.

EPPO, 2011. Alert List: Tomato apical stunt pospiviroid. Updated 26 August 2011 http://www.eppo.org/QUARANTINE/Alert_List/viruses/TASVD0.htm.

Fekih-Hassen, I., Kummert, J., Marbot, S., 2004. First report of pear blister canker viroid, peach latent mosaic viroid, and hop stunt viroid Infecting fruit trees in Tunisia. Plant Dis. 88, 1164.

Flores, R., Gago-Zacher, S., Serra, P., De la Peña, M., Navarro, B., 2017. Chapter 31: Chrysanthemum chlorotic mottle viroid. In: Viroids and Satellites. Academic Press, ISBN: 9780128014981, pp. 331–338, https://doi.org/10.1016/B978-0-12-801498-1.00031-0.

Gorsane, F., Elleuch, A., Hamdi, I., Salhi-Hannachi, A., Fakhfakh, H., 2010. Molecular detection and characterization of hop stunt viroid sequence variants from naturally infected pomegranate (Punica granatum L.) in Tunisia. Phytopathol. Mediterr. 49, 152–162. https://doi.org/10.14601/Phytopathol_Mediterr-3430.

Hadidi, A., Mazyad, H.H., Madkour, M.A., Bar-Joseph, M., 2003. Viroids in the Middle East. In: Hadidi, A., Flores, R., Randles, J.W., Semancik, J.S. (Eds.), Viroids. CSIRO Publishing, Collingwood, VIC, Australia, pp. 275–278.

Hamdi, I., Elleuch, A., Bessaies, N., Grubb, C.D., Fakhfakh, H., 2015. First report of citrus viroid V in North Africa. J. Gen. Plant Pathol. 81, 87. https://doi.org/10.1007/s10327-014-0556-9.

Hamdi, I., Soltani, R., Baraket, G., et al., 2022. First report of apple hammerhead viroid infecting 'Richard Delicious' apple (Malus domestica) in Tunisia. J. Plant Pathol. 104, 811–812. https://doi.org/10.1007/s42161-022-01027-z.

Ito, T., Namba, N., Ito, T., 2003. Distribution of citrus viroids and apple stem grooving virus on citrus trees in Japan using multiplex reverse transcription polymerase chain reaction. J. Gen. Plant Pathol. 69, 205–207. https://doi.org/10.1007/s10327-002-0036-5.

Jooste, A.E.C., Zwane, Z.R., 2018. Report on avocado sunblotch viroid studies in South Africa. In: South African Avocado Growers' Association Yearbook. 41, pp. 106–109.

Koflavi, S., Marcos, J., Cañizares, C., Pallás, V., Candresse, T., 1997. Hop stunt viroid (HSVd) sequence variant from prunus species: evidence for recombination between HSVd isolates. J. Gen. Virol. 78, 3177–3186.

Ladipo, J.L., 1977. Seed transmission of cowpea aphid-borne virus in some cowpea cultivars. Nigerian J. Plant. Prot. 3, 3–10.

Mahfoudhi, N., Salleh, W., Djelouah, K., 2010. First report of hop stunt viroid in apricot in Tunisia. J. Plant Pathol. 92 (Supp 4), 116. https://www.jstor.org/stable/41998907.

Malfitano, M., Alioto, D., Ragozzino, A., Di Serio, F., Flores, R., 2004. Experimental evidence that Apple dimple fruit viroid does not spread naturally. Acta Hortic. 657, 357–360. https://doi.org/10.17660/ActaHortic.2004.657.56.

Meziani, S., Rouag, N., Milano, R., Kheddam, M., Djelouah, K., 2010. Assessment of the main stone fruit viruses and viroids in Algeria. In: 21st International Conference on Virus and Other Graft Transmissible Diseases of Fruit Crops (2009-07-05/10, Neustadt, DE). Julius-Kühn-Archiv. 427, pp. 289–292.

Mohamed, M.E., Hashemian, S.M.B., Dafalla, G., Bove, Duran-Vila, N., 2009. Occurrence and identification of citrus viroids from Sudan. J. Plant Pathol. 91, 185–190.

Morgan, S.W., Read, D.A., Burger, J.T., Pietersen, G., 2022. Diversity of viroids infecting grapevines in South Africa Vitis germplasm collection. Virus Genes 59 (2), 244–253. https://doi.org/10.21203/rs.3.rs-1742186/v1.

Najar, A., Duran-Vila, N., 2004. Viroid prevalence in Tunisian citrus. Plant Dis. 88, 1286. https://doi.org/10.1094/PDIS.2004.88.11.1286B.

Najar, A., Hamdi, I., Varsani, A., Duran-Vila, N., 2017. Citrus viroids in Tunisia, prevalence and molecular characterization. J. Plant Pathol. 99, 781–786. https://doi.org/10.4454/jpp.v99i3.3989.

Nasr-Eldin, M.A., El-Dougdoug, K.A., Othman, B.A., Ahmed, S.A., Abdel-Aziz, S.H., 2012. Three viroids frequency naturally infecting grapevine in Egypt. Int. J. Virol. 8, 1–13. https://doi.org/10.3923/ijv.2012.1.13.

Navarro, B., Flores, R., 1997. Chrysanthemum chlorotic mottle viroid: unusual structural properties of a subgroup of self-cleaving viroids with hammerhead ribozymes. Proc. Natl. Acad. Sci. U. S. A. 94 (21), 11262–11267. https://doi.org/10.1073/pnas.94.21.11262.

Önelge, N., Çinar, A., Szychowski, J.S., Vidalakis, G., Semancik, J.S., 2004. Citrus viroid II variants associated with 'Gummy Bark' disease. Eur. J. Plant Pathol. 110, 1047–1052. https://doi.org/10.1007/s10658-004-0815.

B. Prospecting for viroid

Opoku, A.A., 1972. Incidence of exocortis virus disease of citrus in Ghana. Ghana J. Agric. Sci. 5, 65–71.

Owens, R.A., Yang, G., Gundersen-Rindal, D., Hammond, R.W., Candresse, T., Bar Joseph, M., 2000. Both point mutation and RNA recombination contribute to citrus viroid III sequence diversity. Virus Genes 20, 243–252. https://doi.org/10.1023/a:1008144712837.

Read, D.A., Pietersen, G., Slippers, B., et al., 2022. Chrysanthemum virus B and chrysanthemum chlorotic mottle viroid infect chrysanthemum in South Africa. Australasian Plant Dis. Notes 17, 29. https://doi.org/10.1007/s13314-022-00478-8.

Reanwarakorn, K., Semancik, J.S., 1999. Correlation of hop stunt viroid variants to cachexia and xyloporrosis diseases of citrus. Phytopathology 89, 568–574. https://doi.org/10.1094/PHYTO.1999.89.7.568.

Roistacher, C.N., Calavan, E.C., Navarro, L., Gonzalez, L., 1977. A new more sensitive indicator for detection of mild isolates of exocortis virus in dwarfed citrus trees citrus exocortis viroid (CEV). Plant Dis. Rep. 61, 135–139.

Rouag, N., Guechi, A., Matic, S., Myrta, A., 2008. Viruses and viroids of stone fruits in Algeria. J. Plant Pathol. 90, 393–395.

Semancik, J.S., Szychowski, J.A., Rakowski, A.G., Symons, R.H., 1993. Isolates of citrus exocortis viroid recovered by host and tissue selection. J. Gen. Virol. 74, 2427–2436.

Singh, R.P., Dilworth, A.D., 2009. Tomato chlorotic dwarf viroid in the ornamental plant *Vinca minor* and its transmission through tomato seed. Eur. J. Plant Pathol. 123, 111–116. https://doi.org/10.1007/s10658-008-9344-8.

Skelton, A., Buxton-Kirk, A., Fowkes, A., Harju, V., Forde, S., Ward, R., Fox, A., 2019. Potato spindle tuber viroid detected in seed of uncultivated Solanumanguivi, *S. coagulans* and *S. dasyphyllum* collected from Ghana, Kenya and Uganda. New. Dis. Rep. 39, 23. https://doi.org/10.5197/j.2044-0588.2019.039.023.

Sofy, A.R., El-Dougdoug, K.A., 2014. First record of a hop stunt viroid variant associated with gumming and stem pitting on Citrus volkamerianatrunk rootstock in Egypt. New. Dis. Rept. 30, 11. https://doi.org/10.5197/j.2044-0588.2014.030.011.

Sofy, A.R., Soliman, A.M., Mousa, A.A., Ghazal, S.A., El-Dougdoug, K.A., 2010. First record of citrus viroid II (CVd-II) associated with gummy bark disease in sweet orange (*Citrus sinensis*) in Egypt. New. Dis. Rep. 21, 24. https://doi.org/10.5197/j.2044-0588.2010.021.024.

Sofy, A.R., Mousa, A.A., Soliman, A.M., El-Dougdoug, K.A., 2012a. The limiting of climatic factors and predicting of suitable habitat for citrus gummy bark disease occurrence using GIS. Int. J. Virol. 8, 165–177. https://doi.org/10.3923/ijv.2012.165.177.

Sofy, A.R., Soliman, A.M., Mousa, A.A., El-Dougdoug, K.A., 2012b. Molecular characterization and bioinformatics analysis of viroid isolate associated with citrus gummy bark disease in Egypt. Int. J. Virol. 8, 133–150. https://doi.org/10.3923/ijv.2012.133.150.

Verhoeven, J.Th.J., Jansen, C.C.C., Roenhorst, J.W., 2006. First report of Tomato apical stunt viroid in tomato in Tunisia. Plant Dis. 90, 528.

Watermeyer, S.R., 1984. Detection of chrysanthemum stunt viroid in South Africa by polyacrylamide gel electrophoresis and bioassays. Plant Dis. 68, 485–488.

Yakoubi, S., Elleuch, A., Besaies, N., Marrakchi, M., Fakhfakh, H., 2007. First report of hop stunt viroid and citrus exocortis viroid on fig with symptoms of fig mosaic disease. J. Phytopathol. 155, 125–128. https://doi.org/10.1111/j.1439-0434.2007.01205.x.

Zhang, Z., Qi, S., Tang, N., Zhang, X., Chen, S., et al., 2014. Discovery of replicating circular RNAs by RNA-Seq and computational algorithms. PLoS Pathog. https://doi.org/10.1371/journal.ppat.1004553.

Zongoma, A.M., Dangora, D.B., Al Rwahnih, M., Bako, S.P., Alegbejo, M.D., Alabi, O.J., 2018. First report of grapevine yellow speckle viroid 1, grapevine yellow speckle viroid 2, and hop stunt viroid infecting grapevines (Vitis spp.) in Nigeria. Plant Dis. 102, 259. https://doi.org/10.1094/pdis-07-17-1133-pdn.

Finding the coconut cadang-cadang and tinangaja viroids, naturally occurring pathogens of tropical monocotyledons of Oceania

J.W. Randles[a], C.A. Cueto[b], G. Vadamalai[c], and D. Hanold[a]

[a]School of Agriculture, Food and Wine, The University of Adelaide, Adelaide, SA, Australia
[b]Philippine Coconut Authority-Albay Research Center, Guinobatan, Albay, Philippines
[c]Department of Plant Protection, Faculty of Agriculture, Institute of Plantation Studies, Universiti Putra Malaysia, Serdang, Selangor, Malaysia

Graphical representation

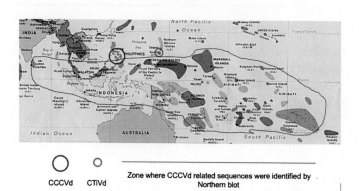

CCCVd CTiVd   Zone where CCCVd related sequences were identified by Northern blot

Definitions

Biological indicator: a plant experimentally infected by a viroid (or other pathogens) as a secondary host displaying characteristic symptoms that can be specifically linked to that pathogenic agent and sometimes even to the variants of that agent.
High-pressure injection: use of a hand-held needle-free jet injector which delivers a small dose of liquid inoculum via a fine orifice at a very high local pressure, leading to subepidermal intercellular infiltration of plant tissue.
Razor slashing: application of inoculum by wetting a stainless steel blade with inoculum and repeatedly cutting the cuticle and epidermis.

Chapter outline

The devastating cadang-cadang epidemic of the coconut palm (*Cocos nucifera*) in the Philippines required the development of a program to identify the cause. The pursuit of a "virus" hypothesis led to the identification of disease-associated small RNA, which was found to have molecular properties typical of the recently recognized viroids, and named coconut cadang-cadang viroid (CCCVd). Uniquely infecting monocotyledonous species, its mode of natural spread and epidemiology are yet to be determined.

This chapter summarizes the internationally supported studies done in the Philippines and Guam where CCCVd and coconut tinangaja viroid (CTiVd) occur, respectively. The discovery that CCCVd is common in oil palm plantations has led to a research and development program in Malaysia. Surveys in Oceania and elsewhere have identified CCCVd-related sequences in coconut and other monocotyledons, which might provide evidence for a source of CCCVd. It is suggested that existing representatives of early plant life should be analyzed for viroid-related sequences, in support of Diener's speculation that viroids are fossils of an RNA world.

Learning objectives

On completion of this chapter, you should be able to:

- Understand how coconut cadang-cadang was identified as a viroid disease.
- Understand the experimental approach and the research programs that implicated CCCVd as the causal pathogen of coconut cadang-cadang disease.
- Explain why the mode(s) of the spread of CCCVd need to be determined.
- Suggest the types of experiments that should be done using oil palm as an experimental model.

Fundamental introduction

Viroids recorded in the tropical zone of Oceania have been listed by Geering and Randles (2012) and their geographical distribution in Oceania and the neighboring Asian region has been reviewed (Geering, 2017; Parakh et al., 2017). All except the known palm-infecting viroids have been introduced to the region with their commercially important dicotyledonous host species and hence their geographical distribution and epidemiology can be traced to introductions of vegetative propagules and seed before the adoption of secure quarantine screening procedures. These include viroids in avocado, chrysanthemum, citrus, grapevine, hops, peach, pear, potato, and tomato (Geering, 2017).

In contrast, the palm-infecting viroids occur naturally in the region, having caused major epidemics in coconut palm plantations in the Philippines and Guam, and economic losses in oil palm (Rodriguez et al., 2017; Anuar and Ali, 2022). The first evidence of their existence was obtained in 1973, following the discovery of PSTVd and CEVd, when low molecular weight RNAs were associated with cadang-cadang disease (Randles, 1975a). The palm viroids differ from all other known viroids in many ways. They are the only viroids known to be pathogenic to monocotyledonous species (Hadidi et al., 2017; Yang et al., 2020), they are lethal in their country of first detection, spread by unknown means to produce epidemics, and their only known hosts cannot be grafted or easily clonally propagated. The coconut cadang-cadang viroid (CCCVd), in its minimum infectious form comprising 246 nt, is the smallest known infectious pathogen. It is classified as a viroid in the genus *Cocadviroid* in the *Pospiviroidae* together with coconut tinangaja viroid (CTiVd), and both have molecular properties in common with the dicotyledon-infecting viroids in this genus (citrus bark cracking viroid and hop latent viroid). Despite these similarities, their unique ecologies have left many questions about their origin, evolution, modes of spread, and replication cycle.

The need to control diseases drives the discovery of pathogens. In viroidology, the economic impact of potato spindle tuber disease led to the discovery of the first known viroid, PSTVd (Diener, 2001). Similarly, the impact of cadang-cadang disease led to the international funding of attempts to determine its etiology and the discovery of firstly, CCCVd, then CTiVd (Rodriguez et al., 2017). This chapter describes the experimental approach and the research programs that implicated these viroids as plant pathogens. It is presented as a history because the discoveries were determined by the circumstances and facilities available and developed at the time and illustrates the phases of disease recognition, isolation of putative causal agents, proof of pathogenicity, and determination of epidemiology required to eventually develop a management scheme for a plant disease. Future research goals are mentioned, such as addressing the source of pathogenic viroids in monocotyledons using structural and sequence properties characteristic of viroids.

A short history of cadang-cadang disease

The cadang-cadang disease of coconut palm was regarded as the most serious threat to the coconut industry in the Philippines from the 1940s (Rillo and Rillo, 1981). By the early 1950s, certain symptoms had been described that were constantly associated with the lethal disease that was spreading within the central Philippines region. In 1949, a national research station then known as Guinobatan Experimental Station (GES) at Guinobatan, Albay, was tasked to deal primarily with the cadang-cadang disease. Early advice came from the US Government and FAO-UN supported by experts and equipment (Anonymous, 1982). By 1971, projects on the symptoms, epidemiology, relationship between palm age and disease incidence, fertilizer studies, and selection of disease-resistant coconut palms had been established. Hypotheses on causation such as nutrition, toxicities, volcanic eruptions, and radioactive fallout existed. Little progress had been achieved in determining the cause, as the disease had not been transmitted artificially, the mode of field spread was unknown, and no control methods were available. Any search for genetic resistance was dependent on natural infections, and even if it existed, a single selection cycle could take 20–50 years and many such cycles would be needed to achieve an acceptable level of resistance. It was considered that there was no solution to the cadang-cadang problem.

Under a national coconut research and development project designed to replant aging coconut plantations with modern varieties and hybrids, using improved fertilizer and agronomic regimes, and assisted by the FAO, UNDP, and the University of Adelaide, many breakthroughs in cadang-cadang disease were achieved in the period 1971–82 (Rillo and Rillo, 1981; Anonymous, 1982; Rodriguez et al., 2017).

These were:

– finding the viroid cause of cadang-cadang (CCCVd);
– showing that phytoplasma infection was not associated with cadang-cadang;
– demonstrating mechanical transmissibility of CCCVd;
– developing various diagnostic methods for viroid detection;
– demonstrating molecular similarity between CTiVd and CCCVd;
– determining the sequence and structure of CCCVd and CTiVd;
– expanding the host range to other palm species;
– developing inoculation methods to test for resistance;
– mapping the boundaries of disease distribution;
– finding low rates of seed and pollen transmission;
– calculating annual losses of palms.

However, the main means of the natural spread of the disease remained unknown, and the removal of infected palms appeared not to slow the spread of the disease.

After the termination of FAO support, collaboration with the Philippines continued from 1985 to 1993 with support from the Australian Centre for International Agricultural Research (ACIAR) and the University of Adelaide (Hanold and Randles, 1998). Studies were expanded and further work led to:

– detection of CCCVd in oil palm plantations in Oceania;
– a comparison of CCCVd structure and thermodynamics with PSTVd;
– development of one- and two-dimensional gel electrophoretic systems for viroid detection;
– discovery of sequence variation of CCCVd in coconut and oil palm;
– diagnosis by sequence detection (hybridization, PCR, RPA, and LAMP);
– conducting a survey for CCCVd-related sequences in coconut palm and a search for alternate hosts.

The Philippine Coconut Authority (PCA) surveyed the incidence of cadang-cadang disease in the Philippines in 2012–13 with the aim of identifying disease-free areas from which coconut germplasm and products could be sourced (Rodriguez et al., 2017). Affected areas are being rehabilitated by the removal of infected palms and replacement with early bearing high-yielding seedling palms. This approach relies on the slow rate of natural spread of the disease and rapid turnover of palms in a shortened replanting cycle. There has been a decline in disease incidence since 2006 (PCA, 2014).

Orange leaf spotting (OS) of oil palm (*Elaies guineensis*) is common in plantations. It has a yield penalty, and the evidence that it has a CCCVd etiology (Hanold and Randles, 1991, 2003) emphasizes the need to eradicate the viroid from new plantations worldwide. Universiti Putra Malaysia (UPM) (Randles et al., 2009) and the Malaysian Palm Oil Board (Anuar and Ali, 2022) are the current centres in Oceania for basic research on the diagnosis and replication of oil palm variants. The oil palm is a more convenient plant than the coconut palm for infection, replication, and pathogenicity studies, as demonstrated by Thanarajoo (2014), where a cDNA clone of a 246 nt oil palm variant of CCCVd could induce typical orange spotting symptoms in seedlings within 6 months of inoculation (Wu et al., 2013). Improvement in the oil palm industry can now be directed at the selection of viroid-free elite germplasm sources, tissue culture, cloning, and use of viroid-tested ramets.

Attempts to inoculate mature flowering palms with viroid have been unsuccessful, so the process from infection to the various stages of cadang-cadang disease in bearing palms, that is, the appearance of leaf symptoms, decline in flower and fruit production, inflorescence decline, and reduced growth of the apical meristem remain to be determined. Since graft inoculation cannot be done with palms, mechanical methods such as slashing and injection of inoculum into the crown of established palms, and artificial pollination have been attempted, but without achieving significant infection above the background incidence in the study area. Epidemiological plots set up to test for the palm-to-palm spread in coconuts found that this had not occurred in the first 13 years of growth (E. Pacumbaba, M.J.B. Rodriguez and J.W. Randles, unpublished).

Discovery of coconut cadang-cadang viroid in the Philippines

The viroid studies began with the recruitment of a plant virologist from the University of Adelaide as a FAO Consultant in Plant Pathology, travel to Manila, Legaspi City, and Guinobatan, and preparation and transport of material for electron microscopy and nucleic acid analysis to Adelaide. This section addresses the situation and experimental approach adopted from the time of the virologist's first consultancy in April 1973.

Location and resources

In 1971, the Guinobatan Experiment Station (GES) of the Bureau of Plant Industry, now the Philippine Coconut Authority-Albay Research Station (PCA-ARC) was in a rural area over 40 km from Legaspi City on an unsealed public highway. It was developed for field studies such as agronomy, fruit breeding, and pest analysis. To commence the project, a secure laboratory and screenhouses had to be designed, built, and developed, and equipment purchased and brought from the United States and Australia. Unreliable water and power, lack of telegraph or telephone communication, and regular typhoons had to be addressed by numerous improvisations. The appointment of FAO consultants to supplement local staff on a series of visits over the 10 year term of the project allowed experiments to be planned, equipped, and set up in the appropriate place. Thus, electron microscopy was done by bringing coated grids and negative stains to GES for application of samples, and examination in Australia. Tissue samples fixed at GES were embedded and sectioned in Australia. Ultracentrifugation was not available at GES, so alternative nucleoprotein concentrating methods were required. Interruptions to gel electrophoresis were overcome by the purchase of a portable generator. Redistillation of water and reagents such as ethanol and phenol was standard practice. Sterilization was achieved with either an oven or a pressure cooker. Air conditioning and refrigeration requirements were limited and disposable consumables did not exist. Chemicals, instruments, and spare parts were generally hand-carried from abroad for the experiments that were planned at the time.

State of the art for characterization of plant viruses

Testing a "virus" hypothesis for plant diseases in the 1960s depended on having symptomatic host species for transmission, bioassay and host range studies, systems and buffers for tissue and sap extracts, differential centrifugation, and electron microscopy by thin sectioning and negative staining. Thus, the purification of viruses (Francki, 1972) to the point where they could be classified by their size and shape using an electron microscope, depended on access to refrigerated centrifuges and ultracentrifuges, and rate zonal and isopycnic density gradient ultracentrifugation. An invaluable finding at this time was that plant viruses can be selectively precipitated with polyethylene glycol (PEG) (Hebert, 1963) allowing virion-enriched pellets to be obtained from large volumes of buffered sap extract in situations when an ultracentrifuge was not available. Further analysis of viral nucleic acid and protein components from such "crude" preparations could be done by gel electrophoresis to assist in virus characterization.

Among the diseases of suspected viral etiology where no virus particles could be found were phytoplasma and viroid diseases. The former was found by visualizing mycoplasma-like cells in phloem tissue by electron microscopy. Viroids were found because their sedimentation coefficient (around 7S) was less than that of viruses (>80S).

Recognition of disease-associated RNA

Without a refrigerated centrifuge at GES, the problem of obtaining a putative "virus" enriched component for further analysis was overcome by optimizing buffer composition for stabilizing coconut palm leaf extracts against endogenous polyphenol oxidase activity, their clarification by low-speed centrifugation at ~2000g, then adding PEG 6000 to 5%, and collecting the resulting precipitate by a second centrifugation at ~2000g (Randles, 1975a). This pellet fraction was resuspended in the buffer for both electron microscope examination and nucleic acid analysis by polyacrylamide gel electrophoresis (PAGE).

Rod-like particles with a superficial resemblance to virus particles were discounted as putative agents of cadang-cadang because they were outside the size range of known virions and were not associated with cadang-cadang disease (Randles, 1975b).

PAGE of nucleic acid extracts from diseased palms showed two low molecular weight RNA bands, with estimated sedimentation coefficients of about 7S and 11S, in addition to the normal 4-5S, 16-25S RNAs, and DNA components. Fig. 1 shows the earliest results of PAGE as done at GES (Randles, 1974). This method was used for disease diagnosis, and the bands, described as ccRNAs, were extracted from gels for various infectivity and molecular assays.

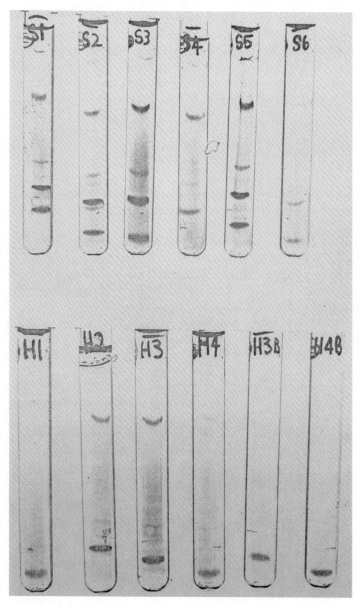

FIG. 1 Analysis by 2.5% poly-acrylamide tube gel electrophoresis of nucleic acids prepared from six diseased (S1 to S6; upper) and four normal (H1 to H4; lower) coconut palms using a portable electrophoresis unit set up for on-site analysis at GES. The position of the bromophenol blue marker dye is indicated by an injection of Indian ink; the lower of the two disease-specific toluidine blue stained bands above the marker represents the monomeric CCCVd, and the lighter stained band above represents the dimeric form. *Reproduced from Randles, J.W., 1974. Report on the Third Visit to the Guinobatan Experiment Station. Coconut Research and Development Project, 8 pp.*

B. Prospecting for viroid

Rejection of a phytoplasma etiology

The economic significance of cadang-cadang dictated that an unambiguous target for research had to be identified, and therefore that known subcellular disease agents other than a viroid had to be excluded. The lethal yellowing epidemic of coconut palms in the Caribbean prompted a trial to test whether a mycoplasma-like organism (phytoplasma) was implicated in cadang-cadang disease. Even though electron microscopy had failed to detect any virus or phytoplasma, these results had to be supported by a test of antibiotic sensitivity.

Progress of disease was shown to be unaffected by treatment with antibiotics known to suppress phytoplasmas (tetracycline) and rickettsia-like organisms (penicillin), supporting the view that cadang-cadang was not a phytoplasma disease. While this trial was in progress, the results of a mechanical inoculation trial with RNA preparations from affected palms showed that symptoms had been induced in young seedlings. These two lines of evidence led to confirmation of the viroid hypothesis for cadang-cadang (Randles et al., 1977).

Host range

Host range studies were done both by testing species of palm in the field with unexplained stunting or failure to bear fruit for the presence of ccRNA, as well as by inoculating seedlings of a range of species with ccRNA.

PAGE assays combined with a cDNA probe showed that symptomatic oil palm (*E. guineensis*) and buri palm (*Corypha elata*) were naturally infected (Randles et al., 1980).

Mechanical inoculation by combined high-pressure injection and razor slashing (Randles et al., 1977) led to palm species other than coconut (oil palm, buri palm, betel nut, date palm, manila palm, palmera, and royal palm) acquiring ccRNA and developing leaf-spotting symptoms (Imperial et al., 1985). No nonpalm species were successfully inoculated (*Nicotiana* spp., tomato, corn, soybean, mungbean, cowpea, *Elephantopus mollis*, *E. spicatus*, *Stachytarpeta* sp., *Commelina* sp., *Imperata cylindrica*, *Calopogonium* sp., *Pueraria* sp., *Centrosema* sp., arrowroot, sitao, talahib, calauag, tagbak) (Randles, 1984).

Coconut palm accessions were tested for resistance by high-pressure inoculation of germinating shoots (Bonaobra et al., 1998). No resistant populations from the 82 tested were identified after 7 years. While some individual palms remained negative, no statistically significant differences were noted. The establishment of routine diagnostic tests for ccRNA at GES/PCA-ARC was essential for work on many biological aspects of cadang-cadang.

Development of methods

Many questions needed to be addressed after the ccRNAs were found and their infectivity had been demonstrated. Their size was approximately the same as PSTVd, but the detection of two ccRNAs, now named coconut cadang-cadang viroid (CCCVd), and other differences, required a study of their molecular properties by PAGE, electron microscopy, thermodynamic methods, sequencing, and mutational analysis.

Field diagnosis of cadang-cadang by PAGE

An early field-adapted version of tube PAGE used agarose-strengthened polyacrylamide gel polymerized in Perspex tubes, closed at the bottom with a dialysis membrane (Randles, 1974). These tubes were mounted in an upper buffer tank, the sample containing bromophenol marker dye and sucrose was loaded on top, and nucleic acids migrated by electrophoresis toward the anode in a lower buffer tank. A simple DC power supply was connected, the run was stopped when the marker dye was close to the bottom of the tube, the gel was removed, and fixed and stained with toluidine blue (Fig. 1).

PAGE was developed for use in a mobile field laboratory (Rodriguez et al., 1989; Randles and Rodriguez, 2003). Leaf nucleic acids were prepared using a benchtop centrifuge, samples were run in 20% nondenaturing polyacrylamide slab gels, and bands were detected with silver stains. Picogram amounts of CCCVd were detectable.

PAGE was also used to detect CTiVd in Guam (Boccardo et al., 1981). The method available at the time at GES/PCA-ARC did not detect the presence of viroid during a survey of other islands in Micronesia (Saipan, Tinian, Rota, Yap, Pelau) (Randles, 1984). Other diseases of unknown etiology in India, Kerala wilt and Tatipaka disease, were not associated with detectable CCCVd (Anonymous, 1982).

Modifications of PAGE

PAGE systems for viroid research have been reviewed (Hanold and Vadamalai, 2017). For CCCVd, 20% PAGE under nondenaturing conditions resolves four molecular sizes, 246, 247, 296, and 297 nt, as well as their dimers. Palms with an unusually severe form of lamina depletion contain additional bands, indicative of sequence mutants (Rodriguez and Randles, 1993) which were resolved because of a change in native molecular structure. Because the molecules are rod-like partly base-paired structures under nondenaturing conditions but melt to open circular structures under denaturing conditions, 2D and bi-directional PAGE were developed for the detection of viroids and virusoids based on these structural changes (Schumacher et al., 1983). A 2D PAGE system in which the crosslinking ratio of the gel matrix was changed between the first and second dimensions (Feldstein et al., 1997) allowed both the linear and circular forms of the monomeric and dimeric forms to be resolved in a single electropherogram. This method identified an additional heterodimeric form.

Labeled cDNA diagnostic probes

Hybridization analyses with [32]P- and ^{3}H-labeled complementary DNA probes for CCCVd (Randles and Palukaitis, 1979) demonstrated that the two disease-specific bands had the same sequence, that they were uniquely associated with diseased palms, and that the sequence was not represented in higher molecular weight RNA or DNA from diseased palms. The ^{3}H-labeled cDNA was developed for use as a stable probe in diagnostic studies of CCCVd after a scintillation counter was purchased for GES, while the short-lived ^{32}P-labeled DNA had to be synthesized and transported to GES for each consultancy. Molecular hybridization analysis was used to test plants and insects for the presence of CCCVd at GES. Nonradioactive colorimetric probes were unreliable because endogenous activity in plant extracts gave false positives. Following the cloning and sequencing of CCCVd (Haseloff et al., 1982), labeled cRNA probes were transcribed from a plasmid containing a monomeric insert of the 246 nt sequence and used for dot-blot or northern blot analyses before the general availability of PCR for diagnosis (Hodgson et al., 1998; Rodriguez, 1993, 2003).

Diagnostic structural properties

CCCVd shares the unique secondary structural properties of other viroids and differs from the virusoids which have a similar size and circular structure (Randles and Hatta, 1979; Randles et al., 1982; Steger et al., 1984). Thus, viroid thermal denaturation profiles show a distinctive cooperative melting pattern and stable intermediate formation resulting from the formation of new hairpin structures. This reversible melting behavior distinguishes viroids from similar-sized plant host, virus, or virusoid RNAs in 2D and bi-directional PAGE formats. For diagnosis, 2D PAGE can test multiple samples in a few hours, and detect low viroid concentrations without the use of radioactivity or highly specialized equipment (Schumacher et al., 1983). Evidence for single base additions and progress to the later stage of disease development when large forms of the viroid appear with a reiterated right terminal sequence (Randles et al., 1988) is readily obtained with this system.

Sequences and taxonomy

Sequencing (Haseloff et al., 1982; Keese et al., 1988) explained the structures of the monomeric and dimeric, small and large forms, of the ccRNAs and led to the classification of CCCVd as the type viroid in the genus *Cocadviroid*. The significance of the duplication of the right-hand terminal sequence at the later stages of disease in coconut is not known.

Ecology

Studies of the ecology of cadang-cadang in the Philippines have emphasized symptomatology, disease distribution, seeking of vectors or other modes of distribution, and whether eradication of infected palms could slow spread. As the natural mode of spread remains unknown, no direct control method has been found (Anonymous, 1982). There has been a decline in the incidence of cadang-cadang from 2.5M infected palms in 2006 to about 0.7M, and hot spots of incidence remain as supported by the generated GIS-maps (PCA, 2014; Rodriguez et al., 2017). It is currently recommended that infected palms should be cut and replaced to rehabilitate affected coconut growing areas.

Tinangaja disease on Guam appears to be clustered, not random, and its eradication was not achieved by a governmental direction to destroy infected trees (Wall and Randles, 2003).

CCCVd beyond the Philippines

Survey in the South Pacific

The discovery of a CCCVd-related sequence in an oil palm with "genetic orange spotting" in the Solomon Islands and preliminary evidence that it spreads between palms (Hanold and Randles, 1991) led to a survey for CCCVd and similar viroids elsewhere in South Pacific countries. Leaf samples were assayed by PAGE and molecular hybridization methods. To allow for likely sequence differences between the CCCVd cRNA probe and candidate viroids, the stringency of washing in hybridization assays was varied. This approach was validated by showing that both CCCVd and CTiVd (with 64% sequence similarity) were detectable by northern blot assay under the low and high stringency conditions adopted, whereas CSVd (with 44% similarity) was not (Hanold and Randles, 1991).

Samples were collected in 28 countries. Sampling was directed at abnormal coconut and other palms, and monocotyledonous species from the *Pandanaceae*, *Commelinaceae*, *Zingiberaceae*, and *Marantaceae*. Individual specimens from each family gave positive signals under low or high stringency conditions using both dot-blot and northern blot assays (Hanold and Randles, 1994; Hanold, 1998a,b). No candidate bands have been cloned or sequenced.

Arranging field surveys in the Pacific Islands region included management of travel, making local contacts, financing field visits and sampling, observing quarantine restrictions, preservation and storage, and access to suitable laboratory facilities. Improvisations were frequently required. For example, Fig. 2 shows a portable kit that allowed coconut fronds to be cut and sampled when a tree climber was not available.

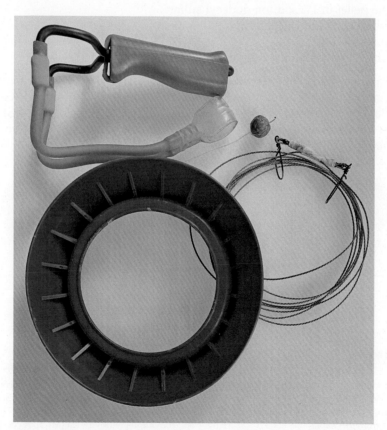

FIG. 2 Components of a portable frond sampling kit for tall coconut palms. A spherical lead weight attached to nylon fishing line is directed with a slingshot over a frond from the ground, an embryotomy wire is tied into loops in the line, the wire is raised until it lies over the midrib of the frond, then it is used in a reciprocating motion to saw through the midrib. Leaflets are collected from the cut frond.

B. Prospecting for viroid

To preserve samples collected in the field, they were sealed in plastic bags, shaded, and held at ambient temperature until refrigerated. They were either frozen at the final laboratory destination or dehydrated. CCCVd was detectable by PAGE after leaf samples were dried by sequential washing in either acetone or ethanol and vacuum drying before storage. A trial at a remote site in the Pacific (Rotuma Island) showed that a mixture of equal parts of commercially available industrial alcohol and unleaded petrol was as effective as acetone for dehydration. For transport and preservation, samples were air-dried and sealed in plastic bags with activated silica gel desiccant (Hanold, 1998b).

The following summarizes the method used for extracting fresh or frozen field samples (Hanold and Randles, 1991).

- Triturate 10–20 g of leaf in blender with 120 mL of cold 100 mM sodium sulfite.
- Strain through muslin, add polyvinylpolypyrrolidone to 2%, and stir for 30 min at 4°C.
- Extract for 5 min with an equal volume of chloroform.
- Add PEG 6000 to 8% and collect the precipitate by centrifugation.
- Dissolve the pellet in 1% sodium dodecyl sulfate.
- Extract with phenol and chloroform.
- Precipitate with CTAB.
- Wash in ethanol with sodium acetate, and dry.

The only samples with bands resembling CCCVd or CTiVd markers by size and sequence similarity in northern blots came from oil and coconut palm (Hanold and Randles, 1991; Hanold, 1998a) but other species showed different-sized bands (Fig. 3). 2D PAGE of selected samples from species of *Alpinia*, *Strelitzia*, *Pandanus*, and *Zingiber* also provided evidence that they contained a viroid-like molecule (Hanold, 1998b) but they have not been further characterized.

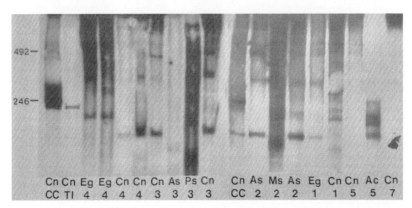

FIG. 3   Autoradiograph of a northern blot assay showing CCCVd-related components from Oceania: in coconut palm from Indonesia (Cn,1) and Solomon Islands (Cn,3,4); in oil palm from Indonesia (Eg,1) and Solomon Islands (Eg,4); *Alpinia* sp. from Papua New Guinea (As,2) and Solomon Islands (As,3); *Areca catechu* from Solomon Islands (Ac,5); *Pandanus* from the Solomon Islands (Ps,3); *Maranta* sp. from Papua New Guinea (Ms,2). A coconut palm from Vanuatu (Cn,7) shows no CCCVd signal. Note positions of monomeric 246 nt and dimeric 492 nt CCCVd forms (Cn, CC) and CTiVd (Cn,TI). *Modified from Hanold, D., 1998a. CCCVd-related sequences in species other than coconut. In: Hanold, D., Randles, J.W. (Eds.), Report on ACIAR-Funded Research on Viroids and Viruses of Coconut Palm and Other Tropical Monocotyledons 1985–1993, ACIAR Working Paper No. 51. Australian Centre for International Agricultural Research Canberra Canberra, Australia, pp. 153–159.*

The method was suitable for use in regional laboratories such as one developed at Dodo Creek in the Solomon Islands (Hanold and Randles, 1991). Alternative methods were not adopted for the survey because they were not adaptable to local conditions or economically viable.

The oil palm model for CCCVd

Current knowledge of nucleotide sequence variation and rapid diagnosis of CCCVd-related molecules have resulted from work done with the oil palm orange leaf spotting-associated CCCVd (Vadamalai, 2005; Vadamalai et al., 2017). Some of the difficulties associated with laboratory studies on the coconut palm in the Philippines such as a suitable laboratory host species have been circumvented by access to a dedicated oil palm research facility in Malaysia where seedlings, tissue culture, clones, and a molecular plant pathology research laboratory are available for research on the pathogenicity, replication, and diagnosis of CCCVd.

Among the diagnostic tests evaluated for oil palm was the ribonuclease protection assay (RPA). Because RPA also detects mismatches between a labeled clone and targeted RNAs, it identifies potential quasispecies of CCCVd. RPA identified CCCVd in coconut palm in Sri Lanka (Vadamalai et al., 2009), and has the potential to be used in future surveys for viroids in monocotyledons. In addition, a range of diagnostic techniques such as RT-PCR (Vadamalai et al., 2006; Wu et al., 2013), RT-LAMP (Thanarajoo, 2014; Thanarajoo et al., 2014; Madihah et al., 2020), and real-time qPCR (Roslan et al., 2022) are employed for rapid and routine detection of CCCVd variants in oil palm.

Geopolitical challenges

The conduct of surveys in Oceania and Asia was actively supported by many governments and their staff (Hanold and Randles, 1998). Discoveries relevant to plant health are frequently unpopular and the International Plant Protection Convention can be in conflict with both sovereign and commercial interests. Occasional experiences include the following:

- Inability to collect official samples from one of the parents of the South-East Asian oil palm germplasm in Bogor Botanic Garden, now dead.
- Disappointment expressed by plantation management on the international announcement of the CCCVd sequence in orange-spotted oil palm.
- A collection from a Central American country being confiscated for unexplained reasons.
- Refusal of entry to plantations.

Prospective: Are there viroid sequences in progenitors of the palms?

The origin of viroids is unknown (Diener, 2001). They were discovered because of their pathogenicity in domesticated plants. The palm-infecting cocadviroids have a different natural history from those infecting cultivated dicots, and present an opportunity to speculate on viroid evolution. The tools available to study viroid evolution include their pathogenicity, size, and sequence. In the survey to seek viroids in Oceania, the only definite examples are CCCVd and CTiVd. Both of the economic hosts, coconut and oil palm, as members of the *Arecaceae* which first appeared in the fossil record about 80 Mya, represent a more ancient group than the dicotyledon hosts of all other known viroids.

In our survey, collections were biased toward plants with abnormalities, expecting that they could represent a symptom of infection. Evidence was obtained that some samples from other monocotyledon families contained nucleic acids similar in size and sequence to CCCVd. While no sequences have been obtained from the viroid-like nucleic acids, one way to study whether they do represent a more primitive group of viroids would be to look for them in representatives of early plant evolution. Cycads first emerged about 280 Mya (Condamine et al., 2015). To test whether a CCCVd-related sequence might be found in currently available descendants of the *Cycadales*, individuals of several species were tested by 2D PAGE and northern blot assay. As shown in Fig. 4, one of three species of cycads tested, *Ceratozamia mexicana*, contained a radioactive spot in the region expected for a viroid-like molecule with sequence similarity to CCCVd. This result suggests that a search for the progenitors of viroids could be directed at the living representatives of the oldest plant families.

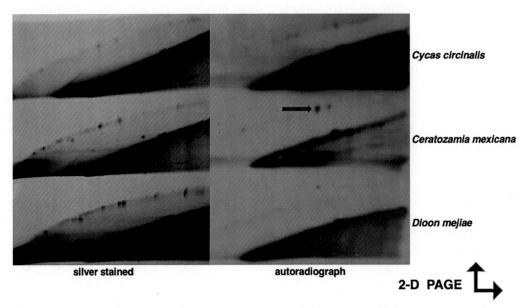

FIG. 4 2D PAGE analysis of nucleic acid extracts from three cycad species showing hybridization of CCCVd ^{32}P-cRNA to a spot in the viroid zone from *Ceratozamia mexicana (red arrow)*. First dimension (from bottom to top) was 5% nondenaturing PAGE, second dimension (from left to right) was 5% denaturing PAGE containing 8 M urea. The viroid zone was confirmed using ASBVd as a marker (J.W. Randles, R. Flores, J.A. Daròs, unpublished).

Chapter summary

The palm viroids, which differ from other known viroids in pathogenicity, size, and structure, emerged because of their lethality and economic importance. Their detection in coconut palm resulted from the application of methods available at the time for plant virus characterization. These differed from the methods developed for the direct purification of nucleic acids from plants. Fortuitously, PEG precipitated the ccRNA from leaf sap extracts, allowing its detection by the primitive methods available at the time.

An important gap in knowledge is the development of a viroid disease model for coconut palms. The processes of spread, infection, and disease development from CCCVd and CTiVd infection have been inferred from field descriptions but there have been no successful attempts to inoculate established palms and follow disease development.

Discovering the wide distribution of CCCVd in oil palm, a much more adaptable species than coconut palm for research, has led to detailed studies on variation in sequence and opened the way to questioning whether other tropical monocotyledons might be infected. No more host species of typical CCCVd have been found, but the detection of CCCVd-related nucleic acids by the hybridization assays employed so far suggests that the suite of wild monocotyledonous plants in Oceania should be assessed for viroid infection by current high-throughput sequencing methods.

The known hosts of CCCVd and CTiVd are more primitive phylogenetically than the dicotyledonous hosts of all other known viroids. A simplistic way to look again at whether viroids are representatives of an early RNA world would be to look for viroid-related molecules or motifs in the most primitive living plants. Horsetails, *Ginkgo*, and cycads fall into this grouping, and the result of a trial with the cycad, *C. mexicana*, suggests that they may provide a platform for studying viroid evolution.

Study question

1. What type of plant is most affected by viroid diseases in Oceania?
2. In a search for vectors of CCCVd, should palm-infecting fungi be tested for CCCVd?
3. Should wild plants of the African oil palm be tested for viroid infection?

Further reading

Geering, A.D.W. (2017). Geographical distribution of viroids in Oceania. In: Hadidi, A., Flores, R., Randles, J.W., Palukaitis, P (Eds.) Viroids and satellites. Elsevier, Academic Press. pp. 497–506.

Geering, A.D.W., Randles, J.W. (2012). Virus diseases of tropical crops. In: eLS. John Wiley & Sons, Ltd.: Chichester.

Hanold, D., Randles, J.W. (Eds.) (1998). Report on ACIAR-funded research on viroids and viruses of coconut palm and other tropical monocotyledons 1985–1993, ACIAR Working Paper No. 51. Canberra, Australia, 222 pp.

Hanold, D., Randles, J.W. (1991). Coconut cadang-cadang disease and its viroid agent. Plant Disease 75, 330–335.

Randles, J.W. (1985). Coconut cadang-cadang viroid. In: Maramorosch, K., McKelvey, J.J. (Eds.). Subviral pathogens of plants and animals: viroids and prions. Academic Press. pp. 39–74.

Boccardo, G. (1985). Viroid etiology of tinangaja and its relationship with cadang-cadang disease of coconut. In: Maramorosch, K., McKelvey, J.J. (Eds.). Subviral pathogens of plants and animals: viroids and prions. Academic Press. pp. 75–99.

Acknowledgments

We gratefully acknowledge the contributions, support, and cooperation of the many countries, institutions, and individuals involved in these studies.

References

Anonymous, 1982. Research on the Cadang-Cadang Disease of Coconut Palm. Food and Agriculture Organization of the United Nations, Rome, p. 75.

Anuar, M.A.S.S., Ali, N.S., 2022. Significant oil palm diseases impeding global industry: a review. Sains Malays. 51, 707–721.

Boccardo, G., Beaver, R.G., Randles, J.W., Imperial, J.S., 1981. Tinangaja and bristle top, coconut diseases of uncertain etiology in Guam, and their relationship to cadang-cadang disease of coconut in the Philippines. Phytopathology 71, 1104–1107.

Bonaobra, Z.S., Rodriguez, M.J.B., Estioko, L.P., Baylon, G.B., Cueto, C.A., Namia, M.T.I., 1998. Screening of coconut populations for resistance to CCCVd using coconut seedlings. In: Hanold, D., Randles, J.W. (Eds.), Report on ACIAR-Funded Research on Viroids and Viruses of Coconut Palm and Other Tropical Monocotyledons 1985–1993, ACIAR Working Paper No. 51. Australian Centre for International Agricultural Research Canberra, Canberra, Australia, pp. 69–75.

Condamine, F.L., Nagalingum, N.S., Marshall, C.R., Morlon, H., 2015. Origin and diversification of living cycads: a cautionary tale on the impact of the branching process prior in Bayesian molecular dating. BMC Evol. Biol. 15, 65. https://doi.org/10.1186/s12862-015-0347-8.

Diener, T.O., 2001. The viroid: biological oddity or evolutionary fossil. Adv. Virus Res. 57, 137–183.

Feldstein, P.A., Levy, L., Randles, J.W., Owens, R.A., 1997. Synthesis and two-dimensional electrophoretic analysis of mixed populations of circular and linear RNAs. Nucleic Acids Res. 25, 4850–4854.

Francki, R.I.B., 1972. Purification of viruses. In: Kado, C.I., Agrawal, H.O. (Eds.), Principles and Techniques in Plant Virology. Van Nostrand Reinhold, New York, pp. 295–335.

Geering, A.D.W., 2017. Geographical distribution of viroids in Oceania. In: Hadidi, A., Flores, R., Randles, J.W., Palukaitis, P. (Eds.), Viroids and Satellites. Elsevier, Academic Press, pp. 497–506.

Geering, A.D.W., Randles, J.W., 2012. Virus diseases of tropical crops. In: eLS. John Wiley & Sons, Chichester.

Hadidi, A., Flores, R., Randles, J.W., Palukaitis, P. (Eds.), 2017. Viroids and Satellites. Elsevier, Academic Press. 716 pp.

Hanold, D., 1998a. CCCVd-related sequences in species other than coconut. In: Hanold, D., Randles, J.W. (Eds.), Report on ACIAR-Funded Research on Viroids and Viruses of Coconut Palm and Other Tropical Monocotyledons 1985–1993, ACIAR Working Paper No. 51. Australian Centre for International Agricultural Research Canberra, Canberra, Australia, pp. 153–159.

Hanold, D., 1998b. Investigation of the characteristics of the CCCVd-related molecules. In: Hanold, D., Randles, J.W. (Eds.), Report on ACIAR-Funded Research on Viroids and Viruses of Coconut Palm and Other Tropical Monocotyledons 1985–1993, ACIAR Working Paper No. 51. Australian Centre for International Agricultural Research Canberra, Canberra, Australia, pp. 160–171.

Hanold, D., Randles, J.W., 1991. Detection of coconut cadang-cadang viroid-like sequences in oil and coconut palm and other monocotyledons in the South-West Pacific. Ann. Appl. Biol. 118, 139–151.

Hanold, D., Randles, J.W., 1994. A new viroid family infecting tropical monocotyledons. In: Foale, M.A., Lynch, P.W. (Eds.), Coconut Improvement in the South Pacific, ACIAR Proceedings No. 53, pp. 55–61.

Hanold, D., Randles, J.W., 2003. CCCVd-related molecules in oil palms, coconut palms and other monocotyledons outside the Philippines. In: Hadidi, A., Flores, R., Randles, J.W., Semancik, J.S. (Eds.), Viroids. CSIRO Publishing, Collingwood, VIC, pp. 336–340.

Hanold, D., Randles, J.W. (Eds.), 1998. Report on ACIAR-Funded Research on Viroids and Viruses of Coconut Palm and Other Tropical Monocotyledons 1985–1993, ACIAR Working Paper No. 51, Canberra, Australia. 222 pp.

Hanold, D., Vadamalai, G., 2017. Gel electrophoresis. In: Hadidi, A., Flores, R., Randles, J.W., Palukaitis, P. (Eds.), Viroids and Satellites. Elsevier, Academic Press, pp. 357–367.

Haseloff, J., Mohamed, N.A., Symons, R.H., 1982. Viroid RNAs of cadang-cadang disease of coconuts. Nature 299, 316–321.

Hebert, T.T., 1963. Precipitation of plant viruses by polyethylene glycol. Phytopathology 53, 362.

Hodgson, R.A.J., Wall, G.C., Randles, J.W., 1998. Specific identification of coconut tinangaja viroid (CTiVd) for differential field diagnosis of viroids in coconut palm. Phytopathology 88, 774–781.

Imperial, J.S., Bautista, R.M., Randles, J.W., 1985. Transmission of the coconut cadang-cadang viroid to six species of palm by inoculation with nucleic acid extracts. Plant Pathol. 34, 391–401.

Keese, P., Osorio-Keese, M.E., Symons, R.H., 1988. Coconut tinangaja viroid: sequence homology with coconut cadang-cadang viroid and other potato spindle tuber viroid related RNAs. Virology 162, 508–510.

Madihah, A., Maizatul-Suriza, M., Idris, A., 2020. Reverse transcription loop-mediated isothermal amplification (RT-LAMP) for detection of coconut cadang-cadang viroid (CCCVd) variants in oil palm. J. Oil Palm Res. 32, 453–463.

Parakh, D.B., Zhu, S., Sano, T., 2017. Geographical distribution of viroids in South, Southeast, and East Asia. In: Hadidi, A., Flores, R., Randles, J.W., Palukaitis, P. (Eds.), Viroids and Satellites. Elsevier, Academic Press, pp. 507–518.

PCA, 2014. Terminal Report. Coconut Cadang-Cadang Disease Surveillance Survey 2012–2013. Philippine Coconut Authority, Quezon City, Philippines. 70 pp.

Randles, J.W., 1974. Report on the Third Visit to the Guinobatan Experiment Station. Coconut Research and Development Project. 8 pp.

Randles, J.W., 1984. Report on 19th and 20th Visits to the Albay Research Center. Coconut Pest and Diseases Project. 10 pp.

Randles, J.W., 1975a. Association of two ribonucleic acid species with cadang-cadang disease of coconut palm. Phytopathology 65, 163–167.

Randles, J.W., 1975b. Detection in coconut of rod-shaped particles which are not associated with disease. Plant Dis. Rep. 59, 349–352.

Randles, J.W., Hatta, T., 1979. Circularity of the ribonucleic acids associated with cadang-cadang disease. Virology 96, 47–53.

Randles, J.W., Palukaitis, P., 1979. In vitro synthesis and characterization of DNA complementary to cadang-cadang-associated RNA. J. Gen. Virol. 43, 649–662.

Randles, J.W., Rodriguez, M.J.B., 2003. Coconut cadang-cadang viroid. In: Hadidi, A., Flores, R., Randles, J.W., Semancik, J.S. (Eds.), Viroids. CSIRO Publishing, Collingwood VIC, pp. 233–241.

Randles, J.W., Boccardo, G., Imperial, J.S., 1980. Detection of the cadang-cadang associated RNA in African oil palm and Buri palm. Phytopathology 70, 185–189.

Randles, J.W., Boccardo, G., Retuerma, M.L., Rillo, E.P., 1977. Transmission of the RNA species associated with cadang-cadang of coconut palm and the insensitivity of the disease to antibiotics. Phytopathology 67, 1211–1216.

Randles, J.W., Rodriguez, J.M.B., Hanold, D., Vadamalai, G., 2009. Coconut cadang-cadang viroid infection of African oil palm. Planter 85 (995), 93–101.

Randles, J.W., Rodriguez, M.J.B., Imperial, J.S., 1988. Cadang-cadang disease of coconut palm. Microbiol. Sci. 5, 18–22.

Randles, J.W., Steger, G., Riesner, D., 1982. Structural transitions in viroid-like RNAs associated with cadang-cadang disease, velvet tobacco mottle virus, and Solanum nodiflorum mottle virus. Nucleic Acids Res. 10, 5569–5586.

Rillo, E.P., Rillo, A.R., 1981. Abstracts on the Cadang-Cadang Disease of Coconut 1937–1980. Philippine Coconut Authority, Quezon City. 50 pp.

Rodriguez, M.J.B., 1993. Molecular Variation in Coconut Cadang-Cadang Viroid (CCCVd) (Ph.D. thesis). University of Adelaide.

B. Prospecting for viroid

Rodriguez, M.J.B., 2003. Coconut and other palm trees: viroid diseases. In: Loebenstein, G., Thottapilly, G. (Eds.), Plant Virus Diseases of Major Crops in Developing Countries. Kluwer Academic Publishers, Dordrecht, pp. 567–582.

Rodriguez, M.J.B., Randles, J.W., 1993. Coconut cadang-cadang viroid (CCCVd) mutants associated with severe disease vary in both the pathogenicity domain and the central conserved region. Nucleic Acids Res. 21, 2771.

Rodriguez, M.J.B., Ignacio-Namia, M.R.T., Estioko, L.P., 1989. A mobile diagnostic laboratory for cadang-cadang disease of coconut. Philipp. J. Coconut Stud. 14, 21–23.

Rodriguez, M.J.B., Vadamalai, G., Randles, J.W., 2017. Economic significance of palm tree viroids. In: Hadidi, A., Flores, R., Randles, J.W., Palukaitis, P. (Eds.), Viroids and Satellites. Elsevier, Academic Press, pp. 39–49.

Roslan, N.D., Vadamalai, G., Idris, A.B., Kong, L.L., Sundram, S., 2022. Comparison of real-time PCR, conventional PCR and RT-LAMP for the detection of coconut cadang-cadang viroid variant in oil palm. J. Oil Palm Res. https://doi.org/10.21894/jopr.2022.0030.

Schumacher, J., Randles, J.W., Riesner, D., 1983. A two-dimensional electrophoretic technique for the detection of circular viroids and virusoids. Anal. Biochem. 135, 288–295.

Steger, G., Hofmann, H., Fortsch, J., Gross, H.J., Randles, J.W., Sanger, H.L., Riesner, D., 1984. Conformational transitions in viroids and virusoids: comparison of results from energy minimization algorithm and from experimental data. J. Biomol. Struct. Dyn. 2, 543–571.

Thanarajoo, S.S., 2014. Rapid Detection, Accumulation and Translocation of Coconut Cadang-Cadang Viroid (CCCVd) Variants in Oil Palm (Ph.D. thesis). Universiti Putra Malaysia.

Thanarajoo, S.S., Kong, L.L., Kadir, J., Lau, W.H., Vadamalai, G., 2014. Detection of coconut cadang-cadang viroid (CCCVd) in oil palm by reverse transcription loop-mediated isothermal amplification (RT-LAMP). J. Virol. Methods 202, 19–23.

Vadamalai, G., 2005. An Investigation of Orange Spotting Disorder in Oil Palm (Ph.D. thesis). University of Adelaide.

Vadamalai, G., Perera, A.A.F.L.K., Hanold, D., Rezaian, M.A., Randles, J.W., 2009. Detection of coconut cadang-cadang viroid sequences in oil and coconut palm by ribonuclease protection assay. Ann. Appl. Biol. 154, 117–125.

Vadamalai, G., Hanold, D., Rezaian, M., Randles, J., 2006. Variants of coconut cadang-cadang viroid isolated from an African oil palm (Elaies guineensis Jacq.) in Malaysia. Arch. Virol. 151, 1447–1456.

Vadamalai, G., Thanarajoo, S.S., Joseph, H., Kong, L.L., Randles, J.W., 2017. Coconut cadang-cadang viroid and coconut tinangaja viroid. In: Hadidi, A., Flores, R., Randles, J.W., Palukaitis, P. (Eds.), Viroids and Satellites. Elsevier, Academic Press, pp. 263–273.

Wall, G.C., Randles, J.W., 2003. Coconut tinangaja viroid. In: Hadidi, A., Flores, R., Randles, J.W., Semancik, J.S. (Eds.), Viroids. CSIRO Publishing, Collingwood, VIC, pp. 242–245.

Wu, Y.H., Cheong, L.C., Meon, S., Lau, W.H., Kong, L.L., Joseph, H., Vadamalai, G., 2013. Characterization of coconut cadang-cadang viroid variants from oil palm affected by orange spotting disease in Malaysia. Arch. Virol. 158, 1407–1410.

Yang, Y., Xing, F., Li, S., Che, H.-Y., Wu, Z.-G., Candresse, T., Li, S.-F., 2020. Dendrobium viroid, a new monocot-infecting apscaviroid. Virus Res. 282, 197958. https://doi.org/10.1016/j.virusres.2020.197958.

Viroids and their distribution in North America

Xianzhou Nie

Fredericton Research and Development Centre, Agriculture and Agri-Food Canada, Fredericton, NB, Canada

Graphical representation

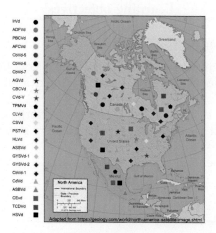

Adapted from https://geology.com/world/north-america-satellite-image.shtml

Definitions

NAPPO: Abbreviation of the North American Plant Protection Organization, a phytosanitary standard setting organization that was created in 1976 and is recognized by the North American Free Trade Agreement (NAFTA). Member countries include Canada, Mexico, and the United States of America.

Fundamentals of Viroid Biology
https://doi.org/10.1016/B978-0-323-99688-4.00003-1

Ornamental plants: Plants that are grown for decoration purposes in gardens and landscapes.

Phytosanitary measures: Practices or protocols and procedures that are used for the control of plant diseases in agricultural (including horticultural and ornamental) crops.

Pome fruit plants: Plants that produce apple-like fruits such as apple, pear, and quince. Botanically, they include members of the flowering plants in the subtribe *Malinae* of the family *Rosaceae*.

Solanaceous plants: Plants of species that belong to the *Solanaceae* family, which include many important horticultural crops such as potato, tomato, eggplant, and chili pepper.

Stone fruit plants: Flowering plants of different families that produce pit-containing fruits (or drupes botanically) that are characterized by an outer fleshy part surrounding a shell (the pit, stone, or pyrena; in botany, hardened endocarp) with a seed inside. Exemplary members include peach, plum, cherry, nectarine, and apricot.

Chapter outline

This chapter provides information and insights on the occurrence, natural hosts, and impacts of different viroids in North America. Various aspects such as geographic distribution, economic significance, and regulatory status in North American Plant Protection Organization countries (namely, Canada, Mexico, and the United States of America) are discussed.

Learning objectives

On completion of this chapter, you should acquire knowledge on:

- the distribution of different viroids in North American countries
- the contributions from pioneer research in North America leading to the discovery of viroids
- the significance of viroid diseases in various crops in North America
- sources of information on the occurrence and regulatory status of various viroids in North American countries
- typical symptoms of some viroid diseases

Introduction

Viroids are the smallest plant pathogens with a genome of 234–430 nucleotides in length. The single-stranded, covalently closed RNA genome does not encode any proteins and relies on host factors for its replication (Flores et al., 2017a). Despite the small size of the genome and the non-protein encoding nature, viroids can cause severe and devastating diseases in many herbaceous and woody horticultural crops such as tomato, potato, citrus, avocado, apple, and pear if unmanaged. Even with fundamental differences with viruses in various aspects such as structure, function, and origin, viroids share similar symptomology and transmission mode with viruses to a large degree. Therefore, several classic viroid diseases such as "spindle tuber" in potatoes, "dapple apple" and "apple scar skin" in apples were believed to have an etiology of the virus (Singh, 2014) before the eventual revelation of "low molecular weight RNA" (i.e., viroid) as the responsible cause of potato spindle tuber disease in the early 1970s by researchers from the USA and Canada (Diener, 1971; Singh and Clark, 1971).

Currently, there are 33 viroid species belonging to two families, *Pospiviroidae* and the *Avsunviroidae*, that have been officially recognized by the taxonomic nomenclature governing body for viroids, the International Committee on Taxonomy of Viruses (ICTV), and at least another 12 whose clarifications are pending (Flores et al., 2021). In North America, at least 21 recognized and three tentative viroids have been reported (Table 1). Among them, potato spindle tuber viroid (PSTVd), previously one of the most devastating pathogens in potato production in general and seed potato production in particular in the USA and Canada, has been eradicated in potatoes in both countries (De Boer and DeHaan, 2005; Sun et al., 2004) and is currently listed as one of the regulated pests of potato in the North American Plant Protection Organization's (NAPPO) member countries, namely, Canada, Mexico, and the USA (NAPPO, 2011). In addition to PSTVd, apple scar skin viroid (ASSVd) and pear blister canker viroid (PBCVd) are currently considered regulated pests in Canada [Canadian Food Inspection Agency (CFIA), https://inspection.canada.ca/plant-health/invasive-species/regulated-pests/eng/1363317115207/1363317187811#z]. In the USA, apple dimple fruit viroid (ADFVd) is on the quarantine list along with PSTVd (https://www.aphis.usda.gov/aphis/ourfocus/planthealth/import-information/rppl/rppl-table).

TABLE 1 Viroids in North America.

Family	Genus	Viroid (Abbr.)	Country	Host	References
Pospiviroidae	*Pospiviroid*	Chrysanthemum stunt viroid (CSVd)	USA	Chrysanthemum spp.	Diener and Lawson (1973); Lawson (1987)
			Canada	*Vinca major*	Bostan et al. (2004); Nie et al. (2005)
		Citrus exocortis viroid (CEVd)	Canada	Impatiens	Bostan et al. (2004); Nie et al. (2005)
				Verbena	Bostan et al. (2004); Nie et al. (2005)
			Mexico	Citrus spp.	Guerrero Gámez et al. (2013)
			USA	Citrus spp.	Kunta et al. (2007)
		Columnea latent viroid (CLVd)	USA	*Columnea erythrophae*	Owens et al. (1978)
			Canada	*Nematanthus wettsteinii*	Singh et al. (1992)

Continued

B. Prospecting for viroid

TABLE 1 Viroids in North America—cont'd

Family	Genus	Viroid (Abbr.)	Country	Host	References
		Iresine viroid (IrVd)	Canada	*V. major*	Bostan et al. (2004); Nie et al. (2005)
				Verbena	Bostan et al. (2004); Nie et al., 2005
		Potato spindle tuber viroid (PSTVd)	Canada	Potato[a]	Singh (2014); De Boer and DeHaan (2005)
			USA	Potato[a]	Diener (1987); Sun et al. (2004)
				Tomato	Ling and Sfetcu (2010); Ling et al. (2014)
				Cestrum sp.	Chitambar (2015)
		Tomato chlorotic dwarf viroid (TCDVd)	Canada	Tomato	Singh et al. (1999)
				Vinca minor	Singh and Dilworth (2009)
			Mexico	Tomato	Ling and Zhang (2009)
			USA	*Petunia hybrida*	Verhoeven et al. (2007)
				Tomato	Ling et al. (2009)
		Tomato planta macho viroid (TPMVd) [synonym, Mexican papita viroid (MPVd)]	Mexico	Tomato	Belalcazar and Galindo (1974); Ling and Zhang (2009)
				Various wild *Solanaceae* spp.	Orozco Vargas and Galindo Alonso (1986)
				Papita	Martínez-Soriano et al. (1996)
			Canada	Tomato	Ling and Bledsoe (2009)
	Cocadviroid	Citrus bark cracking viroid (CBCVd)	USA	Citrus spp.	Duran-Vila et al. (1988); Kunta et al. (2007)
		Hop latent viroid (HLVd)	USA	Hop	Eastwell and Nelson (2007); Nelson et al. (1997)
				Cannabis sativa	Bektaş et al. (2019); Warren et al. (2019)
			Canada	*C. sativa*	A&L Canada Laboratory Inc (2021)

B. Prospecting for viroid

TABLE 1 Viroids in North America—cont'd

Family	Genus	Viroid (Abbr.)	Country	Host	References
	Hostuviroid	Hop stunt viroid (HSVd)	Canada	Apricot	Michelutti et al. (2004)
				Grapevine	Xiao et al. (2019); Fall et al. (2020)
			Mexico	Citrus spp.	Guerrero Gámez et al. (2013)
			USA	Citrus spp.	Kunta et al. (2007)
				Hop	Eastwell and Nelson (2007); Nelson et al. (1997)
				Grapevine	Sano et al. (1986)
				Prunus spp.	Osman et al. (2012)
	Apscaviroid	Apple scar skin viroid (ASSVd) [synonym, Dapple apple viroid (DAVd); pear rusty skin viroid (PRSVd)]	USA	Apple	Hadidi et al. (1991)
			Canada	Apple	Hadidi et al. (1991)
		Apple dimple fruit viroid (ADFVd)	Canada	*Malus/Pyrus/ Prunus*	Rott et al. (2017)
		Apple fruit crinkle viroid (AFCVd)	Canada	*Malus/Pyrus/ Prunus*	Rott et al. (2017)
		Australian grapevine viroid (AGVd)	USA	Grapevine	Rezaian et al. (1992)
		Citrus dwarfing viroid (CDVd)	Mexico	Citrus spp.	Almeyda-León et al. (2002); Guerrero Gámez et al. (2013)
			USA	Citrus spp.	Kunta et al. (2007)
		Citrus viroid V (CVd-V)	USA	Etrog citron	Serra et al. (2008)
		Grapevine yellow speckle viroid-1 (GYSVd-1)	USA	Grapevine	Rezaian et al. (1992)
			Canada	Grapevine	Xiao et al. (2019); Xu et al. (2021)
		Grapevine yellow speckle viroid-2 (GYSVd-2)	USA	Grapevine	Rezaian et al. (1992)
			Canada	Grapevine	Xiao et al. (2019)
		Pear blister canker viroid (PBCVd)	Canada	*Malus/Pyrus/ Prunus*	Rott et al. (2017)

Continued

B. Prospecting for viroid

TABLE 1   Viroids in North America—cont'd

Family	Genus	Viroid (Abbr.)	Country	Host	References
	Coleviroid	*Coleus blumei* viroid 1 (CbVd-1)	Canada	*C. blumei*	Singh et al. (1991); Smith et al. (2018)
			USA	*C. blumei*	Orellana and Karasev (2021)
		C. blumei viroid 5 (CbVd-5)	Canada	*C. blumei*	Smith et al. (2018)
		C. blumei viroid 6 (CbVd-6)	Canada	*C. blumei*	Smith et al. (2019)
		C. blumei viroid 7 (CbVd-7)	Canada	Colums blumei	Smith et al. (2021)
Avsunviroidae	Avsunviroid	Avocado sunblotch viroid (ASBVd)	Mexico	Avocado	De La Torre et al. (2009)
			USA	Avocado	Schnell et al. (2011)
	Pelamoviroid	Chrysanthemum chlorotic mottle viroid (CChMVd)	USA	*Chrysanthemum morifolium*	Horst (1975)
		Peach latent mosaic viroid (PLMVd)	Canada	*Prunus* spp.	Michelutti et al. (2004)
				Malus/Pyrus/ Prunus	Rott et al. (2017)
			USA	*Prunus* spp.	Osman et al. (2012)
			Mexico	Peach	De La Torre-Almaráz et al. (2015)
		Apple hammerhead viroid (AHVd)	Canada	Apple	Messmer et al. (2017)
			USA	Apple	Szostek et al. (2018)

[a] *Eradicated.*

In this chapter, viroids found in North America to date are discussed based on their occurrence in their main hosts. With the advancement and wide application of sequencing technologies, new viroids and/or new hosts to existing viroids will most likely be found. Interested readers are advised to check various databases such as the Centre for Agriculture and Bioscience International (CABI, https://www.cabi.org/isc), European and Mediterranean Plant Protection Organization (EPPO) Global Database (https://gd.eppo.int), International Standards for Phytosanitary Measures (ISPMs) (https://www.ippc.int/en/core-activities/standards-setting/ispms/), and NAPPO (https://www.nappo.org/).

Viroids of solanaceous plants

Three viroids, namely, potato spindle tuber viroid, tomato chlorotic dwarf viroid (TCDVd), tomato planta macho viroid (TPMVd, including Mexican papita viroid), all members of the genus *Pospiviroid*, family *Pospiviroidae*, have been found in solanaceous plants including potato, tomato, and petunia in North America. PSTVd has been eradicated in potatoes in both USA and Canada through strict certification programs and phytosanitary measures (De Boer and DeHaan, 2005; Sun et al., 2004; NAPPO, 2011). A couple of outbreaks of the viroid, one in California and the other in North Carolina, have been reported in commercial greenhouse tomatoes in the USA (Ling and Sfetcu, 2010; Ling et al., 2014). The disease resulted in a significant yield loss due to reduced fruit size in the affected greenhouses. While PSTVd is listed as present in Mexico by CABI (CABI/EPPO, 2014), no details are available. In NAPPO's Regional Standards for Phytosanitary Measures (RSPM) datasheet for potatoes (RSPM3), the viroid is listed as Ab1, namely, "Absent: no pest records." A major concern is the seed-transmissibility of the viroid in both potato and tomato industries as it can be introduced from infected seeds from overseas, especially from regions/countries where outbreaks occur. For potatoes, with the increasing interest in diploid potato breeding and the use of true seeds of diploid hybrid potato in production (Beumer and Stemerding, 2021), careful measures need to be taken to safeguard the crop in both USA and Canada, particularly in seeds imported from overseas. Stakeholders of the potato industry in the USA have expressed their objection to the importation of true potato seed for planting from the Netherlands into the continental United States due to the high risk of disease transmission in a USDA-Animal Plant Health Inspection Service's (APHIS) call for a consultation (https://www.aphis.usda.gov/aphis/newsroom/stakeholder-info/stakeholder-messages/plant-health-news/dried-true-potato-seed-netherlands; https://www.nationalpotatocouncil.org/wp-content/uploads/2020/11/20200525-Letter-True-Seed-Netherlands-APHIS.pdf).

The outbreak of what is now known as TCDVd in tomatoes in a greenhouse in Canada led to the discovery of this viroid (Singh et al., 1999). The viroid likely has a global distribution as it has been reported in many countries of different continents. The occurrence of TCDVd in greenhouse tomatoes has been recorded in all NAPPO member countries, including the USA (Verhoeven et al., 2004; Ling et al., 2009), Canada (Singh et al., 1999), and Mexico (Ling and Zhang, 2009). Similar to PSTVd, TCDVd infection can lead to tremendous yield losses and quality degradation (Singh et al., 1999; Ling et al., 2009) (Fig. 1). It is noteworthy that, in addition to tomato, several petunia (*Petunia hybrida*, a popular solanaceous ornamental species) plants originating from the USA tested positive for TCDVd in the Netherlands (Verhoeven et al., 2007).

TPMVd, including the papita strain (formerly Mexican papita viroid or MPVd), was first identified in Mexico in 1974 (Belalcazar and Galindo, 1974). In 1986, natural infection with TPMVd was found in seven solanaceous species (Orozco Vargas and Galindo Alonso, 1986). This finding prompted the authors to suggest that TPMVd infection in field tomato plants was via solanaceous weeds. The TPMVd papita strain (= MPVd) was found in Mexico infecting papita (*Solanum cardiophyllum*) (Martínez-Soriano et al., 1996). In 2009, the same study that reported TCDVd in tomatoes in Mexico for the first time (Ling and Zhang, 2009), also revealed the occurrence of TPMVd papita strain in the affected greenhouses

FIG. 1 Symptoms induced by potato spindle tuber viroid (PSTVd) and tomato chlorotic dwarf viroid (TCDVd) in tomato plants. Tomato (cultivar Rutgers) plants at the 4-leaf stage were mechanically inoculated with PSTVd or TCDVd (leaf extract from a PSTVd- or TCDVd-infected tomato plant) or inoculation buffer (Mock) and grown in the greenhouse. Pictures shown were taken six (6) weeks after inoculation.

outside of Mexico city during the outbreak from early 2008 to 2009. In British Columbia, Canada, Ling and Bledsoe (2009) reported TPMVd papita strain as the causal agent leading to a viroid-like disease outbreak in greenhouse tomatoes.

While the occurrence of PSTVd, TCDVd, or TPMVd in field tomatoes has not been recorded in North American countries, a severe PSTVd-caused disease outbreak has occurred in a tomato field in Dominic (Ling et al., 2014).

Viroids of ornamentals

An increasing body of evidence demonstrates that ornamentals are natural hosts to various viroids, including all of those that bear a name of an ornamental and some whose first identified hosts are non-ornamentals. Eleven viroids have been found in ornamental plants in North America to date. Except for chrysanthemum chlorotic mottle viroid (CChMVd), which is a member of the genus *Pelamoviroid* of the family *Avsunviroidae*, all are members of the family *Pospiviroidae*. Among the members of *Pospiviroidae*, six, namely, chrysanthemum stunt viroid (CSVd), citrus exocortis viroid (CEVd), columnea latent viroid (CLVd), iresine viroid 1 (IrVd-1), PSTVd, and TCDVd are members of the genus *Pospivorid*, and four, namely, *coleus blumei* viroid 1 (CbVd-1) and CbVd-5 to -7, are members (or tentative members) of the genus *Coleviroid*.

CChMVd has a very narrow host range limited to chrysanthemums. The viroid has been found in major chrysanthemum growing areas in Europe and Asia. In North America, it has only been reported in the USA (Flores et al., 2017b). Nevertheless, the classic CChMVd disease, chrysanthemum chlorotic mottle disease, was first described in the USA in 1971 (Dimock et al., 1971), several years before the causal agent CChMVd was identified (Horst, 1975).

In North America, ornamental species such as impatiens, vinca, verbena, petunia, chrysanthemum, and cestrum are naturally infected by the above-mentioned pospiviroids (Bostan et al., 2004; Nie et al., 2005; Singh et al., 2009; Singh and Dilworth, 2009; Verhoeven et al., 2007; Chitambar, 2015). For CSVd, although the diseases caused by it in chrysanthemum was documented in the USA as early as 1947 (Dimock, 1947), it was not until 1973 that the viroid was determined to be the causal agent of the disease (Diener and Lawson, 1973). In addition to chrysanthemum in the USA (Lawson, 1987), the viroid has been detected in *Vinca major* and trailing *Verbena* spp. in a local commercial nursery in New Brunswick, Canada (Bostan et al., 2004; Nie et al., 2005). In the same study by Bostan et al. (2004) and its follow-up study (Nic et al., 2005), CEVd was detected in trailing *Verbena* and double *Impatiens* using degenerate primers specific to most *Pospiviroid*. In a further study by Singh et al. (2009), CEVd was found to be widely distributed in *Verbena* and *Impatiens* plants and seeds. A high seed-transmission rate (66% in *Impatiens walleriana* and 45% in *Verbena* × *hybrida*) of CEVd was found in both species. CLVd was detected in *Columnea erythrophae* in a local commercial nursery in Maryland, USA, in 1978 (Owens et al., 1978). In 1992, the viroid was isolated from *Nematanthus wettsteinii* plants in Canada (Singh et al., 1992). In both cases, the plants were asymptomatic. IrVd-1 has been found in many ornamental species worldwide. In North America, it was found in trailing *Verbina* and *Vinca major* in the survey for *Pospiviroid* in the local nursery in New Brunswick, Canada (Bostan et al., 2004; Nie et al., 2005). TCDVd was found by Verhoeven et al. (2007) in symptomless *Petunia* × *hybrida* plants imported into the Netherlands in 2007. In 2009, Singh and Dilworth (2009) reported the finding of TCDVd in *Vinca minor* plants that originated from the US and were obtained from a local commercial nursery, and planted outdoors in New Brunswick, Canada. It is particularly worth mentioning that the harsh winter conditions (as low as −12°C) did not affect the survival of the plant hosts nor the viroid. PSTVd was found in three symptomless Cestrum plant samples that were raised in a nursery greenhouse in San Diego County of California in 2013 (Chitambar, 2015). The viroid was eradicated soon thereafter.

Members of the genus *Coleviroid* have only been found in *Coleus blumei* (synonym *Plectranthus scutellarioides*; vernacular name, coleus). To date, seven *Coleviroid*, namely, CbVd-1 through -7, have been discovered worldwide, and among them, three (CbVd-1 to -3) have been officially classified by ICTV. CbVd-1 and -5 to -7 have been found in various varieties of coleus from a local nursery in New Brunswick, Canada (Singh et al., 1991; Smith et al., 2018, 2019, 2021), and CbVd-1 has recently been found in the USA (Orellana and Karasev, 2021). While none of the viroids appear to cause clear-cut symptoms in coleus or have any economic significance, they do offer an interesting perspective for understanding viroid diversification and evolution. The most striking feature of *Coleviroid* is that, in their secondary structure, the left and right arms are shared by at least one member within the genus (Nie and Singh, 2017; Smith et al., 2021). If this trend remains true, it is anticipated that two more viroids resulting from genome recombination among the already identified *Coleviroid* are yet to emerge or be discovered.

B. Prospecting for viroid

Viroids of *Humulus* plants

Hop (*Humulus lupulus* L.) is widely cultivated in the world and used for the brewing industry and beer production. Four viroids, namely, hop stunt viroid (HSVd, *Hostuviroid*), hop latent viroid (HLVd, *Cocadviroid*), apple fruit crinkle viroid (AFCVd, tentatively, *Apscaviroid*), and citrus bark cracking viroid (CBCVd, *Cocadviroid*), all in the *Pospiviroidae* family, have been reported to infect hop naturally. In North America (specifically in the USA), only HLVd and HSVd have been found in the crop to date.

As suggested by its name, HLVd appears to have a latent infection judged by symptoms in foliage in most cultivars. The viroid is widely distributed in hop gardens in many countries, including the USA (Nelson et al., 1997; Eastwell and Nelson, 2007), and the impact of the viroid on hop yield is relatively small on most cultivars, especially in comparison to other hop-infecting viroids. Nevertheless, the viroid can cause a serious reduction in cone yield (Barbara et al., 1990) in certain cultivars and alter the content of secondary metabolites (e.g., α- and β-acids) associated with flavor (Barbara et al., 1990; Patzak et al., 2001). It is noteworthy that HLVd has recently been reported in *Cannabis sativa*, which, like hop, belongs to the Cannabaceae family, exhibiting various symptoms such as stunting, leaf malformation/chlorosis, increased stem brittleness, and yield reduction in California (Bektaş et al., 2019; Warren et al., 2019). In Canada, laboratory testing on cannabis samples in 2020/2021 revealed an HLVd positive rate of over 25% (A&L Canada Laboratory Inc, 2021).

HSVd, AFCVd, and CBCVd cause significantly more severe symptoms and yield reduction on hop crops than HLVd. The distribution of these viroids appears to be different: HSVd has been reported in Asia (mainly Japan) and North America (USA) and Europe; AFCVd has only been found in hop in Asia (Japan); and CBCVd has only been reported in Europe. In the USA, a study in Washington state demonstrated a common occurrence of HSVd in all surveyed hop production regions in the state (Eastwell and Nelson, 2007; Kappagantu et al., 2017). Depending on cultivars, the impact of HSVd infection on dry cone yield and α- and β-acid content vary, ranging from a minor in some (e.g., Nugget, Columbus, and Galena) to significant in some others (e.g., Cascade, Willamette, and Glacier) (Kappagantu et al., 2017). While the distribution and occurrence in other hop production areas in the USA and Canada entirely are unknown, the presence of the viroid in other plant species such as *Citrus* spp. in Texas (Kunta et al., 2007) and Georgia (Stackhouse et al., 2021), *Prunus* spp. in California (Osman et al., 2012), and grapevines in Canada (Xiao et al., 2019; Fall et al., 2020) has been reported. It would not be surprising if the viroid is also present in hop crops in those regions.

Viroids of fruit trees

There are diverse fruit trees that are planted in North America. Three members in the family *Avsunviroidae*, namely, avocado sunblotch viroid (ASBVd), peach latent mosaic viroid (PLMVd), and apple hammerhead viroid (AHVd), and nine members in the family *Pospiviroidae*, namely, CEVd, CBCVd, citrus viroid V (CVd-V), HSVd, ASSVd, Australian grapevine viroid (AGVd), citrus dwarfing viroid (CDVd), grapevine yellow speckle viroid-1 (GYSVd-1), and grapevine yellow speckle viroid-2 (GYSVd-2), have been found in various fruit trees in the continent.

Two viroids, avocado sunblotch viroid (ASBVd, genus *Avsunviroid*, family *Avsunviroidae*) and PSTVd, have been known to infect avocado (*Persea americana* Miller). PSTVd infection in avocados has only ever been recorded once worldwide (Querci et al., 1995), and the infection is likely an isolated case resulting from avocado and potato intercropping in the region of Peru where the incidence occurred (Geering, 2018). ASBVd, on the other hand, has a narrow host range, infecting only avocadoes, but occurs in many countries (including the USA and Mexico) where avocados are grown (De La Torre et al., 2009; Kuhn et al., 2017; Lotos et al., 2018). ASBVd infection can cause significant yield losses even in asymptomatic trees (Saucedo-Carabez et al., 2014). Despite the fact California is where avocado sunblotch was first described (Coit, 1928) and serious outbreaks of the disease occurred in the past, it is currently considered a minor problem in the state where 90% of US avocados are produced. The disease is mainly managed by the use of registered disease-free trees and by the rouging of trees with symptoms (Faber et al., 2001). Florida is another state that produces avocados, and the occurrence of ASBVd has been reported in the National Germplasm Repository in the state (Schnell et al., 2011). However, sunblotch disease is rare in Florida (Crane et al., 2007). ASBVd was first officially reported in 2009 in Mexico (De La Torre et al., 2009), and it is not currently considered a regulated pest in the country. A survey of 70 Mexican race avocado trees (*Persea americana* var. *drymifolia*) from different regions and 35 "Hass" trees from Michoacán state indicated that ASBVd was present in 14% of the "Hass" trees in Michoacán and absent in all Mexican race avocado trees (Pérez et al., 2017).

Viroids in pome and stone fruit trees

Seven viroids, namely, AHVd and PLMVd from the family *Avsunviroidae* and ADFVd, ASSVd, AFCVd, PBCVd, and HSVd from *Pospiviroidae* family, are currently listed by NAPPO in its RSPM datasheet (NAPPO, 2022) for pome fruits. Two ASSVd-caused disease conditions (Hadidi et al., 1991), i.e., dapple apple disease and apple scar skin disease, have been described in the USA (dapple apple, Smith et al., 1956; apple skin scar, Millikan and Martin, 1956). Apple scar skin disease has also been reported in Canada (Welsh and Keene, 1961). Nevertheless, these diseases rarely occur in North America (Hadidi et al., 1991). Indeed, although the exact fruit tree species was not revealed, ASSVd was detected in only one out of 178 specimens of *Malus/Pyrus/Prunus* species collected since the mid-1960s and maintained in the Canadian Food Inspection Agency's (CFIA) Sydney Laboratory in British Columbia using high-throughput sequencing (HTS) for viruses/viroids (Rott et al., 2017). This survey also demonstrated the rarity of AFCVd (1/178), PBCVd (1/178), and ADFVd (4/178) in the specimens. PLMVd, on the other hand, occurred in 21/178 samples. AHVd has been detected in apples in Canada (variety Pacific Gala, Messmer et al., 2017) and the USA (variety Honeycrisp, Szostek et al., 2018). It remains unknown whether this viroid is associated with any specific symptoms.

Three stone fruit viroids (PLMVd, ASSVd, and HSVd) are listed by NAPPO in its RSPM35 (NAPPO, 2022). PLMVd occurs at a high rate in peach germplasm worldwide. A survey of

Prunus species conducted at the USDA's National Clonal Germplasm Repository in California, USA, revealed a PLMVd infection rate of 49.2% in peach, accounting for the majority (ca. 94%) of the infection for all *Prunus* spp. (Osman et al., 2012). A similar level of PLMVd infection was also found in commercial peach and nectarine plants from several states in the USA (Skrzeczkowski et al., 1996). In Canada, a PLMVd infection rate of 24.1% was found in *Prunus* spp. accessions in the Canadian Clonal Genebank located in Harrow (Ontario) (Michelutti et al., 2005). In Mexico, De La Torre-Almaráz et al. (2015) reported the detection of PLMVd in peach plants from three states (i.e., Puebla, Morelos, and Mexico). In the survey conducted on the *Prunus* species in California (Osman et al., 2012), HSVd was found in all tested species including apricot (28.9%), cherry (10%), peach (17.5%), plum (5%), and almond (4.3%), with an overall infection rate of 14.5% in all *Prunus* trees. An infection rate of 4.5% for HSVd was found in apricots in Canada during the survey of *Prunus* accessions in the Canadian Clonal Genebank in 2004 (Michelutti et al., 2004).

Viroids in citrus

Citrus is a natural host to at least seven viroids of four genera in the family *Pospiviroidae*. In North America, specifically in Mexico and the USA, five viroids, namely, CEVd (genus *Pospiviroid*), HSVd (including the non-cachexia-inducing strain, formerly citrus viroids IIa and the cachexia-inducing strains, formerly citrus viroids IIb and IIc, genus *Hostuviroid*), citrus bent leaf viroid (CBLVd, genus *Apscaviroid*), citrus dwarfing viroid (CDVd, formerly citrus viroid III or CVd-III, genus *Apscaviroid*), CBCVd (formerly citrus viroid IV, genus *Cocadviroid*), and citrus viroid V (CVd-V, genus *Apscaviroid*), have been reported infecting citrus (Almeyda-León et al., 2002; Guerrero Gámez et al., 2013; Duran-Vila et al., 1988; Kunta et al., 2007; Serra et al., 2008; Stackhouse et al., 2021). In NAPPA's RSPM for citrus (NAPPO, 2013), all of them are listed as occurring in the USA with a presence level of P4 (= Present: in all parts of the area where host crop(s) are grown). In Mexico, CEVd and the cachexia-inducing HSVd are listed at level P4 whereas the rest are listed at Ab1 (= Absent: no pest records). Nevertheless, Almeyda-León et al. (2002) reported a positive rate of ca. 37% for CDVd in Tahiti lime trees collected from Yucatan, Veracruz, Tabasco, and Colima states in Mexico, suggesting a common occurrence of this viroid in Mexico as well.

Viroids in grapevine

Five recognized viroids, namely, AGVd, CEVd, GYSVd-1, GYSVd-2, and HSVd, and three yet-to-be-recognized viroids (GYSVd-3 and grapevine latent viroid) are reported to infect grapevines (Di Serio et al., 2017; Jiang et al., 2009, 2012), all members (or tentative members) of the family *Pospiviroidae*. In addition, grapevine hammerhead viroid-like RNA has also been identified in grapevines (Wu et al., 2012). Among them, HSVd, GYSVd-1, GYSVd-2, and AGVd have been reported in grapevines in the USA (Sano et al., 1986; Rezaian et al., 1992), and HSVd, GYSVd-1, and GYSVd-2 have been reported in Canada (Xiao et al., 2019; Fall et al., 2020; Xu et al., 2021). In addition, an early study by Semancik et al. (1987) revealed the presence of three viroid-like RNAs designated grapevine viroid 1, 2, and 3 in grapevines in California. None of the viroids are listed by NAPPO in its RSPM for grapevines (NAPPO, 2022) due to the fact that viroids in grapevine are not known to cause any significant agronomic effects.

Chapter summary

Viroids are highly structured, small circular RNAs that do not encode proteins but can replicate and cause diseases in plants. Viroids were not discovered until the beginning of 1970 even though several major diseases including the spindle tuber disease in potatoes, sunblotch disease in avocados, and scar skin in apples, had been known for decades prior. In this chapter, viroids found in North America to date are reviewed according to their occurrence in host species. Also discussed are viroids' geographic distribution, economic significance, and regulatory status in North American Plant Protection Organization countries.

Study questions

1. What are viroids?
2. What was the first viroid discovered in the world and why is it categorized as a quarantine pest in the USA and Canada?
3. Where can you find information about the regulatory status of a viroid in North American countries?
4. Why is it important to study and understand the responses of a host, particularly an economically important host, to a specific viroid?
5. Could you please name five (5) viroids that are considered economically important and explain why you think so?

Further reading

Podleckis, E.V. 2017. Chapter 43: Geographical distribution of viroids in the Americas. In: Hadidi, A., et al., (Eds.), Viroids and Satellites. Academic Press, pp. 459–472. https://doi.org/10.1016/B978-0-12-801498-1.00043-7.
Aviña-Padilla, K., Zamora-Macorra, E.J., Ochoa-Martínez, D.L., Alcántar-Aguirre, F.C., Hernández-Rosales, M., Calderón-Zamora, L., Hammond, R.W. 2022. Mexico: a landscape of viroid origin and epidemiological relevance of endemic species. Cells, 11, 3487. https://doi.org/10.3390/cells11213487.

References

A&L Canada Laboratory Inc, 2021. A&L Cannabis & Hemp Newsletter September 2021. https://www.google.com/url?sa=t&rct=j&q=&esrc=s&source=web&cd=&cad=rja&uact=8&ved=2ahUKEwj6-MHaybz4AhVNa80KHbZxC1UQFnoECAgQAQ&url=https%3A%2F%2Fwww.alcanada.com%2Fpdf%2Fnewsletters%2FAL%2520Cannabis%2520Sept%25202021%2520Newsletter.pdf&usg=AOvVaw1d8MjME1pVzUZh6o7okiVg.

Almeyda-León, I.H., Iracheta-Cárdenas, M.M., Jasso-Argumedo, J.J., Curti-Díaz, S.A., Ruiz-Beltrán, P., Rocha-Peña, M.A., 2002. Reexamination of citrus viroids of Tahiti lime in Mexico. Rev. Mex. Fitopatol. 20, 152–160.

Barbara, D.J., Morton, A., Adams, A.N., Green, C.P., 1990. Some effects of hop latent viroid on two cultivars of hop (*Humulus lupulus*) in the UK. Ann. Appl. Biol. 117, 359–366.

Bektaş, A., Hardwick, K.M., Waterman, K., Kristof, J., 2019. Occurrence of hop latent viroid in *Cannabis sativa* with symptoms of cannabis stunting disease in California. Plant Dis. 103, 2699.

Belalcazar, C.S., Galindo, A.J., 1974. Estudio sobre el virus de la "planta macho" del jitomate. Agrociencia 18, 79–88.

Beumer, K., Stemerding, D., 2021. A breeding consortium to realize the potential of hybrid diploid potato for food security. Nat. Plants 7, 1530–1532.

Bostan, H., Nie, X., Singh, R.P., 2004. An RT-PCR primer pair for the detection of *Pospiviroid* and its application in surveying ornamental plants for viroids. J. Virol. Methods 116, 189–193.

CABI/EPPO, 2014. Potato Spindle Tuber Viroid. Distribution Maps of Plant Diseases (No. April), Map 729 (Edition 3). CABI, Wallingford, UK.

Chitambar, J., 2015. California Plant Pest Rating, Plant Pathogens, Viruses and Viroids, Potato Spindle Tuber Viroid. Calif. Depart. Food Agric. http://blogs.cdfa.ca.gov/Section3162/?p5384. (Accessed 7 November 2022).

Coit, J.E., 1928. Sunblotch of the avocado. In: California Avocado Society Yearbook 20. avocadosource.com, pp. 27–32.

Crane, J.H., Balerdi, C.F., Maguire, I., 2007. Avocado Growing in the Florida Home Landscape. EDIS Publication CIR1034.

De Boer, S.H., DeHaan, T.L., 2005. Absence of *potato spindle tuber viroid* within the Canadian potato industry. Plant Dis. 89, 910.

De La Torre, A.R., Téliz-Ortiz, D., Pallás, V., Sánchez-Navarro, J.A., 2009. First report of *Avocado sunblotch viroid* in avocado from Michoacán, México. Plant Dis. 93, 202.

De La Torre-Almaráz, R., Pallás, V., Sánchez-Navarro, J.A., 2015. First report of *Peach latent mosaic viroid* in peach trees from Mexico. Plant Dis. 99, 899.

Di Serio, F., Izadpanah, K., Hajizadeh, M., Navarro, B., 2017. Viroids infecting the grapevine. In: Meng, B., Martelli, G., Golino, D., Fuchs, M. (Eds.), Grapevine Viruses: Molecular Biology, Diagnostics and Management. Springer, Cham, https://doi.org/10.1007/978-3-319-57706-7_19.

Diener, T.O., 1971. Potato spindle tuber "virus". IV. A replicating, low molecular weight RNA. Virology 45, 411–428.

Diener, T.O., 1987. Potato spindle tuber. In: Diener, T.O. (Ed.), The Viroids. Springer, Boston, MA, pp. 221–233.

Diener, T.O., Lawson, R.H., 1973. Chrysanthemum stunt: a viroid disease. Virology 51, 94–101.

Dimock, A.W., 1947. Chrysanthemum stunt. N. Y. State Flower Growers' Bull. 26, 2.

Dimock, A.W., Geissinger, C.M., Horst, R.K., 1971. Chlorotic mottle: a newly recognized disease of chrysanthemum. Phytopathology 61, 415–419.

Duran-Vila, N., Roistacher, C.N., Rivera-Bustamante, R., Semancik, J.S., 1988. A definition of citrus viroid groups and their relationship to the exocortis disease. J. Gen. Virol. 69, 3069–3080.

Eastwell, K.C., Nelson, M.E., 2007. Occurrence of viroids in commercial hop (*Humulus lupulus* L.) production areas of Washington State. Plant Health Prog. 8, 1.

Faber, B.A., Wilen, C.A., Eskalen, A., Morse, J.G., Hanson, B.R., Hoddle, M.S., 2001. Revised continuously. In: UC IPM Pest Management Guidelines: Avocado. UC ANR Publication 3436, Oakland, CA.

Fall, M.L., Xu, D., Lemoyne, P., Moussa, I.E.B., Beaulieu, C., Carisse, O., 2020. A diverse virome of leafroll-infected grapevine unveiled by dsRNA sequencing. Viruses 12, 1142.

Flores, R., Minoia, S., López-Carrasco, A., Delgado, S., de Alba, Á.E.M., Kalantidis, K., 2017a. Viroid replication. In: Hadidi, A., Randles, J.W., Flores, R., Palukaitis, P. (Eds.), Viroids and Satellites. Academic Press, pp. 71–81.

Flores, R., Gago-Zachert, S., Serra, P., De la Peña, M., Navarro, B., 2017b. Chrysanthemum chlorotic mottle viroid. In: Hadidi, A., Randles, J.W., Flores, R., Palukaitis, P. (Eds.), Viroids and Satellites. Academic Press, pp. 331–338.

Flores, R., Di Serio, F., Navarro, B., Duran-Vila, N., Owens, R.A., 2021. Viroids and viroid diseases of plants. In: Studies in Viral Ecology. John Wiley & Sons, Inc, pp. 231–273.

Geering, A.D., 2018. A review of the status of *Avocado sunblotch viroid* in Australia. Australas. Plant Pathol. 47, 555–559.

Guerrero Gámez, C.E., Alvarado Gómez, O.G., Gutiérrez Mauleón, H., González Garza, R., Álvarez Ojeda, M.G., Luna Rodríguez, M., 2013. Detection of three citrus viroids species from Nuevo Leon and Tamaulipas, Mexico by conventional and real time RT-PCR. Rev. Mex. Fitopatol. 31, 20–28.

Hadidi, A., Hansen, A.J., Parish, C.L., Yang, X., 1991. Scar skin and dapple apple viroids are seed-borne and persistent in infected apple trees. Res. Virol. 142, 289–296.

Horst, R.K., 1975. Detection of a latent infectious agent that protects against infection by chrysanthemum chlorotic mottle viroid. Phytopathology 65, 1000–1003.

B. Prospecting for viroid

Jiang, D., Guo, R., Wu, Z., Wang, H., Li, S., 2009. Molecular characterization of a member of a new species of grapevine viroid. Arch. Virol. 154, 1563–1566.

Jiang, D., Sano, T., Tsuji, M., Araki, H., Sagawa, K., Purushothama, C.R.A., Zhang, Z., Guo, R., Xie, L., Wu, Z., Wang, H., 2012. Comprehensive diversity analysis of viroids infecting grapevine in China and Japan. Virus Res. 169, 237–245.

Kappagantu, M., Nelson, M.E., Bullock, J.M., Kenny, S.T., Eastwell, K.C., 2017. Hop stunt viroid: effects on vegetative growth and yield of hop cultivars, and its distribution in Central Washington State. Plant Dis. 101, 607–612.

Kuhn, D.N., Geering, A.D., Dixon, J., 2017. Avocado sunblotch viroid. In: Hadidi, A., Randles, J.W., Flores, R., Palukaitis, P. (Eds.), Viroids and Satellites. Academic Press, pp. 297–305.

Kunta, M., Da Graca, J.V., Skaria, M., 2007. Molecular detection and prevalence of citrus viroids in Texas. HortScience 42, 600–604.

Lawson, R.H., 1987. Chrysanthemum stunt. In: Diener, T.O. (Ed.), The Viroids. Springer, Boston, MA, pp. 247–259.

Ling, K.S., Bledsoe, M.E., 2009. First report of *Mexican papita viroid* infecting greenhouse tomato in Canada. Plant Dis. 93, 839.

Ling, K.S., Sfetcu, D., 2010. First report of natural infection of greenhouse tomatoes by *potato spindle tuber viroid* in the United States. Plant Dis. 94, 1376.

Ling, K.S., Zhang, W., 2009. First report of a natural infection by *Mexican papita viroid* and *Tomato chlorotic dwarf viroid* on greenhouse tomatoes in Mexico. Plant Dis. 93, 1216.

Ling, K.S., Verhoeven, J.T.J., Singh, R.P., Brown, J.K., 2009. First report of *tomato chlorotic dwarf viroid* in greenhouse tomatoes in Arizona. Plant Dis. 93, 1075.

Ling, K.S., Li, R., Groth-Helms, D., Assis-Filho, F.M., 2014. First report of *Potato spindle tuber viroid* naturally infecting field tomatoes in the Dominican Republic. Plant Dis. 98, 701.

Lotos, L., Kavroulakis, N., Navarro, B., Di Serio, F., Olmos, A., Ruiz-García, A.B., Katis, N.I., Maliogka, V.I., 2018. First report of Avocado sunblotch viroid (ASBVd) naturally infecting Avocado (*Persea americana*) in Greece. Plant Dis. 102, 1470.

Martínez-Soriano, J.P., Galindo-Alonso, J., Maroon, C.J., Yucel, I., Smith, D.R., Diener, T.O., 1996. *Mexican papita viroid*: putative ancestor of crop viroids. Proc. Natl. Acad. Sci. U. S. A. 93, 9397–9401.

Messmer, A., Sanderson, D., Braun, G., Serra, P., Flores, R., James, D., 2017. Molecular and phylogenetic identification of unique isolates of hammerhead viroid-like RNA from 'Pacific Gala' apple (*Malus domestica*) in Canada. Can. J. Plant Pathol. 39, 342–353.

Michelutti, R., Al Rwahnih, M., Torres, H., Gomez, G., Luffman, M., Myrta, A., Pallas, V., 2004. First record of *Hop stunt viroid* in Canada. Plant Dis. 88, 1162.

Michelutti, R., Myrta, A., Pallás, V., 2005. A preliminary account on the sanitary status of stone fruits at the clonal genebank in Harrow, Canada. Phytopathol. Mediterr. 44, 71–74.

Millikan, D.F., Martin, W.R., 1956. An unusual fruit dimple symptom in apple. Plant Dis. Rep. 40, 229–230.

NAPPO, 2011. Movement of potatoes into a NAPPO member country. In: NAPPO Regional Standard for Phytosanitary Measures (RSPM) 3.

NAPPO, 2013. Integrated measures for the movement of citrus propagative material. In: NAPPO Regional Standards for Phytosanitary Measures (RSPM) 16.

NAPPO, 2022. Guidelines for the Movement of Propagative Plant Material of Stone Fruit, Pome Fruit, and Grapevine into a NAPPO Member Country – NAPPO Regional Standards for Phytosanitary Measures (RSPM) 35.

Nelson, M.E., Klein, R.E., Skrzeczkowski, L.J., 1997. Occurrence of hop latent viroid (HLVd) in hops in Washington State. Phytopathology 88, S108.

Nie, X., Singh, R.P., 2017. *Coleus blumei* viroids. In: Hadidi, A., Randles, J.W., Flores, R., Palukaitis, P. (Eds.), Viroids and Satellites. Academic Press, pp. 289–295.

Nie, X., Singh, R.P., Bostan, H., 2005. Molecular cloning, secondary structure, and phylogeny of three pospiviroids from ornamental plants. Can. J. Plant Pathol. 27, 592–602.

Orellana, G.E., Karasev, A.V., 2021. First report of *Coleus blumei* viroid 1 in commercial Coleus blumei in the United States. Plant Dis. 105, 4174.

Orozco Vargas, G., Galindo Alonso, J., 1986. Ecology of *tomato planta macho viroid*: I. Natural host plants; agroecosystem effect on viroid incidence and influence of temperature on viroid distribution. Rev. Mex. Fitopatol. 4, 19–28.

Osman, F., Rwahnih, M.A., Golino, D., Pitman, T., Cordero, F., Preece, J.E., Rowhani, A., 2012. Evaluation of the phytosanitary status of the *Prunus* species in the national clonal germplasm repository in California: survey of viruses and viroids. J. Plant Pathol. 94, 249–253.

B. Prospecting for viroid

Owens, R.A., Smith, D.R., Diener, T.O., 1978. Measurement of viroid sequence homology by hybridization with com-plementary DNA prepared in vitro. Virology 89, 388–394.

Patzak, J., Matoušek, J., Krofta, K., Svoboda, P., 2001. Hop latent viroid (HLVd)-caused pathogenesis: effects of HLVd infection on lupulin composition of meristem culture-derived *Humulus lupulus*. Biol. Plant. 44, 579–585.

Pérez, M.R.V., Ortiz, D.T., Almaraz, R.D.L.T., Martinez, J.O.L., Ángel, D.N., 2017. Avocado sunblotch viroid: pest risk and potential impact in México. Crop Prot. 99, 118–127.

Querci, M., Owens, R.A., Vargas, C., Salazar, L.F., 1995. Detection of potato spindle tuber viroid in avocado growing in Peru. Plant Dis. 79, 196–202.

Rezaian, M.A., Krake, L.R., Golino, D.A., 1992. Common identity of grapevine viroids from USA and Australia revealed by PCR analysis. Intervirology 34, 38–43.

Rott, M., Xiang, Y., Boyes, I., Belton, M., Saeed, H., Kesanakurti, P., Hayes, S., Lawrence, T., Birch, C., Bhagwat, B., Rast, H., 2017. Application of next generation sequencing for diagnostic testing of tree fruit viruses and viroids. Plant Dis. 101, 1489–1499.

Sano, T., Ohshima, K., Hataya, T., Uyeda, I., Shikata, E., Chou, T.G., Meshi, T., Okada, Y., 1986. A viroid resembling hop stunt viroid in grapevines from Europe, the United States and Japan. J. Gen. Virol. 67, 1673–1678.

Saucedo-Carabez, J.R., Téliz-Ortiz, D., Ochoa-Ascencio, S., Ochoa-Martínez, D., Vallejo-Pérez, M.R., Beltrán-Peña, H., 2014. Effect of Avocado sunblotch viroid (ASBVd) on avocado yield in Michoacan, Mexico. Eur. J. Plant Pathol. 138, 799–805.

Schnell, R.J., Tondo, C.L., Kuhn, D.N., Winterstein, M.C., Ayala-Silva, T., Moore, J.M., 2011. Spatial analysis of avocado sunblotch disease in an avocado germplasm collection. J. Phytopathol. 159, 773–781.

Semancik, J.S., Rivera-Bustamante, R., Goheen, A.C., 1987. Widespread occurrence of viroid-like RNAs in grapevines. Am. J. Enol. Vitic. 38, 35–40.

Serra, P., Barbosa, C.J., Daròs, J.A., Flores, R., Duran-Vila, N., 2008. Citrus viroid V: molecular characterization and synergistic interactions with other members of the genus *Apscaviroid*. Virology 370, 102–112.

Singh, R.P., 2014. The discovery and eradication of potato spindle tuber viroid in Canada. Virus Dis. 25, 415–424.

Singh, R.P., Clark, M.C., 1971. Infectious low-molecular weight ribonucleic acid from tomato. Biochem. Biophys. Res. Commun. 44, 1077–1082.

Singh, R.P., Dilworth, A.D., 2009. Tomato chlorotic dwarf viroid in the ornamental plant *Vinca* minor and its trans-mission through tomato seed. Eur. J. Plant Pathol. 123, 111–116.

Singh, R.P., Boucher, A., Singh, A., 1991. High incidence of transmission and occurrence of a viroid in commercial seeds of Coleus in Canada. Plant Dis. 75, 184–187.

Singh, R.P., Lakshman, D.K., Boucher, A., Tavantzis, S.M., 1992. A viroid from *Nematanthus wettsteinii* plants closely related to the Columnea latent viroid. J. Gen. Virol. 73, 2769–2774.

Singh, R.P., Nie, X., Singh, M., 1999. Tomato chlorotic dwarf viroid: an evolutionary link in the origin of pospiviroids. J. Gen. Virol. 80, 2823–2828.

Singh, R.P., Dilworth, A.D., Ao, X., Singh, M., Baranwal, V.K., 2009. Citrus exocortis viroid transmission through commercially-distributed seeds of *Impatiens* and *Verbena* plants. Eur. J. Plant Pathol. 124, 691–694.

Skrzeczkowski, L.J., Howell, W.E., Mink, G.I., 1996. Occurrence of peach latent mosaic viroid in commercial peach and nectarine cultivars in the US. Plant Dis. 80, 823.

Smith, W.W., Barratt, J.G., Rich, A.E., 1956. Dapple apple, an unusual fruit symptom of apples in New Hampshire. Plant Dis. Rep. 40, 756–766.

Smith, R.L., Lawrence, J., Shukla, M., Singh, M., Li, X., Xu, H., Gardner, K., Nie, X., 2018. First report of Coleus blumei viroid 5 and molecular confirmation of *Coleus blumei* viroid 1 in commercial *Coleus blumei* in Canada. Plant Dis. 102, 1862.

Smith, R.L., Lawrence, J., Shukla, M., Singh, M., Li, X., Xu, H., Chen, D., Gardner, K., Nie, X., 2019. Occurrence of *Coleus blumei* viroid 6 in commercial *Coleus blumei* in Canada: the first report outside of China. Plant Dis. 103, 782.

Smith, R.L., Shukla, M., Creelman, A., Lawrence, J., Singh, M., Xu, H., Li, X., Gardner, K., Nie, X., 2021. *Coleus blumei* viroid 7: a novel viroid resulting from genome recombination between Coleus blumei viroids 1 and 5. Arch. Virol. 166, 3157–3163.

Stackhouse, T., Waliullah, S., Oliver, J.E., Williams-Woodward, J.L., Ali, M.E., 2021. First report of hop stunt viroid infecting citrus trees in Georgia, USA. Plant Dis. 105, 515.

B. Prospecting for viroid

Sun, M., Siemsen, S., Campbell, W., Guzman, P., Davidson, R., Whitworth, J.L., Bourgoin, T., Axford, J., Schrage, W., Leever, G., Westra, A., 2004. Survey of potato spindle tuber viroid in seed potato growing areas of the United States. Am. J. Potato Res. 81, 227–231.

Szostek, S.A., Wright, A.A., Harper, S.J., 2018. First report of apple hammerhead viroid in the United States, Japan, Italy, Spain, and New Zealand. Plant Dis. 102, 2670.

Verhoeven, J.T.J., Jansen, C.C.C., Willemen, T.M., Kox, L.F.F., Owens, R.A., Roenhorst, J.W., 2004. Natural infections of tomato by *Citrus exocortis viroid*, *Columnea latent viroid*, *Potato spindle tuber viroid* and *tomato chlorotic dwarf viroid*. Eur. J. Plant Pathol. 110, 823–831.

Verhoeven, J.T.J., Jansen, C.C.C., Werkman, A.W., Roenhorst, J.W., 2007. First report of *Tomato chlorotic dwarf viroid* in *Petunia hybrida* from the United States of America. Plant Dis. 91, 324.

Warren, J.G., Mercado, J., Grace, D., 2019. Occurrence of hop latent viroid causing disease in *Cannabis sativa* in California. Plant Dis. 103, 2699.

Welsh, M.F., Keene, F.W.L., 1961. Diseases of apple that are caused by viruses or have characteristics of virus diseases. Can. Plant Dis. Surv. 41, 123–147.

Wu, Q., Wang, Y., Cao, M., Pantaleo, V., Burgyan, J., Li, W.X., Ding, S.W., 2012. Homology-independent discovery of replicating pathogenic circular RNAs by deep sequencing and a new computational algorithm. Proc. Natl. Acad. Sci. U. S. A. 109, 3938–3943.

Xiao, H., Li, C., Al Rwahnih, M., Dolja, V., Meng, B., 2019. Metagenomic analysis of riesling grapevine reveals a complex virome including two new and divergent variants of grapevine leafroll-associated virus 3. Plant Dis. 103, 1275–1285.

Xu, D., Adkar-Purushothama, C.R., Lemoyne, P., Perreault, J.P., Fall, M.L., 2021. First report of grapevine yellow speckle viroid 1 infecting grapevine (*Vitis vinifera*) in Canada. Plant Dis. 105, 4174.

Viroid-associated plant diseases in South America

Nicola Fiore and M. Francisca Beltrán

University of Chile, Faculty of Agricultural Sciences, Department of Plant Health, Santiago, Chile

Graphical representation

Viroid occurrence in South America

Definitions

Symptomless plant: An infected plant that shows no symptoms.

Return-polyacrylamide gel electrophoresis (R-PAGE): A laboratory technique that allows the separation and detects viroids.

Prevalence: The proportion of plant samples in which evidence of infection is detected.

Leaf epinasty: Downward curving of leaves.

Mosaic: Leaves that are mottled with yellow, white, and light or dark green spots or streaks.

Phylogenetic analysis: The study of evolutionary relationships among biological entities.

Epidemiology: It is the study of the factors that influence the appearance of a disease and how it evolves in space and time.

Chapter outline

In South America, the viroid species detected to date are fifteen. Infected hosts are chrysanthemum, potato, pome fruits, stone fruits, citrus, grapevine, and avocado. This chapter provides information about the viroid species found in the different countries of South America, according to information available in the literature.

Learning objectives

- To know the importance of the diseases caused by viroids in the different hosts and countries of South America.
- To know that mixed infections between different species of viroids and between viroids and viruses are common in several plant hosts.

Fundamental introduction

Fifteen viroids have been found in fruit trees, grapevine, herbaceous ornamental plants, and horticultural plants in South America. *Avocado sunblotch viroid* (ASBVd) and *Peach latent mosaic viroid* (PLMVd) are members of the family *Avsunviroidae*, and thirteen, *Citrus bent leaf viroid* (CBLVd), *Citrus dwarfing viroid* (CDVd), *Citrus exocortis viroid* (CEVd), *Pear blister canker viroid* (PBCVd), *Potato spindle tuber viroid* (PSTVd), *Apple hammerhead viroid* (AHVd), *Apple scar skin viroid* (ASSVd), *Australian grapevine viroid* (AGVd), *Chrysanthemum stunt viroid* (CSVd), *Coleus blumei viroid-1* (CbVd-1), *Grapevine yellow speckle viroid 1* (GYSVd-1), *Grapevine yellow speckle viroid 2* (GYSVd-2), and *Hop stunt viroid* (HSVd) are members of the family *Pospiviroidae* (Table 1). These pathogens are transmitted through vegetative propagation when vegetal material from infected plants is used.

Herbaceous ornamentals

Plant viroids have a negative impact on ornamental plants reducing both the decorative value and quality of propagated material and causing economic damage. CSVd and CbVd-1 are the only viroids known to be important in ornamental plants in South America.

TABLE 1 Geographical distribution of viroids reported in South America.

Viroid	Host	Country
Apple hammerhead viroid (AHVd)	Apple	Brazil
Apple scar skin viroid (ASSVd)	Apple	Argentina
Australian grapevine viroid (AGVd)	Grapevine	Chile
Avocado sunblotch viroid (ASBVd)	Avocado	Peru
		Venezuela
Chrysanthemum stunt viroid (CSVd)	Chrysanthemum	Brazil
	Oxalis latifolia	Colombia
Citrus bent leaf viroid (CBLVd)	*Citrus* spp.	Argentina
		Uruguay
Citrus dwarfing viroid (CDVd)	*Citrus* spp.	Argentina
		Brazil
		Uruguay
		Colombia
Citrus exocortis viroid (CEVd)	*Citrus* spp.	Argentina
		Chile
		Brazil
		Uruguay
		Colombia
	Grapevine	Brazil
Coleus blumei viroid-1 (CbVd-1)	Coleus	Brazil
Grapevine yellow speckle viroid 1 (GYSVd-1)	Grapevine	Chile
		Brazil
Grapevine yellow speckle viroid 2 (GYSVd-2)	Grapevine	Chile
Hop stunt viroid (HSVd)	*Citrus* spp.	Argentina
		Chile
		Brazil
		Uruguay
	Grapevine	Chile
		Brazil
	Peach and nectarine	Chile
	Porcelain berry	Brazil

Continued

B. Prospecting for viroid

TABLE 1   Geographical distribution of viroids reported in South America—cont'd

Viroid	Host	Country
Peach latent mosaic viroid (PLMVd)	Peach	Argentina
		Chile
		Brazil
		Uruguay
	Plum	Uruguay
Pear blister canker viroid (PBCVd)	Pear	Argentina
	Apple	Chile
Potato spindle tuber viroid (PSTVd)	Potato	Peru
		Venezuela
	Avocado	Peru

CSVd is a member of the genus *Pospiviroid* and is distributed worldwide as the causal agent of Chrysanthemum Stunt Disease in cultivated chrysanthemum (*Chrysanthemum* spp.). This disease was reported for the first time in the 1940s in the USA (Dimock, 1947) and almost 30 years later, its viroid etiology was identified (Diener and Lawson, 1973). CSVd has been reported in Brazil causing symptoms like dwarfing, reduced size of the flowers, and color-break (Gobatto et al., 2014). There is a high variation in the severity of disease associated with the presence of the viroid, from symptomless plants to those in which the infection may cause growth reduction of up to 65%. In addition, flowers may show bleaching, and on the leaves of some cultivars may appear bright yellow spots of various sizes (Verhoeven et al., 2017). In Brazil, CSVd is widespread in chrysanthemum crops, and, in most cases, infected plants do not express symptoms, ensuring the permanence of the viroid in the cropping systems (Gobatto et al., 2014). Due to its stability, the viroid often reaches high concentrations in the host plant and is easily disseminated by grafting and mechanical transmission (e.g., foliar contact, and cutting tools contaminated). Its transmission via true seeds has not yet been confirmed, as well as its transmission by insect vectors. The spread of the CSVd is related to the marketing of infected chrysanthemum and other plant species (Gobatto et al., 2019; Wierzbicki and Eiras, 2021). CSVd was also found in plants of broadleaf woodsorrel (*Oxalis latifolia* Kunth) and *Chrysantemum* spp. in chrysanthemum-producing fields in Río Negro, Colombia (Gobatto et al., 2019). Broadleaf woodsorrel plants displayed mosaic and foliar deformation symptoms and findings of natural infection indicated an important role as a reservoir and a potential source of viroid inoculum in the field. No natural CSVd infection was detected in any of the other weed species in Brazilian fields, but a study with experimental hosts showed that CSVd was able to infect the following species: *Amaranthus viridis* L., *Cardamine bonariensis* Pers., *Euphorbia hirta* L. (= *Chamaesyce hirta* (L.) Millsp.), *Conyza bonariensis* (L.) Cronquist, *Digitaria sanguinalis* (L.) Scop., *Gomphrena globose* L., *Helianthus annuus* L., *Lupinus polyphyllus* Lindl., *Mirabilis jalapa* L., *Portulaca oleracea* L., and *Catharanthus roseus* (L.) G. Don (Gobatto et al., 2019). CbVd-1, a species in the genus *Coleviroid*, was first reported in Brazil in commercial coleus [*Plectranthus scutellarioides* (L.) Codd (= *C. blumei* Benth.)], cultivar Amarelo (Fonseca et al., 1990). The viroid was isolated from young leaves and detected by return-polyacrylamide gel electrophoresis (R-PAGE). Coleus viroid migrated faster than other viroids, indicating either a small size or a unique secondary structure. Inoculation of viroid-free coleus cultivars resulted in slightly chlorotic leaves in cultivar Frilled Fantasy and no symptoms in the other two infected cultivars. Viroid infection in the coleus is important because the propagation is usually by cuttings (Fonseca et al., 1990). In Brazil, there are still few studies on the occurrence, distribution, and potential for viroid-induced damage in ornamental plants. Nevertheless, a study with plants of Colombian and Brazilian producers of chrysanthemums showed an average of 15% of losses in production due to infections caused by CSVd. This percentage is equivalent to 5,848,200, 2,924,100, and 487,350 stalks per cycle of production in an area of 60, 30, and 5 ha, respectively. On average, a healthy stalk, containing 5–7 flower points, sells for US$0.10–US$0.15, depending on the variety. So when these stalks are sold at US$0.10, the Colombian producer who has 60 ha, the loss is US$584,820.00 per productive cycle (Gobatto, 2018).

Horticultural plants

Potato

Potato Spindle Tuber Disease was first described in the early 1920s (Martin, 1922; Schultz and Folsom, 1923) but the identification of its causal agent, a small, highly structured, covalently closed circular RNA, known as potato spindle tuber viroid (PSTVd), was achieved in 1971 (Diener, 1971). The first report in South America of PSTVd infecting potato plants (*Solanum tuberosum* L.) was in Venezuela (Singh, 1973), followed by Peru (Salazar et al., 1983). PSTVd is the type member of the genus *Pospiviroid* and some strains are capable of inducing either mild or intermediate symptoms. Symptoms on infected potato foliage are difficult to recognize in the season in which infection occurs and may increase in severity with successive generations of infections (Werner, 1926). Foliar symptoms include stunting, uprightness, and small leaflets. Tubers produced by diseased plants are smaller, fewer in number, elongated, with numerous shallow eyes, and have abnormal skin color and texture (Werner, 1926). Tubers infected with a mild-intermediate strain seem stunted, whose appearance suggests the inability to originate a new plant. On the other hand, a severe strain can cause dwarf curly leaf symptoms (Owens, 2007). PSTVd is highly infectious, being spread rapidly by contact, through pollen, true seed, and by aphids, through trans-encapsidation with *Potato leaf roll virus* (PLRV) (Grasmick and Slack, 1986; Salazar, 1995). The viroid can be a serious problem in seed potato certification programs, potato germplasm centers, and potato breeding programs (Pfannesnstiel and Slack, 1980).

Fruit trees and grapevine

Pome fruits

Three viroids have been detected in pome fruits in South America: PBCVd in pear and apple trees, and ASSVd and AHVd in apple trees.

B. Prospecting for viroid

ASSVd is a member of the genus *Apsacaviroid* that induces a serious disease on pome fruit trees named Apple Scar Skin. The viroid was first detected in South America infecting apple trees (*Malus domestica* Borkh.) in several orchards located in Argentina (Nome et al., 2012). Viroid infections cause symptoms in apple fruits that render the fruits unmarketable and cause economic losses. Symptoms induced by ASSVd infections are restricted to fruits and include color dappling, cracking, scarring, and distortion depending upon the cultivar (Koganezawa et al., 2003). Recently, AHVd was reported causing damage to the Brazilian apple industry. AHVd is a member of the genus *Pelamoviroid* and was reported in apple trees cv. Royal Gala. Infected plants show shoot decline/dieback, trunk splitting and mosaic, and/or ringspots (Nickel et al., 2021). PBCVd, a member of the genus *Apscaviroid*, was first reported in Argentina (Nome et al., 2011). It is a viroid responsible for Pear Blister Canker Disease, although the absence of symptoms is very common. In pear trees (*Pyrus communis* L.), PBCVd may cause bark pustules and rounded-scaly cankers varying from 2 to 6cm in diameter on the stems (Nome et al., 2011). Argentina is one of the main exporters of pears worldwide, with the Río Negro Valley concentrating the largest production. Therefore, diseases occupy an important place since they reduce the production and quality of the fruit and the profitability of the exploitation. A study carried out in pear trees cv. Red Sensation Bartlett, to evaluate the impact of this disease, showed symptoms restricted to the bark with superficial cracks that progressively evolved into cankers, scales, or deep fissures. Yield decreased in infected trees by about 30% (kg fruits/tree), other variables such as growth and production ratio were also affected (De Rossi et al., 2015; Frayssinet et al., 2012). PBCVd was also found in Chile in pear trees, with a prevalence of 2.5% (Medina et al., 2017).

Stone fruits

The viroids frequently detected in stone fruit trees are PLMVd, important for their role in decreasing stone fruit production, and HSVd, which infects mainly nectarine and peach trees but is asymptomatic. HSVd is a member of the genus *Hostuviroid* and was detected by RT-PCR in nectarine and peach plants in Chile with a prevalence of 1.2%, and no symptoms were observed in infected plants (Fiore et al., 2016a). Peach Latent Mosaic Disease is economically important and is associated with PLMVd, which belongs to the *Pelamoviroid* genus. It was reported to infect peach trees (*Prunus persica* (L.) Stokes) in commercial orchards in Argentina, Chile, Brazil, and Uruguay. The viroid-affected trees usually are asymptomatic or show vein banding and yellowing symptoms on leaves, chlorotic mosaics and sprouting delay, fruit suture necrosis (Fig. 1) and splitting, and delays in budding, flowering, and fruit ripening (Flores et al., 2006; Herranz et al., 2002; Nieto et al., 2008). PLMVd was first detected in peach trees with mosaic on leaves and fruit tissues with typical mosaic symptoms in Brazil (Hadidi et al., 1997). In 2001, PLMVd was detected during a survey in the main peach producing area in Uruguay (Herranz et al., 2002), and in peach trees with yellowing in commercial orchards in Argentina (Nieto et al., 2008). PLMVd infecting peach trees in Chile was reported in 2003 (Muñoz, 2004). In a survey in Chile, 25.7% of stone fruit plants resulted infected with PLMVd. Peach plants infected with PLMVd were asymptomatic or expressed mild mosaic, necrosis of buds, or fruit suture necrosis (Fiore et al., 2016a). PLMVd is largely favored by their mode of spread in the field, which occurs mainly by contaminated pruning

FIG. 1 Peach fruit from a plant infected by *Peach latent mosaic viroid*. Fruit deformed, discolored, and with suture necrosis.

tools and, although secondarily, through aphids (Fiore et al., 2016a). The co-occurrence of PLMVd with *Prunus necrotic ringspot virus* (PNRSV) has been frequently reported in peach trees (Fiore et al., 2016a; Herranz et al., 2002). A study conducted in Chile shows a synergistic effect between PLMVd and PNRSV on the transcriptome of peach, the down-regulated genes being prevalent. Among the set of repressed genes, the most interesting are those involved in host defense-related pathways. The number of gene expression changes was relatively low in the case of simple infections (Herranz et al., 2013). High incidences of PLMVd at the orchard level can cause significant financial loss given the effect on the vigor of the tree, affecting yields. Also, damage to the fruit (deformations and spots) cause their depreciation with the consequent economic detriment (Muñoz, 2004).

Citrus

The genus *Citrus* is one of the most important horticultural crops with worldwide fruit production (Najar et al., 2018). *Citrus* includes several species that produce important fruits such as oranges, mandarins (*Citrus reticulata* Blanco), limes (*Citrus latifolia* Tan.), lemons (*Citrus limon* (L.) Burm. f.), sour oranges (*Citrus aurantium* L.), and grapefruits (*Citrus × paradisi* Macfad.). Four viroids are responsible for *Citrus* spp. infections in South America: CEVd, CBLVd, CDVd, and HSVd. The three first viroids belong to the genus *Apscaviroid*, and HSVd to the genus *Hostuviroid*. In Brazil, citrus is one of the most important crops, being one of the top five countries in citrus production. Rootstocks used in citrus plants are important to have a profitable production against some limiting factors such as climate conditions, bad soil conditions, and diseases (Najar et al., 2018). Citrus viroid dissemination occurs mainly through the propagation of contaminated bud-wood material (Eiras et al., 2010). Symptoms like bark cracking and scaling followed by severe dwarfing and yield losses occur in viroid-infected citrus plants when commercial species are grafted on sensitive rootstocks such as trifoliate orange (*Poncirus trifoliata* (L.) Raf.), Rangpur lime (*C. limonia* Osb.), and citranges (*C. sinensis* × *P. trifoliata*), and viroid strain severity is high (Pagliano et al., 2013). This is particularly important, considering that more than 90% of citrus plants in Uruguay are grafted on either trifoliate orange or citranges, thus local citrus production may be at risk (Bisio et al., 2004). CEVd is the causal agent of Exocortis Disease and infects *Citrus* spp. in Argentina, Chile, Brazil, Uruguay, and Colombia. Symptoms have been classified into mild with few and small bark-cracks, moderate, with frequent bark-cracks, and severe displaying easily visible and extended bark cracking in the branches (also referred to as "quebra-galho"). The severe phenotype has been related to mixed infection between CEVd, HSVd, and CDVd in "Tahiti" acid lime (Eiras et al., 2010; Murcia et al., 2010a). Observation of symptoms of CEVd on single indicator plants such as Arizona 861 S1 citron has been a good alternative for efficient and cost-effective diagnosis. In plants inoculated with CEVd, the characteristic symptoms of severe epinasty and stunting were observed (Camps et al., 2014). Sensitive rootstock infected with CEVd showed bark scaling in Argentina (Palacios and Figueroa, 2022). HSVd is the causal agent of Cachexia Disease (or Citrus Xyloporosis) and infects *Citrus* spp. in Argentina, Chile, Brazil, and Uruguay. The disease is caused by specific variants of HSVd, formerly named *Citrus viroid II* (CVd-II). There are cachexia-inducing strains (CVd-IIb and CVd-IIc) and non-cachexia-inducing strains (referred to as CVd-IIa) (Najar et al., 2018). The causal agents of Cachexia Disease can incite severe gumming, discoloration, and wood pitting symptoms in alemow (*Citrus macrophylla* Wester), clementines (*Citrus clementina* Hort. ex Tanaka), mandarin, satsumas (*Citrus unshiu* Marcow), kumquats (*Citrus japonica* Lour.), and hybrids like tangelos (*Citrus paradisi* Macf. × *C. tangerina* Hort. ex Tan.) (Eiras et al., 2013). In Brazil, the Cachexia Disease was first described in the 1930s infecting "Barão" sweet orange [*Citrus sinensis* (L.) Osb.] grafted on "Rangpur" lime rootstock, and later was detected in "Pera" sweet orange, "Mexerica-do-rio," and "Dancy" mandarins and in "Red Blush" grapefruit (Müller and Costa, 1993). Symptoms of gumming and browning of phloem tissues and wood pitting in the field were observed in "Navelina ISA 315" trees infected by a cachexia-induced HSVd isolate (Carvalho et al., 2003). Plants of Arizona 861-S1 citron inoculated with HSVd showed symptoms of leaf break (Camps et al., 2014). CDVd (also known as CVd-III) infects Citrus trees in Argentina, Brazil, Uruguay, and Colombia. CDVd does not induce specific symptoms, but they cause a reduction in tree size and fruit crop. CDVd in the indicator plant Etrog citron (*Citrus medica* L.) causes mild stunting and a "leaf-dropping pattern" due to a moderate epinasty resulting from petiole and midvein necrosis (Murcia et al., 2009). Symptoms of infected citron plants showed different symptoms generated for infection with mild (CDVd^{IIIc}), moderate (CDVd^{so}), or severe strains (CDVd$^{IIIa\ it}$ and CDVd$^{IIIb\ sp}$). Severe strains caused young leaves of infected citron scions to show epinasty and old leaves developed petiole necrosis extending along the midrib of the leaves (Murcia et al., 2009). CDVd was found predominant in a survey in Argentina, in 90% of the samples analyzed (Palacios and Figueroa, 2022). CEVd, CBLVd, HSVd, and CDVd are widespread throughout citrus orchards in Uruguay with the highest prevalence (92%) of HSVd. Mixed infections of viroid were observed between HSVd and CDVd (17%), CEVd, HSVd, and CDVd (5%), and CBLVd, HSVd, and CDVd (4%). Single viroid infections were only observed for CBLVd and HSVd, with HSVd being the most common viroid detected (40% of infected plants) (Pagliano et al., 2013). In a survey for citrus viroid in Colombia, 40% of the samples analyzed were infected with viroids, 70% of the Tahiti lime sources were infected with HSVd and CDVd or with CEVd, HSVd, and CDVd, 33% of the citrons tested were infected with CEVd and CDVd, and Valencia sweet orange was viroid-free, the same for Mexican lime sources (Murcia et al., 2010b). Strains of CEVd, CDVd, and HSVd (CVd-IIa) were detected in grapefruit grafted on trifoliate orange rootstock (Targon et al., 2005). CBLVd is present in Argentina and Uruguay. The survey in Uruguay points out that viroid infections occur in many citrus orchards (62%), affecting all the commercial varieties tested and confirming the occurrence of CEVd, CBLVd, HSVd, and CDVd (Pagliano et al., 2013). The first report of CBLVd in Argentina occurred in grapefruit in mixed infection with CEVd, HSVd, and CDVd (Palacios and Figueroa, 2022). Mixed infections were detected in 86% of the samples. A total of five samples were found infected only with CDVd and one with CEVd. Other infecting citrus viroids like *Citrus bark cracking viroid* (CBCVd) have not been detected in Uruguay (Pagliano et al., 2013) or Argentina (Palacios and Figueroa, 2022). Probably, CEVd and HSVd are the most important citrus viroids for the economic losses caused, a concern when dealing with a culture that moves billions of dollars and generates thousands of direct jobs and indirect services in several countries of the five continents.

Avocado

ASBVd, a member of the genus *Avsunviroid*, is the causal agent of Sunblotch Disease in Avocados (*Persea americana* Miller). ASBVd has been reported in Peru and Venezuela (Rondon and Figueroa, 1976; Vargas et al., 1991). Active monitoring of commercial orchards in Chile has not detected this viroid.

Querci et al. (1995) reported the identification of PSTVd infecting avocados in Peru. Avocado trees with single infections of PSTVd (22.4%) or ASBVd (8%) and mixed infections of PSTVd and ASBVd (7.5%) were detected. PSTVd-infected trees from coastal locations showed an erect branching pattern with weak and slender branches. Severely affected trees showed prominent symptoms in leaves, inflorescence, and fruits.

Grapevine

Vitis spp. is susceptible to several graft-transmitted agents that cause relevant diseases. Five viroids have been reported infecting grapevines in South America. HSVd was found not only in *Vitis vinifera* L. and *Vitis labrusca* L. but also in some genotypes of wild grapevines like *V. aestivals* Michx. (= *Vitis gigas* J.H. Fennel) and *Vahlodea flexuosa* Thunb. A high prevalence of HSVd has been reported for grapevine in Chile and Brazil (Fajardo et al., 2018; Fiore et al., 2016b). HSVd was detected in 91% of the samples of Chilean grapevine with no distinction about cultivar or geographic distribution (Zamorano et al., 2015). According to the classification proposed by Amari et al. (2001), HSVd isolates found in Chilean grapevines were associated with P-H/Cit3 and Hop groups. HSVd has also been reported infecting porcelain berries [(*Parthenocissus heterophylla* (Blume) Merr.] (Fajardo et al., 2016, 2018). CEVd has been reported in Brazil. The Brazilian variants did not form specific groups and no relationship between the variants and geographical origin was observed.

Mixed infection of CEVd and HSVd has been reported in *V. vinifera* cv. Cabernet Sauvignon without symptoms and *V. labrusca* cv. Niagara Rosada presents yellow speckles in Brazil (Eiras et al., 2006). Three viroids from the genus *Apscaviroid* have been reported infecting grapevines, AGVd, and GYSVd-2 in Chile, and GYSVd-1 in Chile and Brazil (Fajardo et al., 2016; Zamorano et al., 2015). In Brazil, GYSVd-1 was detected in *V. labrusca* cv. Niagara Rosada, *V. vinifera* cv. Semillon, cv. Moscato Bailey, and cv. Violeta, displaying conspicuous yellow speckles (Fajardo et al., 2016). GYSVd-1 was also detected in asymptomatic vines. The prevalence of GYSVd-1 was 71.4% in Brazilian vineyards. Four types of GYSVd-1 were identified among the Brazilian isolates. Particularly, type 1, previously described as asymptomatic, and type 3, described as symptomatic (Szychowski et al., 1998). The sequence variability of GYSVd-1 isolates was previously noticed for Chilean isolates. GYSVd-1 was found in 20% of samples from Chilean *V. vinifera* plants with high prevalence in the variety "Syrah" being detected in 67% of the samples analyzed, all of them showing declining symptoms (Zamorano et al., 2015). GYSVd-2 was detected in 10.9% of Chilean grapevine samples with more prevalence in table grape varieties (67%) (Zamorano et al., 2015). Mixed infection between one of the GYSVd and *Grapevine fanleaf virus* can cause the symptom known as "vein banding" in the leaves (Fig. 2). Phylogenetic analysis of GYSVd-2 clustered all isolates in one group, closely related to Chinese and Australian isolates. AGVd was detected in 9.1% of Chilean grapevine samples. Chilean isolates of AGVd clustered indistinctly in both reported phylogenetic groups, even when the isolates shared cultivar and geographic origin (Zamorano et al., 2015).

FIG. 2   "Vein banding" in grapevine leaf resulting from a mixed infection between *Grapevine yellow speckle viroid* and *Grapevine fanleaf virus*.

Prospective

Some of the symptoms caused by viroids can be confused with alterations resulting from the action of an abiotic stress factor (for example, nutritional deficiency). Check whether or not there is a presence of viroids in plants, to take the appropriate control measures. Currently, few groups investigate viroids in South America; therefore, not enough information is being generated in this regard. The training of researchers interested in the study of viroids is urgent for South America.

Future implications

It is necessary to produce viroid-free plants. It is urgent to carry out epidemiological studies and investigate the mechanisms of viroid-plant interaction, which allow the development of control strategies against these pathogens. On the other hand, outreach activities to the producers should be increased, to recognize the symptoms and act quickly to reduce the spread of viroids in the crops.

Chapter summary

The status of viroids infection remains unknown in most of the countries of South America. Viroid diseases have not received special attention in this part of the world. It is important to have more researchers whose line of work is oriented towards the study of these pathogens. For the control of diseases caused by viroids, the prevention of infections is essential. Improving detection methods and obtaining viroid-free plant material for propagation purposes, is the most appropriate way to reduce the viroids spread. Next-Generation Sequencing is a diagnostic tool that can significantly increase the knowledge of viroids and their control.

Study questions

1. What are the three most widespread viroid species in South America?
2. What are the three viroid species that cause the most damage in South America?
3. What are the most characteristic symptoms caused by viroids?
4. How is it possible to reduce the spread of viroids in South America?

Further reading

Camps, R., Castro, M., Besoain, X. 2014. Simultaneous detection of CTV, CEVd and HSVd using Arizona 861 S1 citron and RT-PCR. Ciencia e Investigación Agraria, 41 (2), 23–24. https://doi.org/10.4067/S0718-16202014000200012

Fiore, N., Zamorano, A., Pino, A. M., González, F., Rosales, I. M., Sánchez-Navarro, J. A., & Pallás, V. (2016). Survey of stone fruit viruses and viroids in Chile. J. Plant Pathol. 5.

B. Prospecting for viroid

Gobatto, D., de Oliveira, L. A., de Siqueira Franco, D. A., Velásquez, N., Daròs, J.-A., Eiras, M. 2019. Surveys in the chrysanthemum production areas of Brazil and Colombia reveal that weeds are potential reservoirs of chrysanthemum stunt viroid. Viruses 11 (4), 355. https://doi.org/10.3390/v11040355

Owens, R. A. 2007. Potato spindle tuber viroid: The simplicity paradox resolved? Mol. Plant Pathol., 8 (5), 549–560. https://doi.org/10.1111/j.1364-3703.2007.00418.x

Vargas, C. O., Querci, M., Salazar, L. F. 1991. Identificación y diseminación del viroide del manchado solar del Palto (*P. americana* L.) en el Perú y la existencia de otros viroides en palto. Fitopatología, 26, 23–27.

Verhoeven, J., Hammond, R., Stancanelli, G. 2017. Economic significance of viroids in ornamental crops. In: Viroids and Satellites. Elsevier, p. 12.

References

Amari, K., Gomez, G., Myrta, A., Di Terlizzi, B., Pallás, V., 2001. The molecular characterization of 16 new sequence variants of hop stunt viroid reveals the existence of invariable regions and a conserved hammerhead-like structure on the viroid molecule. J. Gen. Virol. 82 (4), 953–962. https://doi.org/10.1099/0022-1317-82-4-953.

Bisio, L., Vignale, B., Lombardi, P., Carrau, F., 2004. Resultados del programa de mejoramiento de patrones cítricos en Uruguay. In: Proceedings of the 10th Congress Citriculture.

Camps, R., Castro, M., Besoain, X., 2014. Simultaneous detection of CTV, CEVd and HSVd using Arizona 861 S1 citron and RT-PCR. Ciencia e Investigación Agraria 41 (2), 23–24. https://doi.org/10.4067/S0718-16202014000200012.

Carvalho, S.A., Machado, M.A., Müller, G.W., 2003. Avaliação de indicadoras e portaenxertos na indexação biológica do viróide da xiloporose em citros. Laranja 24, 145–155.

De Rossi, R.P., Frayssinet, S., Santos López, S., 2015. Nuevo viroide en perales en Patagonia Norte. Fruticultura y Diversificación 21 (75), 40–44.

Diener, T.O., 1971. Potato spindle tuber "virus": IV. A replicating, low molecular weight RNA. Virology 45 (2), 411–428. https://doi.org/10.1016/0042-6822(71)90342-4.

Diener, T.O., Lawson, R.H., 1973. Chrysanthemum stunt: a viroid disease. Virology 51 (1), 94–101. https://doi.org/10.1016/0042-6822(73)90369-3.

Dimock, A.W., 1947. Chrysanthemum stunt. N. Y. State Flower Growers' Bull. 26, 2.

Eiras, M., Targon, M.L.P.N., Fajardo, T.V.M., Flores, R., Kitajima, E.W., 2006. Citrus exocortis viroid and hop stunt viroid isolates doubly-infecting grapevines in Brazil. Fitopatología Brasileira 31, 440–446.

Eiras, M., Silva, S.R., Stuchi, E.S., Flores, R., Daròs, J.-A., 2010. Viroid species associated with the bark-cracking phenotype of 'Tahiti' acid lime in the State of São Paulo, Brazil. Trop. Plant Pathol. 7.

Eiras, M., Silva, S.R., Stuchi, E.S., Carvalho, S.A., Garcêz, R.M., 2013. Identification and characterization of viroids in 'Navelina ISA. Trop. Plant Pathol. 6.

Fajardo, T.V.M., Eiras, M., Nickel, O., 2016. Detection and molecular characterization of grapevine yellow speckle viroid 1 isolates infecting grapevines in Brazil. Trop. Plant Pathol. 41 (4), 246–253. https://doi.org/10.1007/s40858-016-0097-1.

Fajardo, T.V.M., Eiras, M., Nickel, O., 2018. First report of hop stunt viroid infecting Vitis gigas, *V. flexuosa* and *Ampelopsis heterophylla*. Australas. Plant Dis. Notes 13 (1), 3. https://doi.org/10.1007/s13314-017-0287-9.

Fiore, N., Zamorano, A., Pino, A.M., González, F., Rosales, I.M., Sánchez-Navarro, J.A., Pallás, V., 2016a. Survey of stone fruit viruses and viroids in Chile. J. Plant Pathol. 5.

Fiore, N., Zamorano, A., Sánchez-Diana, N., González, X., Pallás, V., Sánchez-Navarro, J., 2016b. First detection of grapevine rupestris stem pitting-associated virus and grapevine rupestris vein feathering virus, and new phylogenetic groups for grapevine fleck virus and hop stunt viroid isolates, revealed from grapevine field surveys in Spain. Phytopathol. Mediterr. 55 (2). https://doi.org/10.14601/Phytopathol_Mediterr-15875.

Flores, R., Delgado, S., Rodio, M.E., Ambrós, S., Hernández, C., Serio, F.D., 2006. Peach latent mosaic viroid: not so latent. Mol. Plant Pathol. 7 (4), 209–221. https://doi.org/10.1111/j.1364-3703.2006.00332.x.

Fonseca, M.E.N., Boiteux, L.S., Singh, R.P., Kitajima, E.W., 1990. A viroid from Coleus species in Brazil. Plant Dis. 74 (80).

Frayssinet, S., Santos López, S., De Rossi, R.P., 2012. Pear Blister Canker Viroid (PBCVd) Y parámetros Productivos en Perales Red Bartlett, Clon Red Sensation. XIV Jornadas Fitosanitarias Argentinas, San Luis.

Gobatto, D., 2018. Chrysanthemum stunt viroid: Diversidade genética, hospedeiras alternativas, interações viroide-hospedeiro e estudo do impacto na produção do crisântemo. Instituto Biológico. http://repositoriobiologico.com. br/jspui/bitstream/123456789/96/1/DANIELLE%20GOBATTO.pdf.

Gobatto, D., Chaves, A.L.R., Harakava, R., Marque, J.M., Daròs, J.A., Eiras, M., 2014. Chrysanthemum stunt viroid in Brazil: survey, identification, biological and molecular characterization and detection methods. J. Plant Pathol. 10.

Gobatto, D., de Oliveira, L.A., de Siqueira Franco, D.A., Velásquez, N., Daròs, J.-A., Eiras, M., 2019. Surveys in the Chrysanthemum production areas of Brazil and Colombia reveal that weeds are potential reservoirs of Chrysanthemum stunt viroid. Viruses 11 (4), 355. https://doi.org/10.3390/v11040355.

Grasmick, M.E., Slack, S.A., 1986. Effect of potato spindle tuber viroid on sexual reproduction and viroid transmission in true potato seed. Can. J. Bot. 64 (2), 336–340. https://doi.org/10.1139/b86-048.

Hadidi, A., Giunchedi, L., Shamloul, A.M., Poggi-Pollini, C., Amer, M.A., 1997. Occurrence of peach latent mosaic viroid in stone fruits and its transmission with contaminated blades. Plant Dis. 81, 154–158.

Herranz, M.C., Maeso, D., Soria, J., Pallás, V., 2002. First report of peach latent mosaic viroid on peach in Uruguay. Plant Dis. 86 (12), 1405. https://doi.org/10.1094/PDIS.2002.86.12.1405C.

Herranz, M.C., Niehl, A., Rosales, M., Fiore, N., Zamorano, A., Granell, A., Pallas, V., 2013. A remarkable synergistic effect at the transcriptomic level in peach fruits doubly infected by prunus necrotic ringspot virus and peach latent mosaic viroid. Virol. J. 10, 164. http://www.virologyj.com/content/10/1/164.

Koganezawa, H., Yang, X., Zhu, S.F., Hashimoto, J., Hadidi, A., 2003. Apple scar skin viroid in apple. In: Hadidi, A., Flores, R., Randles, J.W., Semancik, J.S. (Eds.), Viroids. CSIRO Publishing, pp. 137–141.

Martin, W.H., 1922. «Spindle Tuber», a New Potato Trouble. Hints to Potato Growers, New Jersey State Potato Association, p. 3.

Medina, G., Quiroga, N., Méndez, P., Zamorano, A., Fiore, N., 2017. Identificación y caracterización molecular de los virus y viroides que infectan al peral (*Pyrus communis* L.) en Chile. In: XXV Congreso de la Sociedad Chilena de Fitopatología. XIX Congreso Latinoamericano de Fitopatología. LVIIAPS Caribbean Division Metting, Termas de Chillán, Chile.

Müller, G.W., Costa, A.S., 1993. Doenças causadas por vírus viroides e similares em citros. In: Rosseti, V. (Ed.), Doenças causadas por algas, fungos, bactérias e vírus. Fundação Cargill, pp. 55–84.

Muñoz, M., 2004. Peach latent mosaic viroid (PLMVd). (Informativo Fitosanitario N.° 5).

Murcia, N., Bernad, L., Serra, P., Bani Hashemian, S.M., Duran-Vila, N., 2009. Molecular and biological characterization of natural variants of Citrus dwarfing viroid. Arch. Virol. 154 (8), 1329–1334. https://doi.org/10.1007/s00705-009-0430-9.

Murcia, N., Bernad, L., Caicedo, A., Duran-Vila, N., 2010a. Citrus Viroids in Colombia. Int. Org. Citrus Virol. Conf. Proc. (1957–2010) 17 (17). https://doi.org/10.5070/C5008577NK.

Murcia, N., Hashemian, S.M.B., Bederski, K., Wulff, N.A., Barbosa, C.J., Bové, J.M., Duran-Vila, N., 2010b. Viroids in Tahiti lime scions showing bark cracking symptoms. Int. Org. Citrus Virol. Conf. Proc. (1957–2010) 17 (17). https://doi.org/10.5070/C58TF8M45D.

Najar, A., Hamdi, I., Mahmoud, K.B., 2018. Citrus viroids: Characterization, prevalence, distribution and struggle methods. J. New Sci. 50, 9.

Nickel, O., Fajardo, T.V.M., Candresse, T., 2021. First report on occurrence of apple hammerhead viroid in apples in Brazil. In: XXXII Congreso Brasileiro de Virologia: Virologia em Casa.

Nieto, A.M., Di Feo, L., Nome, C.F., 2008. First report of peach latent mosaic viroid in peach trees in Argentina. Plant Dis. 92 (7), 1137. https://doi.org/10.1094/PDIS-92-7-1137C.

Nome, C.F., Difeo, L.V., Giayetto, A., Rossini, M., Frayssinet, S., Nieto, A., 2011. First report of pear blister canker viroid in pear trees in Argentina. Plant Dis. 95 (7), 882. https://doi.org/10.1094/PDIS-03-11-0154.

Nome, C., Giagetto, A., Rossini, M., Di Feo, L., Nieto, A., 2012. First report of *apple scar skin viroid* (ASSVd) in apple trees in Argentina. New Dis. Rep. 25 (1), 3. https://doi.org/10.5197/j.2044-0588.2012.025.003.

Owens, R.A., 2007. Potato spindle tuber viroid: the simplicity paradox resolved? Mol. Plant Pathol. 8 (5), 549–560. https://doi.org/10.1111/j.1364-3703.2007.00418.x.

Pagliano, G., Umaña, R., Pritsch, C., Rivas, F., Duran-Vila, N., 2013. Occurrence, prevalence and distribution of citrus viroids in Uruguay. J. Plant Pathol. 5.

B. Prospecting for viroid

Palacios, M.F., Figueroa, J., 2022. First report of citrus bent leaf viroid and citrus dwarfing viroid in Argentina. J. Citrus Pathol. 9 (1). https://doi.org/10.5070/C49151296.

Pfannesnstiel, M.A., Slack, S.A., 1980. Response of potato cultivars to infection by the potato spindle tuber viroid. Phytopathology 70, 922–926.

Querci, M., Owens, R.A., Vargas, C., Salazar, L.F., 1995. Detection of potato spindle tuber viroid in avocado grown in Peru. Plant Dis. 79, 196–202.

Rondon, A., Figueroa, M., 1976. Sunblotch of avocado in Venezuela. Agronomía Tropical 26, 463–466.

Salazar, L.F., 1995. Los virus de la Papa y su control. (Centro Internacional de la papa).

Salazar, L.F., Owens, R.A., Smith, D.R., Diener, T.O., 1983. Detection of potato spindle tuber viroid by nucleic acid spot hybridization: evaluation with tuber sprouts and true potato seed. Am. Potato J. 60 (8), 587–597. https://doi.org/10.1007/BF02854108.

Schultz, E.S., Folsom, D., 1923. Transmission, variation, and control of certain degeneration diseases of Irish potatoes. J. Agric. Res. 25, 43–118.

Singh, R.P., 1973. Experimental host range of the potato spindle tuber 'virus'. Am. Potato J. 50 (4), 111–123. https://doi.org/10.1007/BF02857207.

Szychowski, J.A., Credi, R., Reanwarakorn, K., Semancik, J.S., 1998. Population diversity in grapevine yellow speckle viroid-1 and the relationship to disease expression. Virology 248 (2), 432–444. https://doi.org/10.1006/viro.1998.9292.

Targon, M.L.P.N., Carvalho, S.A., Stuchi, E.S., Souza, J.M., Muller, G.W., Borges, K.M., Machado, M.A., 2005. Hybridization techniques for indexing of citrus viroids in Sao Paulo State, Brazil. Laranja 26 (1), 25–38.

Vargas, C.O., Querci, M., Salazar, L.F., 1991. Identificación y diseminación del viroide del manchado solar del Palto (*Persea americana* L.) en el Perú y la existencia de otros viroides en palto. Fitopatología 26, 23–27.

Verhoeven, J., Hammond, R., Stancanelli, G., 2017. Economic significance of viroids in ornamental crops. In: Viroids and Satellites. Elsevier, p. 12.

Werner, H.O., 1926. The spindle-tuber disease as a factor in seed potato production. Bull. Agric. Exp. Station Nebraska 32.

Wierzbicki, R., Eiras, M., 2021. Pospiviroides de importância para hortaliças e ornamentais: Panorama dos aspectos fitopatológicos e dos riscos para os sistemas agrícolas brasileiros. Revisao Anual of Patologia de Plantas 27.

Zamorano, A., González, X., Quiroga, N., Fiore, N., 2015. OP 26—Detection and characterization of Chilean isolates of grapevine viroids. In: Proceedings of the 18th Congress of ICVG, p. 2.

Viroid pathogenesis and viroid-host interaction

Viroid pathogenicity

*Charith Raj Adkar-Purushothama[a], Francesco Di Serio[b],
Jean-Pierre Perreault[a], and Teruo Sano[c]*

[a]RNA Group, Department of Biochemistry and Functional Genomics, Faculty of Medicine
and Health Sciences, Applied Cancer Research Pavilion, University of Sherbrooke,
Sherbrooke, QC, Canada [b]Consiglio Nazionale delle Ricerche (CNR), Istituto per la
Protezione Sostenibile delle Piante (IPSP), Bari, Italy [c]Faculty of Agriculture and
Life Science, Hirosaki University, Hirosaki, Japan

Graphical representation

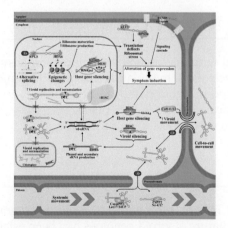

Proposed mechanisms of viroid pathogenesis and intercellular movement. This figure is reused from Fig. 3 of Ortolá
and Daròs, 2023 (Ortolá, B., Daròs, J.-A., 2023. Viroids: non-coding circular RNAs able to autonomously replicate
and infect higher plants. Biology 12 (2), 172. https://doi.org/10.3390/biology12020172) as per the Creative Commons
Attribution (CC BY) license (http://creativecommons.org/licenses/by/4.0/).

Definitions

Noncoding RNAs: RNAs (ribonucleic acids) that do not code for a protein.
Catalytic RNAs: RNAs that have self-cleaving activity.
RNA silencing: Also known as gene silencing or RNA interference (RNAi) is a mechanism of the sequence-specific regulation of gene expression that is highly conserved in most eukaryotic organisms. In addition, it plays various important functions in both the development and the maintenance of the vital activities of living organisms.
Pathogenesis: The mechanism by which a disease is developing.
Viroid structure: A viroid is a noncoding RNA. Their nucleotide sequences (that is, the primary structure of viroid) assume specific conformations that are characterized by Watson-Crick and wobble base pairing between self-complementary regions, forming a secondary structure, which is in turn arranged into a three-dimensional structure (that is, the tertiary structure of viroid).

Chapter outline

In previous chapters, we have seen that viroid infection induces visible symptoms in susceptible host plants despite both their small genome and their noncoding nature. This chapter examines the nucleotides, nucleotide sequences, and structural motifs that play a role in both viroid pathogenicity and the mechanism underlying symptom expression in susceptible host plants.

Learning objectives

- Viroid disease associated with microscopic and macroscopic symptoms in host plants.
- Structural motifs and nucleotides modulating viroid pathogenicity.
- Host components involved in viroid pathogenicity.
- Viroid-induced biochemical changes in host plants.
- Effects of viroid infection on the host's central dogma.

Introduction

Viroids are the smallest known autonomously replicating, noncoding RNA molecules. Even before the noncoding nature of viroid genomes had been established, how such a small RNA molecule could induce disease was a matter of interest (Owens and Hammond, 2009). In 1971, while proposing the new class of pathogens to be known as "viroids," T.O. Diener suggested that the genome of *Potato spindle tuber viroid* (PSTVd) might function as an abnormal regulatory RNA rather than as a functional mRNA as is observed in the case of viruses (Diener, 1971). The noncoding nature of viroid RNAs was confirmed in 1978 by the elucidation of the complete nucleotide sequence of PSTVd (Gross et al., 1978). Subsequent comparative studies of mild and severe isolates of PSTVd using RNA fingerprinting illustrated that only a minor change in sequence could dramatically affect symptom expression (Dickson et al., 1979). These fundamental discoveries lead to studies focusing on identifying the structural motifs involved in disease induction. As a result, our understanding of the molecular properties of viroid RNA, its structure, its pathogenicity, and its interaction with the host factors progressed significantly. In 2004, when researchers discovered that transgenic tomato plants expressing an inverted repeat hairpin construct of PSTVd exhibited viroid-associated symptoms (Wang et al., 2004), the study of viroid pathogenicity took a sudden turn implicating a role for RNA silencing in viroid pathogenicity. In this chapter, we will try to understand the different aspects of both the viroid and the host that play a role in both viroid pathogenicity and symptom expression in susceptible host plants.

Viroid-associated symptoms in host plants

Although viroids are considered as being plant pathogens, not all viroid infections are symptomatic. For example, the Australian grapevine viroid (AGVd) seldom induces visible symptoms in its host plant (grapevine). Contrarily, PSTVd induces an array of symptoms in tomato plant cultivars. Furthermore, higher temperatures are known to enhance viroid-associated disease symptoms in host plants. That said, viroid-associated symptoms are the result of complex viroid-host-environment interactions. The symptom intensity may vary with both the host/cultivar and the sequence of the viroid. Out of all of the known viroids, only the coconut cadang-cadang viroid (CCCVd) and the coconut tinangaja viroid (CTiVd) infections result in the death of host plants (Škorić, 2017).

Stunting is the most conspicuous viroid-associated disease symptom. However, other symptoms that are commonly associated with viroid infections may appear on the leaves (e.g., epinasty, chlorosis, necrosis, distortion, vein clearing, and vein discoloration), the bark (e.g., cracking, scaling, gumming, pitting, and pegging), the reproductive organs (e.g., flower and fruit shape/color, reduction in numbers and sterility) (Flores et al., 2017; Škorić, 2017). Infection often affects the root system by reducing its volume, as well as by causing its malformation (spindle-shaped tubers). At the cellular level, the cytopathic effects associated with viroid infection include the distortion of cell walls, abnormalities in the chloroplast, either the disruption or the proliferation of cytoplasmic membranes leading to the formation of "plasmalemmasomes" and the formation of electron-dense deposits in both the cytoplasm and the chloroplasts (Di Serio et al., 2013).

Structural motifs and nucleotides modulating viroid pathogenicity

Due to its lack of protein coat, the secondary structure of a viroid RNA plays a crucial role in viroid pathogenicity by providing the information necessary for a successful infection. Structural elements of viroid RNAs may interact with host factors for replication, pathogenesis, and transport to occur in the host plant (Navarro et al., 2021). Therefore, the elucidation of both the secondary and the tertiary structures of viroids is critical to be able to understand

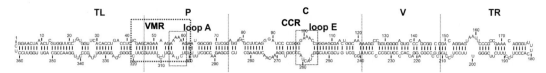

FIG. 1 The most stable secondary structures of PSTVd. The most thermostable structure of PSTVd (GenBank Acc. No. AB623143) was determined using the RNA structure software. The different structural/functional motifs are delimited by the *vertical dotted lines*, and the presence of both loop A and loop E are noted on the PSTVd structure. The nucleotides in the *red box* denote the Virulence Modulating Region (VMR; nucleotides 40–62 and 300–319). Loop A and loop E are shown in the *black box*.

viroid pathogenicity. Based on the nucleotide sequence of PSTVd, it was demonstrated that this viroid adopts a rod-like secondary structure with loops and bulges due to self-complementary intramolecular base-pairing (Gross et al., 1978). This seminal work, which was followed by the elucidation of the secondary structures of other viroids such as the *Citrus exocortis viroid* (CEVd) and the *Chrysanthemum stunt viroid* (CSVd), and those of their sequence variants, formed a basis upon which a structural model containing five structural and functional domains was proposed for PSTVd and all of its related viroids, except the *Avocado sunblotch viroid* (ASBVd) (Keese and Symons, 1985). These five domains are: the Terminal Left (TL), the Pathogenicity (P), the Central (C), the Variable (V), and the Terminal Right (TR) domains (Fig. 1). Two stretches of conserved nucleotides in the upper and lower stands within the C domain form the central conserved region (CCR). The C domain is capable of forming alternative structures that may regulate replication (Owens and Hammond, 2009). Although the V domain exhibited high sequence variability, the P domain is known to be associated with both pathogenicity and disease symptom severity for some viroids. The presence or the absence of the C domain divides viroids into two families, the *Pospiviroidae and the Avsunviroidae*. For details on the classification of viroids, see Chapter 2, *Viroid taxonomy*. Due to intramolecular base-pairing, any alteration of a viroid nucleotide may alter the structure of the viroid molecule. Hence, it is important to know both a viroid's structure and its nucleotide sequence to understand both its pathogenicity and its interaction with the host's components.

Pathogenicity determinants in the members of the *Pospiviroidae* family

After the determination of the structural/functional motifs of a viroid molecule, several studies using different host-viroid combinations were performed to identify the structural motifs modulating viroid pathogenicity. The majority of these studies were focused on the effects of PSTVd and its variants on sensitive tomato cultivars such as cv. Rutgers because tomato plants, as compared to other host plants, are convenient and rapid experimental hosts, only requiring 4 to 6 weeks for completion of the bioassay (Owens and Hammond, 2009). These experiments revealed that the viroid-induced symptoms were mainly determined by the nucleotide sequences in the P domain. For example, a change of 1 or 2 nucleotides in the P domain drastically changed symptom expression in tomato plants (Dickson et al., 1979). The secondary structure elucidation indicated that the observed virulence was correlated with the instability of a region of the P domain, specifically the "virulence modulating region" (VMR) (Fig. 1). It was hypothesized that the change of a nucleotide in the VMR affects symptom severity by modifying the ability of the VMR to interact with unidentified host factors (Schnölzer et al., 1985; Wassenegger et al., 1996). This hypothesis was supported by the drawing of a correlation between PSTVd virulence and the activation of the p68 protein kinase (Diener et al., 1993). However, the correlation between the VMR and disease severity observed with PSTVd was not detected with either CEVd (Visvader and Symons, 1985). To understand the possible contributions of other structural domains to pathogenicity, a series of interspecies chimeras between CEVd and TASVd were constructed and examined, by bioassays, for differences in symptom development (stunting, epinasty, veinal necrosis, etc.) in tomato plants. This experiment highlighted the involvement of the TL, rather than the P domain, in symptom severity with both CEVd and TASVd. This experiment also illustrated that the symptom severity observed is not completely independent of the viroid titer. This observation in turn led to the identification of multiple discrete sequence regions that may correspond to these pathogenicity determinants, namely, TL, P, V, and TR (Sano et al., 1992).

The most employed technique with which to study the role of specific nucleotides in viroid pathogenicity is mutagenic studies. In a mutagenic study, in which a single nucleotide located at nucleotide position 257 (that is, in the lower C domain) of PSTVd was changed from U to A, the conversion of a PSTVd wild-type intermediate strain to a lethal one in tomato was observed. The mutated viroid-infected plants showed severe growth stunting and a flat top of the shoot (Qi and Ding, 2003). Likewise, changing the nucleotide located at position 201 to C from U, G or A affected the replication of the viroid as this mutation changes the loop structure at the right boundary of the RY-motif's consensus sequence. This loop structure was determined to bind the viroid-binding protein, Virp1 (Martínez de Alba et al., 2003).

Subsequently, the presence of several pathogenicity-determining nucleotides located outside of the P domain was reported in both PSTVd and other members of *Pospiviroidae*. One example of this is a detailed and extensive study on two PSTVd isolates (PSTVd-Dahlia [PSTVd-D] and PSTVd-intermediate [PSTVd-I]) that differ by nine nucleotides. In tomato cv. Rutgers, PSTVd-D induces very mild symptoms, while PSTVd-I induces severe symptoms (Tsushima et al., 2011). To understand the role of these nine mutated nucleotides in the pathogenicity of PSTVd, point mutants were created by swapping the nucleotides between the two viroid variants, and these were then subjected to bioassays in tomato cv. Rutgers. Interestingly, two mutants (PSTVd-I:C42U and PSTVd-I:64U) resulted in both reduced viroid titers and symptom attenuation in infected plants. The double mutants created in PSTVd-I at positions 42 and 64 (PSTVd-I:C42U/64U) completely attenuated all disease symptoms (Kitabayashi et al., 2020; see Fig. 2). RT-qPCR analysis on both the pathogenesis-related protein 1b1 and the chalcone synthase genes showed a direct correlation with symptom severity. However, detailed analysis on other mutants demonstrated that the VMR and nearby nucleotides located at positions 42 and 64 work in concert with those located at positions 43, 310, and 311/312 to cause the slower and stable accumulation of PSTVd-D (Dahlia variant) without eliciting any excessive host defense response, thus contributing to the attenuation of the disease symptoms.

C. Viroid pathogenesis and viroid-host interaction

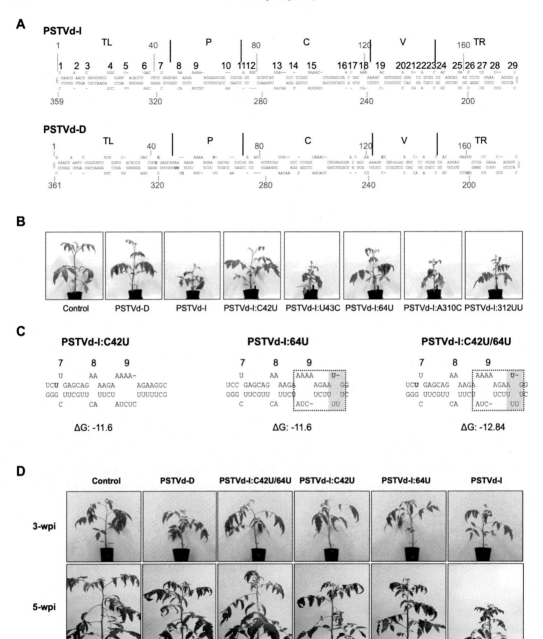

FIG. 2 Effects of single and double mutations on tomato plants. (A) The full-length structures of both PSTVd-I and PSTVd-D are illustrated. The structural/functional domains of PSTVd (Terminal left (TL), Pathogenicity (P), Central (C), Variable (V), and Terminal right (TR)) are delimited by the *vertical solid lines* and are named accordingly. The 29
(Continued)

C. Viroid pathogenesis and viroid-host interaction

Pathogenicity determinants of the members of the *Avsunviroidae* family

To date, at least three structural motifs have been identified in the members of the *Avsunviroidae* family that are known to induce specific disease symptoms in host plants (Fig. 3). First, a single nucleotide change in ASBVd is possibly involved in regulating symptom expression in the host plant, as was the case for PSTVd. Specifically, the insertion of U between nucleotide positions 115 and 118 of ASBVd in the right terminal is associated with leaf variegation and bleaching. This insertion leads to a more open terminal loop that could play a role in pathogenesis (Fig. 3A; Semancik and Szychowski, 1994; Schnell et al., 2001). Second, some variants of *Peach latent mosaic viroid* (PLMVd) possessed an insertion of 12–13 nts in the hairpin loop that caps the stem that includes the hammerhead self-cleaving sequences. These PLMVd variants were shown to induce albinism (Peach calico [PC]; bleaching) on most parts of the leaves of infected pleach trees (Fig. 3B; Malfitano et al., 2003). Third, *Chrysanthemum chlorotic mottle viroid* (CChMVd) sequence variants differing in the nucleotide sequence of a tetraloop formed in their most stable secondary structure (that is, GAAA or UUUC) are associated with either latent or symptomatic (leaf chlorotic mottle) infections in a susceptible chrysanthemum cultivar (Navarro and Flores, 1997). It was shown that by mutating the UUUC sequence to GAAA, the symptomatic form of CChMVd switches from being a severe to a latent variant on susceptible chrysanthemum cultivars independently of viroid titer (De la Peña et al., 1999; De la Peña and Flores, 2002; Fig. 3C).

Role of the host components in viroid pathogenicity

Due to the noncoding nature of viroids, viroid genomes must directly interact with certain host factors for disease induction to occur. These host factors include the host genomes, transcriptomes, proteins, and other regulatory pathways. Consequently, several research groups have examined the ability of viroids to interact with different host factors. The first investigation on the ability of mature PSTVd to interact with host proteins (tomato) was based on the analysis of one-dimensional SDS-PAGE and Northwestern blotting experiments and demonstrated the interaction of PSTVd with all four histones proteins and also with two larger nuclear proteins of approximately 31 and 41 kDa in size (Wolff et al., 1985). Follow-

FIG. 2, CONT'D loops are numbered from left to right on PSTVd-I. The *red-colored* nucleotides on the PSTVd-D structure represent the nucleotides that are different from those in PSTVd-I. (B) Tomato cv. Rutgers exhibited symptoms at 4-weeks post inoculation (wpi) with the PSTVd-D, PSTVd-I, and PSTVd-I single mutants. (C) Enlarged secondary structure of PSTVd-I showing the mutations at positions 42 and 64. The nucleotides in *red* indicate those that are changed to the PSTVd-D type. The structural changes observed at the site of mutation, as compared to PSTVd-I, are indicated by the *gray boxes*. The *dotted boxes* indicate the structural deviations observed in the mutants as compared to the secondary structure of PSTVd-I. The PairFold online tool was used to predict the minimum secondary structure free energy of the pairs of RNA sequences. (D) The tomato plants cv. Rutgers inoculated with the PSTVd double mutant (PSTVd-I:C42U/64U) exhibited symptom attenuation at both 3-wpi and 5-wpi as compared to what was observed with both the single mutants (PSTVd-I:C42U, PSTVd-I:64U) and with PSTVd-I wild type. *Modified from Figs. 1–3 of Kitabayashi, S., et al., 2020. Identification and molecular mechanisms of key nucleotides causing attenuation in pathogenicity of Dahlia isolate of potato spindle tuber viroid. Int. J. Mol. Sci. 21(19). https://doi.org/10.3390/ijms21197352 and was adapted as per the Creative Commons Attribution (CC BY) license (http://creativecommons.org/licenses/by/4.0/).*

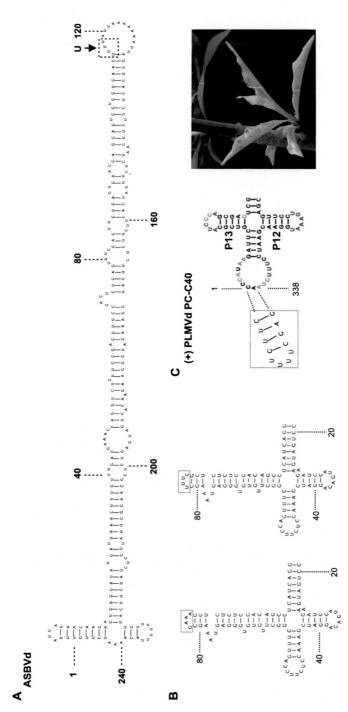

FIG. 3 Motifs of the members of the *Avsunviroidae* family involved in pathogenesis. (A) The secondary structure of the (+) polarity RNA strand of the complete ASBVd is shown. The addition of a U between the bases located at positions 115 and 118 of the right-handed loop is associated with leaf variegation and bleaching in ASBVd-infected avocado plants. (B) The (+) polarity RNA strands of partial CChMVd RNA sequence variants are shown. The change in the tetraloop nucleotide sequence from UUUC to GAAA converts a severe symptomatic variant of CChMVd to a latent one on susceptible chrysanthemum cultivars. (C) The presence of a 12-nt hairpin in PLMVd induces the calico effect (that is, the bleaching of the leaves) in susceptible peach cultivars such as GF305 (inset image). *Modified from Fig. 2 of Adkar-Purushothama, C.R., Perreault, J.P., 2020a. Current overview on viroid-host interactions. Wiley Interdiscip. Rev. RNA 11(2), e1570. https://doi.org/10.1002/wrna.1570; License number: 5521350067514.*

up in vitro studies with either linear or circular PSTVd RNAs, using a combination of UV cross-linking experiments followed by RNase digestion, permitted researchers to isolate a 43 kDa protein from the cellular complexes. However, the function of this protein in the cells was not determined (Klaff et al., 1989). Experimentation with wheat germplasm revealed the interaction of *RNA polymerase* II with both of the terminal loops of PSTVd (Goodman et al., 1984). This was the first illustration of the interaction of a specific host protein with a viroid structural motif. Later, it was shown that the transcription of the PSTVd (−) strand by *RNA polymerase* II likely starts in the left terminal loop of the circular (+) strand template (Kolonko et al., 2006).

To date, several viroid-binding proteins have been characterized. The binding of a tomato bromodomain-containing protein, Virp1, to the 71-nt bulged hairpin of PSTVd was demonstrated by immunoprecipitation assay (Martínez de Alba et al., 2003). This viroid-host protein interaction is important for the long-distance movement of the pospiviroids in the host's vascular system (Maniataki et al., 2003) and is implicated in nuclear transport within the infected cell (Ma et al., 2022). Similarly, a ribonucleoprotein complex formed between the *Hop stunt viroid* (HSVd) RNA and the phloem RNA binding protein PP2 isolated from Cucumis sativus (CsPP2) is found to be involved in the long-distance trafficking of the viroid RNA (Gómez and Pallás, 2001; Gómez and Pallás, 2004). The characterization of the gene encoding cucumber PP2 revealed the presence of a potential double-stranded RNA-binding motif in the encoded protein. However, the viroid-PP2 interaction seems an unlikely trigger for pathogenesis because the viroid-PP2 interaction appears to be nonspecific (that is, PP2 can bind several other RNA molecules) (Owens and Hammond, 2009). In the case of the *Avsunviroidae* family members, UV-irradiated avocado leaves infected with ASBVd revealed ASBVd-host protein interactions (Daros, 2002). More precisely, ASBVd was found to interact with two small chloroplast RNA-binding proteins encoded in the nuclear genome of avocados. The use of tandem mass spectrometry revealed the binding of both PARBP33 and PARBP35 to multimeric ASBVd RNA transcripts. These interactions likely stimulate hammerhead self-cleavage in vitro and are likely to be associated with replication rather than with pathogenicity (Owens and Hammond, 2009).

Viroid-induced biochemical changes

Signal transduction is the process through which cells communicate with the external environment, interpret stimuli, and respond to them. This mechanism is controlled by signaling cascades which play the role of intracellular transmitters and can transmit biochemical information between the cell membrane and the nucleus (Catozzi et al., 2016). At least two protein kinases have been proposed to possibly play a role in viroid pathogenicity by modulating a signaling cascade. Specifically, PSTVd infection assay in tomato plants leads to the autophosphorylation of a protein associated with a double-stranded RNA-stimulated protein kinase activity (Hiddinga et al., 1988). This protein is a 68 kDa analog of PKR, the mammalian double-stranded RNA-dependent protein kinase that is implicated in the regulation of animal RNA virus replication (Langland et al., 1995). The incubation of the RNA transcripts derived from the PSTVd strains differing in pathogenicity with purified PKR leads to differential activation levels, thus supporting an implication for it in the viroid's pathogenicity (Diener et al., 1993; Langland et al., 1995). In other words, the incubation of a severe strain of PSTVd resulted in 10-fold greater activation when compared to that observed with a mild strain of PSTVd. Although these observations suggest a direct interaction between PKR and a PSTVd pathogenicity domain sequence motif, direct evidence for such an interaction is still lacking (Owens and Hammond, 2009).

The second protein kinase that appears to play a role in modulating viroid pathogenicity is a 55 kDa protein kinase known as PKV (protein kinase viroid-induced). The infection of tomato seedlings with PSTVd resulted in the transcriptional activation of PKV. Briefly, the level of PKV transcription was higher in the plants infected by either a severe or an intermediate strain as compared to what was observed in either plants infected with a mild strain of PSTVd or in mock-infected plants. PKV is like the mammalian cyclic nucleotide-dependent kinase, implying involvement in the transduction of extracellular signals. In vitro studies using recombinant PKV protein demonstrated its ability to auto-phosphorylate on serine and tyrosine residues, suggesting that it belongs to the class of dual-specificity protein kinases (Hammond and Zhao, 2000). Overexpression of signal-transducing tomato PKV in tobacco resulted in stunting, modified vascular development, reduced root formation, and male sterility (Hammond and Zhao, 2009). Further studies have shown that PKV may play a role in gibberellic acid (GA) signaling. Although the infection with certain strains of PSTVd resulted in the transcriptional activation of the gene encoding PKV (*pkv*), the direct binding of PSTVd to the PKV protein triggering the pathogenesis has not been shown (Owens and Hammond, 2009).

Similar to PSTVd, CEVd also demonstrated the ability to phosphorylate diverse proteins during infection at the onset of symptom appearance (Vera and Conejero, 1989). Here, researchers noted that the modifications were enhanced in the presence of manganese (Mn^{2+}), indicating a role for the Mn^{2+}-dependent protein kinase in the phosphorylation modifications of the host's proteins. Although is it not known how a viroid can alter protein phosphorylation, these data illustrate that these modifications have a critical effect on several biological pathways. Furthermore, CEVd was reported to alter the expression of two low molecular weight pathogenesis-related (PR) proteins in *Gynura aurantiaca* (Conejero and Semancik, 1977). PR protein synthesis is a component of systemic acquired resistance that is known to coordinate plant's defense responses to infection. Comparative protein and transcript analyzes have revealed that, irrespective of the viroid species, viroids induce PR proteins during infection (Owens and Hammond, 2009). The use of the microarray technique illustrated that PSTVd infection can induce an array of genes in tomato plants that are related to stress, growth, development, and defense. Briefly, to understand the role of RNA silencing in viroid disease induction, microarray analysis, and large-scale RNA sequence analysis were combined to be able to compare changes in both the tomato gene expression levels and the microRNA levels associated with PSTVd infection in two tomato cultivars (cv. Rutgers and cv. Moneymaker) and a transgenic tomato line expressing vd-sPSTVd RNAs in the absence of viroid replication. The results demonstrated that the changes in the mRNA levels for cv. Rutgers were extensive as compared to those observed in the cv. Moneymaker, with more than half of the approximately 10,000 genes present on the array being affected. PSTVd infection downregulated chloroplast biogenesis in both tomato cultivars, irrespective of disease resistance. Furthermore, PSTVd affected the mRNAs that are involved in the biosynthesis of gibberellin, several hormones, and signaling pathways. These results are illustrated in Fig. 4 (Owens et al., 2012).

C. Viroid pathogenesis and viroid-host interaction

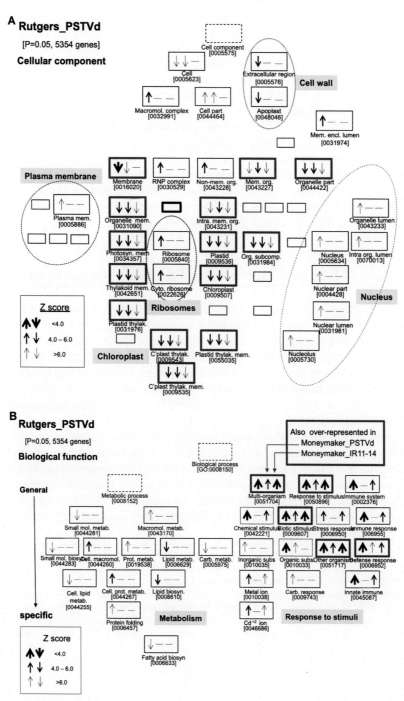

FIG. 4 See caption on next page.

(Continued)

C. Viroid pathogenesis and viroid-host interaction

Like the members of the *Pospiviroidae* family, the members of the *Avsunviroidae* family showed viroid infection-induced biochemical changes in the host proteins. For example, mass spectrometric analysis of the proteins extracted from PLMVd-infected peach trees (Prunus persica) revealed the presence of six putative RNA-binding proteins, including the elongation factor 1-alpha (eEF1A) (Dube et al., 2009). eEF1A was known to be involved in both the replication and the translation of RNA viruses by interacting directly with either the RNA molecule itself or with the viral RNA-dependent RNA polymerase (Bastin and Hall, 1976; Joshi et al., 1986; Yamaji et al., 2006). Furthermore, viroids are also known to interfere with the host's translational machinery, thus contributing to symptom development. Both indirect and direct evidence demonstrating the effects of viroid infection on ribosomal genes and ribosomal stress that are potentially connected with viroid pathogenicity have been demonstrated in tomato plants (Adkar-Purushothama et al., 2017; Cottilli et al., 2019). Chapter 12, *Viroid and protein translation*, provides a detailed account of how viroids influence both host protein expression and the translation machinery.

Effect of viroids on host gene expression and the host response to viroid infection

In molecular biology, the central dogma refers to the flow of genetic information from DNA to RNA and then to protein, or from RNA to protein. Viroid infection is found to interfere with host DNA, RNA, and the protein translation machinery. The effects of viroid infection on host gene expression have been examined at both the transcriptional and posttranscriptional levels by several groups either for a specific individual gene or for a group of genes. These experiments showed that viroid infection results in the increased transcription of defense and stress-related genes, including the genes encoding the pathogenesis-related (PR) proteins and the β-1,3-glucanases (Vera et al., 1989; Domingo et al., 1994). On the contrary, viroid infection is also known to reduce the transcription of certain genes such as the *LeExp2* expansin gene (Qi and Ding, 2003). *LeExp2* expansin is known to be involved in cell expansion. The reduction in *LeExp2* expansin transcription suggests a possible role for this gene in stunting, a common symptom associated with viroid infection. Similarly, CEVd infection resulted in reduced levels of gibberellin 20-oxidase transcripts in citrus plants, another enzyme involved in plant growth (Vidal et al., 2003). Recent developments in both sequencing

---

FIG. 4, CONT'D   The effects of viroid infection on host gene expression. The association of differentially expressed tomato genes with specific (A) cellular components and (B) biological functions are illustrated. Groups of overrepresented genes, as defined by the Gene Ontology (GO) project, were identified by the parametric analysis of gene set enrichment for both Potato spindle tuber viroid (PSTVd)-infected and transgenic plants. The complete results for only the PSTVd-infected tomato cv. "Rutgers" plants are shown, with dark outlines indicating groupings also present in the other two treatments. The *arrows within the boxes* representing individual GO categories denote the relative degree of overrepresentation for, from left to right, infected "Rutgers" PSTVd, "Moneymaker" PSTVd, and "Moneymaker" transgenic plants, as well as the direction of the effect (up- or downregulation). Note that the names of certain GO categories have been modified to save space and that the categories toward the bottom of the figure are more specific and less inclusive than those near the top. *Modified from Figs. 3 and 4 of Owens, R.A. et al., 2012. Global analysis of tomato gene expression during Potato spindle tuber viroid infection reveals a complex array of changes affecting hormone signaling. Mol. Plant Microbe Interact. 25(4), 582–98. https://doi.org/10.1094/MPMI-09-11-0258. (Mol. Plant Microbe Interact. 13(9), 903–910. https://doi.org/10.1094/MPMI.2000.13.9.903).*

technology and microarrays have allowed researchers to study several genes' expression levels simultaneously in a single experiment. Chapter 14, *Transcriptomic analyses provide insights into plant-viroid interactions*, introduces the basic concept of this technology and provides not only a brief history of transcriptomic analysis but also of its application in understanding plant-viroid interactions.

Role of gene silencing in viroid pathogenesis

Gene silencing is a regulatory pathway in cells that prevents the expression of certain genes. Gene silencing can occur either at the transcriptional or the posttranscriptional stage. Depending on the stage where gene silencing occurs, it is called transcriptional gene silencing (TGS) or posttranscriptional gene silencing (PTGS). In TGS, transcription is prevented by modifying the DNA by methylation, while in PTGS (which is also known as RNA silencing) RNA is blocked from being translated. DNA methylation is an epigenetic mechanism involving the transfer of a methyl group onto the C5 position of cytosine, forming 5-methylcytosine. DNA methylation regulates gene expression either by recruiting proteins involved in gene repression or by inhibiting the binding of transcription factor(s) to DNA (Moore et al., 2013). Studies have demonstrated that PSTVd infection in tobacco can induce the de novo methylation of a PSTVd complementary DNA transgene in a sequence-specific manner, thus revealing for the first time the mechanisms of RNA-dependent DNA methylation (RdDM) (Wassenegger et al., 1994). Chapter 13, *Viroid infection and host epigenetic alterations*, describes the relationship between the various epigenetic alterations in the plant genomes induced by viroid infection.

RNA silencing, also known as posttranscription gene silencing (PTGS) or RNA interference (RNAi), is a mechanism of the sequence-specific regulation of gene expression that is highly conserved in most eukaryotic organisms and which plays various important functions in both the development and the maintenance of the vital activities of living organisms. RNA silencing also is a potent antiviral defense mechanism against either double-stranded or structured RNA pathogens (Adkar-Purushothama and Perreault, 2020a,b). RNA silencing results in both the production and the accumulation of 21- to 24-nt-long small RNAs (sRNAs) that are specific to the invading RNA pathogen. As mentioned in both Chapters 3 (Viroid Structure) and 4 (Replication and movement), because viroids are highly base-paired structures and form double-stranded RNA intermediates during replication, they act as both inducers and targets of RNA silencing (Pallas et al., 2012; Flores et al., 2015; Navarro et al., 2021). Upon infection, viroids elicit the RNA silencing machinery of the host plant, and the viroid RNAs are processed by either the DICER or the DCL RNase III-type ribonucleases, resulting in the production of siRNAs of 21- to 24-nt that are called viroid derived small RNAs (vd-sRNAs) (Dadami et al., 2013). The process of viroid-induced RNA silencing in host plants is outlined in Fig. 5. In early 2000, vd-sRNAs were detected in plants infected with viroids belonging to the two families, suggesting that these viroids are the targets of RNA silencing irrespective of their subcellular localization during replication (Itaya et al., 2001; Papaefthimiou et al., 2001; Martínez de Alba et al., 2002; Markarian et al., 2004). These *in-planta* experiments demonstrated that vd-sRNAs are produced from both the (+) and (−) strands of the viroid RNA

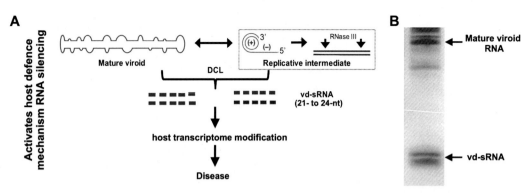

FIG. 5 Viroid-induced RNA silencing and its secondary effect on the host transcriptome. (A) After invading the host plant's cells, the circular viroid replicates in either the nuclei (members of the family *Pospiviroidae*; presented in this figure) or in the chloroplast (members of the family *Avsunviroidae*), via a rolling circle mechanism. During this process, using the host's transcriptional machinery, the (+) strand (represented in blue) viroid molecule is transcribed into the (−) strand (represented in red), and then the (−) strand is transcribed to produce a greater than unit length (+) strand. These transcribed multimeric (+) strands are then processed into unit-length and are then circularized for the production of the mature viroid molecule. The highly base-paired mature circular viroid molecule and the double-stranded RNA structures formed between the (+) and (−) strand molecules during viroid replication, both might trigger the host's RNA silencing machinery and be cleaved into small RNAs of 21- to 24-nt called viroid-derived small RNAs (vd-sRNA). The presence of a large number of accumulated vd-sRNAs in infected cells can extensively alter the host's transcriptome and adversely affect its growth. (B) Total RNA isolated from viroid-infected plants separated by electrophoresis through a denaturing 8% polyacrylamide gel showing both the mature viroid RNA and the vd-sRNA.

(reviewed by Adkar-Purushothama and Perreault, 2020a,b). The accumulation of vd-sRNAs has been extensively studied in different viroid-host combinations using next-generation sequencing technologies. The profiling of vd-sRNA on their respective viroid RNAs suggested that the (+) strand produced more sRNA than did the (−) strand. To date, the RNA silencing mediated downregulation of host transcriptomes, and its associated symptoms, have been demonstrated for the members of both the *Pospiviroidae* and *Avsunviroidae* families (see Chapter 15, *Viroid-induced RNA silencing and its secondary effect on the host transcriptome*, for a review).

Prospective

From the beginning of the discovery of viroids, the mechanism of viroid pathogenicity has been the major theme of viroid research because they were identified as being pathogens. Since viroids are the smallest known RNA pathogens and the fact that they do not encode any protein information, it was thought that all the factors involved in both their autonomous replication and their pathogenicity exist within the molecules themselves. Various structural domains and RNA motifs present in their highly folded RNA molecules (such as the pathogenicity domain, the virulence modulating region, loop E, the hammerhead-self-cleaving motif, etc.), and their possible functions, have been extensively analyzed so far. Analysis of nucleotide sequences and molecular structures, and the virulence of natural or artificial viroid mutants with different pathogenicity, have revealed structural domains directly involved in virulence and specific nucleotide changes that cause particularly marked virulence changes. However, it became clear that the viroid pathogenicity is determined not only by the pathogenicity determinants, and that it cannot be separated from the functions of the other domains that regulate replication and the movement from both cell to cell and cell to tissue. In addition, host factors that interact with these viroid motifs, and the host defense mechanisms targeting viroid infection are also found to be important. Interestingly, the pathogenesis of the *Avsunviroidae* family is straightforward. Specific viroid-derived small RNAs generated by viroid-induced RNA silencing were found to cause the posttranscriptional degradation of partially homologous host mRNAs involved in chloroplast biogenesis via the RNA silencing pathway and to cause specific disease symptoms such as chlorosis and the albinism of both the leaves and the fruit pericarp. In contrast, the pathogenicity of the *Pospiviroidae* family is more complex. Large-scale transcriptome analyzes of various combinations of viroids and hosts revealed that viroid infection activates the host's defense mechanisms, including RNA silencing, and induces either the up- or the downregulation of hundreds of genes involved in defense, stress, and in the plant's innate immunity. Finally, the extensive reprogramming of the host's genes regulation, including things such as hormone signaling, the miRNA pathway, and so on, can lead to the development of viroid-specific disease symptoms (e.g., leaf epinasty, chlorosis, necrosis, and systemic stunting). An overview of the molecular mechanisms underlying the onset of the pospiviroids' pathogenesis also appears to have started to emerge. The goal is to be able to explain both the viroid's pathogenicity and the molecular mechanisms underlying the development of viroid diseases based on either the specific gene-to-gene or gene-to-protein interactions involved in the host's gene expression network.

Future implications

Being a molecular pathogen, which is unable to code for any proteins, viroids serve as a unique model with which to study noncoding RNAs at different levels. Although they are 10-fold smaller than a smallest known virus, they contain all the information required for both their survival and their induction of disease symptoms in susceptible host plants. Since their discovery, their pathogenicity has been studied using different viroid-host combinations to dissect the mechanism through which viroids affect the host's physiology. Studying the molecular mechanism of viroid pathogenicity can provide essential information for understanding how RNA structural motifs function in vivo and influence normal biological processes within cells. Given all the information on viroid that we have gathered to date, namely, their capability of interfering with the host's structural components, proteins, genomes, metabolic pathways, and so on, it becomes clearer that viroids have a multifaced pathogenicity. Although RNA silencing has been implicated in the pathogenesis as it both directly and indirectly affects the host's overall gene expression, it is important to understand how viroids adapt and escape the host's defense mechanisms before triggering the RNA silencing machinery. Because viroids are very simple molecules, both genetically and structurally, the induced pathogenicity is much more complex to understand. Since there are no effective measures for controlling viroid diseases other than diagnosis, elucidation of the basic mechanisms underlying viroid pathogenesis is an urgent issue for developing strategies for controlling viroid diseases.

Chapter summary

This chapter explains the importance of both a viroid's structure and its sequence in inducing disease symptoms in host plants. We noticed how the structural motifs of viroids interact with host factors, and how single nucleotide changes can alter symptom expression. First, it became evident that specific vd-sRNAs generated by RNA silencing targeting viroids are the direct cause of the chlorosis induced by the members of the family *Avsunviroidae*. Generation of the specific viroid-sRNAs that are derived from the stem-loop located at the end of the rod-like domain is involved in the specific symptom-inducing variants. However, the pathogenicity of the viroids belonging to the family *Pospiviroidae* is not only caused by the suppression the of host's genes' expression via specific viroid-sRNAs but also by the reprogramming of the transcriptional regulation in the host's defense reaction against viroid infection. The resultant alterations, such as those in hormone biosynthesis/signaling, microRNA pathways, and so on, are now thought to be more complex, genome-wide, and far-reaching mechanisms that lead to pathogenesis. At the cellular level, viroids can induce biochemical change, while at the molecular level, they can also alter gene expression by interfering with both transcription and translation. Additionally, viroids use the host's RNAi mechanism in their pathogenicity. In simple terms, being simple molecular entities, viroids have all the information required to modulate the host's architecture and to cause specific symptoms.

Protocols/procedures/methods

Method 1: Generation of viroid mutants

The generation of mutagenic viroid constructs is a fundamental requirement to demonstrate the role of specific nucleotides in pathogenicity. A simple viroid mutant generation technique is described below.

1. Determine the nucleotide under study to be mutated.
2. Design a forward primer so that the nucleotide under study is changed in the primer and that this nucleotide is in the middle of the primer.
3. Design the reverse primer so its 5′ sequence overlaps with the forward primer. This overlap region must include a single restriction enzyme site (a sticky end restriction site is preferable when manufacturing dimeric viroid constructs).
4. Perform PCR amplification.
5. Clone the PCR amplicon and confirm the introduction of the mutation by sequencing.
6. Digest the clone using the restriction enzyme sites located on both the forward and the reverse primers.
7. Purify both the digested viroid and desired plasmid DNAs.
8. Ligate the purified viroid DNA to multimerize it.

9. Separate the multimerized viroid DNAs on a 1.5% agarose gel in the presence of a DNA marker.
10. Elute dimeric viroid DNA from the agarose gel.
11. Ligate the dimeric viroid DNA into the desired plasmid digested with the same restriction enzyme (step 6).
12. Sequence the ligated plasmid to confirm the dimeric form of viroid, the mutation, and the direction of the insert in the plasmid.
13. Follow method 2 to study the role of the mutated nucleotide in pathogenicity.

Method 2: Viroid bioassay with which to study the pathogenesis

Bioassay is the primary technology used in viroid biology with which to study the effect of the viroid on both the host's phenotype and genotype. This fundamental principle is applicable for other viroids with or without minor modifications. Several indicator plants are used for the bioassay based on the type of viroid as well as on the study type. For example, to study the pathogenic effect of specific nucleotide variations of PLMVd, peach GF305 is commonly used as an indicator plant. In contrast, tomato cv. Rutgers is used for studying PSTVd-host interactions as cv. Rutgers exhibits an array of symptoms compared to what other tomato cultivars do. Below is the detailed description of the bioassay of PSTVd in tomato cv. Rutgers.

1. Requirement
   a. Dimeric head-to-tail viroid constructs in a cloning vector having a transcription initiation sequence at either end of the viroid construct.
   b. Appropriate indicator plant. For example, the tomato cv. Rutgers used to study PSTVd.
   c. Environmentally controlled plant incubator.
2. Generation of viroid RNA
   a. Determine the direction of the dimeric viroid construct based on sequence information.
   b. Single digest the plasmid after the dimeric viroid sequence, in such a way that transcription generates positive-strand dimeric viroid RNA.
   c. Transcribe the positive strand viroid RNA by incubating at 37°C for 1 h in the presence of transcription reagents and the appropriate transcription enzyme.
   d. Separate the transcribed RNA by 1.0% agarose gel electrophoresis.
   e. Elute the appropriate transcribed RNA from the agarose gel and dissolve it in RNase-free water.
   f. Quantify the RNA by spectrometry.
3. Preparation of the indicator plants
   a. Sow tomato cv. Rutgers seeds in sterilized soil.
   b. The seedling will emerge 7–10 days postsowing.
   c. Transfer 2-week old seedlings (at the time of emergence of first true leaves) to new pots.
   d. Plants are ready for inoculation when the first true leaves are fully opened (approximately 5 days posttransfer to new pots).
4. Inoculation and incubation
   a. Prepare the viroid RNA inoculum with a final concentration of 50 ng/μL of RNA and 1% bentonite in 0.1 M phosphate buffer saline (PBS).

C. Viroid pathogenesis and viroid-host interaction

b. Pipette 10 μL of the viroid RNA preparation onto each of the top two fully opened leaves that are dusted with abrasive such as carborundum.

c. Using a 200 μL pipette tip, gently rub the RNA preparation against each leaf 15 times.

d. After 5 min, rinse the inoculated leaves by spraying them with distilled water.

e. Incubate plants overnight at 23°C in the dark.

f. Continue the incubation at 27°C with 16 h of light and 8 h of dark.

g. Depending on the PSTVd variant, the symptoms become visible 10–21 days postinoculation (dpi).

Study questions

1. What are the visible symptoms of a viroid infection?
2. Name two structural motifs that are involved in viroid pathogenicity?
3. Briefly explain two pathogenicity determinants found in the members of the *Avsunviroidae* family?
4. Briefly explain the biochemical changes caused by viroid infection?
5. How does viroid infection affect both a host gene and its expression?

Further reading

Adkar-Purushothama, C. R. and Perreault, J. P. (2020) 'Current overview on viroid-host interactions', *Wiley interdisciplinary reviews. RNA*, 11(2), e1570. https://doi.org/10.1002/wrna.1570https://doi.org/10.1002/wrna.1570.

Navarro, B., Flores, R. and Di Serio, F. (2021) 'Advances in Viroid-Host Interactions', *Annual review of virology*, 8(1), 305–325. https://doi.org/10.1146/annurev-virology-091919-092331.

Ortolá, B. and Daròs, J. A. (2023) 'Viroids: Non-Coding Circular RNAs Able to Autonomously Replicate and Infect Higher Plants', Biology, 12(2):172. https://doi.org/10.3390/biology12020172

Owens, R. A. and Hammond, R. W. (2009) 'Viroid pathogenicity: One process, many faces', *Viruses*, 1(2), 298–316. https://doi.org/10.3390/v1020298.

References

Adkar-Purushothama, C.R., Perreault, J.P., 2020a. Current overview on viroid-host interactions. Wiley Interdiscip. Rev. RNA 11 (2), e1570. https://doi.org/10.1002/wrna.1570.

Adkar-Purushothama, C.R., Perreault, J.P., 2020b. Impact of nucleic acid sequencing on viroid biology. Int. J. Mol. Sci. 21 (15). https://doi.org/10.3390/ijms21155532.

Adkar-Purushothama, C.R., Lyer, P.S., Perreault, J.P., 2017. Potato spindle tuber viroid infection triggers degradation of chloride channel protein CLC-b-like and ribosomal protein S3a-like mRNAs in tomato plants. Sci. Rep. 7 (1), 8341. https://doi.org/10.1038/s41598-017-08823-z.

Bastin, M., Hall, T.C., 1976. Interaction of elongation factor 1 with aminoacylated brome mosaic virus and tRNA's. J. Virol.

Catozzi, S., et al., 2016. Signaling cascades transmit information downstream and upstream but unlikely simultaneously. BMC Syst. Biol. 10 (1), 84. https://doi.org/10.1186/s12918-016-0303-2.

Conejero, V., Semancik, J.S., 1977. Exocortis viroid: alteration in the proteins of Gynura aurantiaca accompanying viroid infection. Virology 77 (1), 221–232. https://doi.org/10.1016/0042-6822(77)90420-2.

Cottilli, P., et al., 2019. Citrus exocortis viroid causes ribosomal stress in tomato plants. Nucleic Acids Res. 47 (16), 8649–8661. https://doi.org/10.1093/nar/gkz679.

Dadami, E., et al., 2013. DICER-LIKE 4 but not DICER-LIKE 2 may have a positive effect on potato spindle tuber viroid accumulation in nicotiana benthamiana. Mol. Plant. https://doi.org/10.1093/mp/sss118.

Daros, J.-A., 2002. A chloroplast protein binds a viroid RNA in vivo and facilitates its hammerhead-mediated self-cleavage. EMBO J. 21 (4), 749–759. https://doi.org/10.1093/emboj/21.4.749.

Martínez de Alba, A.E., Flores, R., Hernández, C., 2002. Two chloroplastic viroids induce the accumulation of small RNAs associated with posttranscriptional gene silencing. J. Virol. 76 (24), 13094–13096. https://doi.org/10.1128/JVI.76.24.13094-13096.2002.

De la Peña, M., Flores, R., 2002. Chrysanthemum chlorotic mottle viroid RNA: dissection of the pathogenicity determinant and comparative fitness of symptomatic and non-symptomatic variants. J. Mol. Biol. https://doi.org/10.1016/S0022-2836(02)00629-0.

de la Peña, M., Navarro, B., Flores, R., 1999. Mapping the molecular determinant of pathogenicity in a hammerhead viroid: a tetraloop within the in vivo branched RNA conformation. Proc. Natl. Acad. Sci. USA 96 (17), 9960–9965. https://doi.org/10.1073/pnas.96.17.9960.

Di Serio, F., De Stradis, A., Delgado, S., Flores, R., Navarro, B., 2013. Cytopathic effects incited by viroid RNAs and putative underlying mechanisms. Front. Plant Sci. 3, 288. https://doi.org/10.3389/fpls.2012.00288.

Dickson, E., et al., 1979. Minor differences between nucleotide sequences of mild and severe strains of potato spindle tuber viroid. Nature 277 (5691), 60–62. https://doi.org/10.1038/277060a0.

Diener, T.O., 1971. Potato spindle tuber "virus". Virology 45 (2), 411–428. https://doi.org/10.1016/0042-6822(71)90342-4.

Diener, T.O., et al., 1993. Mechanism of viroid pathogenesis: differential activation of the interferon-induced, double-stranded RNA-activated, Mr 68 000 protein kinase by viroid strains of varying pathogenicity. Biochimie 75 (7), 533–538. https://doi.org/10.1016/0300-9084(93)90058-Z.

Domingo, C., Conejero, V., Vera, P., 1994. Genes encoding acidic and basic class IIIβ-1,3-glucanases are expressed in tomato plants upon viroid infection. Plant Mol. Biol. 24 (5), 725–732. https://doi.org/10.1007/BF00029854.

Dube, A., Bisaillon, M., Perreault, J.-P., 2009. Identification of proteins from Prunus persica that interact with peach latent mosaic viroid. J. Virol. 83 (23), 12057–12067. https://doi.org/10.1128/JVI.01151-09.

Flores, R., et al., 2015. Viroids, the simplest RNA replicons: how they manipulate their hosts for being propagated and how their hosts react for containing the infection. Virus Res. 209, 136–145. https://doi.org/10.1016/j.virusres.2015.02.027.

Flores, R., et al., 2017. Viroid pathogenesis. In: Viroids and Satellites. Elsevier, pp. 93–103, https://doi.org/10.1016/B978-0-12-801498-1.00009-7.

Gómez, G., Pallás, V., 2001. Identification of an in vitro ribonucleoprotein complex between a viroid RNA and a phloem protein from cucumber plants. Mol. Plant Microbe Interact. 14 (7), 910–913. https://doi.org/10.1094/MPMI.2001.14.7.910.

Gómez, G., Pallás, V., 2004. A long-distance translocatable phloem protein from cucumber forms a ribonucleoprotein complex in vivo with Hop stunt viroid RNA. J. Virol. 78 (18), 10104–10110. https://doi.org/10.1128/JVI.78.18.10104-10110.2004.

Goodman, T.C., et al., 1984. Viroid replication: equilibrium association constant and comparative activity measurements for the viroid-polymerase interaction. Nucleic Acids Res. 12 (15), 6231–6246. https://doi.org/10.1093/nar/12.15.6231.

Gross, H.J., et al., 1978. Nucleotide sequence and secondary structure of potato spindle tuber viroid. Nature 273 (5659), 203–208. https://doi.org/10.1038/273203a0.

Hammond, R.W., Zhao, Y., 2000. Characterization of a tomato protein kinase gene induced by infection by potato spindle tuber viroid. Mol. Plant-Microbe Interact. 13 (9), 903–910. https://doi.org/10.1094/MPMI.2000.13.9.903.

Hammond, R.W., Zhao, Y., 2009. Modification of tobacco plant development by sense and antisense expression of the tomato viroid-induced AGC VIIIa protein kinase PKV suggests involvement in gibberellin signaling. BMC Plant Biol. https://doi.org/10.1186/1471-2229-9-108.

Hiddinga, H., et al., 1988. Viroid-induced phosphorylation of a host protein related to a dsRNA-dependent protein kinase. Science 241 (4864), 451–453. https://doi.org/10.1126/science.3393910.

Itaya, A., et al., 2001. Potato spindle tuber viroid as inducer of RNA silencing in infected tomato. Mol. Plant Microbe Interact. 14 (11), 1332–1334. https://doi.org/10.1094/MPMI.2001.14.11.1332.

C. Viroid pathogenesis and viroid-host interaction

Joshi, R.L., Ravel, J.M., Haenni, A.L., 1986. Interaction of turnip yellow mosaic virus Val-RNA with eukaryotic elongation factor EF-1 [alpha]. Search for a function. EMBO J. https://doi.org/10.1002/j.1460-2075.1986.tb04339.x.

Keese, P., Symons, R.H., 1985. Domains in viroids: evidence of intermolecular RNA rearrangements and their contribution to viroid evolution. Proc. Natl. Acad. Sci. USA 82 (14), 4582–4586. https://doi.org/10.1073/pnas.82.14.4582.

Kitabayashi, S., et al., 2020. Identification and molecular mechanisms of key nucleotides causing attenuation in pathogenicity of Dahlia isolate of potato spindle tuber viroid. Int. J. Mol. Sci. 21 (19). https://doi.org/10.3390/ijms21197352.

Klaff, P., et al., 1989. Reconstituted and cellular viroid-protein complexes. J. Gen. Virol. 70 (9), 2257–2270. https://doi.org/10.1099/0022-1317-70-9-2257.

Kolonko, N., et al., 2006. Transcription of potato spindle tuber viroid by RNA polymerase II starts in the left terminal loop. Virology. https://doi.org/10.1016/j.virol.2005.11.039.

Langland, J.O., et al., 1995. Identification of a plant-encoded analog of PKR, the mammalian double-stranded RNA-dependent protein kinase. Plant Physiol. 108 (3), 1259–1267. https://doi.org/10.1104/pp.108.3.1259.

Ma, J., Dissanayaka Mudiyanselage, S.D., Park, W.J., Wang, M., Takeda, R., Liu, B., Wang, Y., 2022. A nuclear import pathway exploited by pathogenic noncoding RNAs. Plant Cell 34, 3543–3556. https://doi.org/10.1093/plcell/koac210.

Malfitano, M., et al., 2003. Peach latent mosaic viroid variants inducing peach calico (extreme chlorosis) contain a characteristic insertion that is responsible for this symptomatology. Virology. https://doi.org/10.1016/S0042-6822(03)00315-5.

Maniataki, E., et al., 2003. Viroid RNA systemic spread may depend on the interaction of a 71-nucleotide bulged hairpin with the host protein VirP1. RNA (New York, N.Y.) 9 (3), 346–354. Available at: http://www.ncbi.nlm.nih.gov/pubmed/12592008.

Markarian, N., et al., 2004. RNA silencing as related to viroid induced symptom expression. Arch. Virol. 149 (2), 397–406. https://doi.org/10.1007/s00705-003-0215-5.

Martínez de Alba, A.E., et al., 2003. A bromodomain-containing protein from tomato specifically binds potato spindle tuber viroid RNA in vitro and in vivo. J. Virol. 77 (17), 9685–9694. Available at: http://www.ncbi.nlm.nih.gov/pubmed/12915580.

Moore, L.D., Le, T., Fan, G., 2013. DNA methylation and its basic function. Neuropsychopharmacology 38 (1), 23–38. https://doi.org/10.1038/npp.2012.112.

Navarro, B., Flores, R., 1997. Chrysanthemum chlorotic mottle viroid: unusual structural properties of a subgroup of self-cleaving viroids with hammerhead ribozymes. Proc. Natl. Acad. Sci. U. S. A. 94 (21), 11262–11267. https://doi.org/10.1073/pnas.94.21.11262.

Navarro, B., Flores, R., Di Serio, F., 2021. Advances in viroid-host interactions. Annual Review of Virology 8 (1), 305–325. https://doi.org/10.1146/annurev-virology-091919-092331.

Owens, R.A., Hammond, R.W., 2009. Viroid pathogenicity: one process, many faces. Viruses 1, 298–316. https://doi.org/10.3390/v1020298.

Owens, R.A., et al., 2012. Global analysis of tomato gene expression during Potato spindle tuber viroid infection reveals a complex array of changes affecting hormone signaling. Mol. Plant Microbe Interact. 25 (4), 582–598. https://doi.org/10.1094/MPMI-09-11-0258.

Pallas, V., Martinez, G., Gomez, G., 2012. The interaction between plant viroid-induced symptoms and RNA silencing. Methods Mol. Biol. 894, 323–343.

Papaefthimiou, I., et al., 2001. Replicating potato spindle tuber viroid RNA is accompanied by short RNA fragments that are characteristic of post-transcriptional gene silencing. Nucleic Acids Res. 29 (11), 2395–2400. https://doi.org/10.1093/nar/29.11.2395.

Qi, Y., Ding, B., 2003. Inhibition of cell growth and shoot development by a specific nucleotide sequence in a noncoding viroid RNA. Plant Cell 15 (6), 1360–1374. https://doi.org/10.1105/tpc.011585.

Sano, T., et al., 1992. Identification of multiple structural domains regulating viroid pathogenicity. Proc. Natl. Acad. Sci. USA 89 (21), 10104–10108. https://doi.org/10.1073/pnas.89.21.10104.

Schnell, R.J., et al., 2001. Sequence diversity among avocado sunblotch viroids isolated from single avocado trees. Phytoparasitica 29 (5), 451–460. https://doi.org/10.1007/BF02981864.

Schnölzer, M., et al., 1985. Correlation between structure and pathogenicity of potato spindle tuber viroid (PSTV). EMBO J. 4 (9), 2181–2190.

C. Viroid pathogenesis and viroid-host interaction

Semancik, J.S., Szychowski, J.A., 1994. Avocado sunblotch disease: a persistent viroid infection in which variants are associated with differential symptoms. J. Gen. Virol. 75 (7), 1543–1549. https://doi.org/10.1099/0022-1317-75-7-1543.

Škorić, D., 2017. Viroid biology. In: Viroids and Satellites. Elsevier, pp. 53–61, https://doi.org/10.1016/B978-0-12-801498-1.00005-X.

Tsushima, T., et al., 2011. Molecular characterization of Potato spindle tuber viroid in dahlia. J. Gen. Plant Pathol. 77 (4), 253–256. https://doi.org/10.1007/s10327-011-0316-z.

Vera, P., Conejero, V., 1989. The induction and accumulation of the pathogenesis-related P69 proteinase in tomato during citrus exocortis viroid infection and in response to chemical treatments. Physiol. Mol. Plant Pathol. 34 (4), 323–334. https://doi.org/10.1016/0885-5765(89)90029-5.

Vera, P., Hernandez-Yago, J., Conejero, V., 1989. "Pathogenesis-related" P1(p14) protein. Vacuolar and Apoplastic localization in leaf tissue from tomato plants infected with Citrus Exocortis viroid; in vitro synthesis and processing. J. Gen. Virol. 70 (8), 1933–1942. https://doi.org/10.1099/0022-1317-70-8-1933.

Vidal, A.M., et al., 2003. Regulation of gibberellin 20-oxidase gene expression and gibberellin content in citrus by temperature and citrus exocortis viroid. Planta 217 (3), 442–448. https://doi.org/10.1007/s00425-003-0999-2.

Visvader, J.E., Symons, R.H., 1985. Eleven new sequence variants of citrus exocortis viroid and the correlation of sequence with pathogenicity. Nucleic Acids Res. 13 (8), 2907–2920. Available at: http://www.ncbi.nlm.nih.gov/pubmed/2582367.

Wang, M.-B., et al., 2004. On the role of RNA silencing in the pathogenicity and evolution of viroids and viral satellites. Proc. Natl. Acad. Sci. USA 101 (9), 3275–3280. https://doi.org/10.1073/pnas.0400104101.

Wassenegger, M., et al., 1994. RNA-directed de novo methylation of genomic sequences in plants. Cell 76 (3), 567–576. https://doi.org/10.1016/0092-8674(94)90119-8.

Wassenegger, M., et al., 1996. A single nucleotide substitution converts potato spindle tuber viroid (PSTVd) from a noninfectious to an infectious RNA for Nicotiana tabacum. Virology 226 (2), 191–197. https://doi.org/10.1006/viro.1996.0646.

Wolff, P., et al., 1985. Complexes of viroids with histones and other proteins. Nucleic Acids Res. 13 (2), 355–367. https://doi.org/10.1093/nar/13.2.355.

Yamaji, Y., et al., 2006. In vivo interaction between Tobacco mosaic virus RNA-dependent RNA polymerase and host translation elongation factor 1A. Virology 347 (1), 100–108. https://doi.org/10.1016/j.virol.2005.11.031.

C. Viroid pathogenesis and viroid-host interaction

12

Viroids and protein translation

Purificación Lisón[a], Francisco Vázquez-Prol[a], Irene Bardani[b], Ismael Rodrigo[a], Nikoleta Kryovrysanaki[c], and Kriton Kalantidis[b,c]

[a]Institute for Plant Molecular and Cellular Biology, Universitat Politècnica de València-Consejo Superior de Investigaciones Científicas, Valencia, Spain [b]Department of Biology, University of Crete, Heraklion, Greece [c]Institute of Molecular Biology and Biotechnology, Foundation for Research and Technology, Heraklion, Greece

Graphical representation

(A) **Central dogma of molecular biology**. Genomic DNA uncondensed and replicates to form new cells (1) with the help of several DNA polymerases. The RNA polymerases read DNA as a template to transcribe RNA (2), which can be translated into proteins (3) by the ribosomes. (B) **Non-coding nature of viroids**. Viroids replicate through DNA-dependent RNA polymerases (DdRP). Viroid RNA would not be recognized by the ribosome as translatable information and would not produce proteins.

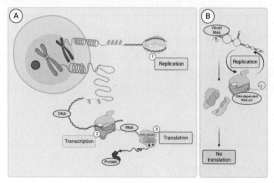

Definitions

Transcription is the process of copying the genetic information stored in DNA into different sorts of RNA molecules. If that information encodes for proteins, then a messenger RNA (mRNA) is produced. DNA also has the sequences to build other RNAs such as transfer RNA (tRNA) or ribosomal RNA (rRNA).

Translation is the process by which ribosomes synthesize proteins. The ribosome *reads* the sequence of the messenger RNA and *translates* a string of characters made by RNA bases into a sequence of amino acids, by using the genetic code.

Genetic code is the set of guidelines used by living cells to translate information encoded within genetic material (four nucleotide bases) into proteins (21 amino acids). The genetic code includes 64 entries, each one corresponding with a combination of three RNA bases (codon) that codes either for one amino acid (aa) or one start/stop codon. This correspondence is embodied through the transfer RNA. The genetic code is degenerated since one aa can be encoded by several codons, but it is unequivocal, since one codon codes only for one aa.

Start codon is the first codon of a messenger RNA transcript translated by a ribosome. The start codon always codes for methionine in eukaryotes. The most common start codon is AUG, which corresponds with ATG in the DNA sequence.

Ribosome is a cellular particle made of RNA and protein that constitutes the translating machinery of the cell. Ribosomes are classified according to their sedimentation coefficient (S). Eukaryotes have 80S ribosomes, consisting of a small (40S) and a large (60S) subunit. Ribosomes are composed of approximately 60% rRNA and 40% ribosomal proteins by mass.

Ribosomal RNA (rRNA) is the primary component of ribosomes and the predominant form of RNA found in most cells. In plants, the 18S rRNA is present in the small ribosome subunit, whilst 25S, 5.8S, and 5S rRNAs are the main constituents of the large ribosome subunit.

Ribosome biogenesis is the process of making ribosomes and it starts in the nucleolus. While the 5S rRNA is transcribed independently, the 18S, 5.8S, and 25S rRNAs are transcribed as a single primary transcript called 35S pre-RNA. This pre-rRNA is processed into the individual rRNAs which then associate with the ribosomal proteins, forming the two types of ribosomal subunits (large and small). These will later assemble in the cytosol to make functional ribosome particles.

Chapter outline

In this chapter, we will study the relationship between viroids and protein translation. The non-coding nature of viroids will be discussed. Changes in protein expression and translation machinery produced by viroids will also be addressed, including the described interactions between viroids and host proteins.

Learning objectives

- To address the non-coding nature of viroids.
- To ponder the important alterations in host gene expression caused by these small, non-coding pathogens.

C. Viroid pathogenesis and viroid-host interaction

- To consider the interactions between viroids and the host proteins.
- To understand the changes that viroids cause on translational machinery.

Fundamental introduction

The "central dogma" of molecular biology explains the flow of genetic information that occurs in living organisms. The normal flow of biological information includes the following processes: DNA can be copied to DNA (DNA replication), DNA information can be copied into mRNA (transcription*), and proteins can be synthesized using mRNA as a template in the ribosomes* (translation*). Particularly, translation* proceeds in three phases: initiation, elongation, and termination. Additionally, ribosome* recycling could be considered as a fourth step. During the initiation, the 40S ribosomal subunit assembles around the target mRNA and scans until a start codon* is found. Then, the first tRNA, which brings the first amino acid (methionine) according to genetic code*, is attached at the initiation codon*. Following initiation, the 60S ribosomal subunit joins to form the 80S ribosome* and the elongation starts. During this stage, following the instructions of the mRNA codons, amino acids are brought to the ribosome by tRNAs and linked together to form a polypeptide. When a stop codon* is reached, the ribosome* releases the polypeptide, leading to the termination. Finally, during the ribosome* recycling phase, the mRNA is released and the 80S ribosome* is separated into its 40S and 60S components, remaining the ribosomal components intact to move on to the next mRNA (Fig. 1).

Viroids are the simplest and smallest known agents of infectious disease, nevertheless having the capacity to cause disease in several economically important crop plants. They produce a wide variety of symptoms in the infected plants that often resemble those provoked by viral infections. Unlike viruses, viroids consist exclusively of naked, covalently closed single-stranded RNA molecules with a high degree of secondary structure and a small size ranging from 239 to 401 nucleotides. Furthermore, viroid RNA appears not to encode for any known protein, indicating that they depend completely on the replication machinery of the host cell.

Since viroids could not be translated into proteins, alterations in protein accumulation during infection should involve only host-encoded proteins. This makes the viroid-plant interaction system a very convenient model to study the plant response to pathogen attack in terms of protein changes.

In this chapter, we will review the non-coding nature of viroids and the changes that viroids produce on host protein expression and the translation machinery.

(*) See definitions.

Non-coding nature of viroids

All known entities with a biological cycle have their genomic information encoded on nucleic acid, in most cases being DNA. However, some very simple biological entities, such as viruses, viroids, and viral satellites have their "genome" based on RNA. We tend to regard nucleic acids as molecules that encode information in the form of a sequence of nitrogenous

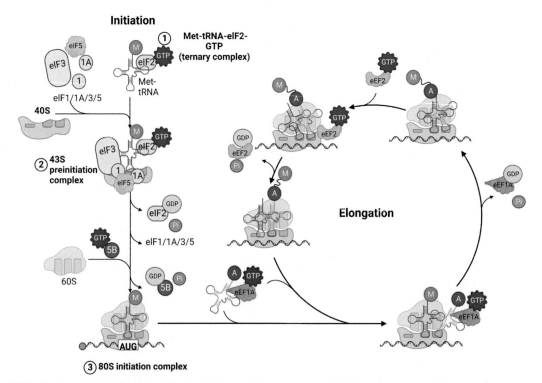

FIG. 1 Translation process. To initiate translation, the tRNA that carries the amino acid methionine (Met-tRNA) forms a complex with the Eukaryotic Initiation Factor 2 (eIF2) bound to GTP (ternary complex). Then, the 43S preinitiation complex is formed among the ternary complex, the 40S subunit of the ribosome, and the Eukaryotic Initiation Factors 1, 1A, 3, and 5. eIF2 consumes GTP to release the Initiation Factors from the ribosome and bind mRNA. Then, the 60S subunit binds to the complex through GTP consumption by the Initiation Factor 5B (5B) to form the 80S initiation complex, which begins elongation. During elongation, the Eukaryotic Elongation Factor 1A (eEF1A) brings the tRNA carrying the corresponding amino acid to the aminoacyl site of the ribosome and consumes GTP to link the amino acids, thus elongating the peptide chain. Then, the Eukaryotic Elongation Factor 2 (eEF2) binds to the ribosome and consumes GTP to translocate the new tRNA to the peptidyl site, pushing the previous tRNA to the exit site of the ribosome. This cycle repeats until the peptide is fully formed.

bases, the majority of which is eventually translated into a sequence of amino acids that form peptides, polypeptides, and proteins. However, we already early on realized that there are some RNA molecules with unarguable biological functions even in the absence of translation capability. Such examples have been housekeeping RNAs (tRNAs and rRNAs), regulatory RNAs (miRNAs and siRNAs), and more. However, none of these non-protein coding RNAs (often for simplicity known as non-coding RNAs which is inaccurate) replicate.

Therefore, when viroids were first discovered in the early 70s (Diener, 1971; Semancik and Weathers, 1972), it was reasonable to presume that their RNA "genome" would be protein-encoding. Nevertheless, attempts to detect viroid translation activity and protein synthesis either in vitro (Davies et al., 1974; Hall et al., 1974) or in vivo failed. More specifically, both Diener and Semancik groups addressed this seminal question with similar approaches; they tested various in vitro translation systems from plants, bacteria, and yeast for their ability to translate viroid RNA and whether viroid RNA could bind amino acids. Each work used a different pospiviroid in the experiments (potato spindle tuber viroid, PSTVd, and citrus exocortis viroid, CEVd), but in all cases the evidence toward translation of viroid RNA was negative.

In vitro approaches are powerful to understand phenomena without obstructions from the complex environment. However, at the same time, there is always the possibility that there is an essential element missing in the in vitro experiment, and the in vivo situation is not accurately simulated. Therefore, it is always important for in vitro experiments to be complemented with in vivo approaches.

It is of interest that soon after the coining of the term "viroid" by Diener, Zaitlin, and Hariharasubramanian used radioactively labeled leucine incorporation in conjugation with polyacrylamide gel electrophoresis to identify small polypeptides that could be generated by PSTVd (at that time known as the PST Virus) (Diene, 1971; Zaitlin and Hariharasubramanian, 1972). Even though the experiment was specifically set up to help identify very small peptides that would fit the size of the viroid genome, no such peptides were detected in their experiments.

Later, improved electrophoresis approaches were applied toward the same aim. CEVd infected and control *Gynura aurantiaca* leaves were incubated with a mixture of ^{14}C-labeled amino acids (arginine, leucine, lysine, valine), and following an incubation period proteins were extracted and run on a polyacrylamide gel of a density adjusted to separate small polypeptides (14% acrylamide). Again, no novel viroid-related proteins were detected on the gel. The authors detected some quantitative differences in the concentration of specific proteins between infected and non-infected samples. However, these proteins were present in both samples, albeit with different intensities, and therefore the authors concluded that they are endogenous proteins affected by CEVd infectivity rather than viroid proteins (Conejero and Semancik, 1977).

The group of Semancik applied an additional ex-planta in vivo approach to testing the ability of viroid RNA to be translated. Injection of CEVd RNA into oocytes of *Xenopus laevis* was performed under conditions that demonstrated translation of bromegrass mosaic virus (BMV) RNA and rabbit hemoglobin mRNA. Again, neither viroid RNA translation activity nor viroid-specific protein intermediate could be observed (Conejero and Semancik, 1977).

Altogether, the above experimentation convinced researchers in the field that viroid (PSTVd and CEVd) RNA is not translated and viroid pathogenesis is probably accomplished by host proteins regulated by viroid RNA.

Changes in protein expression associated with viroid infection

The identification of proteins that differentially accumulated in viroid-infected plants when compared with control non-infected plants has been performed by different procedures. The classical methods, including sodium dodecyl sulfate-polyacrylamide gel electrophoresis (SDS-PAGE), density gradient, and in vivo labeling, allow the identification of a limited number of differentially accumulated proteins, whilst high-throughput proteomic approaches offer a holistic view of the plant response.

By using classic methods, changes in protein accumulation caused by viroids were first described, involving only host-encoded proteins. In this sense, protein alterations following the onset of symptoms of the citrus exocortis viroid (CEVd) disease in *G. aurantiaca* were detected in protein extracts (Conejero and Semancik, 1977). Comparing protein profiles of control and viroid-infected plants, changes in the abundance of proteins were also detected in several viroid-host combinations (Conejero et al., 1979; Henriquez and Sänger, 1982; Hadidi, 1988).

Some of the proteins differentially accumulated in these viroid-infected plants were identified as pathogenesis-related (PR) proteins, which are defensive proteins produced in plants upon a pathogen attack. PR proteins have a wide range of functions, acting as chitinases, peroxidases, anti-microbial agents, hydrolases, protease inhibitors, and other activities (Zribi et al., 2021). Additional proteins that accumulate upon viroid infection were also identified, but their roles in pathogenesis and defense reactions remain unknown.

With the advent of proteomics (high-throughput, large-scale "shotgun" analysis of proteins), changes in the tomato proteome in response to CEVd were studied (Lisón et al., 2013). Using two-dimensional difference gel electrophoresis (2D-DIGE) followed by mass spectrometry to study the response in tomato plant cells to CEVd infection, a total of 45 differentially accumulated proteins were characterized. The validation of the results by RT-PCR, a technique used to study differences in mRNA levels, allowed the classification of these proteins into two expression groups. The first group included genes showing changes at the transcriptional level upon CEVd infection, thus correlating transcription with translation. This group included defensive proteins such as an endochitinase, β-glucanase, and pathogenesis-related proteins PR10 and P69G. The second group of proteins displayed no changes at the transcriptional level, therefore non-correlating with the alterations in protein levels. This suggests viroids seem to have the ability to interfere with the host translational machinery. Members of this group included several ribosomal proteins such as S3, S5, and L10 as well as translation factors, including the elongation factors one (eEF1A) and two (eEF2), and the translation initiation factor 5-alpha (eIF5A).

Interaction of viroids with host proteins

Viroids do not encode proteins but still can replicate, spread cell-to cell and systemically infect several hosts and often cause symptoms. The above makes it clear that these tiny agents not only affect the host by using up resources but are also able to accomplish all these functions by hijacking parts of the host machinery. This became evident quite early and led to a hunt for host factors that are exploited by viroids to complete their replication cycle. Of course, these host factors are not there just to enable viroid infectivity but have their importance for the host. In addition, we should keep in mind that the role these host factors hold for viroid biology is not necessarily the same as the role the viroids hold for the host. There are two basic approaches to identifying host factors that interact with a pathogen: (a) genetic screens that identify host factors whose mutation decreases, increases, or even abolishes pathogen infectivity and (b) looking for host factors—usually proteins—that directly or indirectly interact with the pathogen. In the case of viroids, the genetic screening approach, so widely and successfully used in *Arabidopsis* after *Drosophila* and yeast, was hard to apply since viroid host plants are not convenient models, due to difficult genetics, too long generation time, large genomes, lack of tools, etc. Thus, studies have been focusing on the identification of host proteins significantly affected by viroid infectivity and even more so, proteins that interact directly with viroid RNA. These time-consuming experiments have helped to identify some host proteins with critical roles in viroid infectivity.

The search for host factors that are significant for the biological cycle of viroids has been based on both direct and indirect assays. A classical and very important indirect approach provided strong evidence to suggest that pospiviroids are replicated by DNA-dependent RNA polymerase II (DdRP II), which in this case is hijacked to function as an RNA-dependent RNA polymerase (RdRP). Mühlbach and Sänger (1979) applied α-amanitin to tomato protoplasts infected with what it was called in the paper, the cucumber pale fruit viroid (CPFV), known today as HSVd (Mühlbach and Sänger, 1979). Replication of the viroid was only blocked at a concentration that this chemical inhibits specifically the function of DdRP II. It was only two decades later when a direct interaction between this polymerase and another pospiviroid was published (Warrilow and Symons, 1999).

The first attempt to look for proteins directly interacting with the viroid was performed in vitro reconstitution using nuclear extracts. This approach yielded a small number of viroid-protein complexes and only one of these complexes, based on size hypothesized to be histones, was then also verified by in vivo analysis from the nucleosomal fraction (Wolff et al., 1985). Chloroplast preparations were tested for their ability to support the replication of avocado sunblotch viroid (ASBVd) in vitro. These experiments provided strong evidence that a nuclear-encoded RNA polymerase (NEP) is the chloroplast polymerase involved in avsunviroid replication (Navarro and Flores, 2000). Other early approaches utilized either in vitro binding between viroid RNAs and host proteins extracted from infected tissue (Wolff et al., 1985; Hadidi, 1986, 1988), or expressed from cDNA libraries (Sägesser et al., 1997), or in vitro analysis of subcellular fractions (Hadidi, 1988; Klaff et al., 1996). However, at the time it was not possible to evaluate the functional significance of the suggested interaction.

Technological advances including better, more specific binding protocols and/or peptide analysis methods allowed the identification of a small number of specific proteins, which could now be tested for specific and direct interactions with viroids as well as for their importance as host factors (Daròs and Flores, 2002; Maniataki et al., 2003; Martínez de Alba et al., 2003). Daròs and Flores used UV irradiation to fix protein-viroid complexes in vivo. Their system involved the chloroplast-replicating avocado sunblotch viroid in avocado leaves. Tandem mass spectrometry analysis allowed the characterization of two closely similar viroid-binding proteins, named PARBP33 and PARBP35 (Daròs and Flores, 2002). Experiments in the same work with recombinant PARBP33 showed that this protein enhanced the efficiency of the viroid ribozyme for self-cleavage. Later studies using RNA silencing as a reverse genetics tool allowed testing the physiological significance of proteins identified this way. Suppressing the viroid binding protein VIRP1 identified previously revealed this factor to be essential for viroid infectivity (Kalantidis et al., 2007; Chaturvedi et al., 2012). Somewhat surprisingly, VIRP1-suppressed plants did not show any other phenotypic aberration.

Bojić and colleagues demonstrated that RNA Pol II associates with the left terminal domain of PSTVd (+) RNA, using a co-immunoprecipitation strategy. However, RNA Pol II did not interact with any PSTVd (−) RNAs. Wang and colleagues showed that the Transcription Factor IIIA from *N. benthamiana* interacts with PSTVd RNA. More specifically, they found that the canonical 9-zinc finger TFIIIA (TFIIIA-9ZF), as well as its variant TFIIIA-7ZF, interact with (+)-PSTVd, but only TFIIIA-7ZF interacts with (−)-PSTVd. Furthermore, suppression of TFIIIA-7ZF reduces PSTVd replication, and overexpression of TFIIIA-7ZF enhanced PSTVd replication *in planta*. Consistently, footprinting assays showed that TFIIIA-7ZF bound to a PSTVd region critical for transcription initiation. All of the above reveal that TFIIIA-7ZF directly guides RNA Pol II-driven transcription with PSTVd RNA as a template (Bojić et al., 2012; Wang et al., 2016).

Given the lack of a convenient model for genetic screening, viroid scientists often search for host factors by intuition or educated guesses: there are specific processes that take place in the viroid biological cycle such as replication, cleavage of concatemeric intermediates and ligation of linear monomeric molecules that suggest the type of protein functions to look for. Having a clue as to what to look for aids researchers then spot down candidate proteins before they are actually in vitro and in vivo testing for their function in viroid biology. Following such strategies, several factors very important to viroid biology have been identified. The systemic movement of viroids through phloem led scientists to search within the phloem sap for viroid interacting proteins. Indeed, a strong viroid binding protein was found in the sap of *Cucumis sativus*, a plant particularly amenable to phloem work due to the large amount and ease in the extraction of its phloem sap (Gómez and Pallás, 2001; Owens et al., 2001). Although these were originally in vitro experiments, it was later shown that this is likely a real in vivo interaction enabling viroid mobility, both for hop stunt viroid (Gómez and Pallás, 2004) and for apple scar skin viroid (Walia et al., 2015). Both groups of viroids replicating in the chloroplast and the nucleus undergo a ligation step, where the monomeric linear genomic RNAs of the viroid go circular. For another group of viroids, avsunviroids (for a review on this distinct group of viroids, see Flores et al., 2000) replicating in the chloroplast a likely candidate was tRNA ligase that is targeted to the chloroplast to catalyze there the circularization of chloroplastic tRNA. Using first in vitro approaches, researchers could provide strong evidence that indeed the chloroplast-targeted tRNA ligase can catalyze viroid circularization. Then, they could show the physiological significance of their findings through in vivo experiments (Nohales et al., 2012b). For nuclear-replicating viroids, the same group undertook a somewhat different approach; they first identified the protein fraction of the host (tomato) protein extract that was catalyzing the circularization reaction, and then with a combination of biochemical and molecular and database search steps came down to a single candidate gene, the DNA ligase I gene. The candidate gene was then tested for its function in ligating the viroid RNA through a reverse genetics approach strongly supporting their hypothesis (Nohales et al., 2012a).

Alterations in the translational machinery caused by viroids

Despite their non-coding nature, viroids have been described to produce changes in the translational machinery. As indicated above, CEVd produces alterations in the accumulation of several proteins involved in translation, including the ribosomal proteins S3, S5, and L10, or the eukaryotic translation factors eEF1A, eEF2, and eIF5A. Moreover, post-translational modifications in these translation factors caused by the viroid were also detected by 2D-Western blot, since antisera revealed new spots in the protein extracts from infected plants (Lisón et al., 2006).

Some of these proteins involved in the translation process have also been described to directly interact with viroids. In this sense, the ribosomal protein L5 from *Arabidopsis thaliana* also binds specifically PSTVd RNA in vitro (Eiras et al., 2011). Besides, eEF1A interacts with both peach latent mosaic viroid (Dube et al., 2009) and with CEVd (Lisón et al., 2013). The precise contribution of these interactions to the molecular biology of viroids and plant pathogenesis remains unclear.

Apart from the modification of the expression of proteins that are involved in translation, viroids also produce changes in ribosomal RNAs or mRNAs coding for ribosomal proteins. Cucumber plants infected with hop stunt viroid (HSVd) display an increase in the transcription of rRNA precursors upon infection, which correlates with a modification of DNA methylation in their promoter region (Martinez et al., 2014). Particularly, HSVd provokes dynamic demethylation of repetitive regions in the cucumber genome that include rRNA genes and transposable elements in gametic cells (Castellano et al., 2016). In addition, PSTVd triggers the degradation of the ribosomal protein S3a-like mRNAs in tomato plants (Adkar-Purushothama et al., 2017).

Finally, CEVd produces alterations in ribosome biogenesis (Fig. 2), particularly in the 18S rRNA maturation process. This non-coding pathogen is present in the ribosomal fractions of infected tomato plants, provoking changes in the global polysome profiles. Particularly, CEVd infection leads to alterations in the accumulation of the 40S ribosomal subunit, constituted by 18S rRNA whose maturation is compromised. Moreover, the ribosomal stress mediator NAC082 is induced following viroid infection, confirming that viroids produce ribosomal stress, which also correlates with viroid symptomatology (Cottilli et al., 2019). This correlation was further confirmed in the hypersusceptible ethylene-insensitive *never ripe* tomato mutants (Prol et al., 2020). Finally, the alteration in ribosome biogenesis, evidenced by both the upregulation of the tomato ribosomal stress marker *SlNAC082* and the impairment in 18S rRNA processing, was extended to other viroids, including tomato chlorotic dwarf viroid (TCDVd) and PSTVd (Prol et al., 2021).

Revisiting viroid translation capacity: are negative results conclusive?

The non-coding nature of viroids has troubled scientists for a long time. This is possible because molecular biology was viewing nucleic acids until the late nineties as molecules carrying mainly translational information. In this light, a parasite being non-coding seemed odd to say the least, and certainly stood alone. After all, a human circular RNA pathogen of hepatitis delta virus (HDV) was identified soon after viroids and was reminiscent of them in some aspects, such as structure, folding, and replication mechanism, although does encode for two proteins. As mentioned above viroid researchers addressed this question early on with all the methodology available at the time. The verdict was unanimous: with the methods available, no translation of viroid can be detected in vitro or in vivo. Since then, the scientific community came to terms with ncRNAs, with the discovery of a multitude of such molecules serving

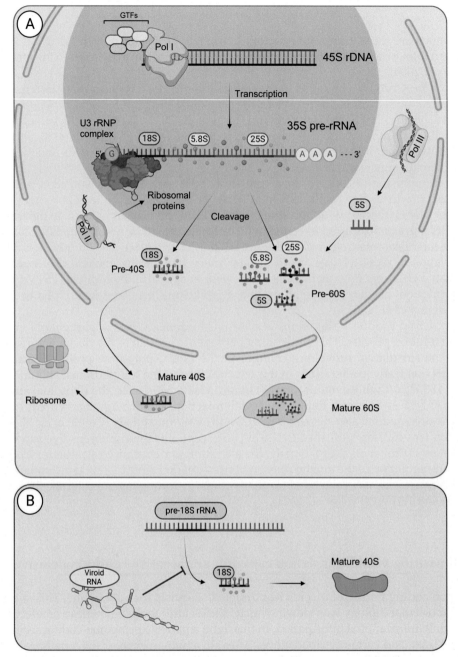

FIG. 2 Ribosome biogenesis in plants. (A) In the nucleolus, several General Transcription Factors (GTFs) are involved in the transcription of the 45S ribosomal DNA (45S rDNA) by the RNA Polymerase I into a 35S pre-rRNA, which contains the information for most of the RNA present in the ribosome. In the nucleus, RNA Polymerase II transcribes the genes of ribosomal proteins, which are translocated to the nucleolus after translation. Then, the U3 rRNP complex and several ribosomal proteins catalyze the cleavage of the pre-rRNA 35S into the 18S, 5.8S, and 25S fragments, which are translocated outside of the nucleolus. In the nucleus, RNA Polymerase III transcribes the 5S fragment. Finally, the rRNA fragments bind ribosomal proteins to form the 40S and 60S subunits of the ribosome and are translocated to the cytosol, where they can start the translation. (B) Some viroid RNAs interfere with the correct cleavage of the 18S fragment, impairing the proper biogenesis of the 40S subunit and the ribosome assembly.

important functions in almost all biological systems. So now the community felt that viroids once again were the first of their kind to be found rather than the odd one out.

The field became even more fascinating with the discovery of circular RNAs (circRNAs), across species from Archaea to humans, animals, and plants (Salzman et al., 2012; Jeck et al., 2013; Memczak et al., 2013; Wang et al., 2014; Ye et al., 2015). Conceptually, viroids were not the odd one out anymore, and it was easier to come to terms with the fact that their function comes from their structure, not their translational capacity. But then, there was a series of works reporting that some of these "ncRNAs" were indeed coding: first came a report in plants that pri-miRNAs can translate regulatory peptides (Ruiz-Orera et al., 2014; Lauressergues et al., 2015; Legnini et al., 2017) and then from circRNA some of which was also shown to be translated (Granados-Riveron and Aquino-Jarquin, 2016; Legnini et al., 2017; Pamudurti et al., 2017; Yang et al., 2017; Yin et al., 2017). These findings immediately opened the possibility that viroids could also hold some translational potential that possibly the methodology at the time of the earlier experimentations was not able to uncover.

Looking at the effect CEVd had on translation, Cottilli and colleagues found that the viroid genomic RNA and its derived viroid small RNAs co-sedimented with tomato ribosomes in vivo, and provoked changes in the global polysome profiles, particularly in the 40S ribosomal subunit accumulation. In the same study, they reported viroid causing alterations in ribosome biogenesis in the infected tomato plants, affecting the translation machinery and eventually leading to symptom development. Apart from the important information this work conveys on symptom development caused by a non-protein coding agent, it also sheds light on viroid interaction with ribosomes (Cottilli et al., 2019).

Having evidence for ribosome viroid interaction made again tempting the hypothesis that viroid RNA may under some circumstances be translated. There are reports from two groups that have tested this hypothesis: first, Marquez-Molins, and co-workers analyzed HSVd and eggplant latent viroid (ELVd) for putative ORFs. They could show that not only are such ORFs present in these viroids but also in addition they are potentially able to encode peptides carrying subcellular localization signals coincident with the corresponding replication-specific organelle (Marquez-Molins et al., 2021). In addition, they addressed the functional importance of these sequences. Bioassays in established hosts revealed that mutations in these ORFs diminish their biological efficiency although these mutations of course did not affect solely the potential peptide produced. Based on this evidence the authors suggested that under certain cellular conditions, as non-canonical translatable transcripts, viroids could potentially be translated. This possibility was revisited by Katsarou et al. (2022). The authors indeed identified potential ORFs practically in all the viroids included in the analysis. They found that for some viroid strains the presence of these "ORFs" was significantly higher than would have been for random sequences of the same size. Again, they could show that circular viroids would co-sediment with polysomes indicating a possibility of them being translated. However, neither classic techniques such as in vitro translation nor MS proteome analysis revealed any trace of the expected or any other viroid-encoded peptide (Katsarou et al., 2022). Taken together using a mixture of old and recent techniques the authors confirmed that at least the pospiviroids tested do not produce peptides, suggesting that the association of viroid RNAs to ribosomes should be due to reasons other than translation. One can only speculate whether such an interaction is related to a favourable (for the viroid) effect on the translation or protection of the viroid RNA.

C. Viroid pathogenesis and viroid-host interaction

Chapter summary

In this chapter, we have described classic and recent experiments supporting that viroids are RNA pathogens with no coding capacity. However, they can produce changes not only in host-protein expression but also in the translation machinery, interacting with some host proteins.

Study questions

1. Do viroids produce changes in protein expression and the translation machinery of the plant? Describe some examples.
2. Name three proteins which have been identified to interact with viroids.
3. Briefly describe how viroids alter the ribosome biogenesis.
4. Indicate two recent works where the presence of ORFs in viroids have been explored.
5. Are viroids translated? Justify the answer.

Further reading

Cottilli, P., Belda-Palazón, B., Adkar-Purushothama, C.R., Perreault, J.P., Schleiff, E., Rodrigo, I., Ferrando, A., Lisón, P., 2019. Citrus exocortis viroid causes ribosomal stress in tomato plants. Nucleic Acids Res. 47, 8649–8661.

Kalantidis, K., Denti, M.A., Tzortzakaki, S., Marinou, E., Tabler, M., Tsagris, M., 2007. Virp1 Is a host protein with a major role in potato spindle tuber viroid infection in nicotiana plants. J. Virol. 81, 12872–12880.

Katsarou, K., Adkar-Purushothama, C.R., Tassios, E., Samiotaki, M., Andronis, C., Nikolaou, C., Perreault, J., 2022. Revisiting the non-coding nature of pospiviroids. Cells 11, 265.

Lisón, P., Tárraga, S., López-Gresa, P., Saurí, A., Torres, C., Campos, L., Bellés, J.M., Conejero, V., Rodrigo, I., 2013. A noncoding plant pathogen provokes both transcriptional and posttranscriptional alterations in tomato. Proteomics 13, 833–844.

References

Adkar-Purushothama, C.R., Iyer, P.S., Perreault, J.P., 2017. Potato spindle tuber viroid infection triggers degradation of chloride channel protein CLC-b-like and ribosomal protein S3a-like mRNAs in tomato plants. Sci. Rep. 7, 8341.

Bojić, T., Beeharry, Y., Da Zhang, J., Pelchat, M., 2012. Tomato RNA polymerase II interacts with the rodlike conformation of the left terminal domain of the potato spindle tuber viroid positive RNA genome. J. Gen. Virol. 93, 1591–1600.

Castellano, M., Martinez, G., Marques, M.C., Moreno-Romero, J., Kohler, C., Pallas, V., Gomez, G., 2016. Changes in the DNA methylation pattern of the host male gametophyte of viroid-infected cucumber plants. J. Exp. Bot. 67, 5857–5868.

Chaturvedi, S., Jung, B., Gupta, S., Anvari, B., Rao, A.L.N., 2012. Simple and robust in vivo and in vitro approach for studying virus assembly. J. Vis. Exp. (61), 1–5.

Conejero, V., Semancik, J.S., 1977. Exocortis viroid: alteration in the proteins of Gynura aurantiaca accompanying viroid infection. Virology 77, 221–232.

Conejero, V., Picazo, I., Segado, P., 1979. Citrus exocortis viroid (CEV): protein alterations in different hosts following viroid infection. Virology 97, 454–456.

Cottilli, P., Belda-Palazón, B., Adkar-Purushothama, C.R., Perreault, J.P., Schleiff, E., Rodrigo, I., Ferrando, A., Lisón, P., 2019. Citrus exocortis viroid causes ribosomal stress in tomato plants. Nucleic Acids Res. 47, 8649–8661.

Daròs, J.A., Flores, R., 2002. A chloroplast protein binds a viroid RNA in vivo and facilitates its hammerhead-mediated self-cleavage. EMBO J. 21, 749–759.

Davies, J.W., Kaesberg, P., Diener TO, 1974. Potato spindle tuber viroid. XII. An investigation of viroid RNA as a messenger for protein synthesis. Virology 61, 281–286.

Diener TO, 1971. Potato spindle tuber '"virus"' IV. A replicating, low molecular weight RNA. Virology 45, 411–428.

Dube, A., Bisaillon, M., Perreault, J.-P., 2009. Identification of proteins from *Prunus persica* that interact with peach latent mosaic viroid. J. Virol. 83, 12057–12067.

Eiras, M., Nohales, M.A., Kitajima, E.W., Flores, R., Daròs, J.A., 2011. Ribosomal protein L5 and transcription factor IIIA from *Arabidopsis thaliana* bind in vitro specifically potato spindle tuber viroid RNA. Arch. Virol. 156, 529–533.

Flores, R., Daròs, J.A., Hernandez, C., 2000. Avsunviroidae family: viroids containing hammerhead ribozymes. Adv. Virus Res. 55, 271–323.

Gómez, G., Pallás, V., 2001. Identification of an in vitro ribonucleoprotein complex between a viroid RNA and a phloem protein from cucumber plants. Mol. Plant-Microbe Interact. 14, 910–913.

Gómez, G., Pallás, V., 2004. A long-distance translocatable phloem protein from cucumber forms a ribonucleoprotein complex in vivo with hop stunt viroid RNA. J. Virol. 78, 10104–10110.

Granados-Riveron, J.T., Aquino-Jarquin, G., 2016. The complexity of the translation ability of circRNAs. Biochim. Biophys. Acta, Gene Regul. Mech. 1859, 1245–1251.

Hadidi, A., 1986. Relationship of viroids and certain other plant pathogenic nucleic acids to group I and II introns. Plant Mol. Biol. 7, 129–142.

Hadidi, A., 1988. Synthesis of disease-associated proteins in viroid-infected tomato leaves and binding of viroid to host proteins. Phytopathology 78, 575.

Hall, T.C., Wepprich, R.K., Davies, J.W., Weathers, L.G., Semancik, J.S., 1974. Functional distinctions between the ribonucleic acids from citrus exocortis viroid and plant viruses: cell-free translation and aminoacylation reactions. Virology 61, 486–492.

Henriquez, A.C., Sänger, H.L., 1982. Analysis of acid-extractable tomato leaf proteins after infection with a viroid, two viruses, and a fungus and partial purification of the 'pathogenesis-related' protein p 14. Arch. Virol. 74, 181–196.

Jeck, W.R., Sorrentino, J.A., Wang, K., Slevin, M.K., Burd, C.E., Liu, J., Marzluff, W.F., Sharpless, N.E., 2013. Circular RNAs are abundant, conserved, and associated with ALU repeats. RNA 19, 141–157.

Kalantidis, K., Denti, M.A., Tzortzakaki, S., Marinou, E., Tabler, M., Tsagris, M., 2007. Virp1 is a host protein with a major role in potato spindle tuber viroid infection in Nicotiana plants. J. Virol. 81, 12872–12880.

Katsarou, K., Adkar-purushothama, C.R., Tassios, E., Samiotaki, M., Andronis, C., Nikolaou, C., Perreault, J., 2022. Revisiting the non-coding nature of pospiviroids. Cell 11, 265.

Klaff, P., Riesner, D., Steger, G., 1996. RNA structure and the regulation of gene expression. Plant Mol. Biol. 32, 89–106.

Lauressergues, D., Couzigou, J.M., San Clemente, H., Martinez, Y., Dunand, C., Bécard, G., Combier, J.P., 2015. Primary transcripts of microRNAs encode regulatory peptides. Nature 520, 90–93.

Legnini, I., Di Timoteo, G., Rossi, F., et al., 2017. Circ-ZNF609 is a circular RNA that can be translated and functions in Myogenesis. Mol. Cell 66, 22–37.e9.

Lisón, P., Rodrigo, I., Conejero, V., 2006. A novel function for the cathepsin D inhibitor in tomato. Plant Physiol. 142, 1329–1339.

Lisón, P., Tárraga, S., López-Gresa, P., Saurí, A., Torres, C., Campos, L., Bellés, J.M., Conejero, V., Rodrigo, I., 2013. A noncoding plant pathogen provokes both transcriptional and posttranscriptional alterations in tomato. Proteomics 13, 833–844.

Maniataki, E., Tabler, M., Tsagris, M., 2003. Viroid RNA systemic spread may depend on the interaction of a 71-nucleotide bulged hairpin with the host protein VirP1. RNA 9, 346–354.

Marquez-Molins, J., Navarro, J.A., Seco, L.C., Pallas, V., Gomez, G., 2021. Might exogenous circular RNAs act as protein-coding transcripts in plants? RNA Biol. 18, 98–107.

Martínez de Alba, A.E., Sägesser, R., Tabler, M., Tsagris, M., 2003. A bromodomain-containing protein from tomato specifically binds potato spindle tuber viroid RNA in vitro and in vivo. J. Virol. 77, 9685–9694.

Martinez, G., Castellano, M., Tortosa, M., Pallas, V., Gomez, G., 2014. A pathogenic non-coding RNA induces changes in dynamic DNA methylation of ribosomal RNA genes in host plants. Nucleic Acids Res. 42, 1553.

Memczak, S., Jens, M., Elefsinioti, A., et al., 2013. Circular RNAs are a large class of animal RNAs with regulatory potency. Nature 495, 333–338.

Mühlbach, H.-P., Sänger, H., 1979. Viroid replication is inhibited by a-amanitin. Nature 278, 185–188.

C. Viroid pathogenesis and viroid-host interaction

Navarro, Â., Flores, R., 2000. Characterization of the initiation sites of both polarity strands of a viroid RNA reveals a motif conserved in sequence and structure. EMBO J. 19, 2662–2670.

Nohales, M.-Á., Flores, R., Daros, J.-A., 2012a. Viroid RNA redirects host DNA ligase 1 to act as an RNA ligase. PNAS 109, 13805–13810.

Nohales, M.-Á., Molina-Serrano, D., Flores, R., Daròs, J.-A., 2012b. Involvement of the chloroplastic isoform of tRNA ligase in the replication of viroids belonging to the family Avsunviroidae. J. Virol. 86, 8269–8276.

Owens, R.A., Blackburn, M., Ding, B., 2001. Possible involvement of the phloem lectin in long-distance viroid movement. Mol. Plant-Microbe Interact. 14, 905–909.

Pamudurti, N.R., Bartok, O., Jens, M., et al., 2017. Translation of CircRNAs. Mol. Cell 66, 9–21.

Prol, F.V., López-Gresa, M.P., Rodrigo, I., Bellés, J.M., Lisón, P., 2020. Ethylene is involved in symptom development and ribosomal stress of tomato plants upon citrus exocortis viroid infection. Plants (Basel) 9, 582.

Prol, F.V., Márquez-Molins, J., Rodrigo, I., López-Gresa, M.P., Bellés, J.M., Gómez, G., Pallás, V., Lisón, P., 2021. Symptom severity, infection progression and plant responses in solanum plants caused by three pospiviroids vary with the inoculation procedure. Int. J. Mol. Sci. 22, 6189.

Ruiz-Orera, J., Messeguer, X., Subirana, J.A., Alba, M.M., 2014. Long non-coding RNAs as a source of new peptides. elife 3, 1–24.

Sägesser, R., Martinez, E., Tsagris, M., Tabler, M., 1997. Detection and isolation of RNA-binding proteins by RNA-ligand screening of a cDNA expression library. Nucleic Acids Res. 25, 3816–3822.

Salzman, J., Gawad, C., Wang, P.L., Lacayo, N., Brown, P.O., 2012. Circular RNAs are the predominant transcript isoform from hundreds of human genes in diverse cell types. PLoS One 7, e30733.

Semancik, J., Weathers, L., 1972. Exocortis disease: evidence for a new species of '"infectious"' low molecular weight RNA in plants. Nat. New Biol. 237, 226–229.

Walia, Y., Dhir, S., Zaidi, A.A., Hallan, V., 2015. Apple scar skin viroid naked RNA is actively transmitted by the whitefly trialeurodes vaporariorum. RNA Biol. 12, 1131–1138.

Wang, P.L., Bao, Y., Yee, M.C., Barrett, S.P., Hogan, G.J., Olsen, M.N., Dinneny, J.R., Brown, P.O., Salzman, J., 2014. Circular RNA is expressed across the eukaryotic tree of life. PLoS One 9, e90859.

Wang, Y., Qu, J., Ji, S., Wallace, A.J., Wu, J., Li, Y., Gopalan, V., Ding, B., 2016. A land plant-specific transcription factor directly enhances transcription of a pathogenic noncoding RNA template by DNA-dependent RNA polymerase II. Plant Cell 28, 1094–1107.

Warrilow, D., Symons, R.H., 1999. Citrus exocortis viroid RNA is associated with the largest subunit of RNA polymerase II in tomato in vivo. Arch. Virol. 144, 2367–2375.

Wolff, P., Gilz, R., Schumacher, J., Riesner, D., 1985. Complexes of viroids with histones and other proteins. Nucleic Acids Res. 13, 355–367.

Yang, Y., Fan, X., Mao, M., et al., 2017. Extensive translation of circular RNAs driven by N 6-methyladenosine. Cell Res. 27, 626–641.

Ye, C.Y., Chen, L., Liu, C., Zhu, Q.H., Fan, L., 2015. Widespread noncoding circular RNAs in plants. New Phytol. 208, 88–95.

Yin, M., Wang, Y., Zhang, L., Li, J., Quan, W., Yang, L., Wang, Q., Chan, Z., 2017. The Arabidopsis Cys2/His2 zinc finger transcription factor ZAT18 is a positive regulator of plant tolerance to drought stress. J. Exp. Bot. 68, 2991–3005.

Zaitlin, M., Hariharasubramanian, V., 1972. A gel electrophoretic analysis of proteins from plants infected with tobacco mosaic and potato spindle tuber viruses. Virology 47, 296–305.

Zribi, I., Ghorbel, M., Brini, F., 2021. Pathogenesis related proteins (PRs): from cellular mechanisms to plant defense. Curr. Protein Pept. Sci. 22, 396–412.

13

Viroid infection and host epigenetic alterations

Joan Marquez-Molins[a,b], German Martinez[b], Vicente Pallás[c], and Gustavo Gomez[a]

[a]Institute for Integrative Systems Biology (I2SysBio), Consejo Superior de Investigaciones Científicas (CSIC)—Universitat de València (UV), Paterna, Spain [b]Department of Plant Biology, Uppsala BioCenter, Swedish University of Agricultural Sciences and Linnean Center for Plant Biology, Uppsala, Sweden [c]Instituto de Biología Molecular y Celular de Plantas (IBMCP), Consejo Superior de Investigaciones Científicas (CSIC)—Universitat Politècnica de València (UPV), Valencia, Spain

Graphical representation

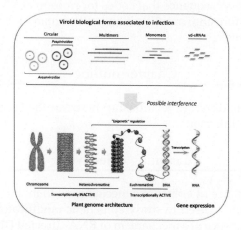

Effects of the viroid infection on the host epigenetic-landscape. Viroid derived biological forms (circular monomers, multimers, linear monomers, and viroid derived small RNAs [vd-sRNAs]) produced during viroid life cycle accumulate in infected plants. As a consequence of this canonical infection process, diverse alterations are induced in the host epigenome. The biological pathways subverted by viroids to promote these changes remain, in general, in a conundrum. (+) *plus* strand. (−) *minus* strand.

Definitions

Epigenetics: is the study of changes that do not occur in the DNA sequence but are heritable. Therefore, it refers to chemical modifications of DNA or histone proteins placed around DNA that do not change the base sequence.

Chromatin accessibility: refers to the degree to which nuclear macromolecules are able to physically contact chromatinized DNA and is determined by the occupancy and topological organization of nucleosomes as well as other chromatin-binding factors that occlude access to DNA. Compacted chromatin is inaccessible to the transcriptional machinery in contrast to open chromatin in which active transcription occurs.

DNA Cytosine methylation: is a form of post-replicative DNA modification. In plants, it can occur in all three different contexts (CG, CHG, CHH; where H = A, C, or T) in the fifth carbon of cytosine residues (5mC).

Histone modifications: are covalent post-translational m (PTM) to histone proteins that include methylation, phosphorylation, acetylation, ubiquitylation, and sumoylation. The PTMs made to histones can impact gene expression by altering chromatin structure or recruiting histone modifiers. Histone proteins act to package DNA, which wraps around the eight histones, into chromosomes. Histone modifications act in diverse biological processes such as transcriptional activation/inactivation, chromosome packaging, and DNA damage/repair.

RNA directed DNA methylation: is a biological process in which non-coding RNA molecules direct the addition of DNA methylation to specific DNA sequences. The RdDM pathway is unique to plants, although other mechanisms of RNA-directed chromatin modification have also been described in fungi and animals.

Functional subversion: consists of the use of host factors (usually proteins) by a cellular parasite that is able to alter the canonical activity of this host factor.

Ribosomal gene: is a DNA sequence that codes for ribosomal RNA. These sequences regulate transcription initiation and amplification and contain both transcribed and non-transcribed spacer segments.

Chapter outline

In this chapter, we will study the relationship between the epigenetic alterations in the plant genomes and viroid infection

Learning objectives

- To highlight the importance of epigenetic mechanisms to maintain plant homeostasis
- To be aware of the methods used to study epigenetic modifications
- To highlight that the existence of a mechanism of RNA-directed DNA methylation which is crucial for plant genome stability, was discovered while studying viroids
- Understand the influence of viroid infection as a perturbation of plant homeostasis and the dynamism of the phenomenon
- To comprehend how viroids may subvert host mechanisms for their benefit

Fundamental introduction

Epigenetic mechanisms are behind the control of gene expression in eukaryotes. Plant responses to biotic stress involve a series of coordinated signaling networks to identify and respond to the pathogen invasion by minimizing the damage at the expense of development (Pumplin and Voinnet, 2013). The switch from the regular developmental program to the stress response program requires changes in gene expression that produce proteins to cope with stress and inhibit growth-related pathways. This is termed transcriptional reprogramming and is the consequence of dynamic changes at a deeper level in the genome (Tsuda and Somssich, 2015). Plants have widespread epigenetic marks that are responsible for controlling gene expression (Slotkin, 2016). These modifications can be classified as microstructural (at the DNA level) and macrostructural (at the chromatin level) (Köhler and Springer, 2017; Annacondia et al., 2018) (Fig. 1). DNA modifications in plant genomes mostly consist of methylation of the fifth carbon of cytosine residues (5mC) and can occur in all three different contexts (CG, CHG, CHH; being H=A, C or T), (Zhang et al., 2018). Overall, the DNA methylation state is the outcome of dynamic regulation by de novo methylation, maintenance mechanisms, and active demethylation, processes that are catalyzed by various enzymes and targeted by distinct regulatory pathways (Zhang et al., 2018). Specifically, cytosine methylation is established de novo by siRNAs through the pathway denominated RNA-directed DNA methylation (RdDM), and generally results (but not always) in compacted chromatin and transcriptional gene silencing (TGS) (Matzke et al., 2015). RdDM is guided by small RNAs (sRNAs) predominantly of 24 nucleotides (nt) in size, derived from double-stranded precursors synthesized by RNA Polymerase IV (Pol IV), in close partnership with the RNA-dependent RNA polymerase 2 (RDR2) (Wendte and Pikaard, 2017). However,

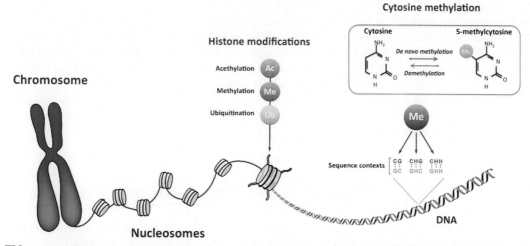

FIG. 1 Graphical representation of the epigenetic modifications that can be present in the plant genome. The three histone-modification described in the text (acetylation, methylation, and ubiquitination) are in magenta, red and green, respectively. A depiction of the cytosine methylation and the DNA sequence context (CG, CHG, CHH being H=A, C, or T) in which it can occur are detailed in the right part.

the chromatin structure is additionally influenced by histone marks such as acetylation, methylation, and ubiquitination. Interestingly, these histone modifications can have different effects on the transcriptional status despite their chemical similarity. For instance, in *Arabidopsis*, histone H3 lysine 4 (H3K4) mono-/di- or tri-methylation is associated with highly transcribed genes (Zhang et al., 2009) while H3K27 tri-methylation is found in silenced genes (Zhang et al., 2007), and H3K9 di-methylation is predominantly present in silenced transposable elements (TE) (Zhou et al., 2010). In short, host transcriptional reprogramming is a central part of plant defense upon pathogen recognition and it is tightly regulated by epigenetic mechanisms that modify the chromatin accessibility (Tsuda and Somssich, 2015; Annacondia et al., 2018).

The study of epigenetics has been possible because of the development of methods to assess the methylation status or the enrichment in histone modifications of genomic sequences (Li, 2021). Invented in early 1990s, DNA bisulfite treatment is considered the fundamental technology for detection of DNA methylation status (Frommer et al., 1992). This method is based on the treatment of genomic DNA with sodium bisulfite resulting in deamination of unmethylated cytosines to uracil, leaving methylated cytosines intact that allows 5mCs to be distinguished from unmethylated cytosines throughout the genomic DNA (Jones, 2012). Sequencing of bisulfite treated-based DNA provides a highly quantitative and efficient approach to identify 5mC at single base-pair resolution. For locus-specific analysis, several variants of PCR amplification have been developed (Rand et al., 2002; Derks et al., 2004). However, the availability of next generation sequencing technologies has made possible the profiling at a genome-wide context.

The most commonly used method to detect and quantify chromatin modifications and interaction patterns is chromatin immunoprecipitation (ChIP) technique (Collas, 2010). ChIP assay relies on specific antibodies that can recognize particular histone modification markers or epigenetic modulators in conjunction with specific DNA fragments, which allows for assigning locus-specific functions of histone modifications or transcriptional factor complexes that may directly or indirectly influence chromatin structure and subsequent transcriptional machinery efficiency (Gade and Kalvakolanu, 2012). If the target histone modification and DNA regulatory region are specific, ChIP followed by conventional PCR or quantitative real-time PCR (qPCR) will reveal the enrichments of specific histone modifications or binding ability of a remodeling complex to the specific DNA region. However, if the specific modification patterns are unknown or just to obtain a profile at the whole genome scale, sequencing-based ChIP (termed ChIP-seq) is the approach of choice to determine histone modification enrichment (Furey, 2012).

Discovery of RNA directed DNA methylation

The first association between viroid infection and DNA methylation was reported at the end of the past century by researchers studying potato spindle tuber viroid (PSTVd) infection in tobacco plants transformed with a partial non-infectious cDNA of the PSTVd genome (Wassenegger et al., 1994). Although this analysis did not describe a direct relationship between viroid infection and epigenetic changes associated with the pathogen, it led to the

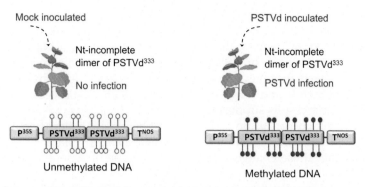

FIG. 2 Details of the experimental approaches employed by Wassenegger et al. (1994) to describe the RdDM phenomenon. Cytosines in the transgene sequence corresponding to partial (333 nt in length) PSTVd dimeric constructions remain unmethylated in mock-inoculated *N. benthamiana* transgenic plants *(left)*. However, the similar cytosines in this transgene sequence are specifically methylated as a consequence of PSTVd infection *(right)*. Methylated cytosines are represented in *red bold*.

discovery of the RNA-directed DNA-methylation (RdDM) phenomenon. In their analyses, Wassenegger et al. (1994) showed that PSTVd transgenes became methylated at the DNA level only when the transgenic plants were infected with the full viroid sequence (Fig. 2). This was the first evidence that RdDM existed in any organism and it was crucial to the research on this topic (Wassenegger and Dalakouras, 2021). The current RdDM model propose that RNA polymerase IV (Pol IV)-derived 24-nt small interference RNAs (siRNAs; canonical RdDM) or, in exceptional cases, Pol II-derived 21/22-nt siRNAs (non-canonical RdDM), are loaded onto AGO4/6 or 9 and guide domains-rearranged methyltransferase 2 (DRM2) to methylate cognate DNA, most likely through a process wherein siRNAs directly interact with DNA or interact with the nascent transcripts produced by Pol V (Erdmann and Picard, 2020; Gallego-Bartolomé, 2020).

Viroid infection and host epigenetic changes

As mentioned earlier, the initial experimental evidence supporting the link between viroid infection and DNA methylation was provided by the observation that PSTVd replication was associated with specific de novo methylation of a PSTVd-cDNA integrated as a transgene in tobacco plants (Wassenegger et al., 1994). However, it was not until the year 2014 when the direct interference of viroid infection with the methylation of host sequences was demonstrated. Employing hop stunt viroid (HSVd) as a pathogenic model it was established that viroid infection induces the hypo-methylation of the promoter region of the ribosomal genes in cucumber plants (Martinez et al., 2014). This alteration of the methylation pattern, predominantly affected cytosines in the CG context and was associated with an increased transcription and accumulation of ribosomal RNA (rRNA) precursors. Similar results were obtained by analyzing transgenic *Nicotiana benthamiana* (*N. benthamiana*) plants constitutively expressing dimeric transcripts of HSVd (Castellano et al., 2015). Furthermore, the sequencing of the full ribosomal gene from bisulfite-treated DNA of cucumber plants confirmed the hypomethylation at ten-days post inoculation (dpi) that turned into hypermethylation as the infection progressed, evidencing the dynamism of the process (Márquez-Molins et al., 2022). Additionally, the analysis of HSVd effects in the host male gametophyte of infected cucumber plants, showed that this viroid induced a dynamic demethylation of repetitive DNA regions (rRNA genes and TEs) in cucumber gametophytic cells (Castellano et al., 2016a). Altogether, these results support the notion that the plant epigenetic changes associated with HSVd infection constitute a regulatory phenomenon that is not restricted to a unique host-viroid interaction (cucumber or *N. benthamiana*) or plant tissue (vegetative or reproductive) and that those changes affect different type of repetitive sequences (rRNA and TEs).

Surprisingly, aiming to elucidate the molecular basis of the alterations in the host DNA methylation associated with HSVd infection, it was demonstrated, that in infected cucumber plants, mature forms of HSVd are able to interact in vivo with Histone Deacetylase 6 (HDA6) (Castellano et al., 2016b). Furthermore, HDA6 expression was increased in HSVd-infected plants and the transient overexpression of HDA6 reverted the hypomethylation status of rDNA in infected plants.

In arabidopsis, HDA6 is recognized as a component of the RdDM pathway, involved in the maintenance and de novo DNA methylation of TEs, rRNA genes, and transgenes (Aufsatz et al., 2002; Probst et al., 2004; Hristova et al., 2015) via interactions with Methyltransferase 1 (MET1). Based on these results, it was proposed that the host epigenetic alterations associated to the infection were related to the recruitment and functional inactivation of cucumber HDA6 by HSVd (Castellano et al., 2016b). Interestingly, the lack of HDA6 activity has been associated with spurious Pol II transcription of nonconventional rDNA templates (usually transcribed by RNA Pol I) (Earley et al., 2010). Similarly, in HSVd-infected plants, it was reported an increased accumulation of rRNAs precursors (pre-rRNAs) and sRNAs derived from ribosomal RNA (rb-sRNAs) indicating an unusual transcriptional environment (Martinez et al., 2014; Castellano et al., 2015, 2016a,b). Therefore, it was proposed that the recruitment of HDA6 mediated by HSVd may promote a spurious Pol II activity of rRNA repeats and in parallel favor the transcription of non-canonical templates such as viroid genomic RNA (Castellano et al., 2016b).

Further studies of the relationship between epigenetic regulation and viroid infection have shown different scenarios. For example, in the potato-PSTVd pathosystem, infection with PSTVd was associated with the hypermethylation of the promoter regions of certain potato genes and their consequent transcriptional repression in *N. benthamiana* (Lv et al., 2016). In this work, using transgenic *N. benthamiana* plants overexpressing GFP transcripts as experimental host, it was demonstrated that PSTVd replication enhanced GFP silencing by the hypermethylation of the promoter region of this transgene (Lv et al., 2016). The authors suggested that the interference of PSTVd in the host methylation pathway might be attributable to Virp1, a bromodomain-containing viroid-binding protein (Martínez de Alba et al., 2003) required for PSTVd replication (Kalantidis et al., 2007). Similarly, in the tomato-PSTVd pathosystem, strong activation of host DNA methylation was observed (Torchetti et al., 2016). In particular, viroid replication up-regulated the expression of key genes involved in the maintenance of DNA methylation (such as Methyltransferase 1 (MET1), Chromomethylase 3 (CMT3) and Domains Rearranged Methylase 2 (DRM2) and histone methylation (like the

H3K9 histone methyltransferase KRYPTONITE/SUVH4, KYP). Altogether these results support the notion that (at least in the analyzed host) PSTVd infection may be associated to a general increase of the DNA methylation in the host genome at both, histones and cytosines, levels. However, it remains unknown if the overexpression of methylation-related genes occurs indirectly as a host response or if the viroid might interfere by interacting with any host factor to promote these alterations. Future analysis of the global DNA-methylation and histone profile landscapes and their dynamism during PSTVd, HSVd or other viroid infection will help to shed light into the overall dynamism of viroid-induced epigenetic changes and their potential connection to transcriptional changes.

Potential ways of promoting viroid-associated host epigenetic changes

A growing body of evidence link pathogen invasion with alteration of epigenetic modifications in the genome of animal and plant cells (Gómez-Díaz et al., 2012; Annacondia et al., 2018; Wang et al., 2019; Tsai and Cullen, 2020). In the case of plants, the interaction of other molecular pathogens such as viruses with the epigenetic pathways of their hosts has provided the field with good examples of how viroids might be mediating their induced epigenetic changes. For example, plant DNA viruses of the family *Geminiviridae* are known to interfere with the host DNA methylation machinery (Rodríguez-Negrete et al., 2013; Yang et al., 2013) to avoid the methylation of their genome, and the consequent transcriptional repression (Raja et al., 2008). This alteration is mediated by the activity of viral proteins (termed viral RNA silencing suppressors, VSRs) that suppress the host epigenetic silencing activity against the viral genomes. As a side-effect of this avoidance of epigenetic silencing in their own viral genomes, Geminivirus-VSRs induce the transcriptional reactivation of TEs in the host genome (Rodríguez-Negrete et al., 2013; Yang et al., 2013). Similarly, RNA viruses generally code for VSRs (Csorba et al., 2015) and, some of these, for example, the 2b protein of cucumber mosaic virus, have been shown to interact with siRNAs from the RdDM pathway altering the methylation levels of its target genes (Hamera et al., 2012). Nevertheless, other viruses such as pelargonium line pattern virus (PLPV), which do not have any known interaction with the RdDM pathway through its silencing suppressor, was also able to induce methylation changes in its host genome (Pérez-Cañamás et al., 2020). Devoid of the ability to encode any protein homologous to VSRs, it is reasonable to assume that viroid might redesign the host epigenome during infection through alternative mechanisms. According to the experimental evidence obtained from diverse viroid-host interactions, we envision two alternative mechanisms to explain the interference of these pathogenic RNAs with the host epigenetic pathways, (i) Direct Interaction: through the involvement of viroid RNAs (mature forms and/or viroid derived sRNAs [vd-sRNAs]) and (ii) Indirect Interaction: by means of the unknown functional regulation of certain host factors (Fig. 3).

Involvement of viroid RNAs (direct interaction)

The early detection of vd-sRNAs recovered from plants infected by members of both *Pospiviroidae* (Itaya et al., 2001; Papaefthimiou et al., 2001) and *Avsunviroidae* (Martinez de Alba et al., 2002) families, supported the notion that the vd-sRNAs could guide the silencing

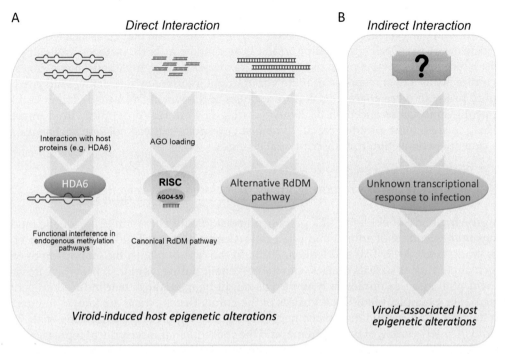

FIG. 3 Potential mechanisms of interference with epigenetic routes in viroid infected plants. (A) The HSVd-mediated recruitment of host-HDA6 is related to epigenetic changes observed in infected cucumber plants (Martinez et al., 2014). The specific de novo methylation (mediated by vd-sRNAs) of PSTVd-cDNA sequences inserted in host-genome was originally described in PSTVd-infected plants (Wassenegger et al., 1994). Alternatively, it has been recently proposed that "de novo" methylation step would be triggered by long dsRNAs rather than canonical siRNAs (Wassenegger and Dalakouras, 2021). (B) The possibility that the epigenetic changes observed during viroid infection might be part of the transcriptional defense program of response to stress cannot be discarded.

of plant-endogenous transcripts (Papaefthimiou et al., 2001; Wang et al., 2004; Gómez et al., 2009). Although this hypothesis was validated in diverse viroid-host interactions (Wang et al., 2011; Navarro et al., 2012, 2021; Adkar-Purushothama et al., 2015, 2017, 2018; Avina-Padilla et al., 2015; Adkar-Purushothama and Perreault, 2018; Bao et al., 2019), it cannot be excluded the possibility that DCL4/2-synthesized 21–22 nt vd-sRNAs could be incorporated into non-canonical RdDM pathways or that spurious DCL3 activity could generate 24-nt length vd-sRNAs that could be incorporated into the endogenous RdDM pathway, and that both events could guide the specific methylation of complementary regions in the host genome (Fig. 3A, right). This idea is supported by the first observation of the specific de novo methylation mediated by vd-sRNAs (generated during PSTVd replication) of full or partial length sequences of viroid cDNA inserted in the genome of tobacco plants (Wassenegger et al., 1994). Additionally, a plausible explanation for epigenetic changes taking place during viroid infection could be the colonization of the host AGO proteins, which are known to be influenced by anomalous presence of siRNAs (McCue et al., 2013; Minoia et al., 2014; Annacondia and

Martinez, 2021). Hence, alteration of host RNA silencing pathways, including the RdDM pathway, might lead to altered DNA methylation levels and their associated histone marks.

Alternatively, diverse observations (Dalakouras and Wassenegger, 2013) have prompted the proposal of an extended model in which the de novo methylation step would be triggered by long (>50 bp) double stranded RNAs (dsRNAs) rather than canonical siRNAs (Wassenegger and Dalakouras, 2021). Although the involvement of longer viroid-derived sequences, might be containing specific structural motifs (Wüsthoff and Steger, 2022), significantly reduces the probability that homologous host DNA regions can exist, this last possibility cannot be completely ruled out (Fig. 3A, middle). It must be noted that both DCL4/2- and DCL3-produced 21/22- and 24-nt TE-derived sRNAs are known to initiate silencing of "retrovirus-like" transcriptionally active TEs (Marí-Ordóñez et al., 2013; Nuthikattu et al., 2013).

Several studies performed in diverse viroid-hosts pathosystems have evidenced that viroid of both families are capable to subvert host factors to adapt the cell environment to their functional needs predominantly related to the intracellular compartmentalization (Martínez de Alba et al., 2003), replication (Mühlbach and Sänger, 1979; Navarro et al., 2000; Gas et al., 2007; Nohales et al., 2012a,b; Seo et al., 2021) and export to vascular tissue for long-distance trafficking (Gómez and Pallás, 2001, 2004; Owens et al., 2001; Morozov et al., 2014). Considering this experimental evidence, it can be envisioned a scenario in which viroids might modulate the host epigenetic landscape by interfering with the functional activity of certain host factors involved in the DNA methylation pathway (Fig. 3A, left). The mechanisms exploited by a viroid to functionally alter a host factor may be the direct physical interaction, as it was described for the interaction HSVd-HDA6 in cucumber plants (Castellano et al., 2016b), or by inducing the differential accumulation of host proteins directly (Torchetti et al., 2016) or indirectly (Lv et al., 2016) involved in the epigenetic pathways that maintain the stability of the plant genome.

Unknown regulation of host factors (indirect interaction)

Another plausible scenario would be that the epigenetic changes observed during viroid infection were just part of the transcriptional defense program orchestrated during stress (Fig. 3B). Defense genes such as nucleotide binding site and leucine-rich repeat domains (NBS-LRR) proteins are localized in heterochromatic regions of plant genomes (Lee and Yeom, 2015) and are activated at the transcriptional level in epigenetic mutants (Stokes et al., 2002; López Sánchez et al., 2016). Additionally, DNA methylation changes are connected to the accessibility of transcriptional regulators such as transcription factors to genomic regions (O'Malley et al., 2016), which are well-known players in the transcriptional response to stresses (Fujita et al., 2006; Chang et al., 2020). It is interesting to note that actually several genes from diverse epigenetic pathways that respond to bacterial infection contain transcription factor binding domains in their promoter regions (Agnès et al., 2013). Future research will help to solve these unsolved questions that are basic to understand the interplay between epigenetic regulation and viroid biology.

Remarks and future perspectives

The results obtained in the recent years show a close interaction between the pathogenesis process induced by HSVd and PSTVd (two representative members of the *Pospiviroidae* family) and the DNA methylation level of the host genome. However, to date, only viroid-induced changes affecting particular plant genes have been analyzed and the global impact on DNA methylation at a genomic scale is yet unknown. Consequently, deciphering the effects exerted by these plant-pathogenic RNAs on the plant epigenome (at both DNA methylation and histone modification levels) as well as identifying subverted host-endogenous mechanisms that might regulate these epigenetic changes emerge as relevant challenges for viroid research in the near future.

Another aspect to be elucidated is the specificity of the viroid-host interactions that produce epigenetic alterations, or in other words, whether some DNA methylation changes might be a common phenomenon triggered by nuclear-replicating viroids, or specific to certain viroid/host combinations. For example, the current results in the field (from a very limited number of studies) indicate that while HSVd induced the demethylation of cucumber (Martinez et al., 2014; Castellano et al., 2016a) and *N. benthamiana* (Castellano et al., 2015) genes, an opposite situation (strong hypermethylation of the analyzed genes) was observed in PSTVd infected tomato (Torchetti et al., 2016) and potato (Lv et al., 2016) plants. However, both effects may not be necessarily contradictory, since due to the absence of information about the temporal evolution of the host-epigenome during the infection and the intrinsically different target regions analyzed, it cannot be discarded that the different methylation effects associated to these two nuclear viroids might be consequence of a specific temporal stage of a dynamic host response. Alternatively, it should be considered the possibility that this functional inconsistence may be due, as mentioned earlier, to the limited number of host genes analyzed in both pathosystems. Additionally, it would be interesting to compare the potential methylation effects between severe and mild variants of a specific viroid to try to correlate these epigenomic alterations with their corresponding pathogenic processes. It is well known that induction of the viroid infection responses occurs earlier and is stronger in plants infected with a severe variant (HSVd-g54; accession No. AB219944) than in those infected with a mild variant (HSVd-h; accession No. AB039271) (Xia et al., 2017).

Another interesting issue to be considered is whether the infectious process triggered by viroids of the *Avsunviroidae* family could be associated to host epigenetic changes. Although these pathogenic RNAs replicate in the chloroplast, where no RdDM activity has been reported (Wang et al., 2020), it is thought that they possess a nuclear processing step before being targeted to the chloroplasts (Gómez and Pallás, 2012a,b; Baek et al., 2017) and *Avsunviroidae* infection is associated to the recovery of vd-sRNAs (Martinez de Alba et al., 2002).

Finally, it is important to remark that the first evidence linking viroid infection and epigenetic alterations in the endogenous host-DNA was only 8 years ago (Martinez et al., 2014) and the road traveled so far is short. Future technical advances such as increased depth of sequencing technologies and availability of genome sequences from non-model species will help to understand these alterations to a broader extent. Consequently, much more work is needed to clarify the molecular basis and the functional impact (for both the pathogen and the plant) of the remodeling of the host epigenome in response to viroid infection.

Simplified strategy to analyze epigenetic changes induced by viroid infection*

Extract total DNA from viroid-infected and non-infected plants

Subject the denatured DNA to bisulfite conversion (unmethylated Cytosines are transformed to Uracil

Stop reaction (Sulfonation) and remove residual sodium bisulfite

Desalt and purify treated DNA

Sequence the bisulfite treated and untreated DNA (after sequencing Uracil is recognized as Thymine)

Compare the conversion rate C/T of sequences derived from infected and non infected samples respect to untreated control

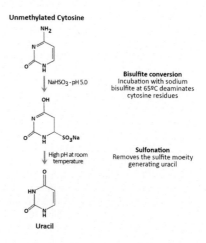

* In general, this process could be performed entirely using commercially available kits.

Study questions

1. In what DNA sequence context occurs the cytosine methylation process?
2. What host protein has been proposed to be involved in the alteration of the plant-epigenetic landscape during HSVd infection?
3. What experimental approach should be employed to analyze the changes in cytosine methylation?
4. What Viroid was used as experimental model in the discovering of the RNA-directed DNA-methylation (RdDM) phenomenon?

Further reading

The Epigenetics Revolution (2013). Nessa Carey. Columbia University Press, NY, USA. ISBN: 9780231161176

Reinberg Epigenetics (2015). David Allis, Marie-laure Caparros, Thomas Jenuwein and Danny Reinberg. Cold Spring Harbor Laboratory Press, USA. ISBN: 9781936113590.

Plant Epigenetics (2017). Nikolaus Rajewsky, Stefan Jurga, Jan Barciszewski Eds. Springer. ISBN: 978-3-319-55519-5

References

Adkar-Purushothama, C.R., Perreault, J.-P., 2018. Alterations of the viroid regions that interact with the host defense genes attenuate viroid infection in host plant. RNA Biol. 15, 955–966. https://doi.org/10.1080/15476286.2018.1462653.

Adkar-Purushothama, C.R., Brosseau, C., Giguère, T., Sano, T., Moffett, P., Perreault, J.-P., 2015. Small RNA derived from the virulence modulating region of the potato spindle tuber viroid silences callose synthase genes of tomato plants. Plant Cell 27, 2178–2194. https://doi.org/10.1105/tpc.15.00523.

Adkar-Purushothama, C.R., Iyer, P.S., Perreault, J.-P., 2017. Potato spindle tuber viroid infection triggers degradation of chloride channel protein CLC-b-like and ribosomal protein S3a-like mRNAs in tomato plants. Sci. Rep. 7, 8341. https://doi.org/10.1038/s41598-017-08823-z.

Adkar-Purushothama, C.R., Sano, T., Perreault, J.P., 2018. Viroid-derived small RNA induces early flowering in tomato plants by RNA silencing. Mol. Plant Pathol. 19, 2446–2458. https://doi.org/10.1111/mpp.12721.

Agnès, Y., Gersende, L., Florence, J., Jingyu, W., Laure, B., Yu, W., et al., 2013. Dynamics and biological relevance of DNA demethylation in Arabidopsis antibacterial defense. Proc. Natl. Acad. Sci. 110, 2389–2394. https://doi.org/10.1073/pnas.1211757110.

Annacondia, M.L., Martinez, G., 2021. Reprogramming of RNA silencing triggered by cucumber mosaic virus infection in Arabidopsis. Genome Biol. 22, 1–22. https://doi.org/10.1186/s13059-021-02564-z.

Annacondia, M.L., Magerøy, M.H., Martinez, G., 2018. Stress response regulation by epigenetic mechanisms: changing of the guards. Physiol. Plant. 162, 239–250. https://doi.org/10.1111/ppl.12662.

Aufsatz, W., Mette, M.F., van der Winden, J., Matzke, M., Matzke, A.J., 2002. HDA6, a putative histone deacetylase needed to enhance DNA methylation induced by double-stranded RNA. EMBO J. 21, 6832–6841. https://doi.org/10.1093/emboj/cdf663.

Avina-Padilla, K., Martinez de la Vega, O., Rivera-Bustamante, R., Martinez-Soriano, J.P., Owens, R.A., Hammond, R.W., et al., 2015. In silico prediction and validation of potential gene targets for pospiviroid-derived small RNAs during tomato infection. Gene 564, 197–205. https://doi.org/10.1016/j.gene.2015.03.076.

Baek, E., Park, M., Yoon, J.-Y., Palukaitis, P., 2017. Chrysanthemum chlorotic mottle viroid-mediated trafficking of foreign mRNA into chloroplasts. Res. Plant Dis. 23, 288–293. https://doi.org/10.5423/RPD.2017.23.3.288.

Bao, S., Owens, R.A., Sun, Q., Song, H., Liu, Y., Eamens, A.L., et al., 2019. Silencing of transcription factor encoding gene StTCP23 by small RNAs derived from the virulence modulating region of potato spindle tuber viroid is associated with symptom development in potato. PLoS Pathog. 15, e1008110. https://doi.org/10.1371/journal.ppat.1008110.

Castellano, M., Martinez, G., Pallás, V., Gómez, G., 2015. Alterations in host DNA methylation in response to constitutive expression of hop stunt viroid RNA in Nicotiana benthamiana plants. Plant Pathol. 64, 1247–1257. https://doi.org/10.1111/ppa.12358.

Castellano, M., Martinez, G., Marques, M.C., Moreno-Romero, J., Köhler, C., Pallas, V., et al., 2016a. Changes in the DNA methylation pattern of the host male gametophyte of viroid-infected cucumber plants. J. Exp. Bot. 67, 5857–5868. https://doi.org/10.1093/jxb/erw353.

Castellano, M., Pallas, V., Gomez, G., 2016b. A pathogenic long noncoding RNA redesigns the epigenetic landscape of the infected cells by subverting host histone deacetylase 6 activity. New Phytol. 211, 1311–1322. https://doi.org/10.1111/nph.14001.

Chang, Y.-N., Zhu, C., Jiang, J., Zhang, H., Zhu, J.-K., Duan, C.-G., 2020. Epigenetic regulation in plant abiotic stress responses. J. Integr. Plant Biol. 62, 563–580. https://doi.org/10.1111/jipb.12901.

Collas, P., 2010. The current state of chromatin immunoprecipitation. Mol. Biotechnol. 45, 87–100. https://doi.org/10.1007/S12033-009-9239-8.

Csorba, T., Kontra, L., Burgyán, J., 2015. Viral silencing suppressors: tools forged to fine-tune host-pathogen coexistence. Virology 479–480. https://doi.org/10.1016/j.virol.2015.02.028.

Dalakouras, A., Wassenegger, M., 2013. Revisiting RNA-directed DNA methylation. RNA Biol. 10, 453–455. https://doi.org/10.4161/rna.23542.

Derks, S., Lentjes, M.H.F.M., Hellebrekers, D.M.E.I., De Bruïne, A.P., Herman, J.G., Van Engeland, M., 2004. Methylation-specific PCR unraveled. Cell. Oncol. 26, 291–299. https://doi.org/10.1155/2004/370301.

Earley, K.W., Pontvianne, F., Wierzbicki, A.T., Blevins, T., Tucker, S., Costa-Nunes, P., Pontes, O., Pikaard, C.S., 2010. Mechanisms of HDA6-mediated rRNA gene silencing: suppression of intergenic pol II transcription and differential effects on maintenance versus siRNA-directed cytosine methylation. Genes Dev. 24, 1119–1132.

Erdmann, R.M., Picard, C.L., 2020. RNA-directed DNA methylation. PLoS Genet. 16, e1009034. https://doi.org/10.1371/JOURNAL.PGEN.1009034.

Frommer, M., McDonald, L.E., Millar, D.S., Collis, C.M., Watt, F., Grigg, G.W., et al., 1992. A genomic sequencing protocol that yields a positive display of 5-methylcytosine residues in individual DNA strands. Proc. Natl. Acad. Sci. U. S. A. 89, 1827–1831. https://doi.org/10.1073/PNAS.89.5.1827.

Fujita, M., Fujita, Y., Noutoshi, Y., Takahashi, F., Narusaka, Y., Yamaguchi-Shinozaki, K., et al., 2006. Crosstalk between abiotic and biotic stress responses: a current view from the points of convergence in the stress signaling networks. Curr. Opin. Plant Biol. 9, 436–442. https://doi.org/10.1016/j.pbi.2006.05.014.

Furey, T.S., 2012. ChIP–seq and beyond: new and improved methodologies to detect and characterize protein–DNA interactions. Nat. Rev. Genet. 13, 840–852. https://doi.org/10.1038/nrg3306.

Gade, P., Kalvakolanu, D.V., 2012. Chromatin immunoprecipitation assay as a tool for analyzing transcription factor activity. Methods Mol. Biol. 809, 85. https://doi.org/10.1007/978-1-61779-376-9_6.

Gallego-Bartolomé, J., 2020. DNA methylation in plants: mechanisms and tools for targeted manipulation. New Phytol. 227, 38–44. https://doi.org/10.1111/nph.16529.

Gas, M.-E., Hernández, C., Flores, R., Daròs, J.-A., 2007. Processing of nuclear viroids in vivo: an interplay between RNA conformations. PLoS Pathog. 3, e182. https://doi.org/10.1371/journal.ppat.0030182.

Gómez, G., Pallás, V., 2001. Identification of an in vitro ribonucleoprotein complex between a viroid RNA and a phloem protein from cucumber plants. Mol. Plant-Microbe Interact. 14, 910–913. https://doi.org/10.1094/MPMI.2001.14.7.910.

Gómez, G., Pallás, V., 2004. A long-distance translocatable phloem protein from cucumber forms a ribonucleoprotein complex in vivo with hop stunt viroid RNA. J. Virol. 78, 10104–10110. https://doi.org/10.1128/JVI.78.18.10104-10110.2004.

Gómez, G., Pallás, V., 2012a. A pathogenic non coding RNA that replicates and accumulates in chloroplasts traffics to this organelle through a nuclear-dependent step. Plant Signal. Behav. 7, 882–884.

Gómez, G., Pallás, V., 2012b. Studies on subcellular compartmentalization of plant pathogenic noncoding RNAs give new insights into the intracellular RNA-traffic mechanisms. Plant Physiol. 159, 558–564. https://doi.org/10.1104/pp.112.195214.

Gómez, G., Martínez, G., Pallás, V., 2009. Interplay between viroid-induced pathogenesis and RNA silencing pathways. Trends Plant Sci. 14, 264–269. https://doi.org/10.1016/j.tplants.2009.03.002.

Gómez-Díaz, E., Jordà, M., Peinado, M.A., Rivero, A., 2012. Epigenetics of host–pathogen interactions: the road ahead and the road behind. PLoS Pathog. 8, e1003007. https://doi.org/10.1371/journal.ppat.1003007.

Hamera, S., Song, X., Su, L., Chen, X., Fang, R., 2012. Cucumber mosaic virus suppressor 2b binds to AGO4-related small RNAs and impairs AGO4 activities. Plant J. 69, 104–115. https://doi.org/10.1111/j.1365-313X.2011.04774.x.

Hristova, E., Fal, K., Klemme, L., Windels, D., Bucher, E., 2015. HISTONE DEACETYLASE6 controls gene expression patterning and DNA methylation-independent euchromatic silencing. Plant Physiol. 168, 1298–1308. https://doi.org/10.1104/pp.15.00177.

Itaya, A., Folimonov, A., Matsuda, Y., Nelson, R.S., Ding, B., 2001. Potato spindle tuber viroid as inducer of RNA silencing in infected tomato. Mol. Plant-Microbe Interact. 14, 1332–1334. https://doi.org/10.1094/MPMI.2001.14.11.1332.

Jones, P.A., 2012. Functions of DNA methylation: islands, start sites, gene bodies and beyond. Nat. Rev. Genet. 13, 484–492. https://doi.org/10.1038/nrg3230.

Kalantidis, K., Denti, M.A., Tzortzakaki, S., Marinou, E., Tabler, M., Tsagris, M., 2007. Virp1 is a host protein with a major role in potato spindle tuber viroid infection in Nicotiana plants. J. Virol. 81, 12872–12880. https://doi.org/10.1128/JVI.00974-07.

Köhler, C., Springer, N., 2017. Plant epigenomics-deciphering the mechanisms of epigenetic inheritance and plasticity in plants. Genome Biol. 18, 132. https://doi.org/10.1186/s13059-017-1260-9.

Lee, H.-A., Yeom, S.-I., 2015. Plant NB-LRR proteins: tightly regulated sensors in a complex manner. Brief. Funct. Genomics 14, 233–242. https://doi.org/10.1093/bfgp/elv012.

Li, Y., 2021. Modern epigenetics methods in biological research. Methods 187, 104–113. https://doi.org/10.1016/J.YMETH.2020.06.022.

López Sánchez, A., Stassen, J.H.M., Furci, L., Smith, L.M., Ton, J., 2016. The role of DNA (de)methylation in immune responsiveness of Arabidopsis. Plant J. 88, 361–374. https://doi.org/10.1111/tpj.13252.

Lv, D.Q., Liu, S.W., Zhao, J.H., Zhou, B.J., Wang, S.P., Guo, H.S., et al., 2016. Replication of a pathogenic non-coding RNA increases DNA methylation in plants associated with a bromodomain-containing viroid-binding protein. Sci. Rep. 6, 35751. https://doi.org/10.1038/srep35751.

C. Viroid pathogenesis and viroid-host interaction

Marí-Ordóñez, A., Marchais, A., Etcheverry, M., Martin, A., Colot, V., Voinnet, O., 2013. Reconstructing de novo silencing of an active plant retrotransposon. Nat. Genet. 45, 1029–1039. https://doi.org/10.1038/ng.2703.

Márquez-Molins, J., Villalba-Bermell, P., Corell-Sierra, J., Pallás, V., Gómez, G., 2022. Multiomic analisys reveals that viroid infection induces a temporal reprograming of plant-defence mechanisms at multiple regulatory levels. bioRxiv. https://doi.org/10.1101/2022.01.06.475203. 2022.01.06.475203.

Martinez de Alba, A.E., Flores, R., Hernandez, C., 2002. Two chloroplastic viroids induce the accumulation of small RNAs associated with posttranscriptional gene silencing. J. Virol. 76, 13094–13096. https://doi.org/10.1128/jvi.76.24.13094-13096.2002.

Martínez de Alba, A.E., Sägesser, R., Tabler, M., Tsagris, M., 2003. A bromodomain-containing protein from tomato specifically binds potato spindle tuber viroid RNA in vitro and in vivo. J. Virol. 77, 9685–9694. https://doi.org/10.1128/jvi.77.17.9685-9694.2003.

Martinez, G., Castellano, M., Tortosa, M., Pallas, V., Gomez, G., 2014. A pathogenic non-coding RNA induces changes in dynamic DNA methylation of ribosomal RNA genes in host plants. Nucleic Acids Res. 42, 1553–1562. https://doi.org/10.1093/nar/gkt968.

Matzke, M.A., Kanno, T., Matzke, A.J.M., 2015. RNA-directed DNA methylation: the evolution of a complex epigenetic pathway in flowering plants. Annu. Rev. Plant Biol. 66, 243–267. https://doi.org/10.1146/annurev-arplant-043014-114633.

McCue, A.D., Nuthikattu, S., Slotkin, R.K., 2013. Genome-wide identification of genes regulated in trans by transposable element small interfering RNAs. RNA Biol. 10, 1379–1395. https://doi.org/10.4161/rna.25555.

Minoia, S., Carbonell, A., Di Serio, F., Gisel, A., Carrington, J.C., Navarro, B., et al., 2014. Specific Argonautes selectively bind small RNAs derived from potato spindle tuber viroid and attenuate viroid accumulation in vivo. J. Virol. 88, 11933–11945. https://doi.org/10.1128/jvi.01404-14.

Morozov, S., Makarova, S., Erokhina, T., Kopertekh, L., Schiemann, J., Owens, R., et al., 2014. Plant 4/1 protein: potential player in intracellular, cell-to-cell and long-distance signaling. Front. Plant Sci. 5, 26.

Mühlbach, H.P., Sänger, H.L., 1979. Viroid replication is inhibited by α-amanitin. Nature 278, 185–188. https://doi.org/10.1038/278185a0.

Navarro, J.A., Vera, A., Flores, R., 2000. A chloroplastic RNA polymerase resistant to tagetitoxin is involved in replication of avocado sunblotch viroid. Virology 268, 218–225. https://doi.org/10.1006/viro.1999.0161.

Navarro, B., Gisel, A., Rodio, M.E., Delgado, S., Flores, R., Di Serio, F., 2012. Small RNAs containing the pathogenic determinant of a chloroplast- replicating viroid guide the degradation of a host mRNA as predicted by RNA silencing. Plant J. 70, 991–1003. https://doi.org/10.1111/j.1365-313X.2012.04940.x.

Navarro, B., Gisel, A., Serra, P., Chiumenti, M., Di Serio, F., Flores, R., 2021. Degradome analysis of tomato and nicotiana benthamiana plants infected with potato spindle tuber viroid. Int. J. Mol. Sci. 22, 3725. https://doi.org/10.3390/ijms22073725.

Nohales, M.-A., Molina-Serrano, D., Flores, R., Daros, J.-A., 2012a. Involvement of the chloroplastic isoform of tRNA ligase in the replication of viroids belonging to the family avsunviroidae. J. Virol. 86, 8269–8276. https://doi.org/10.1128/jvi.00629-12.

Nohales, M.Á., Flores, R., Daròs, J.A., 2012b. Viroid RNA redirects host DNA ligase 1 to act as an RNA ligase. Proc. Natl. Acad. Sci. U. S. A. 109, 13805–13810. https://doi.org/10.1073/pnas.1206187109.

Nuthikattu, S., McCue, A.D., Panda, K., Fultz, D., DeFraia, C., Thomas, E.N., et al., 2013. The initiation of epigenetic silencing of active transposable elements is triggered by RDR6 and 21-22 nucleotide small interfering RNAs. Plant Physiol. 162, 116–131. https://doi.org/10.1104/pp.113.216481.

O'Malley, R.C., Huang, S.C., Song, L., Lewsey, M.G., Bartlett, A., Nery, J.R., et al., 2016. Cistrome and epicistrome features shape the regulatory DNA landscape. Cell 165, 1280–1292. https://doi.org/10.1016/j.cell.2016.08.063.

Owens, R.A., Blackburn, M., Ding, B., 2001. Possible involvement of the phloem lectin in long-distance viroid movement. Mol. Plant-Microbe Interact. 14, 905–909. https://doi.org/10.1094/MPMI.2001.14.7.905.

Papaefthimiou, I., Hamilton, A., Denti, M., Baulcombe, D., Tsagris, M., Tabler, M., 2001. Replicating potato spindle tuber viroid RNA is accompanied by short RNA fragments that are characteristic of post-transcriptional gene silencing. Nucleic Acids Res. 29, 2395–2400. https://doi.org/10.1093/nar/29.11.2395.

Pérez-Cañamás, M., Hevia, E., Hernández, C., 2020. Epigenetic changes in host ribosomal DNA promoter induced by an asymptomatic plant virus infection. Biology 9, 91. https://doi.org/10.3390/BIOLOGY9050091.

Probst, A.V., Fagard, M., Proux, F., Mourrain, P., Boutet, S., Earley, K., Lawrence, R.J., Pikaard, C.S., Murfett, J., Furner, I., Vaucheret, H., Mittelsten Scheid, O., 2004. Arabidopsis histone deacetylase HDA6 is required for maintenance of transcriptional gene silencing and determines nuclear organization of rDNA repeats. Plant Cell 16, 1021–1034. https://doi.org/10.1105/tpc.018754.

Pumplin, N., Voinnet, O., 2013. RNA silencing suppression by plant pathogens: defence, counter-defence and counter-counter-defence. Nat. Rev. Microbiol. 11, 745–760. https://doi.org/10.1038/nrmicro3120.

Raja, P., Sanville, B.C., Buchmann, R.C., Bisaro, D.M., 2008. Viral genome methylation as an epigenetic defense against geminiviruses. J. Virol. 82, 8997–9007. https://doi.org/10.1128/jvi.00719-08.

Rand, K., Qu, W., Ho, T., Clark, S.J., Molloy, P., 2002. Conversion-specific detection of DNA methylation using real-time polymerase chain reaction (ConLight-MSP) to avoid false positives. Methods 27, 114–120. https://doi.org/10.1016/S1046-2023(02)00062-2.

Rodríguez-Negrete, E., Lozano-Durán, R., Piedra-Aguilera, A., Cruzado, L., Bejarano, E.R., Castillo, A.G., 2013. Geminivirus Rep protein interferes with the plant DNA methylation machinery and suppresses transcriptional gene silencing. New Phytol. 199, 464–475. https://doi.org/10.1111/nph.12286.

Seo, H., Kim, K., Park, W.J., 2021. Effect of VIRP1 protein on nuclear import of citrus exocortis viroid (CEVd). Biomol. Ther. 11, 95. https://doi.org/10.3390/biom11010095.

Slotkin, R.K., 2016. Plant epigenetics: from genotype to phenotype and back again. Genome Biol. 17, 57. https://doi.org/10.1186/s13059-016-0920-5.

Stokes, T.L., Kunkel, B.N., Richards, E.J., 2002. Epigenetic variation in Arabidopsis disease resistance. Genes Dev. 16, 171–182. https://doi.org/10.1101/gad.952102.

Torchetti, E.M., Pegoraro, M., Navarro, B., Catoni, M., Di Serio, F., Noris, E., 2016. A nuclear-replicating viroid antagonizes infectivity and accumulation of a geminivirus by upregulating methylation-related genes and inducing hypermethylation of viral DNA. Sci. Rep. 6, 35101. https://doi.org/10.1038/srep35101.

Tsai, K., Cullen, B.R., 2020. Epigenetic and epitranscriptomic regulation of viral replication. Nat. Rev. Microbiol. 18, 559–570. https://doi.org/10.1038/s41579-020-0382-3.

Tsuda, K., Somssich, I.E., 2015. Transcriptional networks in plant immunity. New Phytol. 206, 932–947. https://doi.org/10.1111/nph.13286.

Wang, M.B., Bian, X.Y., Wu, L.M., Liu, L.X., Smith, N.A., Isenegger, D., et al., 2004. On the role of RNA silencing in the pathogenicity and evolution of viroids and viral satellites. Proc. Natl. Acad. Sci. U. S. A. 101, 3275–3280. https://doi.org/10.1073/pnas.0400104101.

Wang, Y., Shibuya, M., Taneda, A., Kurauchi, T., Senda, M., Owens, R.A., et al., 2011. Accumulation of potato spindle tuber viroid-specific small RNAs is accompanied by specific changes in gene expression in two tomato cultivars. Virology 413, 72–83. https://doi.org/10.1016/j.virol.2011.01.021.

Wang, C., Wang, C., Zou, J., Yang, Y., Li, Z., Zhu, S., 2019. Epigenetics in the plant–virus interaction. Plant Cell Rep. 38, 1031–1038. https://doi.org/10.1007/s00299-019-02414-0.

Wang, L., Leister, D., Kleine, T., 2020. Chloroplast development and genomes uncoupled signaling are independent of the RNA-directed DNA methylation pathway. Sci. Rep. 10, 15412. https://doi.org/10.1038/s41598-020-71907-w.

Wassenegger, M., Dalakouras, A., 2021. Viroids as a tool to study RNA-directed DNA methylation in plants. Cell 10, 1187. https://doi.org/10.3390/cells10051187.

Wassenegger, M., Heimes, S., Riedel, L., Sänger, H.L., 1994. RNA-directed de novo methylation of genomic sequences in plants. Cell 76, 567–576. https://doi.org/10.1016/0092-8674(94)90119-8.

Wendte, J.M., Pikaard, C.S., 2017. The RNAs of RNA-directed DNA methylation. Biochim. Biophys. Acta, Gene Regul. Mech. 1860, 140–148. https://doi.org/10.1016/j.bbagrm.2016.08.004.

Wüsthoff, K.-P., Steger, G., 2022. Conserved motifs and domains in members of Pospiviroidae. Cell 11, 230. https://doi.org/10.3390/cells11020230.

Xia, C., Li, S., Hou, W., Fan, Z., Xiao, H., Lu, M., et al., 2017. Global transcriptomic changes induced by infection of cucumber (*Cucumis sativus* L.) with mild and severe variants of hop stunt viroid. Front. Microbiol. 8, 1–16. https://doi.org/10.3389/fmicb.2017.02427.

Yang, L.-P., Fang, Y.-Y., An, C.-P., Dong, L., Zhang, Z.-H., Chen, H., et al., 2013. C2-mediated decrease in DNA methylation, accumulation of siRNAs, and increase in expression for genes involved in defense pathways in plants infected with beet severe curly top virus. Plant J. 73, 910–917. https://doi.org/10.1111/tpj.12081.

Zhang, X., Clarenz, O., Cokus, S., Bernatavichute, Y.V., Pellegrini, M., Goodrich, J., et al., 2007. Whole-genome analysis of histone H3 lysine 27 trimethylation in arabidopsis. PLoS Biol. 5, e129. https://doi.org/10.1371/journal.pbio.0050129.

Zhang, X., Bernatavichute, Y.V., Cokus, S., Pellegrini, M., Jacobsen, S.E., 2009. Genome-wide analysis of mono-, di- and trimethylation of histone H3 lysine 4 in Arabidopsis thaliana. Genome Biol. 10, R62. https://doi.org/10.1186/gb-2009-10-6-r62.

C. Viroid pathogenesis and viroid-host interaction

Zhang, H., Lang, Z., Zhu, J.K., 2018. Dynamics and function of DNA methylation in plants. Nat. Rev. Mol. Cell Biol. 19, 489–506. https://doi.org/10.1038/s41580-018-0016-z.

Zhou, J., Wang, X., He, K., Charron, J.B.F., Elling, A.A., Deng, X.W., 2010. Genome-wide profiling of histone H3 lysine 9 acetylation and dimethylation in arabidopsis reveals correlation between multiple histone marks and gene expression. Plant Mol. Biol. 72, 585–595. https://doi.org/10.1007/s11103-009-9594-7.

Transcriptomic analyses provide insights into plant-viroid interactions

Jernej Jakše[a], Ying Wang[b], and Jaroslav Matoušek[c]

[a]Department of Agronomy, Biotechnical Faculty, University of Ljubljana, Ljubljana, Slovenia
[b]Plant Pathology Department, University of Florida, Gainesville, FL, United States [c]Biology Centre of the Czech Academy of Sciences, Department of Molecular Genetics, Institute of Plant Molecular Biology, České Budějovice, Czech Republic

Graphical representation

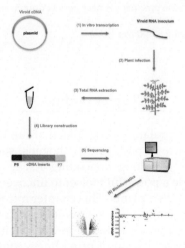

A flow chart showing a process for transcriptomic analyses on viroid-infected samples.

Abbreviations

AFCVd	apple fruit crinkle viroid
CBCVd	citrus bark cracking viroid
CDPK	calcium-dependent protein kinase
CEVd	citrus exocortis viroid
circRNA	endogenous circular RNA that forms a lariat structure
dRNA-Seq	high-throughput sequencing of mRNAs with a poly(A) tail but without a 5′ cap
EST	Express Sequence Tags derived from cDNAs
ET	ethylene
ETI	effector-triggered immunity
GA	gibberellic acid
HLVd	hop latent viroid
HSVd	hop stunt viroid
HTS	highly parallel sequencing technologies
JA	jasmonic acid
lncRNAs	long noncoding RNAs that are often longer than 200 nt
miRNA	microRNA whose precursor can fold into a hairpin structure
ncRNA	noncoding RNA
NGS	next generation sequencing
PAMP	pathogen-associated molecular patterns
PCA	principal component analysis
Pol II	DNA-dependent RNA polymerase II
PSTVd	potato spindle tuber viroid
PTI	PAMPs-triggered immunity
RIN	RNA integrity number
RNA-Seq	high-throughput sequencing of mRNAs
ROS	reactive oxygen species
RPKM	reads per kilobase of exon per million mapped reads
SA	salicylic acid
SAGE	serial analysis of gene expression
siRNA	short interfering RNA whose precursor is near perfect double-stranded RNA
sRNA	small RNA
sRNA-Seq	high-throughput sequencing of sRNAs

Chapter outline

Plants have evolved elegant regulatory programs that can rapidly elicit the expression of defense-related genes in response to pests and pathogens, such as viroids. These defense-related genes include small RNAs, long noncoding RNAs, protein-coding mRNAs, etc. Transcriptomic analyses provide a powerful toolset to uncover those genes that have altered expression profiles in viroid-infected plants. This chapter introduces the basic concept and a brief history of transcriptomic analysis as well as its application in understanding plant-viroid interactions. This chapter intends to focus on introducing the background and methodology of transcriptomic analyses instead of providing a comprehensive literature review. While the chapter focuses on the description of the most popular RNA deep sequencing analyses, readers are referred to other resources in the Further Reading section to gain a better grasp on small RNA deep sequencing, degradome RNA deep sequencing, etc.

Learning objectives

- Objective 1: understand the history of transcriptomic analyses
- Objective 2: understand the general methodology for analyzing viroid-infected samples
- Objective 3: understand general transcriptomic knowledge regarding plant-viroid interactions
- Objective 4: provoke novel ideas in understanding viroid pathogenesis using transcriptomic analyses

Fundamental introduction

Viroids

Viroids are single-stranded circular RNAs that do not encode proteins. No DNA intermediate is present during viroid replication. There are two viroid families, *Avsunviroidae* and *Pospiviroidae*. Members of *Avsunviroidae* adopt a highly branched RNA structure in one of the terminal regions and replicate in chloroplasts. They also possess ribozyme activity, meaning that they can cleave their own RNA intermediates during replication (Wang, 2021; Navarro et al., 2021a). Members of *Pospiviroidae* form rod shape structures and replicate in the nucleus. They do not possess ribozyme activities (Wang, 2021; Navarro et al., 2021a). All viroids infect plants and sometimes cause devastating consequences in production loss. The physiological changes in infected plants reflect the changes in gene expression with active defense.

Definition of transcriptome and transcriptomics

The transcriptome is defined as a complete set of RNA molecules in a single cell, tissue, or organism at a given time, developmental stage, or condition influenced by biotic or abiotic factors. It includes all currently known RNA species such as coding RNAs like messenger RNAs (mRNAs) as well as noncoding RNAs (ncRNAs) including transfer RNAs (tRNAs), ribosomal RNAs (rRNAs), small RNAs (sRNAs) including microRNAs (miRNAs) and short interfering RNAs (siRNAs), circular RNAs (circRNA), long noncoding RNAs (lncRNAs), and many exogenous RNAs including viroid RNA molecules. The term transcriptome was introduced in the mid-1990s in an analysis of the genome-wide expression of yeast genes (Velculescu et al., 1997). Transcriptomics thus combines various experimental approaches and bioinformatic tools to answer questions regarding the presence and abundance of RNA molecules.

Transcriptomics methods development in brief

The methods used for transcriptome research continued to evolve over time, becoming more sophisticated and, with the help of skilled engineers, miniaturized at an unprecedented throughput. The scope of this chapter does not allow us to go into depth, but we will describe the key discoveries and technologies that have revolutionized the field of transcriptomics.

Before the advent of cloning techniques, the analysis of certain RNA molecules relies on the abundance of RNA targets. Therefore, it is not surprising that tRNA was the first nucleic acid molecule whose sequence structure was deduced in 1965 using a lengthy biochemical approach that differs from today's enzyme-based Sanger or next generation sequencing (NGS) methods (Holley et al., 1965). A series of advancements in molecular cloning technologies significantly enhanced the ability to explore transcriptome in various samples. To name a few of the essential cloning technologies, various transcriptomic analyses rely on reverse transcription (RT) by reverse transcriptase (Baltimore, 1970; Temin and Mizutani, 1970), polymerase chain reaction (PCR) (Mullis et al., 1986), DNA ligase-based ligation (Lehman, 1974), and so on. Initial attempts to analyze transcripts based on sequencing without reference genome sequence led to the development of expressed sequence tags (EST) and serial analysis of gene expression (SAGE) that can correlate RNA transcript abundance with specific tissues (Adams Mark et al., 1991; Velculescu Victor et al., 1995). Both methods rely on massive Sanger sequencing of short complementary DNAs (cDNAs). The real breakthrough in transcriptomics, with the ability to measure gene expression of the organism of interest at the whole transcriptome level, began with the development of two techniques that are still popular nowadays, microarray (Schena et al., 1995) and RNA-Seq approach (Mortazavi et al., 2008). Microarray was developed in the mid-1990s, which is based on a hybridization approach. The RNA-Seq was developed about 10 years later, which is based on a sequence counting approach. Both methods provide quantitative, parallel monitoring of many or all genes in biological samples.

Microarray-based transcriptomics

We can think of microarrays as super-miniaturized dot blots of DNA or RNA samples. Ordinary dot-blots are usually DNA or RNA samples blotted onto a membrane with the size of multiwell plates (about $13 \times 9\,cm$). Standard detection techniques such as Southern or Northern blotting are used. Microarrays of the approximate size of microscopic glass slides or closed housings with a surface area of a few cm^2 can contain up to 10,000 printed or synthesized sequences or oligos that act as probes to attract complementary cDNAs from samples of interest. cDNAs or RNAs are fluorescently labeled, allowing laser-based scanning detection of the microarrays. The signal strengths provide means for quantifying expression.

Two main approaches have been developed for microarray transcriptomics: spotted arrays and in situ synthesized high-density short oligoprobe arrays. The first method (DeRisi et al., 1996) uses fine needles to print small amounts of longer cDNAs at defined positions on the surface of glass slides. This approach allows the use of any sequence of interest from various biological sources, even nonmodel species. Typically, the nucleic acids (cDNAs, RNAs) of the control and treated samples are labeled with different fluorescent dyes and then mixed for hybridization. High-density in situ synthesized arrays (e.g., Affymetrix GeneChip arrays)

(Pease et al., 1994) are produced by special in situ procedures manufactured by commercial vendors. Those arrays can carry up to 10,000 probes (genes) but restrain the use of multiple fluorophores. Usually, they are only available for model species.

RNA-Seq transcriptomics

Soon after the introduction of highly parallel sequencing technologies (HTS) (Mardis, 2008), a specialized technology (RNA-Seq) was established to analyze the whole collection of mRNAs from various biological samples (Martin and Wang, 2011). RNA-Seq analyses provide insights into expression differences among samples by quantifying RNA molecules at a single-nucleotide resolution (Mortazavi et al., 2008). From a technical perspective, RNA-Seq technology outcompetes microarrays for transcriptomic expression studies. However, microarray technology has certainly established its niche in diagnostics (Zhang et al., 2013) or capture-based sequencing approaches (Albert et al., 2007).

General types of transcriptomic analyses in viroid research

In the early 2000s, macroarray-based analysis, similar to microarray but with a much lower resolution, has been employed in an attempt to understand the gene expression changes in potato spindle tuber viroid (PSTVd)-infected plants (Itaya et al., 2002). The first microarray-based transcriptomic analysis on PSTVd infection investigated gene expression profiles in susceptible and tolerant tomato cultivars infected with PSTVd, using tomato genome arrays from Affymetrix (Owens et al., 2012). While there is explosive development of various NGS methods, sRNA sequencing (sRNA-Seq), mRNA sequencing (RNA-Seq), and degradome RNA sequencing (dRNA-Seq) are often used. These three approaches can complement each other and provide a global view of plant transcriptomic changes in response to viroid infection. For example, sRNA-Seq data can provide insights into the expression of sRNAs, and their function in guiding the cleavage of target mRNAs can be illustrated by dRNA-Seq data. To confirm the biological significance, the expression level of the cleaved transcripts can be evaluated by RNA-Seq data.

sRNA-Seq and RNA-Seq are relatively straightforward. sRNA-Seq enriches and sequences short transcripts that are 18–30 nucleotides (nt) in length. RNA-Seq usually enriches and sequences transcripts containing a poly(A) tail and longer than 200 nt in length. It is noteworthy that RNA-Seq covers not only protein-coding transcripts but also lncRNAs. dRNA-Seq covers mRNAs that do not possess the 5′-end cap (Addo-Quaye et al., 2009; German et al., 2008, 2009). One major mechanism for removing mRNA caps relies on the sRNA-guided mRNA cleavage. The cleaved products, collectively called degradome, contain 3′-end poly(A) tail but lack the 5′-end cap. Therefore, it is essential to design a methodology that clones degradome from poly(A)-enriched transcripts, otherwise degraded RNAs from other pathways will interfere with data annotation. The construction of degradome relies on the activity of type III restriction enzymes, which will only retain very short fragments downstream of sRNA-guided cleavage sites. Therefore, dRNA-Seq libraries are similar in length as compared with sRNA-Seq libraries.

Transcriptomic studies of viroid-infected samples

Infection methods

Viroids can naturally infect host plants and can spread mainly by mechanical injury as highly structured and relatively resistant RNA. However, artificial inoculation has a significant advantage in analyzing the molecular basis of plant-viroid interactions in a laboratory with a more controlled and homogenous setup. There are multiple ways to infect plants with diverse types of viroids (Hadidi et al., 2003; Vazquez Prol et al., 2021). In the early 1980s, artificially prepared cDNA and transcribed RNA were found infectious (Cress et al., 1983). Mechanical wounding by cDNA-derived viroid transcripts with the help of Carborundum successfully led to plant infection with PSTVd, hop stunt viroid (HSVd), and citrus exocortis viroid (CEVd) (Täbler and Sänger, 1984; Meshi et al., 1984; Visvader et al., 1985). In most cases, dimeric viroid clones were used to achieve efficient infection for members of *Pospiviroidae* (Hammond et al., 1989; Candresse et al., 1990). Later, the finding that various viroid cDNA clones are infectious (Täbler and Sänger, 1984) greatly facilitated studies of viroid structures and many related biological questions (Owens et al., 1985; Hammond et al., 1989; Ambros et al., 1999; Matoušek et al., 2004). Construction of infectious plant vectors containing viroid sequences was used for *Agrobacterium tumefaciens*-mediated host transformation (Salazar et al., 1988) and transient expression after infiltration into leaves (Gardner et al., 1986). Equimolar mixtures of various unrelated viroids can be used to study the impact of mixed populations on plant transcriptomes and viroid interactions (Matoušek et al., 2017). However, it is noteworthy that the inoculation methods need to be carefully selected to mimic the natural infection process. Different inoculation methods can lead to different degrees of symptom severity, suggesting that some inoculation methods may not faithfully reflect the natural infection process (Vazquez Prol et al., 2021). In particular, *A. tumefaciens*-mediated infection challenges plants with agrobacterium in addition to viroids and introduces an artificial step of transcribing viroid cDNAs to generate viroid RNAs. Therefore, transcriptomic analyses using samples infected via *A. tumefaciens*-mediated infection may not truthfully delineate the plant-viroid interactions.

Cytopathic effects in viroid-infected plants and the implications in alterations of gene expression

While some viroids do not trigger symptom development in plants (Verhoeven et al., 2013; Daros, 2016; Li et al., 2021), many viroids are well-known pathogens to economically important crops (Flores et al., 2020). Viroids of *Avsunviroidae* incite symptoms early, specific, and locally. For example, defects in chloroplast development are often observed. By contrast, viroids of *Pospiviroidae* induce symptoms late, nonspecific, and systemic. For instance, infected plants often have an irregular proliferation of cell membranes, cell wall distortions, and chloroplast malformations. The cytopathology triggered by viroids of *Pospiviroidae* is in line with the phenotypes of alteration of hormone signaling and the activation of innate immunity. Therefore, gene expressions related to innate immunity as well as hormone metabolism and signaling are of particular interest for in-depth analysis.

Gene expression profiles underlying plant-viroid interactions

With the popularizing of NGS technology and cost reduction, RNA-Seq analyses have been widely used to understand the gene expression profiles in viroid-infected plants. The ultimate goal is to delineate the regulatory mechanism underpinning plant defense and viroid pathogenesis. Bearing this in mind, numerous studies have illustrated some common sets of gene modules that have been influenced in infected plants. Below, we briefly introduce some of the essential pathways/gene modules that may be the key to understanding plant-viroid interactions.

Immunity

Plants maintain a balance of gene expression to cope with the needs of normal growth or responses to environmental stimuli. There are two layers of defense from plant innate immunity, namely pathogen-associated molecular patterns (PAMPs)-triggered immunity (PTI) and effector-triggered immunity (ETI) (Jones and Dangl, 2006). PTI is deployed mainly at the cell surface to sense the presence of pathogens near the cells, whereas ETI largely functions within cells to identify effectors and toxins "injected" by pathogens. PTI and ETI work cooperatively to ensure plant survival. Triggering plant innate immunity leads to the activation of calcium-dependent protein kinases (CDPKs) and mitogen-activated protein kinase (MAPK) cascades, induction of reactive oxygen species (ROS), changes in ion efflux and influx, activation of signaling involved in hypersensitive responses, and induction of cell wall fortification.

Host immune responses were observed in PSTVd- and CEVd-infected tomato plants as well as HSVd-infected hop through RNA-Seq analyses (Zheng et al., 2017; Thibaut and Claude, 2018; Kappagantu et al., 2017) (Fig. 1). In PSTVd-infected tomato plants, the

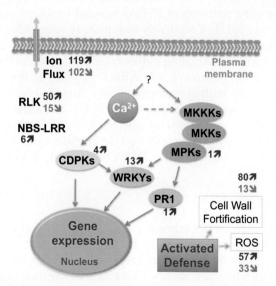

FIG. 1 Activation of plant immune responses upon PSTVd infection. A simplified view of plant immune response. Numbers of genes upregulated (*red arrows*) or downregulated (*green arrows*) by PSTVd infection are listed next to each gene family/category. This figure is a reuse of Figure 9 published in Zheng et al. (2017).

MAPK3-based signaling cascade and four CDPKs are activated. As a consequence of the activation of the MAPK3-based cascade, the prominent marker gene, PR-1, is also activated together with 13 WRKY transcription factors involved in plant defense (Zheng et al., 2017). Key enzymes involved in ROS metabolisms, such as the ascorbate glutathione cycle and L-ascorbate degradation V, are induced in PSTVd-infected tomatoes as well (Zheng et al., 2017). In total, 90 genes involved in ROS metabolism and 107 genes involved in ROS responses showed significant expression changes. In line with the fortification of the cell wall in viroid-infected samples, there were 93 genes involved in cell wall biogenesis showing significant expression changes (Zheng et al., 2017). In addition, significant expression changes were found in 221 genes involved in ion transport (Zheng et al., 2017). Therefore, viroids activating host immune responses is a new framework for explaining the viroid-triggered cytopathic symptoms.

Hormones

Phytohormones are pivotal signaling molecules that regulate plant growth and responses to environmental stimuli, including virus/viroid infections. There are nine major phytohormones: salicylic acid (SA), jasmonic acid (JA), ethylene (ET), abscisic acid (ABA), auxin, gibberellic acid (GA), cytokinins (CKs), brassinosteroids (BRs), and strigolactones (SLs). SA, JA, and ET are well-known defense-related hormones. However, except SLs, all the other eight major phytohormones participate in plant-virus interactions either in a synergistic or an antagonistic fashion (Zhao and Li, 2021). The role of phytohormones in plant-virus interactions is complex and highly variable in different host-virus combinations (Zhao and Li, 2021).

Transcriptomic analyses, either microarray- or RNA-Seq-based, also shed light on the role of phytohormones in plant-viroid interactions. Using microarray analysis, data showed the activation of the ET signaling pathway in PSTVd-infected tomatoes, regardless of the symptom severity (Owens et al., 2012). A relatively recent study using RNA-Seq analysis showed massive changes in SA (93 genes in the biogenesis pathway and 146 genes in the signaling pathway), auxin (83 genes in the signaling pathway), ET (40 genes in the biogenesis pathway and 98 genes in the signaling pathway), and ABA (135 genes in the signaling pathway) in PSTVd-infected tomato plants with symptoms, highlighting the complex hormone signaling underlying plant-viroid interactions (Zheng et al., 2017). Interestingly, GA signaling was found to be significantly repressed in PSTVd-infected tomato plants and potato tubers when severe symptoms developed, which suggests that GA signaling is likely a downstream event for symptom development (Katsarou et al., 2016; Owens et al., 2012; Więsyk et al., 2020).

Transcription factors

Since members of *Pospiviroidae* directly employ nuclear RNA polymerase II (Pol II) for replication, they have a direct impact on host transcription. This feature is distinct from other RNA viruses that code for their own polymerases to catalyze replication. To redirect Pol II to viroid genomic RNA as templates, an RNA-specific transcription factor, TFIIIA-7ZF, is required, at least for the case of PSTVd (Wang et al., 2016). TFIIIA-7ZF is a conserved splicing variant of transcription factor IIIA in land plants. The expression of TFIIIA-7ZF is not significantly changed in PSTVd-infected plants, but binding with PSTVd likely affects its endogenous function as well as influences Pol II availability to DNA templates. In line with this, ectopic expression of TFIIIA-7ZF cloned from hop appears to affect the branching of tobacco plants (Fig. 2) (Matoušek and Steger, 2022).

Mediators are a group of regulatory co-activators associated with Pol II during transcription on DNA templates. They were originally considered as general factors of transcription, but increasing evidence suggests that individual mediator subunits may specifically regulate plant development or responses to biotic and abiotic stresses (Buendia-Monreal and Gillmor, 2016; Samanta and Thakur, 2015; Malik et al., 2017). A recent study analyzed five viroid species (apple fruit crinkle viroid [AFCVd], citrus bark cracking viroid [CBCVd], hop latent viroid [HLVd], HSVd, and PSTVd) in either *Nicotiana* species or hop and found individual plant mediator subunits displayed differential and tailored expression patterns in response to different viroids (Nath et al., 2020). Therefore, mediator activity may also specifically contribute to distinct host-viroid combinations.

Besides TFIIIA-7ZF, WRKY transcription factors, and mediators, other transcription factors are also displaying altered expression in viroid-infected plants. While more transcription factors are showing significant expression changes in tomato plants infected by the severe strain of PSTVd, members in WRKY, NAC, MYB, and AP2/ERF families exhibited consistent expression changes in plants infected by both the mild strain and the severe strain of PSTVd (Więsyk et al., 2018, 2020). AP2/ERF family transcription factors mainly function in the ET signaling pathway (Shoji and Yuan, 2021). NAC transcription factors participate in plant immune responses by impacting plant hormone signaling (Yuan et al., 2019). MYB transcription factors are involved in regulating plant growth and responses to biotic and abiotic responses (Ambawat et al., 2013). Since transcription factors generally play essential roles in regulating target gene expressions, those expression changes in transcription factors likely contribute to the global transcriptomic changes and symptom development underlying plant-viroid interactions.

Changes in noncoding transcripts

Two major types of noncoding transcripts are of particular interest in transcriptomic studies, sRNAs and lncRNAs. sRNAs, including miRNAs and siRNAs, play an essential role in regulating gene expression at both transcriptional and posttranscriptional levels (Axtell, 2013; Borges and Martienssen, 2015). sRNAs using their sequence complementarity to guide Argonaut-based RNA-induced silencing complex either directly destroying target mRNAs or regulating gene transcription robustness (Baulcombe, 2004). Since sRNA-based regulation is dosage-dependent, expression changes in sRNAs may lead to changes in target gene expression (Bartel, 2009). However, it is essential to obtain empirical data supporting the changes in sRNA activity in viroid-infected plants. Numerous research efforts were made to dissect the changes in host sRNA-based regulation during plant-viroid interactions. For instance, a comprehensive transcriptomic analysis including sRNA-Seq, RNA-Seq, and degradome RNA-Seq found that there is a specifically enhanced miR167 activity in guiding cleavage of auxin response factor 8 (ARF8; *Solyc02g037530*) but not ARF8-1 (*Solyc03g031970*) in PSTVd-infected tomatoes (Zheng et al., 2017). In addition, miR393-based regulation over

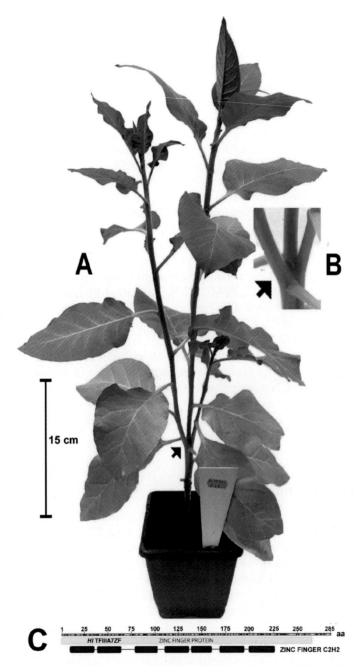

FIG. 2   The unusual branching of the main shoot in tobacco transformed with *Hl*TFIIIA-7ZF isolated from HLVd-infected hop (*Humulus lupulus* L.). (A) Transformed tobacco plant; (B) the detail of splitting of the main shoot, as one of the main symptoms in tobacco transformed with TFIIIA-7ZF. *Arrows* indicate the position of unusual branching. (C) Protein domains of *Hl*TFIIIA-7ZF as calculated using Inter Pro Scan and visualized using Geneious Prime 2020.04.

TIR1 (a critical auxin transporter) is significantly reduced in PSTVd-infected tomato plants (Zheng et al., 2017). These observations are in line with the transcriptomic changes in auxin-related signaling in PSTVd-infected tomatoes. Two studies found that PSTVd replication can lead to upregulation of miR398-regulated production of ROS (Suzuki et al., 2019; Fujibayashi et al., 2021), which provides a potential link between RNA silencing and innate immunity in defending viroids. These studies represent great examples where multiple comparative transcriptomic analyses helped understand the function of sRNAs in plant-viroid interactions.

It is noteworthy that viroid replication alone generates many viroid-derived siRNAs (vd-sRNAs) (Flores et al., 2020). A general hypothesis stands that those vd-sRNAs may alter the expression of plant transcripts, which contributes to pathogenesis (Flores et al., 2020). However, most vd-sRNA-based cleavages of target mRNAs in PSTVd-infected tomatoes are not repeatable in biological replicates, suggesting that they lack biological significance (Zheng et al., 2017; Navarro et al., 2021b). More importantly, none of the vd-sRNA targets, even verified by dRNA sequencing data, showed any significant expression changes in RNA-Seq analyses in tomato and *N. benthamiana* plants infected by PSTVd (Zheng et al., 2017; Navarro et al., 2021b). Therefore, caution is needed to claim any role of vd-sRNAs in viroid pathogenesis. As pointed out by a recent review, it is critical to directly associate the presence of vd-sRNAs in triggering initial molecular lesion or symptom (Flores et al., 2020). In addition, the loss-of-function approach is required to draw any solid conclusion regarding the function of vd-sRNAs. To date, very few vd-sRNAs are well supported as regulators of pathogenesis. A good example is two vd-sRNAs from peach latent mosaic viroid that guide cleavage of the mRNA encoding a chloroplast heat shock protein (Navarro et al., 2012).

Besides sRNAs, studies have also revealed the dynamic expression of lncRNAs in viroid-infected plants (Nath et al., 2021; Zheng et al., 2017). In general, lncRNA expression is stochastic (Zheng et al., 2017). Therefore, carefully designed experiments with multiple biological replicates are required to reveal viroid-responsive lncRNAs. In this case, the standard criteria for identifying differential expressed protein-coding transcripts are no longer suitable. Instead, additional criteria should be implemented to only select those that displayed consistently induced- or repressed-expression patterns in all replicates. By using these stringent criteria, only 44 out of 6726 tomato lncRNAs are qualified as PSTVd-responsive (Zheng et al., 2017). Their functions in plant-viroid interactions remain to be experimentally tested.

Methods for transcriptomic analyses

Collecting RNA samples

Before constructing deep sequencing libraries, the first step is to verify that plants samples are indeed infected with viroids. Generally, total RNAs were extracted either using a commercial kit or Tri Reagent-based purification methods. The purified total RNAs are subjected to urea-PAGE gel electrophoresis and RNA gel blots to detect the presence of viroids. Other detection methods, such as RT-PCR, can be used instead of RNA gel blots. However, RNA gel blots offer insights into the relative abundance of genomic RNA (circular form) and the replication intermediates, which are useful information.

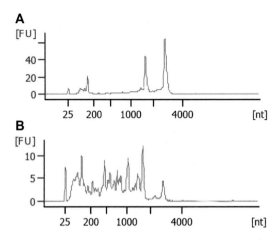

FIG. 3 Examples of integrity analysis of isolated total RNA samples using Agilent Bioanalyzer 2100. (A) An RNA sample with good integrity and well-defined peaks of 28S and 18S rRNA with RIN=8.7; (B) a sample of degraded total RNA electrophoretic profile with RIN below the threshold (4.7). Both samples are total RNA isolated from hops (*Humulus lupulus* L.).

The quality of RNA is important for library constructions and subsequent analyses. Thanks to the advancement of instruments, an Agilent bioanalyzer can perform precise electrophoretic RNA measurements. An established algorithm can use the RNA migration profile to calculate the RNA integrity number (RIN). The RIN is scored from 1 to 10, with 10 as the best quality. Samples with a RIN value >6 are considered good quality, although higher values are often preferred. The RIN number is now widely used to assess the quality of the purified RNAs (Schroeder et al., 2006) (Fig. 3).

The experimental design must include at least one control (e.g., a healthy plant) and one treated condition (e.g., a viroid-infected plant) for comparison, although a time course or multiple treatments are highly preferred. Biological replicates are now a must, with at least three being a minimum for analysis, although more replicates are not uncommon. Biological replicates are samples from different resources, typically different plants. They are necessary to measure biological variation among samples and to perform proper statistical analysis. As for technical replicates, they are replicates of the same biological sample. It is generally believed that technical replicates are not necessary for RNA-Seq experiments because the technical reproducibility of RNA-Seq is considered satisfactory (McIntyre et al., 2011).

Library constructions

RNA library constructions have been made easy these days, at a much affordable cost. In general, the goal of RNA-Seq analyses is to compare the expression changes in gene transcripts. There are two strategies for enriching mRNAs. The most common strategy is to use an oligo dT column to specifically enrich RNAs with a poly(A) tail. Other RNAs, like rRNAs, tRNAs, and RNAs from organelles, will be left out from further procedures. However, this method will also exclude some important transcripts, such as circRNAs. Alternatively, it is possible to use sequence-specific oligonucleotides, which are immobilized in agarose or other types of supporting resins, to deplete rRNAs. Since rRNAs are the most dominant species in samples, the depleted RNAs are then ready for library constructions. There are commercial kits available for both strategies.

The enriched mRNAs are then ready for library construction. Here, we briefly introduce the strand-specific library construction protocol. Strand-specific libraries keep the information on RNA orientation, which is useful when analyzing overlapped genes transcribed from opposite orientations. The enriched mRNAs are fragmented and subjected to reverse transcription using random hexamer primers with the presence of actinomycin D. Actinomycin D specifically inhibits DNA-dependent, but not RNA-template, DNA synthesis. The first-strand cDNA is then used for generating double-stranded DNA. Notably, dUTP replaces dTTP in the second-strand DNA synthesis. After end-repairing through a series of enzymes and adapter ligation, the UTP-containing strand is degraded by uracil DNA glycosylase. The remaining strand is then amplified again via PCR to generate double-stranded libraries, which will be ready for sequencing after removing PCR primers (Zhong et al., 2011).

Sequencing platforms

Most sequencing services offer Illumina sequencing with either single-end or paired-end options, differing by sequencing from one end or both ends of the libraries. Researchers have the option to choose the number of reads needed (M) or the volume of data (GB), which is related to the sequencing depth that has to be sufficient. Various Illumina platforms are the most commonly used instruments in research facilities. Ion Torrent platforms, which introduce a postlight sequencing approach, are also available. The Ion Torrent platforms offer strand-specific RNA sequencing by default, which is an advantage. However, paired-end sequencing is not available from Ion Torrent platforms.

Library quality check

The computing environment for bioinformatic analyses needs more memory and computing power, so stronger servers or even cluster computers are needed. Most analysis is performed in a Linux environment, which is the basis for using open-source software. Galaxy platforms are very popular (Goecks et al., 2010). There are also commercial or paid solutions for RNA-Seq analysis, such as BioBam's OmicsBox, Qiagen's CLC Genomics/Server, Sequentia's cloud solution AIR, or CSC IT Center for Science's Chipster (Kallio et al., 2011). Recently, an integrative toolkit (TBtools) with a user-friendly interface has been developed, which makes such analyses even easier (Chen et al., 2020).

After receiving sequencing data, it is essential to perform a quality check to select high-quality data only for downstream analyses. The most commonly used tool for quality control of NGS data is FastQC software, which provides an easy-to-read output in HTML format about the potential problems in the sequencing data, highlighting quality scores, GC content, duplication level, possible adapter contamination, and others in the analyzed data set. Similar tools are NGS QC Toolkit (Patel and Jain, 2012) and PRINSEQ (Schmieder and Edwards, 2011), both which offer the ability to trim reads in the following steps. The next step is trimming the adapter sequences and low-quality reads. Trimmomatic and CUTADAP are the most popular programs for this purpose (Bolger et al., 2014; Martin, 2011).

Data analyses

The reads are then mapped to the reference genome using Bowtie (Langmead et al., 2009) or STAR (Dobin et al., 2013), or are subjected to de novo assembly if there is no reference genome available. There are several considerations to further assess the quality of sequencing data. First, the read mapping rate is important. After removing reads mapped to organelle genomes and rRNAs, the remaining reads are aligned to the reference genome. A good library should have more than 90% of the reads mapped to the reference genome. Second, sample replicates should be clustered in principal component analysis (PCA). PCA analysis can help estimate whether sample replicates are similar to each other and distinct from other treatments.

High-quality data can then be used for various downstream analyses. The most common application is to find differentially expressed genes. Although genes with essential functions in plant-viroid interactions are not always changing their expression profiles, genes that are significantly changed in expression values are obvious targets for further functional analyses. The mapped reads are then subjected to DESeq2 (Love et al., 2014), which is used to calculate differential gene expression based on the number of reads per kilobase of exon per million mapped reads (RPKM). The differentially expressed genes can then be grouped based on their Gene Ontology annotations. This is the basic analysis of RNA-Seq data. For other analyses, readers should look into specific bioinformatic textbooks or literature.

Prospective

Despite the rapid progress in delineating the gene expression profiles (including protein-coding transcripts, sRNAs, and lncRNAs) in viroid-infected plants, the mechanism regarding viroid pathogenesis, particularly for those of the family *Pospiviroidae*, remains elusive. Many of such changes in gene expressions may only be the consequences but not the causes. It is important to note that gene expression is elegantly controlled by the openness of the chromosome regions, availability of the transcription factors, transcription rates, co-transcriptional processing of mRNAs, nuclear export, RNA stability in the cytoplasm, and translational efficiency (Bailey-Serres, 2013). RNA-Seq alone can only estimate the steady-state of mRNAs, which ignores many key regulations. Perhaps extending the investigation to chromatin level using other sequencing technologies can provide more insights.

Future implications

Most transcriptomics studies were performed on the multicellular levels (i.e., on the level of individual leaves, stems, flower parts, etc.) or the whole plants. Such experiments utilize cells of different specialization and a mixture of infected and noninfected cells. The mixed infected and noninfected cells in plants have been demonstrated by tissue print technique (Stark-Lorenzen et al., 1997). The study showed that PSTVd accumulated in the upper parts of tomato plants newly grown after inoculation and was predominantly in association with the ring formed by the vascular tissue. Although this issue may be alleviated by increasing biological replicates, a better experimental design may be implemented to tease apart the specific principles underlying the plant-viroid interactions in infected cells. One strategy is to use suspension cell cultures. For instance, a previous study (Stöcker et al., 1993) analyzed the impact of PSTVd on tomato suspension cultures that are photosynthetically active. As compared to uninfected cells, PSTVd-infected cells grew slowly, were morphologically different in size and shape, and formed tight cell aggregates. Electron microscopy showed that starch accumulation in chloroplasts, deformation of the chloroplast envelope, and irregular plasmalemmasomes at the cell membrane were associated. Another approach is to use homogenous cells, like pollen (Steinbachová et al., 2021; Shrestha et al., 2020). The pollen system can also help identify factors important for the elimination of AFCVd and CBCVd that are two pollen and seed nontransmissible viroids (Matoušek et al., 2020). To further understand the principles underpinning plant-viroid interactions at the organismal levels, the recently developed single-cell sequencing technology may be harnessed in the near future.

Chapter summary

This chapter explains the development of tools for analyzing transcriptome. Using those tools, some general knowledge has been gained regarding the gene expression profiles underlying viroid infection. The knowledge gained will guide future investigations on illustrating the molecular basis of viroid pathogenesis. Such efforts may help develop effective measures combating viroids.

Study questions

(1) Transcriptomic studies aim at understanding the expression and abundance of RNA populations in given samples. How are the two general types of techniques in analyzing transcriptome? How many different deep sequencing-based approaches are used to analyze viroid-infected samples?
(2) What are the general methods to infect plant materials with viroids? Why should one carefully select the infection method? What are the considerations in designing controls?

(3) What groups of genes have been commonly found as differentially expressed genes in samples infected with viroids (mainly the members of the *Pospiviroidae*)? What are the considerations in studying the role of vd-sRNAs in viroid infection?

(4) In RNA-seq sequencing experiments, RNA strand-specific sequencing protocols are available for library construction that preserve mRNA sequence specificity information. What are the advantages of strand-specific sequencing over nonstrand-specific sequencing in an RNA-Seq experiment?

(5) It has been a practice to submit RNA-Seq sequencing data for public availability to the Sequence Read Archive depository (SRA) (available at https://www.ncbi.nlm.nih.gov/sra), where the raw sequencing data and, if available, alignment information are stored. RNA or microarray gene expression data can also be stored in the Gene Expression Omnibus (GEO) repository (available at https://www.ncbi.nlm.nih.gov/gds/). Try to find out how much viroid-related data is available in each of the repositories. Which viroid species is the most studied?

(6) How to design transcriptomic experiments to further understand the molecular basis of plant-viroid interactions and viroid pathogenesis?

Further reading

While there are so many excellent reviews and recourses for the topics below, we refer to the following reference as introductory reading materials.

1. For general knowledge regarding viroid replication and infection (Ding, 2009; Wang, 2021; Flores et al., 2020; Navarro et al., 2021a)
2. For constructing RNA-Seq, sRNA-Seq, and dRNA-Seq (German et al., 2009; Zhong et al., 2011; Chen et al., 2012)
3. For analyzing RNA-Seq, sRNA-Seq, and dRNA-Seq (Langmead et al., 2009; Dobin et al., 2013; Love et al., 2014; Mohorianu et al., 2017; Addo-Quaye et al., 2009)

Acknowledgment

J.J. is supported by Slovenian Research Agency. Y.W. is supported by US NIGMS and NSF. J.M. is supported by Czech and German Science Foundations.

References

Adams Mark, D., Kelley Jenny, M., Gocayne Jeannine, D., Dubnick, M., Polymeropoulos Mihael, H., Xiao, H., Merril Carl, R., Wu, A., Olde, B., Moreno Ruben, F., Kerlavage Anthony, R., McCombie, W.R., Venter, J.C., 1991. Complementary DNA sequencing: expressed sequence tags and human genome project. Science 252, 1651–1656.

Addo-Quaye, C., Miller, W., Axtell, M.J., 2009. CleaveLand: a pipeline for using degradome data to find cleaved small RNA targets. Bioinformatics 25, 130–131.

Albert, T.J., Molla, M.N., Muzny, D.M., Nazareth, L., Wheeler, D., Song, X., Richmond, T.A., Middle, C.M., Rodesch, M.J., Packard, C.J., Weinstock, G.M., Gibbs, R.A., 2007. Direct selection of human genomic loci by microarray hybridization. Nat. Methods 4, 903–905.

Ambawat, S., Sharma, P., Yadav, N.R., Yadav, R.C., 2013. MYB transcription factor genes as regulators for plant responses: an overview. Physiol. Mol. Biol. Plants 19, 307–321.

Ambros, S., Hernandez, C., Flores, R., 1999. Rapid generation of genetic heterogeneity in progenies from individual cDNA clones of peach latent mosaic viroid in its natural host. J. Gen. Virol. 80 (Pt 8), 2239–2252.

Axtell, M.J., 2013. Classification and comparison of small RNAs from plants. Annu. Rev. Plant Biol. 64, 137–159.

Bailey-Serres, J., 2013. Microgenomics: genome-scale, cell-specific monitoring of multiple gene regulation tiers. Annu. Rev. Plant Biol. 64, 293–325.

Baltimore, D., 1970. Viral RNA-dependent DNA polymerase: RNA-dependent DNA polymerase in virions of RNA tumour viruses. Nature 226, 1209–1211.

Bartel, D.P., 2009. MicroRNAs: target recognition and regulatory functions. Cell 136, 215–233.

Baulcombe, D., 2004. RNA silencing in plants. Nature 431, 356–363.

Bolger, A.M., Lohse, M., Usadel, B., 2014. Trimmomatic: a flexible trimmer for Illumina sequence data. Bioinformatics 30, 2114–2120.

Borges, F., Martienssen, R.A., 2015. The expanding world of small RNAs in plants. Nat. Rev. Mol. Cell Biol. 16, 727–741.

Buendia-Monreal, M., Gillmor, C.S., 2016. Mediator: a key regulator of plant development. Dev. Biol. 419, 7–18.

Candresse, T., Diener, T.O., Owens, R.A., 1990. The role of the viroid central conserved region in cDNA infectivity. Virology 175, 232–237.

Chen, Y.R., Zheng, Y., Liu, B., Zhong, S., Giovannoni, J., Fei, Z., 2012. A cost-effective method for Illumina small RNA-Seq library preparation using T4 RNA ligase 1 adenylated adapters. Plant Methods 8, 41.

Chen, C., Chen, H., Zhang, Y., Thomas, H.R., Frank, M.H., He, Y., Xia, R., 2020. TBtools: an integrative toolkit developed for interactive analyses of big biological data. Mol. Plant 13, 1194–1202.

Cress, D.E., Kiefer, M.C., Owens, R.A., 1983. Construction of infectious potato spindle tuber viroid cDNA clones. Nucleic Acids Res. 11, 6821–6835.

Daros, J.A., 2016. Eggplant latent viroid: a friendly experimental system in the family Avsunviroidae. Mol. Plant Pathol. 17, 1170–1177.

DeRisi, J., Penland, L., Brown, P.O., Bittner, M.L., Meltzer, P.S., Ray, M., Chen, Y., Su, Y.A., Trent, J.M., 1996. Use of a cDNA microarray to analyse gene expression patterns in human cancer. Nat. Genet. 14, 457–460.

Ding, B., 2009. The biology of viroid-host interactions. Annu. Rev. Phytopathol. 47, 105–131.

Dobin, A., Davis, C.A., Schlesinger, F., Drenkow, J., Zaleski, C., Jha, S., Batut, P., Chaisson, M., Gingeras, T.R., 2013. STAR: ultrafast universal RNA-seq aligner. Bioinformatics 29, 15–21.

Flores, R., Navarro, B., Delgado, S., Serra, P., Di Serio, F., 2020. Viroid pathogenesis: a critical appraisal of the role of RNA silencing in triggering the initial molecular lesion. FEMS Microbiol. Rev. 44, 386–398.

Fujibayashi, M., Suzuki, T., Sano, T., 2021. Mechanism underlying potato spindle tuber viroid affecting tomato (Solanum lycopersicum): loss of control over reactive oxygen species production. J. Gen. Plant Pathol. 87, 226–235.

Gardner, R.C., Chonoles, K.R., Owens, R.A., 1986. Potato spindle tuber viroid infections mediated by the Ti plasmid of Agrobacterium tumefaciens. Plant Mol. Biol. 6, 221–228.

German, M.A., Pillay, M., Jeong, D.H., Hetawal, A., Luo, S., Janardhanan, P., Kannan, V., Rymarquis, L.A., Nobuta, K., German, R., De Paoli, E., Lu, C., Schroth, G., Meyers, B.C., Green, P.J., 2008. Global identification of microRNA-target RNA pairs by parallel analysis of RNA ends. Nat. Biotechnol. 26, 941–946.

German, M.A., Luo, S., Schroth, G., Meyers, B.C., Green, P.J., 2009. Construction of Parallel Analysis of RNA Ends (PARE) libraries for the study of cleaved miRNA targets and the RNA degradome. Nat. Protoc. 4, 356–362.

Goecks, J., Nekrutenko, A., Taylor, J., Galaxy Team, T., 2010. Galaxy: a comprehensive approach for supporting accessible, reproducible, and transparent computational research in the life sciences. Genome Biol. 11, R86.

Hadidi, A., Flores, R., Randles, J., Semancik, J., 2003. In: Hadidi, A., Flores, R., Randles, J., Semancik, J. (Eds.), Viroids. CSIRO Publishing, Australia.

Hammond, R.W., Diener, T.O., Owens, R.A., 1989. Infectivity of chimeric viroid transcripts reveals the presence of alternative processing sites in potato spindle tuber viroid. Virology 170, 486–495.

Holley, R.W., Apgar, J., Everett, G.A., Madison, J.T., Marquisee, M., Merrill, S.H., Penswick, J.R., Zamir, A., 1965. Structure of a ribonucleic acid. Science 147, 1462–1465.

Itaya, A., Matsuda, Y., Gonzales, R.A., Nelson, R.S., Ding, B., 2002. Potato spindle tuber viroid strains of different pathogenicity induces and suppresses expression of common and unique genes in infected tomato. Mol. Plant-Microbe Interact. 15, 990–999.

C. Viroid pathogenesis and viroid-host interaction

Jones, J.D., Dangl, J.L., 2006. The plant immune system. Nature 444, 323–329.

Kallio, M.A., Tuimala, J.T., Hupponen, T., Klemelä, P., Gentile, M., Scheinin, I., Koski, M., Käki, J., Korpelainen, E.I., 2011. Chipster: user-friendly analysis software for microarray and other high-throughput data. BMC Genomics 12, 507.

Kappagantu, M., Bullock, J.M., Nelson, M.E., Eastwell, K.C., 2017. Hop stunt viroid: effect on host (Humulus lupulus) transcriptome and its interactions with hop powdery mildew (Podospheara macularis). Mol. Plant-Microbe Interact. 30, 842–851.

Katsarou, K., Wu, Y., Zhang, R., Bonar, N., Morris, J., Hedley, P.E., Bryan, G.J., Kalantidis, K., Hornyik, C., 2016. Insight on genes affecting tuber development in potato upon Potato spindle tuber viroid (PSTVd) infection. PLoS One 11, e0150711.

Langmead, B., Trapnell, C., Pop, M., Salzberg, S.L., 2009. Ultrafast and memory-efficient alignment of short DNA sequences to the human genome. Genome Biol. 10, R25.

Lehman, I.R., 1974. DNA ligase: structure, mechanism, and function. Science 186, 790–797.

Li, S., Wu, Z.G., Zhou, Y., Dong, Z.F., Fei, X., Zhou, C.Y., Li, S.F., 2021. Changes in metabolism modulate induced by viroid infection in the orchid Dendrobium officinale. Virus Res. 308, 198626.

Love, M.I., Huber, W., Anders, S., 2014. Moderated estimation of fold change and dispersion for RNA-seq data with DESeq2. Genome Biol. 15, 550.

Malik, N., Agarwal, P., Tyagi, A., 2017. Emerging functions of multi-protein complex Mediator with special emphasis on plants. Crit. Rev. Biochem. Mol. Biol. 52, 475–502.

Mardis, E.R., 2008. Next-generation DNA sequencing methods. Annu. Rev. Genomics Hum. Genet. 9, 387–402.

Martin, M., 2011. Cutadapt removes adapter sequences from high-throughput sequencing reads. EMBnet. j. 17, 10–12.

Martin, J.A., Wang, Z., 2011. Next-generation transcriptome assembly. Nat. Rev. Genet. 12, 671–682.

Matoušek, J., Steger, G., 2022. The splicing variant TFIIIA-7ZF of viroid-modulated transcription factor IIIA causes physiological irregularities in transgenic tobacco and transient somatic depression of "degradome" characteristic for developing pollen. Cells 11, 784.

Matoušek, J., Orctová, L., Steger, G., Riesner, D., 2004. Biolistic inoculation of plants with viroid nucleic acids. J. Virol. Methods 122, 153–164.

Matoušek, J., Siglová, K., Jakše, J., Radišek, S., Brass, J.R.J., Tsushima, T., Guček, T., Duraisamy, G.S., Sano, T., Steger, G., 2017. Propagation and some physiological effects of Citrus bark cracking viroid and Apple fruit crinkle viroid in multiple infected hop (Humulus lupulus L.). J. Plant Physiol. 213, 166–177.

Matoušek, J., Steinbachová, L., Záveská Drábková, L., Kocábek, T., Potěšil, D., Mishra, A.K., Honys, D., Steger, G., 2020. Elimination of viroids from tobacco pollen involves a decrease in propagation rate and an increase of the degradation processes. Int. J. Mol. Sci. 21, 3029.

McIntyre, L.M., Lopiano, K.K., Morse, A.M., Amin, V., Oberg, A.L., Young, L.J., Nuzhdin, S.V., 2011. RNA-seq: technical variability and sampling. BMC Genomics 12, 293.

Meshi, T., Ishikawa, M., Ohno, T., Okada, Y., Sano, T., Ueda, I., Shikata, E., 1984. Double-stranded cDNAs of hop stunt viroid are infectious. J. Biochem. 95, 1521–1524.

Mohorianu, I., Stocks, M.B., Applegate, C.S., Folkes, L., Moulton, V., 2017. The UEA small RNA workbench: a suite of computational tools for small RNA analysis. Methods Mol. Biol. 1580, 193–224.

Mortazavi, A., Williams, B.A., McCue, K., Schaeffer, L., Wold, B., 2008. Mapping and quantifying mammalian transcriptomes by RNA-Seq. Nat. Methods 5, 621–628.

Mullis, K., Faloona, F., Scharf, S., Saiki, R., Horn, G., Erlich, H., 1986. Specific enzymatic amplification of DNA in vitro: the polymerase chain reaction. Cold Spring Harb. Symp. Quant. Biol. 51 (Pt 1), 263–273.

Nath, V.S., Shrestha, A., Awasthi, P., Mishra, A.K., Kocábek, T., Matoušek, J., Sečnik, A., Jakše, J., Radišek, S., Hallan, V., 2020. Mapping the gene expression spectrum of mediator subunits in response to viroid infection in plants. Int. J. Mol. Sci. 21, 2498.

Nath, V.S., Mishra, A.K., Awasthi, P., Shrestha, A., Matoušek, J., Jakše, J., Kocábek, T., Khan, A., 2021. Identification and characterization of long non-coding RNA and their response against citrus bark cracking viroid infection in Humulus lupulus. Genomics 113, 2350–2364.

Navarro, B., Gisel, A., Rodio, M.E., Delgado, S., Flores, R., Di Serio, F., 2012. Small RNAs containing the pathogenic determinant of a chloroplast-replicating viroid guide the degradation of a host mRNA as predicted by RNA silencing. Plant J. 70, 991–1003.

Navarro, B., Flores, R., Di Serio, F., 2021a. Advances in viroid-host interactions. Annu. Rev. Virol. 8, 305–325.

Navarro, B., Gisel, A., Serra, P., Chiumenti, M., Di Serio, F., Flores, R., 2021b. Degradome analysis of tomato and Nicotiana benthamiana plants infected with potato spindle tuber viroid. Int. J. Mol. Sci. 22, 3725.

Owens, R.A., Kiefer, M.C., Cress, D.E., 1985. Chapter 15 – Construction of infectious potato spindle tuber viroid cDNA clones: implication for investigations of viroid structure-function relationships. In: Maramorosch, K., McKelvey, J.J. (Eds.), Subviral Pathogens of Plants and Animals: Viroids and Prions. Academic Press.

Owens, R.A., Tech, K.B., Shao, J.Y., Sano, T., Baker, C.J., 2012. Global analysis of tomato gene expression during Potato spindle tuber viroid infection reveals a complex array of changes affecting hormone signaling. Mol. Plant-Microbe Interact. 25, 582–598.

Patel, R.K., Jain, M., 2012. NGS QC Toolkit: a toolkit for quality control of next generation sequencing data. PLoS One 7, e30619.

Pease, A.C., Solas, D., Sullivan, E.J., Cronin, M.T., Holmes, C.P., Fodor, S.P., 1994. Light-generated oligonucleotide arrays for rapid DNA sequence analysis. Proc. Natl. Acad. Sci. U. S. A. 91, 5022–5026.

Salazar, L.F., Hammond, R.W., Diener, T.O., Owens, R.A., 1988. Analysis of viroid replication following Agrobacterium-mediated Inoculation of non-host species with potato spindle tuber viroid cDNA. J. Gen. Virol. 69, 879–889.

Samanta, S., Thakur, J.K., 2015. Importance of Mediator complex in the regulation and integration of diverse signaling pathways in plants. Front. Plant Sci. 6, 757.

Schena, M., Shalon, D., Davis Ronald, W., Brown Patrick, O., 1995. Quantitative monitoring of gene expression patterns with a complementary DNA microarray. Science 270, 467–470.

Schmieder, R., Edwards, R., 2011. Quality control and preprocessing of metagenomic datasets. Bioinformatics 27, 863–864.

Schroeder, A., Mueller, O., Stocker, S., Salowsky, R., Leiber, M., Gassmann, M., Lightfoot, S., Menzel, W., Granzow, M., Ragg, T., 2006. The RIN: an RNA integrity number for assigning integrity values to RNA measurements. BMC Mol. Biol. 7, 3.

Shoji, T., Yuan, L., 2021. ERF gene clusters: working together to regulate metabolism. Trends Plant Sci. 26, 23–32.

Shrestha, A., Mishra, A.K., Matoušek, J., Steinbachová, L., Potměšil, D., Nath, V.S., Awasthi, P., Kocábek, T., Jakše, J., Záveská Drábková, L., Zdráhal, Z., Honys, D., Steger, G., 2020. Integrated proteo-transcriptomic analyses reveal insights into regulation of pollen development stages and dynamics of cellular response to Apple Fruit Crinkle Viroid (AFCVd)-infection in Nicotiana tabacum. Int. J. Mol. Sci. 21, 8700.

Stark-Lorenzen, P., Guitton, M.C., Werner, R., Mühlbach, H.P., 1997. Detection and tissue distribution of potato spindle tuber viroid in infected tomato plants by tissue print hybridization. Arch. Virol. 142, 1289–1296.

Steinbachová, L., Matoušek, J., Steger, G., Matoušková, H., Radišek, S., Honys, D., 2021. Transformation of seed non-transmissible hop viroids in Nicotiana benthamiana causes distortions in male gametophyte development. Plants (Basel) 10, 2398.

Stöcker, S., Guitton, M.C., Barth, A., Mühlbach, H.P., 1993. Photosynthetically active suspension cultures of potato spindle tuber viroid infected tomato cells as tools for studying viroid – host cell interaction. Plant Cell Rep. 12, 597–602.

Suzuki, T., Ikeda, S., Kasai, A., Taneda, A., Fujibayashi, M., Sugawara, K., Okuta, M., Maeda, H., Sano, T., 2019. RNAi-mediated down-regulation of Dicer-like 2 and 4 changes the response of 'Moneymaker' tomato to potato spindle tuber viroid infection from tolerance to lethal systemic necrosis, accompanied by up-regulation of miR398, 398a-3p and production of excessive amount of reactive oxygen species. Viruses 11, 344.

Täbler, M., Sänger, H.L., 1984. Cloned single- and double-stranded DNA copies of potato spindle tuber viroid (PSTV) RNA and co-inoculated subgenomic DNA fragments are infectious. EMBO J. 3, 3055–3062.

Temin, H.M., Mizutani, S., 1970. RNA-dependent DNA polymerase in virions of Rous sarcoma virus. Nature 226, 1211–1213.

Thibaut, O., Claude, B., 2018. Innate immunity activation and RNAi interplay in citrus exocortis viroid-tomato pathosystem. Viruses 10, 587.

Vazquez Prol, F., Marquez-Molins, J., Rodrigo, I., Lopez-Gresa, M.P., Belles, J.M., Gomez, G., Pallas, V., Lison, P., 2021. Symptom severity, infection progression and plant responses in Solanum plants caused by three pospiviroids vary with the inoculation procedure. Int. J. Mol. Sci. 22, 6189.

Velculescu Victor, E., Zhang, L., Vogelstein, B., Kinzler Kenneth, W., 1995. Serial analysis of gene expression. Science 270, 484–487.

Velculescu, V.E., Zhang, L., Zhou, W., Vogelstein, J., Basrai, M.A., Bassett Jr., D.E., Hieter, P., Vogelstein, B., Kinzler, K.W., 1997. Characterization of the yeast transcriptome. Cell 88, 243–251.

C. Viroid pathogenesis and viroid-host interaction

Verhoeven, J.T.J., Meekes, E.T.M., Roenhorst, J.W., Flores, R., Serra, P., 2013. Dahlia latent viroid: a recombinant new species of the family Pospiviroidae posing intriguing questions about its origin and classification. J. Gen. Virol. 94, 711–719.

Visvader, J.E., Forster, A.C., Symons, R.H., 1985. Infectivity and in vitro mutagenesis of monomeric cdna clones of citrus exocortis viroid indicates the site of processing of viroid precursors. Nucleic Acids Res. 13, 5843–5856.

Wang, Y., 2021. Current view and perspectives in viroid replication. Curr. Opin. Virol. 47, 32–37.

Wang, Y., Qu, J., Ji, S., Wallace, A.J., Wu, J., Li, Y., Gopalan, V., Ding, B., 2016. A land plant-specific transcription factor directly enhances transcription of a pathogenic noncoding RNA template by DNA-dependent RNA polymerase II. Plant Cell 28, 1094–1107.

Więsyk, A., Iwanicka-Nowicka, R., Fogtman, A., Zagórski-Ostoja, W., Góra-Sochacka, A., 2018. Time-course microarray analysis reveals differences between transcriptional changes in tomato leaves triggered by mild and severe variants of potato spindle tuber viroid. Viruses 10, 257.

Więsyk, A., Lirski, M., Fogtman, A., Zagórski-Ostoja, W., Góra-Sochacka, A., 2020. Differences in gene expression profiles at the early stage of Solanum lycopersicum infection with mild and severe variants of potato spindle tuber viroid. Virus Res. 286, 198090.

Yuan, X., Wang, H., Cai, J., Li, D., Song, F., 2019. NAC transcription factors in plant immunity. Phytopathol. Res. 1, 3.

Zhang, Y., Yin, J., Jiang, D., Xin, Y., Ding, F., Deng, Z., Wang, G., Ma, X., Li, F., Li, G., Li, M., Li, S., Zhu, S., 2013. A universal oligonucleotide microarray with a minimal number of probes for the detection and identification of viroids at the genus level. PLoS One 8, e64474.

Zhao, S., Li, Y., 2021. Current understanding of the interplays between host hormones and plant viral infections. PLoS Pathog. 17, e1009242.

Zheng, Y., Wang, Y., Ding, B., Fei, Z., 2017. Comprehensive transcriptome analyses reveal that potato spindle tuber viroid triggers genome-wide changes in alternative splicing, inducible trans-acting activity of phased secondary small interfering RNAs, and immune responses. J. Virol. 91, e00247–17.

Zhong, S., Joung, J.G., Zheng, Y., Chen, Y.R., Liu, B., Shao, Y., Xiang, J.Z., Fei, Z., Giovannoni, J.J., 2011. High-throughput illumina strand-specific RNA sequencing library preparation. Cold Spring Harb. Protoc. 2011, 940–949.

Viroid-induced RNA silencing and its secondary effect on the host transcriptome

Charith Raj Adkar-Purushothama[a], Jean-Pierre Perreault[a], and Teruo Sano[b]

[a]RNA Group, Department of Biochemistry and Functional Genomics, Faculty of Medicine and Health Sciences, Applied Cancer Research Pavilion, University of Sherbrooke, Sherbrooke, QC, Canada [b]Faculty of Agriculture and Life Science, Hirosaki University, Hirosaki, Japan

Graphical representation

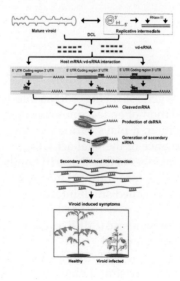

Abbreviations

DCL	DICER-like
dsRNA	double-stranded RNA
hpRNA	hairpin RNA
HSVd	hop stunt viroid
miRNA	microRNA
PSTVd	potato spindle tuber viroid
RdRP	RNA dependent RNA polymerase
RNAi	RNA interference
siRNA	small interfering RNA
sRNA	small RNA
vd-sRNA	viroid-derived small RNA

Working hypothesis: Putative working model for both viroid-induced RNA silencing and viroid-induced global effects on the host transcriptome and the phenotype. Upon infection, the mature viroid RNA replicates in the host plant *via* a rolling circle mechanism. Due to both internal base-pairing and the double-stranded RNA intermediate formed during viroid replication, it induces the host RNA silencing machinery via the activity of the DICER-like (DCL) enzyme and results in the production of viroid-derived small RNA (vd-sRNA). The RNA-induced silencing complex recruits these vd-sRNAs to direct the degradation of any complementary RNAs. This results in the accumulation of aberrant host mRNAs which lack either the 5′-cap or the 3′-poly(A) tail. These aberrant mRNAs might act as templates for the RNA-dependent RNA polymerase 6 (RDR6), resulting in the production of dsRNA. The resulting host dsRNA then serves as a substrate for the RNA silencing machinery. This in turn leads to the production of several secondary siRNAs that are complementary to the host RNA, and the cycle thus continues. In a nutshell, the RNA silencing initiated by the viroid is multiplexed by the host RDR6 to act on several endogenous mRNAs, which results in the induction of visible disease symptoms (modified from Adkar-Purushothama and Perreault, 2020a).

Definitions

RNA silencing: Also known as gene silencing or RNA interference (RNAi), is a mechanism of the sequence-specific regulation of gene expression that is highly conserved in most eukaryotic organisms. In addition, it plays various important functions in both the development and the maintenance of the vital activities of living organisms.

DICER: Endoribonuclease, a member of the RNase III family, that cleaves double-stranded RNA (dsRNA) and pre-microRNA (pre-miRNA) into approximately 20–25 base pairs long double-stranded RNA fragments called small interfering RNA and microRNA, respectively.

RdRP: RNA-dependent RNA polymerase (RdRP) catalyzes the synthesis of an RNA strand complementary to a given RNA template.

RISC: RNA-induced silencing complex (RISC) is a multiprotein complex that incorporates a single strand RNA, called the guide strand, that acts as a template for RISC to recognize complementary messenger RNA (mRNA) transcript. Once found, one of the proteins in RISC, known as Argonaute (AGO), activates and cleaves the mRNA.

Small RNA: Small RNAs are defined as noncoding RNA molecules that are less than 200 nucleotides in length.

vd-sRNA: Viroid-derived small RNAs are small RNAs of 19–26 nucleotides in length that are produced from viroids by the RNA silencing mechanism. These molecules are small noncoding RNAs that can subsequently interact with mRNA species.

secondary siRNA: The double stranded RNA (dsRNA) precursor synthesized as a result of RNA-dependent RNA polymerase (RdRP) activity on the aberrant RNAs produced by sRNA-guided transcript cleavage that act as a target for the RNA silencing machinery. The small RNAs produced from such dsRNAs are called secondary small interfering RNAs (secondary siRNA). Many secondary siRNAs are produced by successive Dicer-Like (DCL) processing and are therefore called phased secondary siRNAs (phasiRNAs).

Chapter outline

The discovery of viroids in the early 1970s opened the door to the study of a very mysterious molecule. Viroids are intriguing because they are long noncoding RNA infectious agents that rely entirely on their structure to survive in both the environment and their host. Hence, it makes one of the best models for RNA biologists to study host-RNA interactions and to develop novel RNA-based techniques. However, they are mysterious because they are noncoding RNA infectious agents, unlike fungus, bacteria, and viruses. Since their discovery, many scientists have struggled to understand how viroids overcome the host's defense mechanism, their mode of pathogenicity and disease management strategies. The previous chapters describe the roles of the various viroid regions and their implication in disease symptom expression; how viroids interact with the host translational machinery and proteins; the effect of viroids on the host DNA methylation pathways; and, how viroids affect both the host transcriptome and miRNA. This chapter provides supporting information on the working model called "role of vd-sRNA in viroid pathogenicity." More specifically, this chapter presents how viroid-triggered RNA silencing and vd-sRNAs both directly and indirectly effect the host transcriptome and morphology.

Learning objectives

- RNA silencing
- Viroid-induced RNA silencing
- Direct effect of viroid-derived small RNAs on host transcriptomes
- RNA silencing and viroid induced disease symptoms
- Indirect and global effect on host transcriptomes caused by viroid-derived small RNAs

Introduction

Viroids are low molecular weight RNA infectious agents that replicate from RNA to RNA in a process beginning with the generation of a multimeric antigenomic (or minus) strand from the circular viroid RNA genomic (or plus) strand through a rolling-circle replication mechanism. In the case of pospiviroids, the multimeric antigenomic strand then serves as a template to produce a multimeric genomic strand. The latter is then processed into monomeric linear molecules that are then circularized. Viroid RNAs have high intramolecular self-complementarity and can form either a rod-shaped or a branched rod-shaped stem-loop structure (Chapter 3) that resembles dsRNA. Due to these characteristics in replication and molecular structure, viroids strongly induce RNA silencing, one of the multilayered defense mechanisms present in their hosts, when they invade and replicate in the host cells. The viroids then become the target and are cleaved into small pieces of 21–24 nucleotides (nt) in length that are called vd-sRNA.

RNA silencing, also known as gene silencing or RNA interference (RNAi), is a sequence-specific mechanism that regulates gene expression. It is highly conserved in most eukaryotic organisms and plays various important functions in both the development and the maintenance of the vital activities of living organisms. RNA silencing is composed of several mechanistically related complex pathways, including both endogenously expressed microRNAs (miRNAs) that regulate endogenous gene expression, and when exogenously induced small interfering RNAs (siRNAs) which are used to defend against invading pathogens such as viruses. For example, siRNAs of 21–24-nt in length produced by RNA silencing that was activated by either double-stranded RNA (dsRNA), hairpin structured RNA (hpRNA), or abnormal RNA derived from invaders, cause the sequence-specific degradation of homologous RNAs.

Viroids are noncoding RNA molecules that do not possess the information that is required for them to be translated into proteins, but which attract and redirect various host factors in the invaded host cells to self-replicate. This, in turn, sometimes causes serious damage to the host metabolism and growth. The exact molecular mechanisms of viroid pathogenicity await further investigation; however, accumulated data provide convincing evidence that: (i) the interactions of specific nucleotide sequences or structural motifs present in viroid RNA with host factors regulating both replication and transportation (Chapter 11); (ii) the changes in protein expression and translation machinery produced by viroids (Chapter 12); (iii) the transcriptional inactivation of host gene expression by viroid-derived DNA methylation (Chapter 13); and (iv) the genome-wide modifications of the transcriptome associated with immunity, defense, hormone signaling, transcription factors, and miRNA biogenesis (Chapter 14) are all involved in either viroid pathogenicity or in the induction of disease symptoms. Furthermore, the notion that the vd-sRNAs, which are produced in large quantities in infected cells by viroid-induced RNA silencing, posttranscriptionally target host gene transcript(s) with homologous sequence(s), thereby modifying the host transcriptome and causing metabolic abnormalities, has recently received experimental support. This chapter outlines the findings obtained to date about the vd-sRNA induced by RNA silencing and its potential secondary effects on the host transcriptome about viroid pathogenicity.

RNA silencing

RNA silencing, or RNAi, regulates gene expression in eukaryotes (Baulcombe, 2004). It is associated with biological responses to either dsRNA or hpRNA. Therefore, it is recognized as an important innate immune mechanism against invading viruses and viroids. The RNAi pathway involves the recognition and processing of either dsRNA or hpRNA into 21–24-nt long small RNA (sRNA) duplexes by either a Dicer or DICER-like (DCL) protein. The processing of the different forms of either dsRNA or hpRNA (pri-miRNA) precursors leads to the production of two main forms of sRNA known as siRNA and miRNA (Lee and Carroll, 2018). The next step in RNA silencing is the incorporation of the sRNA duplex into an ARGONAUTE (AGO) protein, thus forming the RNA-induced silencing complex (RISC) and resulting in the removal of one of the sRNA strands. The RISC incorporated sRNA is called the guide strand, while the sRNA that is removed from the RISC is called the passenger strand. The guide strand helps in the recognition of complementary RNA, resulting in either the AGO-mediated degradation of the RNA or in the translational inhibition of the RNA. The single-strand RNA (ssRNA) containing the complementary region that forms a duplex with the guide strand is known as the target RNA (Baulcombe, 2004; Guo et al., 2016).

The RNA silencing pathways in plants are evolutionarily conserved. Because both the siRNAs and miRNAs regulate gene expression, they are commonly referred to as small regulatory RNAs. Hence, although the regulatory sRNAs function distinctly to cope with the plant functional requirements, they overlap in their pathways. The RNA silencing pathways can be broadly classified into at least three types: (i) the miRNA and *trans*-acting small interfering RNA (tasiRNA) pathways; (ii) the siRNAs and posttranscriptional gene silencing pathway; and, (iii) the RNA-directed DNA methylation pathway (RdDM) (Baulcombe, 2004; Eamens et al., 2008). The major distinctions between the different classes of sRNAs are that they are derived from different dsRNA precursors. Fig. 1 illustrates the biogenesis of the small regulatory sRNAs by three different pathways in plants.

In the miRNA pathway (Fig. 1A), miRNAs negatively regulate gene expression either by RNA cleavage or by blocking the translation of endogenous mRNA. The miRNAs are sRNAs derived from genetic loci known as the MIR genes. The MIR genes are transcribed into pre-miRNAs which form hairpin structures with stems and loops due to the internal base pairing of nucleotides (Eamens et al., 2009). The pre-miRNA molecule is processed by DCL1 in the nucleus, generating a 21-nt long imperfect RNA duplex of mature miRNA (guide strand) and a miRNA* (passenger stand; the symbol "*" is generally used to differentiate the passenger from the guide miRNA strand). The RNA duplexes are methylated at the 3' terminal nucleotides, which is known to protect the duplex RNA from degradation (Li et al., 2005). The duplexes are then exported into the cytoplasm where the miRNA binds to the AGO protein, forming the RISC. Since plant miRNAs share high sequence complementarity with their target mRNA, this leads to RNA cleavage. The primary target genes of plant miRNAs are regulatory genes that play a key role in plant development (Millar and Waterhouse, 2005).

TasiRNAs are generated via the specific miRNA-guided cleavage of the noncoding TAS precursor RNA (that was generated by the transcription of the TAS gene) followed by the conversion of the aberrant RNA into dsRNA by RDR6 (Peragine et al., 2004; Talmor-Neiman et al., 2006). DCL4 then processes the dsRNA into 21-nt siRNAs to produce a phased array of

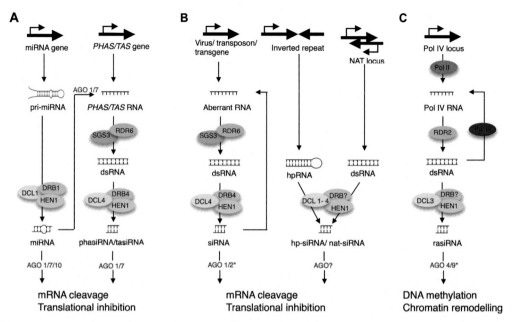

FIG. 1   RNA silencing pathways in plants. (A) In the miRNA biogenesis pathway, pre-miRNA transcribed from the MIR gene forms a hairpin structure that involves imperfect base-pairing. This precursor pre-miRNA stem-loop structure is processed by DCL1 into mature miRNA duplexes (miRNA:miRNA*). The guide strand (miRNA) is then incorporated into an AGO protein, forming the RISC. The resulting complex then guides the posttranscriptional gene silencing (PTGS) of endogenous mRNAs. The endogenous TAS and PHAS transcripts that are cleaved by 21-nt and 22-nt miRNAs, respectively, are used as templates by RDR6, generating dsRNAs which are subsequently processed by DCLs into phased siRNAs called *trans*-acting siRNAs (tasiRNAs) and phasiRNA, respectively, both of which play primary roles in target cleavage. (B) The siRNA biogenesis pathway primarily targets exogenous RNA (viral and subviral), transposons, transgene mRNAs and transcripts derived from natural inverted repeats found in the genome. The aberrant mRNAs, or the transcribed RNAs, are processed by RDR6, producing dsRNA which is subsequently cleaved by DCL4 and DCL1 into 21-nt and 22-nt siRNAs, respectively. The 22-nt siRNA triggers a secondary siRNA biogenesis resulting in the amplification of the RNA silencing. Natural antisense transcripts and inverted repeated sequences form hpRNA or dsRNA, respectively. These hpRNA and dsRNAs are processed by DCLs into siRNAs that target transcripts through an unidentified AGO protein. (C) In the RNA directed DNA methylation pathway (RdDM), RNA polymerase IV (Pol IV) transcripts are converted into dsRNA by RDR2 and then are processed by DCL3 into 24-nt siRNAs. These siRNAs guide RdDM to maintain genome stability. In the figure the symbol "*" indicates an atypical involvement of AGO in RNA silencing pathway.

21-nt siRNAs starting from the miRNA cleavage site (Peragine et al., 2004; Xie et al., 2005; Adenot et al., 2006). PhasiRNAs are also generated by RdRPII from the long, noncoding PHAS precursor RNAs transcribed from both protein-coding genes and from noncoding sequences such as the repetitive DNA called PHAS locus (Zhai et al., 2011; Shivaprasad et al., 2012). The production of phasiRNA begins with a primary cleavage in which a complex of 22-nt-long miRNAs with either AGO1 or AGO7 cleaves the target sequence that is conserved at either the 5′-end, or at both ends, of the PHAS RNA. The cleaved single-stranded PHAS RNA is then rendered double-stranded by the action of RDR6. Then, starting from the cleavage site of the miRNA, either DCL4 or DCL5, a ribonuclease III-like enzyme, secondarily cleaves the double-stranded PHAS RNA at intervals of either 21- (by DCL4) or 24-nt (by DCL5), producing the phasiRNA. The function of the phasiRNA is not yet fully known.

In addition to miRNAs and tasiRNAs or phasiRNAs, the 21–24-nt sRNAs that are involved in the posttranscriptional gene silencing (PTGS) are called siRNA. These siRNAs targets and degrades transcripts generated from viruses or viroids or transposons. These siRNAs are produced from aberrant mRNAs, which are processed by RDR6, producing dsRNA that is cleaved by DCL4 and DCL1 (Fig. 1B). Hence, it is often considered as "exogenic RNA silencing" mechanism. This kind of RNA silencing phenomenon was first observed in studies involving transgenes in which transgenic petunia plants, designed to overexpress a pigmentation enzyme, exhibited a loss of pigmentation in the flowers (van der Krol et al., 1990). Similarly, transgenic tobacco plants expressing either a full-length or a truncated form of the tobacco etch virus (TEV) coat protein induced the sequence-specific target RNA degradation, leading to virus resistance (Lindbo et al., 1993). However, the infections of plants with exogenic RNAs (such as viral and viroid RNAs) are associated with the accumulation of exogenic specific sRNAs, which are processed from double-stranded exogenic RNA by the host RNA silencing machinery and which direct the degradation of single-stranded exogenic RNAs in a manner similar to that of the endogenous siRNA pathway (Wang et al., 2012). Therefore, RNA silencing has been considered as a natural antiviral defence mechanism against invading viruses and viroids.

The third pathway of RNA silencing in plants is associated with de novo DNA methylation (Fig. 1C), and thus in suppressing the transcription of genes in the nucleus (Zhang et al., 2013; Matzke et al., 2015). The RdDM pathway is unique to plants. RdDM is directed by 24-nt siRNAs, which are generated by the combined functioning of the plant-specific RNA polymerase IV (Pol IV), RDR2 and DCL3. The principal function of RdDM is to silence both transposable elements (TEs) and repetitive DNA to maintain genome stability. Indeed, 24-nt siRNAs are also known as either repeat-associated siRNAs or as rasiRNAs as most of these siRNAs are derived from either TEs or repetitive DNA that is present in the plant genome (Guo et al., 2016).

Viroid induced RNA silencing

As described in the previous section, RNA silencing is a natural antiviral defense mechanism against invading exogenous RNA pathogens which results in both the production and the accumulation of invading RNA specific 21–24-nt long sRNAs (Ding, 2010; Adkar-Purushothama and Perreault, 2020a). As mentioned in both Chapter 3 (Structure) and Chapter 4 (Replication and movement), because viroids are highly base-paired structures and form dsRNA intermediates during replication, they act as both inducers and targets of RNA silencing (Pallas et al., 2012; Flores et al., 2015). Upon infection, viroids elicit the RNA silencing machinery of the host plant and the viroid RNAs are processed by either the DCL ribonucleases, resulting in the production of siRNAs of 21–24-nt that are called vd-sRNAs (Dadami et al., 2013). The process of viroid induced RNA silencing in host plants is outlined in Fig. 2.

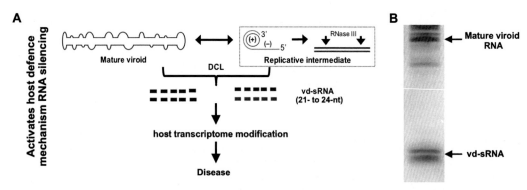

FIG. 2 Viroid-induced RNA silencing and its secondary effect on the host transcriptome. (A) After invading the host plant cells, the circular viroid replicates in either the nuclei (members of the family *Pospiviroidae*; presented in this figure) or in the chloroplast (members of the family *Avsunviroidae*) via a rolling circle mechanism. During this process, using the host's transcriptional machinery, the (+) strand (represented in *blue*) viroid molecule is transcribed into the (−) strand (represented in *red*), and then (−) strand is transcribed to produce a greater than unit length (+) strand. These transcribed multimeric (+) strands are then processed into a unit-length and circularized for production of the mature viroid molecule. The highly base-paired mature circular viroid molecule, and double-stranded RNA structures formed between the (+) and (−) strand molecules during viroid replication, both trigger the host's RNA silencing machinery and are cleaved into small RNAs of 21–24 nt called viroid-derived small RNAs (vd-sRNA). The presence of a large amount of accumulated vd-sRNAs in infected cells can extensively alter the host's transcriptome and adversely affect its growth. (B) Total RNA isolated from viroid infected plant separated in 8% poly-acrylamide gel under denaturation condition showing the mature viroid RNA and vd-sRNA.

In the early 1980s, a comparison of the sequences of PSTVd variants revealed a stretch of 23 nucleotides that possessed sequence similarity with the mammalian U1 small nuclear RNA (snRNA), a member of the snRNA which plays a vital role in the mRNA splicing mechanism (Kiss and Solymosy, 1982). Additionally, a stretch of 36–53-nt of the tomato signal recognition particle (SRP) RNA was shown to exhibit high sequence complementarity to five viroid species that cause disease in tomato plants (Haas et al., 1988). Thermodynamically, this sequence complementarity favored base-pairing in vivo, suggesting that tomato SRP RNA is a possible viroid target (Haas et al., 1988). Although the concept of RNA silencing had not yet been developed, these findings led to the hypothesis that viroids could interfere with the host RNA metabolism and thus be involved in pathogenesis (Diener et al., 1993). The first demonstration of RNA silencing, specifically the de novo methylation of a viroid molecule, was reported in 1994 (Wassenegger et al., 1994). In this study, *Agrobacterium tumefaciens*-mediated leaf-disc transformation of one monomeric and three oligomeric PSTVd molecules into the tobacco genome resulted in the methylation of PSTVd-specific sequences, but not of the flanking T-DNA nor the genomic plant DNA. Furthermore, the methylation of viroid cDNA was observed only after the viroid RNA-RNA replication had occurred in transformed plants (Wassenegger et al., 1994).

In early 2000, vd-sRNAs were detected in plants infected with viroids belonging to the two families, suggesting that these viroids are the targets of RNA silencing irrespective of their subcellular localization during replication (Itaya et al., 2001; Papaefthimiou et al., 2001; Martínez de Alba et al., 2002; Markarian et al., 2004). These *in planta* experiments demonstrated that vd-sRNAs are: (i) produced from both the (+) and (−) strands of the viroid RNA

(Adkar-Purushothama and Perreault, 2020b). The accumulation of vd-sRNAs has been extensively studied in different viroid-host combinations using next-generation sequencing technologies. The profiling of vd-sRNA on their respective viroid RNAs suggested that the (+) strand produced more sRNA than did the (−) strand. This occurs because the (+) strand viroid RNA is more abundant than the (−) strand in infected plants (Symons et al., 1987).

Direct effect of viroid derived small RNAs on the host transcriptome

Peach latent mosaic viroid (PLMVd) induces a variety of symptoms in its natural host (the peach; *Prunus persica*) that includes everything from latent, mosaic symptoms to severe albinism (named peach calico, PC) in leaves, stems and fruits. The PLMVd variants that induce latent and mosaic-symptoms are 335–338-nt long, while the variant associated with PC (hereafter, PLMVd-PC) is 348–351-nt in length and contains an additional insertion of 12–14-nt (Flores et al., 2006). The availability of complete genome sequence data provided a critical resource for genome-wide exploring of the pathogenic determinant of PLMVd-PC-induced alterations in the host transcriptome and in eliciting symptomatology (peach v1.0; International Peach Genome Initiative, http://www.rosaceae.org/peach/genome). Hence, the bioinformatic analysis of vd-sRNAs from the PC-expressing variant against the peach transcriptome revealed the expression of two 21-nt vd-sRNAs (PC-sRNA8a and PC-sRNA8b) from the (−) polarity strand of PLMVd-PC that could potentially hybridize and induce the cleavage of the peach mRNA encoding the chloroplastic heat-shock protein 90 (cHSP90). Interestingly, PC-sRNA8a and PC-sRNA8b predominantly accumulated in the parts of leaves that showed bleaching but did not accumulate in the adjacent green parts. Furthermore, semiquantitative RT-PCR revealed the suppression of *cHSP90* expression in the bleached parts of PLMVd infected plants. This evidence supported the hypothesis that vd-sRNA could potentially induce cleavage and degradation of its RNA target, as was predicted by using bioinformatic tools. The cleavage of the target transcript was validated by 5′ RNA ligase mediated rapid amplification of cDNA ends (5′ RLM-RACE) experiments with RNA preparations from bleached leaf tissues. Specifically, this led to the observation that four out of five cDNA clones possessed a cleavage site on the *cHSP90* transcript with 5′ termini identical to that of the predicted cleavage site. Conversely, the 5′ RLM-RACE products obtained from green tissues lacked similar cleavage sites, indicating that the RISC-mediated degradation of this peach transcript was not occurring in the asymptomatic tissues (Navarro et al., 2012).

The hybridization between a vd-sRNA and its target, followed by cleavage of target RNA by RISC, was conclusively demonstrated *in planta* using artificial miRNA (amiRNA) technology (Adkar-Purushothama et al., 2015a). Specifically, the amiRNA system was developed to transiently and simultaneously express a vd-sRNA derived from the pathogenicity-modulating region of PSTVd and its predicted target sequence found in the tomato callose synthase 11-like mRNA (Cals11-like mRNA). To that purpose, both constructs were agroinfiltrated in *Nicotiana benthamiana* under the control of the cauliflower mosaic virus 35S promoter. Since the target sequence was tagged with green fluorescent protein (GFP) gene, if the vd-sRNA expressed by the amiRNA technology bound the target, the GFP level should then decrease as binding results in RNA silencing. Three days postagroinfiltration the leaves that were co-expressed with vd-sRNA and GFP-tagged target sequence demonstrated reduced fluorescence upon UV illumination as compared to control leaves in which GFP-tagged target and empty vector without vd-sRNA were agroinfiltrated. Immunoblot assays using an anti-GFP antibody confirmed the decrease in GFP accumulation. Furthermore, the 5′-RLM RACE experiment showed that the callose synthase 11-like transcript in PSTVd-infected tomato plants was cleaved at the position expected from the vd-sRNA of interest. The detailed protocol is provided in Protocols/Procedures/Methods (Methods 1 and 3). For evaluation of the direct vd-sRNA's effect on the host's phenotype, at least two methods using amiRNA technology were available: (i) the transgenic expression of vd-sRNA (Eamens et al., 2014); and, (ii) the transient expression of vd-sRNA (Adkar-Purushothama and Perreault, 2018; Adkar-Purushothama et al., 2018). Since developing transgenic plants expressing vd-sRNA as amiRNA is laborious and time consuming, only the transient expression method is explained here. In this method, the vd-sRNA of interest is constructed on an amiRNA backbone (osa-MIR528), cloned into the pGEM-T easy vector, and then transiently expressed by co-agroinfiltration with the binary vectors pCV-A and pCV-B. Vector pCV-A has 35S promoter that helps in expression of the inserted gene of interest whereas pCV-B helps in the systemic movement of both vectors upon co-agroinfiltration in host plant. These binary vectors were developed from *Cabbage leaf curl virus* (CaLCuV) (Tang et al., 2010). Upon agroinfiltration, the former vector can produce vd-sRNAs equivalent to those produced by viroid. These, in turn, bind to their target transcript leading to either RISC-mediated cleavage or translational repression. Finally, this induces visible phenotypic changes if the target transcript is involved in such traits (Fig. 3B). The detailed protocol is provided in "Method 2: amiRNA experiment demonstrating the effect of the vd-sRNA on the plant's phenotype" in this chapter. For instance, viroid infection is often associated with early flowering of the plant. The flowering behavior in Arabidopsis is determined by two genes, FRIGIDA (FRI) and FLOWERING LOCUS C (FLC), which negatively affect flowering (Michaels et al., 2004; Choi et al., 2011). Bioinformatics analysis of vd-sRNA on the tomato genome revealed that the FRIGIDA-like protein 3 (FRL3) mRNA could be the potential target of PSTVd-inducing RNA silencing. To verify the role of putative vd-sRNA on both FRL3 mRNA and on the host's phenotype, an amiRNA experiment was designed to deliver vd-sRNA into the sensitive tomato cultivar "Rutgers." Approximately 35 days later, plants agroinfiltrated with pCV vector expressing the vd-sRNA of interest formed flower buds, while none were observed in the control samples (Fig. 3C). This result shows that PSTVd infection suppresses FLR3 expression *via* vd-sRNA, which in turn accelerates flower bud formation. Similar experiments can be performed to demonstrate the impact of vd-sRNA on host's phenotype.

Generation of secondary siRNAs triggered by viroid infection and their global effect on both the transcriptome and the phenotype—A working hypothesis

Generally, the effect of RNA silencing triggered by siRNA is dose-dependent. That said, a target's downregulation is directly related to the amount of siRNA. In the case of PLMVd-PC infected plants, 171 reads of vd-sRNAs (PC-sRNA8a and PC-sRNA8b) were recovered from the albino regions of the leaves (Navarro et al., 2012). In this scenario, this number of

A

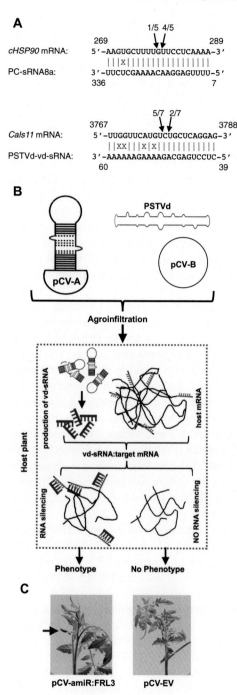

```
                            1/5 4/5
                269          ↓  ↓            289
cHSP90 mRNA:    5'-AAGUGCUUUUGUUCCUCAAAA-3'
                   |||x||||||||||||||||
PC-sRNA8a:      3'-UUCUCGAAAACAAGGAGUUUU-5'
                336                        7


                             5/7 2/7
                3767          ↓  ↓          3788
Cals11 mRNA:    5'-UUGGUUCAUGUCUGCUCAGGAG-3'
                   ||xx|||x|x||||||||||||
PSTVd-vd-sRNA:  3'-AAAAAAGAAAAGACGAGUCCUC-5'
                60                          39
```

B

PSTVd

pCV-A   pCV-B

Agroinfiltration

Host plant

production of vd-sRNA   host mRNA

vd-sRNA:target mRNA

RNA silencing   NO RNA silencing

Phenotype   No Phenotype

C

pCV-amiR:FRL3   pCV-EV

FIG. 3 (A) cHSP90 mRNA/PC-sRNA8a and CalS11-like mRNA/PSTVd-vd-sRNA duplexes predicted to be formed by the sRNAs derived from the PLMVd-PC and PSTVd-I variants, respectively. The *arrows* indicate the 5′ termini of the cHSP90 mRNA and the CalS11-like mRNA fragments isolated from the PLMVd-PC and PSTVd-I infected plants, respectively, as identified by 5′ RLM-RACE products, with the frequencies of the clones being shown (ex. 4/5, indicates that four cleavage products were found out of five analyzed clones). The sequences are shown in the complementary polarity. 5′ RLM RACE results of the cHSP90 mRNA are adapted from Navarro et al. (2012) and of the CalS11-like mRNA from Adkar-Purushothama et al. (2015a), respectively. (B) Flow chart illustrating the details of an amiRNA experiment. The vd-sRNA expressed as an artificial microRNA (amiRNA) in the pCV-A binary vector was agroinfiltrated, together with pCV-B, into tomato plants at the cotyledon stage. The specific interaction of vd-sRNA with the target sequence leads to either RNA-induced silencing complex (RISC)-mediated cleavage or to translational repression, which, in turn, results in the observed phenotype of the plant. (C) At 35 days postinfection, plants agroinfiltrated with pCV-amiR: FRL3 exhibited early flower buds. pCV-amiR: FRL3: plants agroinfiltrated with pCV-amiR:FRL3, which is capable of producing sRNA similar to the vd-sRNA that was predicted to bind to the FRIGIDA-like protein 3 (FRL3) mRNA. pCV-EV: plants agroinfiltrated with pCV-empty vector that was used as a negative control. *Figure B and its legend is reused after modification from Adkar-Purushothama, C.R., Sano, T., Perreault, J.-P., 2018. Viroid derived small RNA induces early flowering in tomato plants by RNA silencing. Mol. Plant Pathol. https://doi.org/10.1111/mpp.12721.*

C. Viroid pathogenesis and viroid-host interaction

vd-sRNAs might have been sufficient to cause the bleaching effect given the fact that cHSP90 is a single-copy gene. However, under multigene conditions, the reduced expression level can be complemented by another gene of the same function (Inoue et al., 2013). In this scenario, how can a viroid's limited number of vs-sRNAs can induce downregulation of a large number of different kinds of host's transcriptome? This led to hypothesis, of the possible production of secondary siRNAs during viroid infection that escalates the downregulation of the host transcriptome thus inducing viroid-associated disease symptoms in the host plant. Please refer "working model" presented at the beginning of this chapter to understand the different types of sRNAs involved in viroid-induced symptom expression.

A global view of host transcript degradation due to viroid infection can be obtained by degradome sequencing. Degradome sequencing (also called, Parallel Analysis of RNA ends (PARE)) is a genome-wide approach that permits the identification of the uncapped 5′ phosphate termini of transcripts in a given sample. Comparison of degradome data obtained from PSTVd infected tomato plants with degradome obtained from mock-inoculated plants allowed identification of more than 1000 positive events of sRNA-guided, but vd-sRNA independent, cleavage of transcripts in viroid infected plants through the production of secondary siRNA and phasiRNA (Zheng et al., 2017), suggesting an important role of secondary siRNAs triggered by viroid infection.

Upon infection, the PSTVd-RG1 variant is known to induce severe disease symptoms in the tomato cultivar "Rutgers." Analysis of deep sequencing data revealed the presence of eight vd-sRNAs derived from the (−) strand of PSTVd-RG1 that have the potential to target the 40S ribosomal protein S3a-like mRNA (RPS3a-like) (Adkar-Purushothama et al., 2017). The latter is a structural constituent of the ribosome; hence, it is directly involved in the protein translation machinery. 5′ RLM-RACE experiments demonstrated the cleavage of the RPS3a-like mRNA (Fig. 4A), and RT-qPCR experiments revealed a significant downregulation of the target transcripts in the PSTVd infected tomato plants as compared to what was seen in the mock inoculated plants. However, amiRNA experiments indicated a low efficiency of target RNA cleavage. To solve this riddle, a parallel analysis of RNA ends (PARE) was performed on the total RNA extracted from both the PSTVd-RG1-infected and the mock-inoculated tomato plants. Data analysis revealed the widespread cleavage of the RPS3a-like transcript in distinct locations other than the predicted vd-sRNA binding site in the viroid-infected plant when compared to the data from the mock inoculated plant (see Fig. 4B).

The above mentioned experimental data provides with attractive idea leading to the working hypothesis on the production of secondary siRNAs in viroid infected plants, that could be expanding the cleavage targets. As shown in Fig. 4C, the aberrant transcript RNAs produced from vd-sRNA-induced primary RNA silencing have to be converted into dsRNA to generate secondary siRNA.

In plants, RNA-dependent RNA polymerases (RdRP) are essential for virus-induced gene silencing. Additionally, the RdRPs are known to play a role in the amplification of the initial RNA silencing reaction through the production of secondary siRNAs (Schwach et al., 2005). In plants, six RdRPs have been isolated (RDR1 to RDR6), of which RDR1 and, particularly, RDR6, have been implicated in virus defense. RDR6 is an important component in RNA silencing for the biogenesis of different siRNAs (Fig. 1). Specifically, the single-stranded aberrant transcript RNAs are used as templates by RDR6, in the presence of the "Suppressor of gene silencing 3" (SGS3), to produce dsRNA. The resulting dsRNAs then become the targets of the RNA silencing machinery, producing the secondary siRNA called transitive siRNAs (Nishikura, 2001) that degrade complementary mRNAs. This process leads to the amplification of secondary siRNAs, like that of phasiRNA or tasiRNA (Sanan-Mishra et al., 2021). Correspondingly, the sequences located outside of the region targeted by vd-sRNA could also result in the further silencing as the secondary targets that are not homologous to the initial silencing trigger of the vd-sRNA (Figs. 1 and 4C).

Three independent studies using two viroids species (HSVd and PSTVd) showed that, under RDR6 knockdown conditions, plants are asymptomatic although they do accumulate higher concentrations of both viroid RNA and vd-sRNA (Gómez et al., 2008; Di Serio et al., 2010; Adkar-Purushothama and Perreault, 2019). This observation indirectly supports the working model "role of vd-sRNA in viroid pathogenicity" by which viroid pathogenicity is exerted, in part, through the production of secondary siRNA and the effect of these secondary siRNA on host's transcriptome and phenotype.

Protocols/procedures/methods

Method 1: amiRNA experiment demonstrating the formation of the vd-sRNA/target duplex *in planta*

To demonstrate vd-sRNA/target duplex formation *in planta*, the use of a transient expression of amiRNA containing vd-sRNA along with a GFP-tagged target sequence is a well-established system. This experiment can be performed as described below:

1. Construction of amiRNA
   (a) Design a reverse primer that contains the vd-sRNA nucleotide sequence with a restriction enzyme site such as *Xba*I.
   (b) Design a forward primer that contains the vd-sRNA* nucleotide sequence with a restriction enzyme site such as *Bam*HI.
   (c) Perform PCR amplification on the osa-MIR528 backbone using the above forward and reverse primers in the presence of a DNA polymerase that lacks proofreading activity.

   Note: WMD3—Web microRNA designer (http://wmd3.weigelworld.org/cgi-bin/webapp.cgi?page=Home;project=stdwmd) is a useful tool for use in designing the primers.
   (d) Digest both the PCR product and the expression vector that possesses a promoter such as CaMV 35S (pBIN61 vector) with the same two restriction enzymes (e.g., *Xba*I and *Bam*HI).
   (e) Ligate the gel purified PCR product to the linearized vector.
   (f) Transform the resulting construct into competent *E. coli* bacteria, culture, and verify the presence of the insert by PCR.
   (g) Extract plasmid DNA from the *E. coli* bacteria and transform it into *A. tumefaciens*.
2. Construction of the GFP-tagged target sequence
   (a) Ligate the target sequence to the 3′ UTR of the GFP gene sequence.
   (b) Insert the ligated GFP-target sequence into the expression vector (pBIN61 vector).

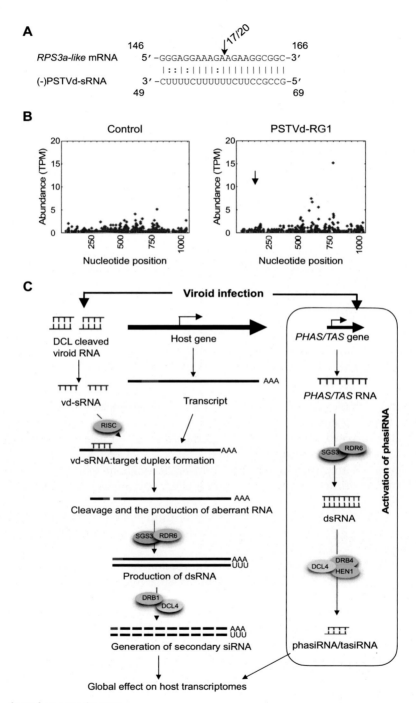

A

	146 ↓17/20 166
RPS3a-like mRNA	5′ -GGGAGGAAAGAAGAAGGCGGC-3′
	|::|:||||:||||||||||||
(-)PSTVd-sRNA	3′ -CUUUUCUUUUUUUCUUCCGCCG-5′
	49 69

B

Control   PSTVd-RG1

Abundance (TPM)

Nucleotide position

C

Viroid infection

DCL cleaved viroid RNA

Host gene

PHAS/TAS gene

vd-sRNA

Transcript

PHAS/TAS RNA

RISC

vd-sRNA:target duplex formation

SGS3   RDR6

Cleavage and the production of aberrant RNA

dsRNA

SGS3   RDR6

Production of dsRNA

DRB4
DCL4   HEN1

DRB1
DCL4

Generation of secondary siRNA

phasiRNA/tasiRNA

Activation of phasiRNA

Global effect on host transcriptomes

FIG. 4   See legend on opposite page.

(c) Transform the construct into competent *E. coli bacteria*, culture, and verify the insert of the insert by PCR.

(d) Extract the plasmid DNA from the *E. coli* bacteria and transform it into *Agrobacterium tumefaciens*.

3. *In planta* experiments

(a) Co-agroinfiltrate the GFP-tagged target constructs along with the amiRNA constructs into fully opened mature *N. benthamiana* leaves.

(b) On the different areas of the same leaves, co-agroinfiltrate the GFP-tagged target constructs along with the empty vector. This serves as the negative control.

4. Data analysis

(a) At 3 days postinfection, observe the agroinfiltrated leaves under UV illumination for the presence of GFP fluorescence. If the vd-sRNA has bound the target, then a reduction in the fluorescence will be noticed as compared to the negative control.

(b) To quantify the amount of GFP downregulation, perform Western blot hybridization using an anti-GFP antibody.

A detailed demonstration of the amiRNA experiment has been reported previously (Adkar-Purushothama et al., 2015a).

Method 2: amiRNA experiment demonstrating the effect of the vd-sRNA on the plant's phenotype

To evaluate the role of the vd-sRNA on the tomato plant's phenotype, amiRNAs for the different vd-sRNA sequences must be initially synthesized on an osa-MIR528 backbone, and then must be ligated into the binary vector pCV-A as described in Method 1. The resulting binary vectors (pCV-A+vd-sRNA) are then transformed into the *A. tumefaciens*

FIG. 4, CONT'D PARE analysis of the RPS3a-like mRNA in viroid infected plants and a putative working model for viroid induced RNA silencing in host plants. (A) RPS3a-like mRNA/vd-sRNA duplexes are predicted to be formed by the sRNA derived from the PSTVd-RG1 variant. The *arrows* indicate the 5′ termini of the RPS3a-like mRNA fragments, isolated from the PSTVd-infected plants, as identified by 5′ RLM-RACE. The observed frequencies of the clones are also shown (ex. 17/20 indicates that 17 cleavage products were found out of 20 analyzed clones). The sequences are shown in the complementary polarity. (B) The PARE reads were profiled against the RPS3a-like mRNAs. The *vertical arrow* denotes the predicted cleavage site. All the reads were normalized to "Transcripts Per Million (TPM)". (C) Due to both internal base-pairing and the double-stranded RNA intermediate formed during the viroid's replication (Flores et al., 2005), the viroid induces the host's RNA silencing machinery through the activity of the DICER-like (DCL) enzyme, and this results in the production of viroid-derived sRNA (vd-sRNA). The RNA-induced silencing complex recruits these vd-sRNAs to direct the degradation of complementary RNAs. This results in the accumulation of aberrant host mRNAs which lack either the 5′-cap or the 3′-poly(A)-tail. These aberrant mRNAs might act as the templates for the polymerase RDR6, resulting in the production of dsRNA. This phased product from host dsRNA serves as a substrate for the RNA silencing machinery. This results in the production of several secondary siRNAs complementary to the host's RNA, and the cycle thus continues. In a nutshell, the RNA silencing initiated by the viroid is multiplexed by the host's polymerase RDR6 to act on the host's mRNAs, resulting in symptom expression. The figure legend is reused after minor editing from Adkar-Purushothama and Perreault (2020a). *Figure A is reproduced and Figure B is modified from Adkar-Purushothama, C.R., Iyer, P.S., Perreault, J.-P., 2017. Potato spindle tuber viroid infection triggers degradation of chloride channel protein CLC-b-like and Ribosomal protein S3a-like mRNAs in tomato plants. Sci. Rep. 7(1), 8341. https://doi.org/10.1038/s41598-017-08823-z.*

strain GV3101. Next, the pCV-B vectors are transformed into the *A. tumefaciens* strain GV3101. Both *A. tumefaciens* samples are then mixed in equal proportion and agroinfiltrated into either tomato or *N. benthamiana* plants. The agroinfiltrated plants are maintained at 23°C with 16h of light and 8h of darkness. A detailed demonstration of this experiment has been reported previously, and showed the effect of vd-sRNA on both early flowering (Adkar-Purushothama et al., 2018) and host morphology (Adkar-Purushothama and Perreault, 2018) in tomato plants, respectively.

Method 3: 5′ RLM RACE experiment confirming the cleavage site on the predicted target transcript

5′ RLM RACE is a commonly used technique with which to confirm the direct cleavage of a target by a miRNA. It can be performed as follows:

1. Ligating an RNA adapter to the free 5′-phosphate end of a cleaved transcript
   (a) To DNase I treated total RNA (1 µg) add an RNA adaptor and 1 unit of T4 RNA ligase.
   (b) Gently tap to mix and incubate at 37°C for 6h.
2. Reverse transcription reaction
   (a) Perform a reverse transcription reaction on the ligated RNA using either an oligo d(T) or a target specific reverse primer.
3. Nested PCR amplification of the target
   (a) Perform a PCR amplification using a proofreading DNA polymerase in the presence of an RNA adapter specific forward and a target specific reverse primer.
   (b) Dilute the PCR products 10–20 times with water.
   (c) Perform a second round of PCR using a forward primer that is present within the adapter sequence and a target specific reverse primer which is located inside of the previous reverse primer on undiluted, 1:10 and 1:20 diluted first PCR products.
   (d) Separate all the nested PCR products by 2.0% agarose gel electrophoresis and verify the amplicons for the presence of ones of the expected size by comparison against standard size markers.
4. Cloning and sequencing
   (a) Elute amplicons of the expected size from the gel.
   (b) Ligate them into a cloning vector using T4 DNA ligase.
   (c) Transform the ligated cloning vector into competent bacteria and grow on selection media.
   (d) Extract the plasmid DNA from selected colonies and verify the presence of the insert by PCR using plasmid specific primers.
   (e) Sequence the plasmids containing inserts.
5. Sequence analysis
   (a) Remove the plasmid sequence from the obtained nucleotide sequence.
   (b) Identify the sequence of the RNA adapter.
   (c) BLAST analyzes the entire nucleotide sequence adjacent to that of the RNA adapter.
   (d) The nucleotide that is adjacent to the RNA adapter is the 5′-phosphate end of an uncapped transcript produced by RISC-mediated cleavage.

Perspective

Several groups of researchers are studying various viroid-host combinations to understand the interaction of viroids with the host components and to elucidate the mechanism behind viroid induced disease symptoms. Although the early days of this research were focused on the role of the structural/functional motifs in causing the viroid-induced symptoms, in recent years it has been redirected with advancements in both molecular and computational biology. For example, the accumulation of vd-sRNAs in viroid infected plants was demonstrated in the early 2000s, while the direct interaction of these vd-sRNAs on host transcriptomes was demonstrated about a decade later. The development of deep sequencing technologies such as sRNA sequencing, RNAseq and degradome analysis has allowed researchers to understand the global effects of vd-sRNA, viroid-activated phasiRNA, or similar secondary siRNAs on host-viroid interactions.

Although the suppression of certain transcripts could result in the induction of visible symptoms in the host plant (ex., the suppression of cHSP90 by PLMVd-PC in peach plants induces bleaching), it is not a universal phenomenon for all of the vd-sRNA/target combinations because such effects are copy number dependent and must allow for the compensatory effect of other genes. For instance, the transient expression of PSTVd-sRNA that was predicted to target the FRIGIDA-like protein 3 (FRL3) mRNA induced early flowering in tomato plants (Adkar-Purushothama et al., 2018). However, *N. benthamiana* plants did not demonstrate early flowering when the same PSTVd-sRNA was transiently expressed (unpublished data; CR Adkar-Purushothama and JP Perreault). To understand the compensation effect, the *FRL3* transcript was amplified and sequenced. The sequence data obtained revealed the presence of four polymorphic forms of the *FRL3* mRNA, being only one identical to the *FRL3* mRNA from tomato. That said, although vd-sRNA can downregulate one *FRL3* transcript, the other three *FRL3* transcripts that are not targeted by the vd-sRNA compensate for the effect of the downregulated transcript and result in an overall null effect on the phenotype. At the present moment, it is worth noting that, as mentioned in other chapters of the section "Viroid pathogenesis and viroid-host interaction," RNA silencing mechanisms are not the only governing phenomenon that play a role in a viroid pathogenicity and disease symptom expression.

As a noncoding RNA infectious agent, viroids should solely rely on their RNA molecule for their survival and its pathogenicity. Therefore, understanding the selection pressure that help in viroid survival could be exploited to develop crop protection strategies. This is detailed in Chapter 17 "Viroid disease control and strategies." The analysis of the deep sequencing data of viroids, and the mapping of them on the viroid (+) and (−) strands, allowed researchers to identify not only the regions of the RNA genome that are more susceptible to RNA silencing but also those that are less susceptible or protected. Targeting of such susceptible regions using RNAi delayed viroid infection in *N. benthamiana* plants (Adkar-Purushothama et al., 2015b). Alternatively, researchers have screened multiple artificial sRNAs targeting sites distributed across the viroid using modern technologies and found them to be effective in controlling PSTVd infection in *N. benthamiana* plants (Carbonell and Daròs, 2017). Without a doubt, having a deeper understanding of viroid-host interactions, and of the role of vd-sRNA in viroid pathogenicity, will help in the development of more efficient crop protection strategies in the near future.

Future implications

Since the first demonstration of vd-sRNA in viroid infected plants (Itaya et al., 2001; Papaefthimiou et al., 2001), many studies have contributed to shedding light on the role of vd-sRNAs in viroid pathogenicity and the host factors that are involved in viroid-induced RNA silencing, as well as permitting the use of RNA silencing as a potential strategy to manage viroid disease. However, there are still a few questions that are completely unanswered and require more attention.

For example, at the beginning of this chapter, we stated that AGO proteins are required for RNA silencing. More specifically, the presence of the 5′ terminal uridine in the PLMVd-PC derived vd-sRNA indicates the involvement of AGO1. However, most of the vd-sRNAs from the three members of family *Pospiviroidae* that were demonstrated to induce the cleavage of target transcripts lack a 5′ terminal uridine, suggesting the possibility of the involvement of an AGO protein other than AGO1. If so, what is the identity of the AGO protein that is involved in pospiviroid-induced RNA silencing? Is PLMVd-sRNA lacking 5′ uridine capable of inducing RNA silencing? What is the efficacity of target cleavage by the two classes of RNA (i.e., those with and without a 5′ terminal uridine)?

A second example is associated with the fact that during viroid infection, both the down- and the upregulation of certain miRNAs are observed. The mechanisms that regulate how miRNAs are either up- or downregulated in the host plant remain elusive. Is the downregulation of miRNA observed in PSTVd-infected plants the result of the vd-sRNA effect, or is it induced by phasiRNA, or by another novel mechanism? Above all, it is important to understand whether viroid-induced RNA silencing triggers the infection, or if is there another mechanism in the host that eventually triggers the RNA silencing mechanism in the host plant. Also, it is not clear how and why different regions of viroids produce different amounts of vd-sRNA. This is another intriguing question that will require attention. More generally, it is interesting to assess the overall contribution of RNA silencing to viroid pathogenicity.

Chapter summary

In this chapter we tried to understand the host special defense mechanism against invading RNA molecules, that is to say RNA silencing. Since it is associated with the biological response to either dsRNA or hpRNA, it is recognized as an important innate immune defense mechanism against invading viruses and viroids. Upon infection, viroids are recognized by the host RNA silencing machinery due to both their structure and RNA/RNA formation during replication. This statement is supported by experiments showing the presence of vd-sRNA in viroid infected plants in different viroid-host combinations. The application of modern molecular biology tools, combined with bioinformatics analyses, has permitted researchers to show that viroids not only produce vd-sRNA, but also that these vd-sRNAs can induce RNA silencing of the host genes containing the target sequence. This includes but is not limited to, the genes that are involved in host defense, transcription factors, and others.

Study questions

1. What is RNA silencing? What are the steps involved in RNA silencing?
2. Even though viroids are single-stranded RNA molecules, how do they induce the host RNA silencing machinery?
3. What are the host factors or enzymes involved in RNA silencing?
4. How are vd-sRNAs generated?
5. Give an example of transcriptional gene silencing caused by viroid infection. How was it proved?
6. Briefly explain the artificial microRNA (amiRNA) experiments designed to prove a specific RNA silencing event.

Further reading

1. Adkar-Purushothama, C. R. and Perreault, J. (2020) Current overview on viroid-host interactions, *Wiley Interdisciplinary Reviews. RNA*, 11(2), p. e1570. https://doi.org/10.1002/wrna.1570.
2. Baulcombe, D. (2004) RNA silencing in plants, *Nature*, 431(7006), pp. 356–363. https://doi.org/10.1038/nature02874.
3. Flores, R. *et al.* (2015) Viroids, the simplest RNA replicons: How they manipulate their hosts for being propagated and how their hosts react for containing the infection, *Virus Research*, 209, pp. 136–145. https://doi.org/10.1016/j.virusres.2015.02.027.
4. Pallas, V., Martinez, G. and Gomez, G. (2012) The interaction between plant viroid-induced symptoms and RNA silencing, *Methods in Molecular Biology*, 894, pp. 323–343.
5. Flores, R., Navarro, B., Delgado, S., Serra, P. and Di Serio, F. (2020) Viroid pathogenesis: a critical appraisal of the role of RNA silencing in triggering the initial molecular lesion, *FEMS Microbiology Reviews*, 44(3), pp. 386–398.
6. Navarro, B., Flores, R. and Di Serio, F. (2021) Advances in viroid-host interactions. *Annual Review of Virology*, 8(1), pp. 305–325.

References

Adenot, X., et al., 2006. DRB4-dependent TAS3 trans-acting siRNAs control leaf morphology through AGO7. Curr. Biol. 16 (9), 927–932. https://doi.org/10.1016/j.cub.2006.03.035.

Adkar-Purushothama, C.R., Perreault, J.-P., 2018. Alterations of the viroid regions that interact with the host defense genes attenuate viroid infection in host plant. RNA Biol., 1–12. https://doi.org/10.1080/15476286.2018.1462653.

Adkar-Purushothama, C.R., Perreault, J.-P., 2019. Suppression of RNA-dependent RNA polymerase 6 favors the accumulation of potato spindle tuber viroid in Nicotiana benthamiana. Viruses 11 (4), 345. https://doi.org/10.3390/v11040345.

Adkar-Purushothama, C.R., Perreault, J., 2020a. Current overview on viroid-host interactions. Wiley Interdiscip. Rev. RNA 11 (2), e1570. https://doi.org/10.1002/wrna.1570.

Adkar-Purushothama, C.R., Perreault, J.P., 2020b. Impact of nucleic acid sequencing on viroid biology. Int. J. Mol. Sci. 21 (15). https://doi.org/10.3390/ijms21155532.

Adkar-Purushothama, C.R., Brosseau, C., et al., 2015a. Small RNA derived from the virulence modulating region of the potato spindle tuber viroid silences callose synthase genes of tomato plants. Plant Cell 27 (8), 2178–2194. https://doi.org/10.1105/tpc.15.00523.

Adkar-Purushothama, C.R., Kasai, A., et al., 2015b. RNAi mediated inhibition of viroid infection in transgenic plants expressing viroid-specific small RNAs derived from various functional domains. Sci. Rep. 5 (1), 17949. https://doi.org/10.1038/srep17949.

Adkar-Purushothama, C.R., Iyer, P.S., Perreault, J.-P., 2017. Potato spindle tuber viroid infection triggers degradation of chloride channel protein CLC-b-like and Ribosomal protein S3a-like mRNAs in tomato plants. Sci. Rep. 7 (1), 8341. https://doi.org/10.1038/s41598-017-08823-z.

Adkar-Purushothama, C.R., Sano, T., Perreault, J.-P., 2018. Viroid derived small RNA induces early flowering in tomato plants by RNA silencing. Mol. Plant Pathol. https://doi.org/10.1111/mpp.12721.

Baulcombe, D., 2004. RNA silencing in plants. Nature 431 (7006), 356–363. https://doi.org/10.1038/nature02874.

Carbonell, A., Daròs, J.-A., 2017. Artificial microRNAs and synthetic trans-acting small interfering RNAs interfere with viroid infection. Mol. Plant Pathol. 18 (5), 746–753. https://doi.org/10.1111/mpp.12529.

Choi, K., et al., 2011. The FRIGIDA complex activates transcription of FLC, a strong flowering repressor in Arabidopsis, by recruiting chromatin modification factors. Plant Cell 23 (1), 289–303.

Dadami, E., et al., 2013. DICER-LIKE 4 but not DICER-LIKE 2 may have a positive effect on potato spindle tuber viroid accumulation in Nicotiana benthamiana. Mol. Plant. https://doi.org/10.1093/mp/sss118.

Di Serio, F., et al., 2010. RNA-dependent RNA polymerase 6 delays accumulation and precludes meristem invasion of a viroid that replicates in the nucleus. J. Virol. 84 (5), 2477–2489. https://doi.org/10.1128/JVI.02336-09.

Diener, T.O., et al., 1993. Mechanism of viroid pathogenesis: differential activation of the interferon-induced, double-stranded RNA-activated, Mr 68 000 protein kinase by viroid strains of varying pathogenicity. Biochimie 75 (7), 533–538. https://doi.org/10.1016/0300-9084(93)90058-Z.

Ding, S.-W., 2010. RNA-based antiviral immunity. Nat. Rev. Immunol. 10 (9), 632–644. https://doi.org/10.1038/nri2824. Nature Publishing Group.

Eamens, A., et al., 2008. RNA silencing in plants: yesterday, today, and tomorrow. Plant Physiol. 147 (2), 456–468. https://doi.org/10.1104/pp.108.117275.

Eamens, A.L., et al., 2009. The Arabidopsis thaliana double-stranded RNA binding protein DRB1 directs guide strand selection from microRNA duplexes. RNA (New York, N.Y.). https://doi.org/10.1261/rna.1646909.

Eamens, A.L., et al., 2014. In Nicotiana species, an artificial microRNA corresponding to the virulence modulating region of Potato spindle tuber viroid directs RNA silencing of a soluble inorganic pyrophosphatase gene and the development of abnormal phenotypes. Virology 450–451, 266–277. https://doi.org/10.1016/j.virol.2013.12.019.

Flores, R., et al., 2005. Viroids and viroid-host interactions. Annu. Rev. Phytopathol. 43 (1), 117–139. https://doi.org/10.1146/annurev.phyto.43.040204.140243.

Flores, R., et al., 2006. Peach latent mosaic viroid: not so latent. Mol. Plant Pathol., 209–221. https://doi.org/10.1111/j.1364-3703.2006.00332.x.

Flores, R., et al., 2015. Viroids, the simplest RNA replicons: how they manipulate their hosts for being propagated and how their hosts react for containing the infection. Virus Res. 209, 136–145. https://doi.org/10.1016/j.virusres.2015.02.027.

Gómez, G., Martínez, G., Pallás, V., 2008. Viroid-induced symptoms in Nicotiana benthamiana plants are dependent on RDR6 activity. Plant Physiol. https://doi.org/10.1104/pp.108.120808.

Guo, Q., et al., 2016. RNA silencing in plants: mechanisms, technologies and applications in horticultural crops. Curr. Genomics 17 (6), 476–489. https://doi.org/10.2174/1389202917666160520103117.

Haas, B., et al., 1988. The 7S RNA from tomato leaf tissue resembles a signal recognition particle RNA and exhibits a remarkable sequence complementarity to viroids. EMBO J. 7 (13), 4063–4074. Available from: http://www.ncbi.nlm.nih.gov/pubmed/2468486.

Inoue, H., Li, M., Schnell, D.J., 2013. An essential role for chloroplast heat shock protein 90 (Hsp90C) in protein import into chloroplasts. Proc. Natl. Acad. Sci. 110 (8), 3173–3178. https://doi.org/10.1073/pnas.1219229110.

Itaya, A., et al., 2001. Potato spindle tuber viroid as inducer of RNA silencing in infected tomato. Mol. Plant Microbe Interact. 14 (11), 1332–1334. https://doi.org/10.1094/MPMI.2001.14.11.1332.

Kiss, T., Solymosy, F., 1982. Sequence homologies between a viroid and a small nuclear RNA (snRNA) species of mammalian origin. FEBS Lett. 144 (2), 318–320. https://doi.org/10.1016/0014-5793(82)80662-5.

Lee, C.H., Carroll, B.J., 2018. Evolution and diversification of small RNA pathways in flowering plants. Plant Cell Physiol. https://doi.org/10.1093/pcp/pcy167.

Li, J., et al., 2005. Methylation protects miRNAs and siRNAs from a 3′-end uridylation activity in Arabidopsis. Curr. Biol. 15 (16), 1501–1507. https://doi.org/10.1016/j.cub.2005.07.029.

Lindbo, J.A., et al., 1993. Induction of a highly specific antiviral state in transgenic plants: implications for regulation of gene expression and virus resistance. Plant Cell, 1749–1759. https://doi.org/10.1105/tpc.5.12.1749.

Markarian, N., et al., 2004. RNA silencing as related to viroid induced symptom expression. Arch. Virol. 149 (2), 397–406. https://doi.org/10.1007/s00705-003-0215-5.

Martínez de Alba, A.E., et al., 2002. Two chloroplastic viroids induce the accumulation of small RNAs associated with posttranscriptional gene silencing. J. Virol. 76 (24), 13094–13096. https://doi.org/10.1128/JVI.76.24.13094-13096.2002.

Matzke, M.A., Kanno, T., Matzke, A.J.M., 2015. RNA-directed DNA methylation: the evolution of a complex epigenetic pathway in flowering plants. Annu. Rev. Plant Biol. 66 (1), 243–267. https://doi.org/10.1146/annurev-arplant-043014-114633.

Michaels, S.D., Bezerra, I.C., Amasino, R.M., 2004. FRIGIDA-related genes are required for the winter-annual habit in Arabidopsis. Proc. Natl. Acad. Sci. U. S. A. 101 (9), 3281–3285.

Millar, A.A., Waterhouse, P.M., 2005. Plant and animal microRNAs: similarities and differences. Funct. Integr. Genomics 5 (3), 129–135. https://doi.org/10.1007/s10142-005-0145-2.

Navarro, B., et al., 2012. Small RNAs containing the pathogenic determinant of a chloroplast-replicating viroid guide the degradation of a host mRNA as predicted by RNA silencing. Plant J. 70 (6), 991–1003. https://doi.org/10.1111/j.1365-313X.2012.04940.x.

Nishikura, K., 2001. A short primer on RNAi. Cell 107 (4), 415–418. https://doi.org/10.1016/S0092-8674(01)00581-5.

Pallas, V., Martinez, G., Gomez, G., 2012. The interaction between plant viroid-induced symptoms and RNA silencing. Methods Mol. Biol. 894, 323–343.

Papaefthimiou, I., et al., 2001. Replicating potato spindle tuber viroid RNA is accompanied by short RNA fragments that are characteristic of post-transcriptional gene silencing. Nucleic Acids Res. 29 (11), 2395–2400. https://doi.org/10.1093/nar/29.11.2395.

Peragine, A., et al., 2004. SGS3 and SGS2/SDE1/RDR6 are required for juvenile development and the production of trans-acting siRNAs in Arabidopsis. Genes Dev. 18 (19), 2368–2379. https://doi.org/10.1101/gad.1231804.

Sanan-Mishra, N., et al., 2021. Secondary siRNAs in plants: biosynthesis, various functions, and applications in virology. Front. Plant Sci. 12. https://doi.org/10.3389/fpls.2021.610283.

Schwach, F., et al., 2005. An RNA-dependent RNA polymerase prevents meristem invasion by potato virus X and is required for the activity but not the production of a systemic silencing signal. Plant Physiol. 138 (4), 1842–1852. https://doi.org/10.1104/pp.105.063537.

Shivaprasad, P.V., Chen, H.-M., Patel, K., 2012. A microRNA superfamily regulates nucleotide binding site – leucine-rich repeats and other mRNAs. Plant Cell 24 (3), 859–874. https://doi.org/10.1105/tpc.111.095380.

Symons, R.H., et al., 1987. Self-cleavage of RNA in the replication of viroids and virusoids. J. Cell Sci. Suppl. 7, 303–318.

Talmor-Neiman, M., et al., 2006. Identification of trans-acting siRNAs in moss and an RNA-dependent RNA polymerase required for their biogenesis. Plant J. 48 (4), 511–521. https://doi.org/10.1111/j.1365-313X.2006.02895.x.

Tang, Y., et al., 2010. Virus-based microRNA expression for gene functional analysis in plants. Plant Physiol. 153 (2), 632–641. https://doi.org/10.1104/pp.110.155796.

van der Krol, A.R., et al., 1990. Flavonoid genes in petunia: addition of a limited number of gene copies may lead to a suppression of gene expression. Plant Cell 2 (4), 291–299. https://doi.org/10.1105/tpc.2.4.291.

Wang, M.-B., et al., 2012. RNA silencing and plant viral diseases. Mol. Plant-Microbe Interact. 25 (10), 1275–1285. https://doi.org/10.1094/MPMI-04-12-0093-CR.

Wassenegger, M., et al., 1994. RNA-directed de novo methylation of genomic sequences in plants. Cell 76 (3), 567–576. https://doi.org/10.1016/0092-8674(94)90119-8.

Xie, Z., et al., 2005. DICER-LIKE 4 functions in trans-acting small interfering RNA biogenesis and vegetative phase change in Arabidopsis thaliana. Proc. Natl. Acad. Sci. 102 (36), 12984–12989. https://doi.org/10.1073/pnas.0506426102.

Zhai, J., Jeong, D.-H., De Paoli, E., et al., 2011. MicroRNAs as master regulators of the plant NB-LRR defense gene family via the production of phased, trans-acting siRNAs. Genes Dev. 25 (23), 2540–2553. https://doi.org/10.1101/gad.177527.111.

Zhang, H., He, X., Zhu, J.-K., 2013. RNA-directed DNA methylation in plants. RNA Biol. 10 (10), 1593–1596. https://doi.org/10.4161/rna.26312.

Zheng, Y., et al., 2017. Comprehensive transcriptome analyses reveal that potato spindle tuber viroid triggers genome-wide changes in alternative splicing, inducible trans-acting activity of phased secondary small interfering RNAs, and immune responses. J. Virol. 91 (11). https://doi.org/10.1128/JVI.00247-17. Edited by A. E. Simon. JVI.00247-17.

C. Viroid pathogenesis and viroid-host interaction

16

Detection of viroids

Zhixiang Zhang and Shifang Li

Institute of Plant Protection, Chinese Academy of Agricultural Sciences, Beijing, China

Graphical representation

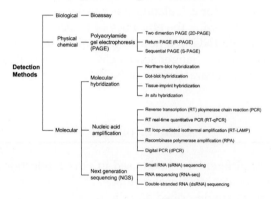

Definitions

Limit of detection (LOD): the lowest concentration of the target in a sample that can be consistently detected with a stated probability (typically at 95% certainty).

Detection sensitivity: the minimum number of copies of the target in a sample that can be accurately measured with an assay. Typically, sensitivity is expressed as LOD.

Detection specificity: the assay detects the appropriate target rather than the other, nonspecific targets in a sample.

Indicator plant: a host plant of a viroid used for bioassays. Indicator plants should be easy to infect and express specific symptoms in a short time.

Viroid structure: refers to the conformations formed by sequences of viroid genomic RNA (primary structure). Single-stranded circular RNA of the viroid genome can form a compact conformation (secondary structure) due to intramolecular Watson-Crick base pairing between self-complementary regions and can further form a three-dimensional arrangement (tertiary structure). Conformation change of viroid RNA is the base for the detection of detecting polyacrylamide gel electrophoresis (PAGE). For more details refer to Chapter 3 "Structure of Viroids."

Molecular hybridization: the process of the formation of a partially or wholly complementary nucleic acid duplex by the association of single strands, usually between DNA and RNA strands, but also between RNA strands; used to detect and isolate specific sequences, measure homology, or define other characteristics of one or both strands. The kinetics of this process are governed by the sequence similarity of two nucleic acid strands, solvent type, solvent ionic strength, temperature, and time.

Melting temperature: the point in a DNA hybridization reaction at which 50% of the nucleotides are annealed (linked) to their complement (i.e., are double-stranded), which is a function of the stringency of the hybridizing conditions.

Threshold cycle (Ct): is the cycle number at which the fluorescent signal of quantitative PCR exceeds the threshold. It is used to calculate the starting DNA copy number because the Ct value is inversely related to the starting amount of the target.

Standard curve: is established by a dilution series of known template concentrations. The log of each known concentration in the dilution series (x-axis) is plotted against the Ct value for that concentration (y-axis). It is used for calculating the starting amount of the target template in experimental samples or for assessing the efficiency of nucleic acid amplification.

Next-generation sequencing (NGS): a new DNA sequencing technology relative to first-generation sequencing technology, Sanger sequencing. It is also called high-throughput sequencing. The most important characteristic of NGS is the ability to sequence millions of DNA simultaneously.

Chapter outline

This chapter introduces different methods used in viroid detection, including biological, physical-chemical, and molecular biological methods. Their basic principles, advantages, disadvantages, and applications are described and discussed. The final part includes considerations for method selection and prospects for viroid detection.

Learning objectives

- To understand available methods of viroid detection
- To understand the basic principle of each method
- To understand the advantages and disadvantages of each method
- The ability to select a proper detection method in a specific situation

Fundamental introduction

The purpose of viroid detection is to examine whether viroids exist in a sample(s) (qualitative manner) or to measure the amount of viroid in a sample(s) (quantitative manner) by certain methods. One assumption underlying viroid detection is that the detected viroid is known, meaning that once it is identified, we can learn about its biological, physical-chemical, and molecular features (see corresponding chapters in this book). To detect the viroid, it is necessary to understand and follow certain methods that include but are not limited to sample collection, preparation of samples for detection, and the selection of appropriate detection methods. Certain specialized detection laboratories must use standard methods to detect viroid(s) from specific samples. The problem then becomes how to perform the detection. This chapter addresses this question and discusses how to collect and prepare samples and the selection of appropriate detection methods depending on the host plant or viroid type. Considerations for sample collection and preparation involve the type of plant tissue, the amount of the collected tissue, and the time needed to collect tissue. To select the proper detection method(s), we must learn about the basic principles, advantages and disadvantages, and application ranges of all available detection methods.

Sample collection and preparation

Once a viroid enters a plant, it establishes a systemic infection, indicating that all tissues of the infected plant contain viroid RNAs. However, concentrations of viroid RNAs in different tissues are not the same. For example, apple scar skin viroid (ASSVd) reaches higher concentrations in apple bark than in fruit pericarp (Li et al., 1995), and hop stunt viroid (HSVd) reaches much higher concentrations in mature apricot fruits than in leaves (Astruc et al., 1996). Certain tissues of infected plants could contain low concentrations of viroid RNAs that are not enough for detection. To ensure the reliability of detection, tissues with higher concentrations of viroid RNAs should be collected.

The amount of collected tissue primarily depends on the sensitivity of the detection method. Generally, polyacrylamide gel electrophoresis (PAGE) and bioassays require a larger amount of plant tissue than molecular detection methods, such as molecular hybridization and reverse transcription polymerase chain reaction (RT-PCR).

Collecting samples in the proper season can also increase the success of viroid detection. Concentrations of viroid RNAs in tissues change depending on the growing season. For instance, monitoring the detection of HSVd in apricot trees during a whole year revealed a significant decrease in the HSVd concentration during September to November (Amari et al., 2001). Seasonal differences in concentration have also been observed for several citrus viroids in citrus.

In most methods, collected plant tissues cannot be directly used to detect viroids but must extract RNAs from plant tissues. RNA extraction aims to separate RNAs (including viroid RNAs) from the cell contents. Cell contents are usually released by mechanical grinding. Importantly, it is harder to grind tough tissues such as dry seeds and branches and bark of woody plants than to grind soft tissues such as flowers and young leaves. Released cell contents are composed of complex constituents, including genomic DNA, proteins, and different metabolites such as polysaccharides and polyphenols, which can interfere with viroid detection. Separating RNAs from these cellular constituents is usually laborious and time-consuming, and sometimes is an important limit in detection methods, especially those aiming for on-site detection. Although many different methods have been developed for RNA extraction, none are suitable for all situations. Currently, RNA extraction methods using cetyltrimethylammonium bromide (CTAB) (Chang et al., 1993) and TRIzol reagents are the most widely used. Modifications or optimizations are usually made for different plants, and many commercial products are also available.

Detection methods

Viroid detection methods can be broadly divided into three classes: biological, physical-chemical, and molecular (nucleic acid). The first two include bioassays and PAGE, respectively; there are three subclasses of molecular methods: molecular hybridization, nucleic acid amplification, and next-generation sequencing (NGS). Each subclass now has several different methods (Fig. 1).

Bioassay

Bioassay on indicator plants is the earliest method developed for the detection of viroids. It was first developed to diagnose potato spindle tuber disease (Raymer and O'Brien, 1962) before the discovery of PSTVd. Bioassay still plays a major role in viroid detection.

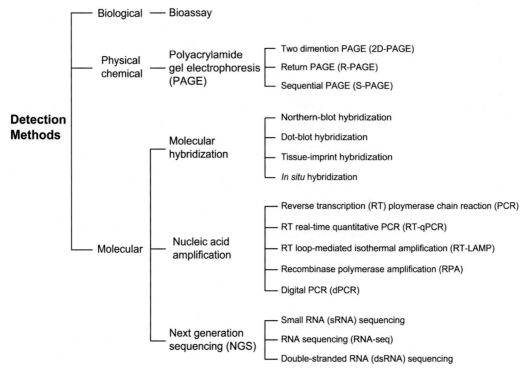

FIG. 1   Methods available for viroid detection. They are classified into three groups (biological, physical-chemical, and molecular) based on their basic principles and features.

A bioassay is based on the specific symptom(s) induced by the viroid that are then detected. Viroids that can be detected by a bioassay must be able to induce visible symptoms on at least one of its host plants, and at least one of these symptoms must be distinguished. Due to these limitations, not all viroids can be detected by the bioassay method. Nie and Singh (2017) summarized the viroids that can be identified by a bioassay.

In bioassays, the detected viroid is inoculated into its indicator plant. Inoculation can be performed through grafting, mechanical rubbing of leaves, or the cutting or slashing of stems. The leaf rubbing method is the most widely used. Except for grafting, other inoculation methods require the preparation of inoculum. While both crude extracts of plant tissue and extracted RNAs can be used as inoculum, crude extracts are simpler and faster to prepare and are thus used more often. Inoculated plants usually require controllable environmental conditions, similar to those in a greenhouse, to grow and express the distinguished symptom, because these symptoms are sensitive to temperature, humidity, light intensity, and day length. The need for specific environmental conditions sometimes limits the use of the bioassay method, especially for large-scale samples.

Disadvantages of the bioassay method include reliability and time required. Except for environmental conditions, other viroids or viruses coexisting in the sample can influence the viroid-specific symptoms, and sometimes, it is difficult to judge detection results based on symptoms. The growing and development of inoculated plants requires time, as does the expression of symptoms, and can take several weeks, months, or even years for woody plants. However, bioassays do not need advanced or expensive facilities and are simple to perform. Although this method is rarely used in routine viroid detection, it plays an important role in some countries. It is important to note that a bioassay is needed to verify the infectivity and biological properties of a new viroid (Owens et al., 2012a).

Polyacrylamide gel electrophoresis (PAGE)

Viroid detection using the PAGE method is based on the structural transformation of genomic RNA. Under natural conditions, viroid genomic RNA forms a compact structure due to its highly self-complementary nature; under denaturation conditions, the compact structure transforms to a loose structure, single-stranded circular or linear RNA, due to the opening of complementary base pairs (Fig. 2A). In PAGE, circular viroid RNAs migrate more slowly than most linear plant RNAs, including rRNA, tRNA, and mRNA. Thus, viroid RNAs will be separated from plant RNAs.

There are three different methods to detect viroid using PAGE (Fig. 2B): sequential PAGE (S-PAGE) (Schumacher et al., 1983), two-dimensional PAGE (2D-PAGE) (Schumacher et al., 1983), and return PAGE (R-PAGE) (Schumacher et al., 1986). All these methods require two electrophoresis treatments, i.e., the first under native conditions and the second under denaturation conditions. Denaturation conditions are achieved by high concentrations of urea for 2D-PAGE and S-PAGE and by high temperatures for R-PAGE.

The S-PAGE and 2D-PAGE methods are similar to operate, but different from that of R-PAGE. Broadly, the former two methods require gel cutting after the first electrophoresis, while the latter method does not. In S-PAGE, the gel containing different sample lanes, between 250 bp and 500 bp of molecular weight marker of DNA, is cut horizontally

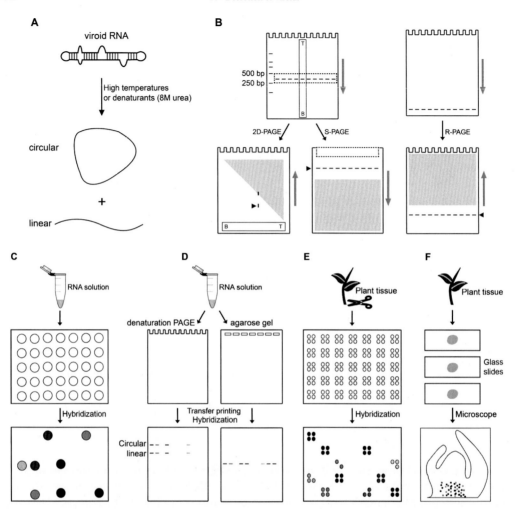

FIG. 2   PAGE detection and molecular hybridization of viroids. (A) Structural change of viroids under denaturation conditions. (B) Diagrams for two-dimensional PAGE (2D-PAGE), sequential PAGE (S-PAGE), and return PAGE (R-PAGE). The long, grey arrows indicate the direction of electrophoresis. "T" and "B" are abbreviations for "top" and "bottom." (C) Dot-blot hybridization; the example at the bottom is the detection result. (D) For Northern-blot hybridization, RNA solutions can be separated by denaturation PAGE and agarose gel electrophoresis, and hybridization results will show two bands (circular and linear viroid RNAs) and one band, respectively. (E) Tissue-imprint hybridization. (F) In situ hybridization, where the schematic drawing is the detection of viroids in the shoot apical meristem of plant.

and then placed, horizontally, at the top of the gel of the second electrophoresis. In 2D-PAGE, the gel of only one sample lane is cut vertically and then placed horizontally at the bottom or top of the gel of the second electrophoresis. In R-PAGE, the whole gel from the first electrophoresis is directly used for the second electrophoresis, which uses a heated electrophoresis buffer to denature the RNA. After the second electrophoresis, the RNA in the gel is visualized by staining with silver or nucleic acid dye such as ethidium bromide. Staining with nucleic acid dye is faster and easier than using silver staining but is less sensitive (Owens et al., 2012b).

Compared to the bioassay method, PAGE detection requires less time, can be used with more viroids, and is more reliable. This detection can be performed in a laboratory without the influence of environmental conditions and can be finished in several hours rather than the weeks or months needed for a bioassay. All viroids can be detected by PAGE, at least in principle. Therefore, the PAGE method has been widely used in the routine detection of different viroids and has played a critical role in controlling viroid diseases and eradicating viroids in some regions. However, since molecular hybridization was first applied to viroid detection in the early 1980s (Owens and Diener, 1981), PAGE has been gradually replaced by molecular methods that are quicker and have both higher sensitivities and specificities. Although the PAGE method is rarely used in routine detection, it remains important for identifying new viroids because it is independent of the viroid sequence.

Molecular hybridization

Serological methods commonly used in plant virus detection, such as enzyme-linked immunosorbent assay (ELISA), are not suitable for viroid detection because viroids are not known to code proteins. The alternative method for viroid detection is molecular hybridization (Owens and Diener, 1981), which is based on the interaction of nucleic acids but not proteins. This method needs a nucleic acid probe, the sequences of which are the same or highly similar to those of viroid RNAs but has an opposite polarity. This probe is labeled with markers that are easy to detect. A viroid can be detected through the probe marker when it specifically interacts with its probe through base complementation.

The strength of the interaction between viroid RNAs and the probes is critical for detection and is primarily determined by the number and type of base pairs formed between them. Naturally, more base pairs will generate a stronger interaction between two nucleic acid molecules. Moreover, RNA/RNA hybrids are more stable than DNA/RNA hybrids because the A/U base pair has a stronger interaction than the A/T base pair. Similarly, hybrids with a higher GC content are more stable because the G/C base pair has a stronger interaction than the A/U base pair. As such, a relatively longer RNA probe with a higher GC content binds to viroid RNAs with a stronger interaction.

There are three settings when preparing a viroid probe: (i) RNA or DNA probe: a DNA probe is more stable and easier to handle and store; an RNA probe is more sensitive due to the stronger interaction of RNA/RNA hybrid. (ii) Radioactive or nonradioactive probe: A radioactive probe, such as ^{32}P-labeled, is sensitive and easy to detect but has strict requirements for handling and a short half-life. Nonradioactive probes, such as digoxigenin (DIG)-labeled, are more stable, do not have special security requirements, and have similar detection sensitivity to radioactive probes (Astruc et al., 1996; Podleckis et al., 1993). (iii) Specific probe or polyprobe/universal probe: A specific probe normally detects one viroid, while a polyprobe and universal probe can simultaneously detect several different viroids, e.g., polyprobes are used to detect viroids infecting citrus, pome, and stone fruit trees, grapevine, and tomato (Pallas et al., 2017) and the universal probe is capable of detecting six coleviroids (Jiang et al., 2013). A polyprobe is constructed by the tandem connection of cDNA sequences of different viroids (Herranz et al., 2005); the universal probe is normally synthesized from the conserved sequences of the viroids of the same genus, such as the central conserved region (CCR).

Currently, the DIG-labeled probe is the most widely used in viroid detection. DIG is detected by an antidigoxigenin antibody conjugated-alkaline phosphatase that can cleave either chromogenic or chemiluminescent substrates to produce color or light. Chromogenic detection based on visible color does not need special detection equipment, has a relatively lower detection sensitivity, and could experience interference due to some colored cellular components. In contrast, chemiluminescent detection based on light requires special optical devices, has a higher detection sensitivity, and experiences little interference from colored cellular components. In practice, chemiluminescent detection is more common.

Four different molecular hybridization methods have been developed for viroid detection (Fig. 2C–F) based on the sample preparation.

(i) Dot-blot hybridization

Drops of prepared RNA solution are directly added to a nitrocellulose or nylon membrane (Fig. 2C). This method is high-throughput and simple to operate, making it suitable for large-scale detection (Owens and Diener, 1981). Both positive and negative controls are necessary because cellular components in prepared RNA solution, such as protein, plant genomic DNA, and secondary metabolites, could interfere with the hybridization reaction resulting in false-positive or false-negative results. Additional or special treatment for purifying RNA can improve detection accuracy. For example, the removal of proteins eliminates false positive results during the detection of peach latent mosaic viroid (PLMVd) (Xu et al., 2009).

(ii) Northern-blot hybridization

RNA preparations are first separated by gel electrophoresis and then transferred to a membrane (Fig. 2D). Gel electrophoresis increases detection accuracy and reliability because of the separation of viroid RNAs from cellular components. Although this method is not suitable for large-scale detection due to the complex nature of gel electrophoresis, it is important for routine detection, especially for verifying the results of other detection methods. Additionally, this method is commonly used in basic viroid research. Using denaturation PAGE, it can distinguish circular and linear forms of viroid RNAs, identify different viroids and different viroid variants (Sano et al., 1985), and detect other viroid-derived RNA molecules including subgenomic RNAs (Minoia et al., 2015), double-stranded replication intermediates (Branch et al., 1981; Owens and Diener, 1982), and small RNAs (Itaya et al., 2001; Papaefthimiou et al., 2001).

(iii) Tissue-imprint hybridization

Plant tissues such as those in the stem, leaf, or fruit are directly pressed onto membranes and viroid RNAs will transfer onto the membrane with plant sap (Fig. 2E). The absence of RNA extraction makes this method simpler and faster, but also reduces detection sensitivity due to the influence of cellular components and the low amount of viroid RNAs in the plant sap. This method is usually used for the preliminary screening of large-scale field samples (Gucek et al., 2017; Pallas et al., 2017).

(iv) In situ hybridization

The reaction of in situ hybridization occurs in tissues or cells (Fig. 2F). The hybridization signal is detected by electrons, light, or confocal laser scanning microscopy. This method is mainly used to study viroid subcellular localization (Harders et al., 1989) and trafficking in plant tissues (Zhu et al., 2001).

Nucleic acid amplification methods

Viroids sometimes accumulate at low levels beyond the detection limits of the PAGE and molecular hybridization methods, especially in some woody plants. Under these conditions, if the low-level viroid RNAs can be copied many times, to amplify the amount of viroid RNAs (like viroid replication), the viroid can be detected. During a bioassay, the amount of viroid RNA increases by autonomous replication in plants. However, the rate of viroid replication is low since it is restricted by host and environmental conditions. Now, the rapid and stable increase of the amount of viroid RNA can be achieved in vitro by nucleic acid amplification methods. The invention of polymerase chain reaction (PCR) (Saiki et al., 1988) has made these methods possible.

(i) PCR

PCR amplifies DNA molecules using thermostable DNA polymerase, a pair of oligonucleotide primers, and deoxyribonucleotide triphosphates (dNTPs). Both double-stranded DNA (dsDNA) and single-stranded complementary DNA (cDNA) of RNA can be used as PCR templates. Amplification starts with a denaturation reaction under high temperatures, in which dsDNA is split into two complementary single-stranded DNA molecules. Next, the reaction temperature decreases, and the pair of primers individually bind to 3'-ends of the two single-stranded DNA molecules (annealing) and extend along the DNA forced by DNA polymerase (primer extension), producing two copies of the dsDNA. These two newly produced dsDNA copies will be used as templates for the next amplification cycle. Finally, millions of dsDNA copies are produced past a certain number of amplification cycles. The amplification products can be detected at the end point of PCR or in real-time during PCR. Endpoint methods detect amplification products when PCR finishes, by gel electrophoresis followed by nucleic acid stain; real-time methods dynamically monitor the amount of amplification products during PCR amplification using optical instruments (see RT-qPCR).

The pair of primers, two short single-stranded DNA molecules, are complementary with 3'-ends of each strand of the dsDNA template. They determine the length of the PCR products and the efficiency and specificity of PCR amplification. Thus, primer design and optimization are critical for PCR. The primer design has been thoroughly discussed (Hadidi and Candresse, 2003). Many professional software packages such as Primer3Plus (Untergasser et al., 2012) are available for primer design.

(ii) RT-PCR

PCR only amplifies double- or single-stranded DNA molecules, but not RNA molecules. However, after reverse transcription (RT) of RNA into cDNA, PCR can be performed. As such, viroid RNAs can be detected by PCR after an RT reaction (RT-PCR).

RT reactions are catalyzed by reverse transcriptase with dNTPs and a primer, including random primers, universal primers, and specific primers. Random primers are applicable to all kinds of RNA. RT products are not only used to detect the viroid of interest but can also be used to detect other viroids or other plant pathogens. For example, a random primer was used in the simultaneous detection of five potato viruses and PSTVd by RT-PCR (Nie and Singh, 2001). A universal primer is usually used to detect the viroids of a genus in the family *Pospiviroidae*, which have conserved sequences such as CCR. For instance, the universal primers of pospi1-RE/FW located in CCR were designed to detect several pospiviroids (Verhoeven et al., 2004). A specific primer is used to detect a particular viroid. To confirm the success of the RT reaction, endogenous control (e.g., a housekeeping gene) is recommended.

RT reactions and PCR can be performed individually (two-step RT-PCR) or simultaneously (one-step RT-PCR). It is easier to develop and optimize the two-step method because there is no interference between the RT reaction and PCR. The first application of RT-PCR in viroid detection used the two-step method (Hadidi and Yang, 1990). Moreover, in the two-step method, the products of an RT reaction using random primers can be used for other purposes such as detecting other coinfected viroids or viruses. One-step RT-PCR is simpler and faster because it has fewer steps, and thus is better for large-scale detection.

By multiplex RT-PCR, different viroids coinfecting are simultaneously detected (Ito et al., 2002; Nie and Singh, 2001; Pallas et al., 2018). During multiplex RT-PCR, different viroids are amplified all at once in a single tube of PCR solution. The amplification products of these viroids have different sizes and can thus be distinguished by gel electrophoresis. Compared with single RT-PCR, multiplex detection is cheaper and faster. However, this method is difficult to develop because of the negative influences between the amplification reactions of viroids.

RT-PCR is performed in a small reaction volume and only requires a few hours. As such, it is simpler and faster than the PAGE and molecular hybridization methods. Importantly, RT-PCR is more sensitive. It can detect trace amounts of the template in small samples because the template is exponentially amplified more than a million-fold. In the RT-PCR assay of ASSVd developed for pome fruit trees, the required total RNA is as low as 1 to 100 pg, and the detection sensitivity is 10- to 100-fold higher than molecular hybridization using cRNA probes and 2500-fold higher than R-PAGE (Hadidi and Yang, 1990).

(iii) RT-qPCR

RT-PCR detection entails gel electrophoresis, which takes time and increases the risk of cross-contamination from PCR products. RT-qPCR avoids these issues by real-time detection of the products directly in a closed tube. Real-time detection relies on fluorescent substances that can label PCR products. Currently, PCR products are mainly labeled by nucleic acid dyes (e.g., SYBR Green 1) and hydrolysis probe (e.g., TaqMan probes).

SYBR Green 1 is a fluorescent dye that can intercalate into the minor groove of dsDNA. Free SYBR Green 1 emits little fluorescence but will emit strong fluorescence upon binding to dsDNA. The strength of fluorescence is positively correlated with the amount of dsDNA binding dyes. Thus, an increase in amplification products can be monitored based on increases in the fluorescence strength during PCR. However, SYBR Green 1 also binds with primer dimers and other nonspecific products generated during PCR and emits fluorescence. Therefore, primers must be well-designed and optimized to avoid nonspecific amplification. In real-time detection using SYBR Green 1, the specificity is typically evaluated by assessing the melting temperature (Tm) of the final products. Specific products have a distinctive Tm.

A hydrolysis probe is an oligonucleotide that is complementary with one strand of dsDNA template, the 5′ and 3′ ends of which are labeled with a fluorescent reporter dye and a fluorescent quencher dye. Intact probes emit little or no fluorescence because the quencher dye covers the fluorescence of the reporter dye. Once the probe is cleaved, the reporter dye will emit strong fluorescence. During PCR, probes anneal to products and are cleaved by DNA polymerase in the extension phase due to the 5′ to 3′ exonuclease activity of DNA polymerase, resulting in strong fluorescence. Compared with the SYBR Green 1 method, the hydrolysis probe method is more expensive but has higher specificity. The first application of RT-qPCR in viroid detection was based on the hydrolysis probe method (Boonham et al., 2004).

In addition to real-time detection, RT-qPCR can quantify the amount of starting template. In principle, the template amount doubles in each PCR amplification cycle and will reach 2^n after n cycles. As such, the amount of starting template can be calculated based on the amount of final products and the number of cycles. In fact, amplification is always affected by PCR inhibitors, reagent consumption, and the accumulation of by-products. Double increase (exponential amplification) cannot occur in each cycle. In late phases, exponential amplification does not occur but reaches a plateau. Thus, the amount of starting template cannot be precisely calculated based on the amount of final PCR products. However, there is a phase of exponential amplification (log or exponential phase) before the plateau phase. In this phase, the amount of starting template can be calculated based on the amount of PCR products that are in proportion to the strength of fluorescence measured in real-time during qPCR.

The amount of starting template is calculated by the threshold cycle (Ct) value that is the cycle number at which the reporter fluorescence exceeds a threshold. The threshold is typically 10 times the standard deviation of the baseline, and is defined as the cycles in which a reporter fluorescence is accumulating, but is still beneath the limits of detection of the instrument (Heid et al., 1996). By default, computer software sets the baseline from cycles 3 to 15. However, this must frequently be manually changed. The Ct value has a negative relationship with the amount of starting template. A lower starting template requires more cycles to generate enough products to emit detectable fluorescence.

Quantification can be achieved in absolute and relative ways. Absolute quantification precisely calculates the copy number of the starting template according to a standard curve (Heid et al., 1996; Higuchi et al., 1993). The standard curve is a straight line showing the linear relationship between the logarithm of copy numbers of a set of calibrators diluted by five- or ten-fold and the Ct values of these calibrators. The calibrator is a solution for the detection target of a known concentration. As such, in the standard curve of the calibrator, each Ct value corresponds to a copy number. If the amplification efficiency is the same for the detected sample and the calibrator, the copy numbers of the starting template in the detected samples will be calculated by their Ct values according to the standard curve.

Relative quantification does not calculate the copy number of the starting template but the change in the amount of starting template between two samples. It does not need a standard curve but requires a housekeeping gene as the reference (reference gene) for normalization, which is performed by subtracting the Ct of the target from the Ct of the reference gene, resulting in ΔCt. The relative amount of target in the two samples is then given by $2^{-\Delta\Delta Ct}$, where $\Delta\Delta Ct = \Delta Ct(sample1) - \Delta Ct(sample2)$. To ensure this calculation is valid and reliable, the PCR amplification efficiencies of the reference gene and the target should be approximately equal to or exceed 90%.

Compared with RT-PCR, RT-qPCR is simpler and faster, has higher detection throughput, and has a lower risk of PCR product crossover contamination. It is more sensitive than

RT-PCR, by 100 times for PSTVd detection (Boonham et al., 2004). Thus, RT-qPCR was quickly adopted after it was first used (Boonham et al., 2004) and is now widely used in viroid detection. It is especially suitable for the large-scale detection of low-concentration viroids. A typical example is the detection of pospiviroids in the seeds of solanaceous crops. Pospiviroids can cause serious diseases in potatoes and tomatoes (Verhoeven et al., 2004) and thus are strictly regulated in many countries. Seed transmission is considered a pathway for the introduction and spread of pospiviroids (Matsushita and Tsuda, 2016). To meet the demand of some countries for the pospiviroid detection of seed lots of solanaceous crops, RT-qPCR assays were developed (Botermans et al., 2020; Yanagisawa et al., 2017). Moreover, fluctuations in the viroid concentration of different tissues or at different times can be monitored by RT-qPCR. For example, a rapid and sensitive RT-qPCR assay based on SYBR Green I was developed to determine the concentrations of citrus viroid III (CVd-III) at different times in the green bark of four different seedlings (Rizza et al., 2009).

(iv) Digital PCR (dPCR)

dPCR is a precise nucleic acid amplification method that can realize real absolute quantification (Sykes et al., 1992; Vogelstein and Kinzler, 1999) and uses a "divide and conquer" strategy (Baker, 2012): a sample is diluted and partitioned into hundreds or even millions of separate reaction units so that each contains one or no copies of the target DNA. The copy number of the target DNA in the original sample is precisely determined by counting the number of positive reaction units (in which the DNA is detected) versus the negative ones (in which it is not). Currently, sample partition is performed either on chips (chip-based dPCR) or through water-in-oil emulsions or droplets (droplet dPCR; ddPCR) (Pinheiro et al., 2012).

dPCR is not only more precise than qPCR (Hindson et al., 2011) but is simpler because it does not need a standard curve, as in qPCR. It can precisely determine extremely low concentrations of target DNA in a sample even in a complex background (Huggett et al., 2013). The most important feature of dPCR is absolute quantification, unlike the relative "absolute quantification" of qPCR. Therefore, dPCR is suitable for situations requiring high precision (Huggett et al., 2013) such as the screening of important virus-free propagating materials. However, dPCR is still in its infancy, is expensive, and has a low detection throughput. Currently, it is only used for detecting peach latent mosaic viroid (PLMVd) (Lee et al., 2021).

(v) Nucleic acid isothermal amplification

Although RT-PCR and RT-qPCR are widely used for viroid detection, they are not suitable for on-site detection in the field because denaturation, annealing, and elongation reactions require various and precise temperatures supplied by a large, heavy, and expensive thermocycler. This limitation has been overcome by nucleic acid isothermal amplification, which is performed under a constant temperature. Since the early 1990s, dozens of isothermal amplification methods have been developed (Craw and Balachandran, 2012; Zhao et al., 2015), most of which have impressive detection sensitivity. Some of these have achieved commercial success. Among them, loop-mediated isothermal amplification (LAMP) and recombinase polymerase amplification (RPA) are most often used in viroid detection.

(a) LAMP

LAMP employs a DNA polymerase with high strand-displacement activity and four primers that recognize six distinct sequences on a template. The four primers contain two inner and two outer primers. The six sequences on one strand of the template covered by the four primers are named F1, F2, and F3 at the 5'-end and B1c, B2c, and B3c at the 3'-end, and their complementary sequences on another strand of the template are F1c, F2c, F3c at the 3'-end, and B1, B2, and B3 at the 5'-end (Fig. 3). There are approximately 40 nucleotides between F2 and F1, as well as B2 and B1. The forward inner primer (FIP) contains F2 and F1c; the backward inner primer (BIP) contains B2 and B1c. F3 and B3 are two outer primers. The PrimerExplorer online software was developed for designing LAMP primers (http://primerexplorer.jp/e/).

The basic principles of LAMP are available on the Eiken Chemical Co., Ltd. website (http://loopamp.eiken.co.jp/e/index.html). In brief, the LAMP reaction occurs at 60–65°C and includes three phases producing starting materials, elongation, and cycling amplification. The starting template is assumed to be dsDNA. Inner primer BIP anneals to B2c of one strand of dsDNA (structure 1) and synthesizes a complementary strand (structure 2), then outer primer B3 anneals to B3C and initiates strand-displacement DNA synthesis, resulting in a dsDNA (structure 3) and a BIP-linked ssDNA. This ssDNA forms a loop at the 5'-end (structure 4) due to the complementary interaction between B1c and B1. Next, structure 4 serves as the template for DNA synthesis initiated by inner primer FIP and the strand-displacement DNA synthesis initiated by the out primer F3, resulting in a dsDNA (structure 5) and an ssDNA. This ssDNA has two loops at its ends, forming a dumbbell-like structure (structure 6). Structure 6 will serve as the starting material for cycling amplification (see the circle formed by structures 6–13 in Fig. 3).

Cycling amplification starts with structure 6, which is quickly converted into a stem-loop DNA (structure 7) due to the DNA synthesis guided by B1 at 3'-end. B2 of BIP anneals to B2c in the loop and guides strand-displacement DNA synthesis. The released ssDNA forms a loop due to the complementary interaction of F1 with F1c at the 3'-end and is linked with the synthesized dsDNA by an open double-stranded loop (structure 8). F1 in the ssDNA guides DNA synthesis, followed by strand-displacement DNA synthesis, producing a compound dsDNA (structure 9), a copy of dsRNA template linked with a stem-loop by a loop. The released ssDNA forms another dumbbell-like structure (structure 10), the sequence of which is complementary to structure 6. Importantly, structure 6 can be produced from structure 10 through structures 11–13 and finally forms an amplification cycle. Structures 9 and 13 then serve as templates for elongation and recycling. Consequently, the final products are stem-loop DNA molecules with several inverted repeats of the template and cauliflower-like structures with multiple loops formed by annealing between alternately inverted repeats of the target in the same strand.

LAMP products can be detected by gel electrophoresis, turbidity, and fluorescence. Electrophoresis shows a typical ladder-like pattern of LAMP products with different sizes. The turbidity of LAMP products is due to magnesium pyrophosphate (Mori et al., 2001). In addition to a large amount of DNA, the LAMP reaction produces a large amount of pyrophosphate ion, which reacts with magnesium ion in the reaction solution, yielding a white precipitate, magnesium pyrophosphate. The turbidity change of the reaction solution can

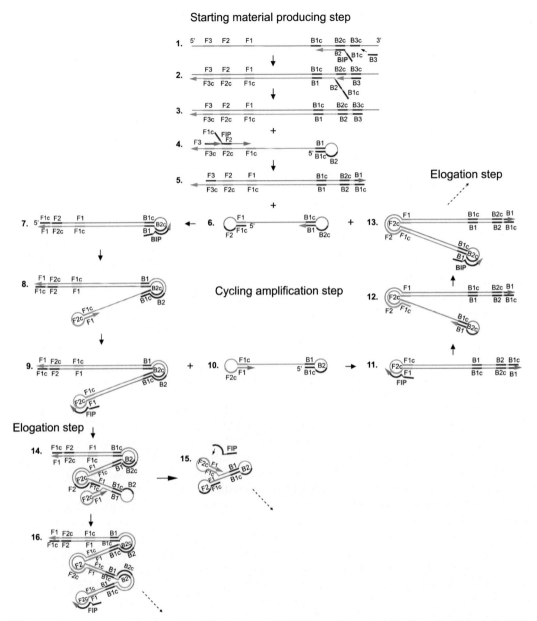

FIG. 3 Schematic representation of LAMP reactions. *Modified from Nagamine, K., Hase, T., Notomi, T. 2002. Accelerated reaction by loop-mediated isothermal amplification using loop primers. Mol. Cell. Probes 16 (3), 223–229.*

be monitored by either the naked eye or with a special instrument. The fluorescent method uses calcein, a metal indicator that yields strong fluorescence upon forming complexes with divalent metallic ions, such as calcium and magnesium. Adding manganous ions to the LAMP reaction solution can realize visible observation based on color change. Magnesium ions cannot compete with manganous ions for complexation with calcein. However, manganous ions will form an insoluble salt with pyrophosphate ions generated as the reaction proceeds and release calcein, which will react with magnesium ions to yield a strong fluorescence (Tomita et al., 2008).

Several factors make LAMP a favorable nucleic acid amplification method (Tomita et al., 2008): (i) It is simple and requires no special reagents or equipment. (ii) It has a fast reaction time of less than an hour and can be less than half an hour if loop primers are used (Nagamine et al., 2002). (iii) It has a high amplification efficiency; a reaction can produce 10^9 copies of the target in less than an hour (Notomi et al., 2000). (iv) It is sensitive; LAMP is tolerant to PCR inhibitors and can detect even a few copies of the target (Notomi et al., 2000). (v) It is highly specific; four primers corresponding to six independent regions of the target DNA ensure high specificity.

RNA can be detected by LAMP after an RT reaction (Notomi et al., 2000). Naturally, viroids can also be detected by RT-LAMP. Different RT-LAMP assays have been developed for PSTVd detection (Lenarcic et al., 2013; Tsutsumi et al., 2010; Verma et al., 2020). They are not only simple, specific, and cost-effective but are 10–100 times more sensitive than RT-PCR. Thus, these assays are suitable for the PSTVd detection of large-scale field samples (Verma et al., 2020). In addition, RT-LAMP is suitable for the detection of seeds with low levels of viroid accumulation. Recently, an RT-LAMP assay was developed to detect pospiviroids using a universal primer (Tseng et al., 2021), which could be used to screen a large number of solanaceous plants and seeds intended for import and export.

(b) RPA

This method was first introduced in 2006 by scientists from ASM Scientific Ltd. (Cambridge, United Kingdom, founded by the Wellcome Trust Sanger Institute) (Piepenburg et al., 2006). Although RPA was introduced relatively recently and has not yet attained a large market share of isothermal amplification technology, it has become one of the fastest-developing methods and has a promising future (Li et al., 2019; Lobato and O'Sullivan, 2018).

Recombinase and polymerase, the first two words in "RPA," are responsible for homologous recombination and DNA synthesis reactions during RPA. As shown in Fig. 4, recombinase initiates RPA by binding to oligonucleotide primers, assisted by accessory proteins and forming nucleoprotein filaments. These filaments scan the dsDNA template and will invade the dsDNA template upon meeting homologous sequences of primers. Once primers hybridize with their complementary sequences, the ssDNA strand will be released and stabilized by ssDNA binding protein (SSB), forming a D-loop structure. Next, recombinase disassembles from the nucleoprotein filament and leaves the 3'-end of primers accessible to DNA polymerase with strand-displacement activity. DNA polymerase drives the primer to extend along the template strand. The extension of a pair of primers synthesizes the copy of the DNA between the primers. Cyclic repetition of this process results in exponential amplification (Daher et al., 2016; Li et al., 2019; Lobato and O'Sullivan, 2018; Piepenburg et al., 2006).

Like other nucleic acid amplification methods, RPA is very sensitive and can detect few DNA copies (Li et al., 2019; Lobato and O'Sullivan, 2018). Moreover, tolerance of certain PCR inhibitors (Kersting et al., 2014) means that RPA has a high amplification efficiency. However, the most distinct advantages of RPA are that it is fast and the reaction conditions are easy to meet. To amplify DNA template to detectable levels, RPA usually takes less than

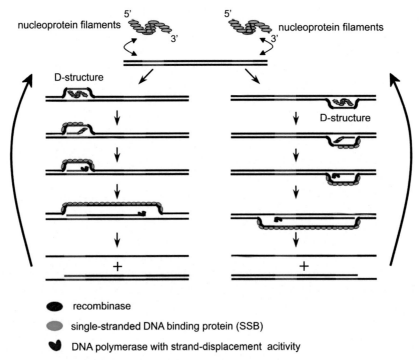

FIG. 4 Schematic of the RPA process. Recombinase *(grey)* binds with primer, forming nucleoprotein filaments that scan template DNA for homologous sequences *(red/blue)*. Following strand exchange, the displaced strand is bound by a single-stranded DNA-binding protein *(green)*. Primers are extended by DNA polymerase with strand-displacement activity *(blue)*. *Modified from Piepenburg, O., Williams, C.H., Stemple, D.L., Armes, N.A. 2006. DNA detection using recombination proteins. PLoS Biol. 4 (7), e204.*

20 min and can be as low as 4–5 min. RPA reactions can occur at temperatures ranging from 22°C to 45°C, though the recommended range is between 37°C and 42°C. This is easy to achieve without special equipment, and can even be supplied by body heat. As such, RPA can be performed using a simple and portable instrument and thus has a promising future for on-site detection in the field.

Like PCR and qPCR, the RPA reaction finally produces a large amount of dsDNA. Thus, the detection methods of amplified products used for PCR and qPCR also can be used for RPA, including gel electrophoresis and real-time check. Additionally, to simplify and quicken operations, different detection methods have been developed, including some that entail microfluidic devices and automation (see details in Lobato and O'Sullivan, 2018). Broadly speaking, RPA products can be detected at the endpoint of amplification or can be monitored in real-time. Probes may be used depending on the detection strategy. Generally, endpoint detection requires fewer instruments than real-time detection and is thus cheaper.

The commonly used endpoint detection of RPA products is lateral flow assays, like lateral flow strips, The results are rapidly generated and can be read by the naked eye. This detection method needs a probe 46–52 nt in length with an antigen at the 5′ end, a polymerase extension blocking group at the 3′ end, and an internal abasic nucleotide analog of at least 30 nucleotides from the 5′ end and 15 nucleotides from the 3′ end. The abasic nucleotide is a tetrahydrofuran residue (THF) that replaces a conventional nucleotide, which can be cleaved by *Nfo* nuclease (the double-strand-specific *Escherichia coli* endonuclease IV) when the probe forms dsDNA with target DNA. The cleavage produces a new 3′ hydroxyl group and thus the probe can continue to extend along the target DNA. In addition to the probe, the primer in an opposite direction from the probe is labeled with another marker (e.g., biotin) at the 5′ end. Thus, RPA products contain two labels, which can be detected in a sandwich assay format by antibodies or antibody/streptavidin.

The fundamental principle of real-time detection of dsDNA is similar for RPA and qPCR, using a quencher and fluorophore to produce fluorescence. To increase detection specificity, a probe is recommended. However, the commonly used probes for qPCR, such as the TaqMan hydrolysis probe, are not compatible with RPA. DNA polymerase for RPA is a strand-displacing DNA polymerase (e.g., a large fragment of *Bacillus subtilis* Pol 1, *Bsu*). Unlike the DNA polymerases used in qPCR, this polymerase has no 5′ to 3′ exonuclease. Thus, the RPA probe is modified and requires an additional exonuclease to be cleaved for emitting fluorescence (Piepenburg et al., 2006; Lobato and O'Sullivan, 2018).

RNA, including viroid RNA, can also be used as the template of RPA followed by RT reaction, namely, RT-RPA. RT-RPA was first used in viroid detection for tomato chlorotic dwarf viroid (TCDVd) (Hammond and Zhang, 2016), followed by several other viroids. The detection sensitivity of RT-RPA for viroids is similar to that of RT-qPCR. The RT-RPA assay developed for tomato apical stunt viroid (TASVd) can detect 8 fg of pure TASVd RNA transcripts (Kovalskaya and Hammond, 2022). In addition, the tolerance to certain PCR inhibitors allows RT-RPA to use plant crude extract for viroid detection (Kovalskaya and Hammond, 2022). With the rapid development of the RPA technique, RT-RPA will be applied to more situations of viroid detection and can enable on-site detection in the field.

Next-generation sequencing (NGS)

If all RNAs in a sample can be sequenced, viroid(s) in the sample will be known. This can be achieved using next-generation sequencing (NGS) (Metzker, 2010; Shendure and Ji, 2008).

NGS detection contains RNA preparation, library construction, sequencing, and bioinformatics analysis (Kutnjak et al., 2021). For viroid detection by NGS, starting materials are usually extracted total RNA or small RNA (sRNA). Total RNAs are normally broken into short fragments due to the limit of the sequencing length of NGS. RNA or small RNA molecules of a certain length are separated and purified for library construction. Adapters are added to the prepared RNA molecules at the 3′ and 5′ ends, transcribed into cDNA, and amplified by PCR. The resulting library is sequenced on proper platforms, yielding millions of reads. Viroid sequences hide in these massive reads and can be screened by bioinformatics analysis. Most of the sequencing reads are derived from the plant genome. Thus, if the plant genome is known, the reads homologous to the plant genome are first removed by mapping. The remaining reads are assembled into longer contigs and then BLAST analysis is performed in a public database such as GenBank to search viroid contigs.

Four types of viroid-derived RNAs exist in viroid-infected plants, single-stranded circular genomic RNA, single-stranded linear subgenomic RNA (Minoia et al., 2015), dsRNA of replication intermediate (Branch et al., 1981; Owens and Diener, 1982), and small RNA (sRNA), which are products of plant defense against viroid infection by RNA silencing (Itaya et al., 2001; Papaefthimiou et al., 2001). Although viroid sRNA molecules are short (normally in sizes of 21–24 nt), they overlap each other and can be assembled into a whole genome (Li et al., 2012). Thus, all four types of viroid-derived RNAs can be used as targets for viroid detection. Different types of RNA are specifically detected by different NGS methods (Hadidi, 2019; Wu et al., 2015), including RNA sequencing (RNA-seq) (for genomic and subgenomic RNAs), dsRNA sequencing (for replication intermediates), and sRNA sequencing (for sRNA).

sRNA sequencing is the most widely used (Wu et al., 2015) because sample preparation and library construction are relatively simpler and more cost-effective. RNA-seq usually requires rRNA deletion to decrease the proportion of plant genomic sequences in sequencing reads. dsRNA sequencing needs a large amount of initial material, and the extraction and purification of dsRNA are laborious and complex. Moreover, the total RNA and dsRNA for sequencing must be randomly broken into short fragments in mechanical ways or digested by enzymes.

Compared with nucleotide acid amplification methods, NGS has no advantages related to cost, time, detection throughput, or ease of operation, due to its complex library construction, the expensive sequencing platform, and the need for professional bioinformatics analysis. As such, NGS is not widely used in routine viroid detection. However, due to its sequence independence and high sensitivity, NGS detection is a powerful tool for discovering viroids (Wu et al., 2012; Zhang et al., 2014), performing viroid surveys in a given country (Choi et al., 2020), viroid detection in mother plants for viroid elimination (Matousek et al., 2020), or screening viroid-free seeds (Fox et al., 2015). NGS is particularly attractive for applications relating to disease diagnosis and quarantine. For example, in 2007, hop growers in Slovenia observed severe stunting of hop plants in several hop gardens. The disease rapidly spread within hop gardens and between farms. The causal agent of this disease could not be identified by traditional detection methods because no known hop-infecting pathogens were detected or excluded. The disease was finally identified by NGS analysis to be a novel variant of citrus bark cracking viroid (CBCVd) (Jakse et al., 2015).

Method selection

As described above, there are many methods for viroid detection. The question then becomes: how to select the appropriate method? All methods have advantages and disadvantages related to sensitivity, specificity, feasibility, rapidness, and cost. The advantages and disadvantages of the viroid detection methods introduced in this chapter are summarized in Table 1, according to descriptions provided by Lopez et al. (2009). A proper method can be selected based on comprehensively considering these advantages and disadvantages.

The appropriate method should be able to be performed under current restraints including those related to budget, equipment, and the availability of a professional operator. Some

TABLE 1   Comparison of different viroid detection methods.

Method	Sensitivity	Specificity	Feasibility	Rapidness	Cost
Bioassay	+++	++++	+++	++	++++
PAGE					
2D-PAGE	+	+	+++	+++	++++
R-PAGE	+	+	+++	+++	++++
S-PAGE	+	+	+++	+++	++++
Molecular hybridization					
Northern-blot	+	++++	++	+	+++
Dot-blot	+	+++	+++	++	++++
In situ	+	+++	+	+	++
Nucleic acid amplification					
RT-PCR	+++	++++	+++	+++	+++
RT-qPCR	+++++	+++++	++++	+++++	+++
RT-LAMP	+++++	+++++	++++	+++++	++++
RT-RPA	+++++	+++++	++++	+++++	++
RT-dPCR	+++++	+++++	++	+++++	+++
Next-generation sequencing					
sRNA	+++++	++++	++	+	+
RNA	+++++	++++	++	+	+
dsRNA	+++++	++++	++	+	+

methods require special equipment such as a thermal cycler for RT-PCR and RT-qPCR and an expensive sequencing platform for NGS. In addition, the proper method must meet detection requirements related to accuracy, reliability, and time. In some situations, the most important requirement is accuracy; in others, it could be time. For instance, if we are testing a propagating material, such as a mother fruit plant used to produce seedlings, the mother plant should be viroid-free. In this case, accuracy is the most important consideration. Thus, method(s) with high sensitivity, such as RT-qPCR and NGS, is optimal. It is wise to avoid selecting the newest or highest quality methods when conventional methods can meet our purpose.

Prospective

Improvements in viroid detection entail making the process simpler, faster, cheaper, and more accurate. Undoubtedly, current methods will be further improved, and new methods will be developed. As these techniques develop, we believe that RT-RPA, RT-LAMP, and NGS will be improved and more widely applied in viroid detection.

RT-RPA and RT-LAMP are independent of the thermal cycler and thus can be performed using small and portable heating devices. Moreover, their amplification products can be detected by miniature devices such as microfluidic chips and lateral-flow strips. As such, real on-site viroid detection in the field can be achieved using these two methods. In the future, with the development of the microfluidic technique, viroid detection could be performed using an all-in-one device that integrates sample preparation, nucleic acid amplification, detection of amplification products, and reporting of the results in easy-to-operate units.

Although NGS has revolutionized viroid detection (Boonham et al., 2014; Wu et al., 2015), it is still not widely used in routine detection due to limitations related to time, cost, and the need for specialized bioinformatics analysis. The breakthrough of these limitations depends on the development of sequencing technology and especially bioinformatics. Currently, in addition to some commercial software, several free and user-friendly automated pipelines such as VirusDetect (Zheng et al., 2017) have been developed for bioinformatics analysis of NGS data. Thus, a researcher with no background in bioinformatics can use NGS methods. In the future, viroid detection by NGS will benefit from the development of certain computer programs used for viroid identification, like progressive filtering of overlapping small RNAs (PFOR) (Wu et al., 2012; Zhang et al., 2014), and the establishment of a special viroid sequence database, like ViroidDB (Lee et al., 2022).

New methods related to clustered regularly interspaced short palindromic repeats (CRISPR)-based technology are also expected. Since the discovery of CRISPR-based detection technology (Gootenberg et al., 2017), the number of different CRISPR–Cas detection systems has rapidly increased (Kaminski et al., 2021). Indeed, a similar method has been applied in viroid detection (Jiao et al., 2021), and new methods such as these will undoubtedly open new avenues for viroid detection.

Chapter summary

1. Different methods are available for viroid detection, including bioassay, PAGE, molecular hybridization, nucleic acid amplification, and NGS methods.
2. A bioassay is based on viroid-induced symptoms on indicator plants and only a few viroids can be detected using this method. PAGE detection is independent of viroid sequence, but depends on the size and structure of viroid RNA. Although these two methods are rarely used in routine detection, they remain essential for viroid discovery and identification.
3. Molecular hybridization is based on base complementation between viroid RNA and the viroid probe and is thus specific, sensitive, and reliable. Dot-blot hybridization can realize high-throughput detection and is suitable for detecting large-scale samples. The simultaneous detection of different viroids by molecular hybridization can be achieved using a polyprobe or universal probes.
4. Nucleic acid amplification methods include RT-PCR, RT-qPCR, RT-LAMP, RT-RPA, and dPCR. They are very sensitive because the amount of starting template is significantly amplified before detection. RT-qPCR and dPCR can quantify the amount of starting

template. RT-LAMP and RT-RPA occur under constant temperatures without the need for a thermal cycler and thus can be used for real on-site detection in the field.

5. NGS detection combines new sequencing technology and bioinformatics. It has significantly revolutionized viroid detection and will continue to play an increasingly important role in the development of automated and user-friendly software or computer programs for bioinformatics analysis.

6. Each detection methods have both advantages and disadvantages. The proper method(s) must be performed based on current detection conditions while meeting detection requirements.

7. Novel methods could be established with the rapid development of CRISPR-based detection technology and microfluidic techniques.

Study questions

1. In what situations is viroid detection required? Please give some examples.
2. How many methods are currently available for viroid detection? Please briefly describe their basic principles, advantages, and disadvantages.
3. Suppose there is a viroid disease outbreak in a country. To control the disease, it is necessary to identify the occurrence and distribution of the viroid, and a large-scale survey of the viroid must be performed. Which method(s) is most appropriate for this large-scale survey? Please provide a detailed solution.
4. If you study a disease and suspect it is caused by a viroid, how would you identify the causal agent? Which detection method(s) is proper for this situation?

Further reading

Protocols for most of the viroid methods described in this chapter are described in the book "Viroids: Methods and Protocols" edited by Ayala L. N. Rao, Irene Lavagi-Craddock, and Georgios Vidalakis, which was published in 2022 by Springer US.

Chapter 13 of "Assay, detection, and diagnosis of plant viruses" in Section III of the book "Plant Virology (5th edition)" edited by Roger Hull is an excellent reference for learning about the detection methods of viral agents.

References

Amari, K., Cañizares, M.C., Myrta, A., Sabanadzovic, S., Di Terlizzi, B., Pallás, V., 2001. Tracking hop stunt viroid (HSVd) infection in apricot trees during a whole year by non-isotopic tissue printing hybridization. Acta Hortic. 550, 315–320.

Astruc, N., Marcos, J.F., Macquaire, G., Candresse, T., Pallas, V., 1996. Studies on the diagnosis of hop stunt viroid in fruit trees: identification of new hosts and application of a nucleic acid extraction procedure based on non-organic solvents. Eur. J. Plant Pathol. 102 (9), 837–846.

Baker, M., 2012. Digital PCR hits its stride. Nat. Methods 9 (6), 541–544.

Boonham, N., Perez, L.G., Mendez, M.S., Peralta, E.L., Blockley, A., Walsh, K., Barker, I., Mumford, R.A., 2004. Development of a real-time RT-PCR assay for the detection of potato spindle tuber viroid. J. Virol. Methods 116 (2), 139–146.

Boonham, N., Kreuze, J., Winter, S., van der Vlugt, R., Bergervoet, J., Tomlinson, J., Mumford, R., 2014. Methods in virus diagnostics: from ELISA to next generation sequencing. Virus Res. 186, 20–31.

Botermans, M., Roenhorst, J.W., Hooftman, M., Verhoeven, J.T.J., Metz, E., van Veen, E.J., Geraats, B.P.J., Kemper, M., Beugelsdijk, D.C.M., Koenraadt, H., Jodlowska, A., Westenberg, M., 2020. Development and validation of a real-time RT-PCR test for screening pepper and tomato seed lots for the presence of pospiviroids. PLoS One 15 (9), e0232502.

Branch, A.D., Robertson, H.D., Dickson, E., 1981. Longer-than-unit-length viroid minus strands are present in RNA from infected plants. Proc. Natl. Acad. Sci. U. S. A. 78 (10), 6381–6385.

Chang, S.J., Puryear, J., Cairney, J., 1993. A simple and efficient method for isolating RNA from pine trees. Plant Mol. Biol. Rep. 11 (2), 113–116.

Choi, H., Jo, Y., Cho, W.K., Yu, J., Tran, P.T., Salaipeth, L., Kwak, H.R., Choi, H.S., Kim, K.H., 2020. Identification of viruses and viroids infecting tomato and pepper plants in Vietnam by metatranscriptomics. Int. J. Mol. Sci. 21 (20), 7565.

Craw, P., Balachandran, W., 2012. Isothermal nucleic acid amplification technologies for point-of-care diagnostics: a critical review. Lab Chip 12 (14), 2469–2486.

Daher, R.K., Stewart, G., Boissinot, M., Bergeron, M.G., 2016. Recombinase polymerase amplification for diagnostic applications. Clin. Chem. 62 (7), 947–958.

Fox, A., Adams, I.A., Hany, U., Hodges, T., Forde, S.M.D., Jackson, L.E., Skelton, A., Barton, V., 2015. The application of next-generation sequencing for screening seeds for viruses and viroids. Seed Sci. Technol. 43 (3), 531–535.

Gootenberg, J.S., Abudayyeh, O.O., Lee, J.W., Essletzbichler, P., Dy, A.J., Joung, J., Verdine, V., Donghia, N., Daringer, N.M., Freije, C.A., Myhrvold, C., Bhattacharyya, R.P., Livny, J., Regev, A., Koonin, E.V., Hung, D.T., Sabeti, P.C., Collins, J.J., Zhang, F., 2017. Nucleic acid detection with CRISPR-Cas13a/C2c2. Science 356 (6336), 438–442.

Gucek, T., Trdan, S., Jakse, J., Javornik, B., Matousek, J., Radisek, S., 2017. Diagnostic techniques for viroids. Plant Pathol. 66 (3), 339–358.

Hadidi, A., 2019. Next-generation sequencing and CRISPR/Cas13 editing in viroid research and molecular diagnostics. Viruses 11 (2), 120.

Hadidi, A., Candresse, T., 2003. Polymerase chain reaction. In: Hadidi, A., Flores, R., Randles, J.W., Semancik, J.S. (Eds.), Viroids. Collingwood, Australia, CSIRO, pp. 115–122.

Hadidi, A., Yang, X.C., 1990. Detection of pome fruit viroids by enzymatic cDNA amplification. J. Virol. Methods 30 (3), 261–269.

Hammond, R.W., Zhang, S.L., 2016. Development of a rapid diagnostic assay for the detection of tomato chlorotic dwarf viroid based on isothermal reverse-transcription-recombinase polymerase amplification. J. Virol. Methods 236, 62–67.

Harders, J., Lukacs, N., Robertnicoud, M., Jovin, T.M., Riesner, D., 1989. Imaging of viroids in nuclei from tomato leaf tissue by in situ hybridization and confocal laser scanning microscopy. EMBO J. 8 (13), 3941–3949.

Heid, C.A., Stevens, J., Livak, K.J., Williams, P.M., 1996. Real time quantitative PCR. Genome Res. 6 (10), 986–994.

Herranz, M.C., Sanchez-Navarro, J.A., Aparicio, F., Pallas, V., 2005. Simultaneous detection of six stone fruit viruses by non-isotopic molecular hybridization using a unique riboprobe or 'polyprobe'. J. Virol. Methods 124, 49–55.

Higuchi, R., Fockler, C., Dollinger, G., Watson, R., 1993. Kinetic PCR analysis: real-time monitoring of DNA amplification reactions. Nat. Biotechnol. 11 (9), 1026–1030.

Hindson, B.J., Ness, K.D., Masquelier, D.A., Belgrader, P., Heredia, N.J., Makarewicz, A.J., Bright, I.J., Lucero, M.Y., Hiddessen, A.L., Legler, T.C., Kitano, T.K., Hodel, M.R., Petersen, J.F., Wyatt, P.W., Steenblock, E.R., Shah, P.H., Bousse, L.J., Troup, C.B., Mellen, J.C., Wittmann, D.K., Erndt, N.G., Cauley, T.H., Koehler, R.T., So, A.P., Dube, S., Rose, K.A., Montesclaros, L., Wang, S.L., Stumbo, D.P., Hodges, S.P., Romine, S., Milanovich, F.P., White, H.E., Regan, J.F., Karlin-Neumann, G.A., Hindson, C.M., Saxonov, S., Colston, B.W., 2011. High-throughput droplet digital PCR system for absolute quantitation of DNA copy number. Anal. Chem. 83 (22), 8604–8610.

Huggett, J.F., Foy, C.A., Benes, V., Emslie, K., Garson, J.A., Haynes, R., Hellemans, J., Kubista, M., Nolan, R.D.M.T., Pfaffl, M.W., Pfaffl, M.W., Shipley, G.L., Vandesompele, J., Wittwer, C.T., Bustin, S.A., 2013. The digital MIQE guidelines: minimum information for publication of quantitative digital PCR experiments. Clin. Chem. 59 (6), 892–902.

Itaya, A., Folimonov, A., Matsuda, Y., Nelson, R.S., Ding, B., 2001. Potato spindle tuber viroid as inducer of RNA silencing in infected tomato. Mol. Plant Microbe Interact. 14 (11), 1332–1334.

Ito, T., Ieki, H., Ozaki, K., 2002. Simultaneous detection of six citrus viroids and apple stem grooving virus from citrus plants by multiplex reverse transcription polymerase chain reaction. J. Virol. Methods 106 (2), 235–239.

Jakse, J., Radisek, S., Pokorn, T., Matousek, J., Javornik, B., 2015. Deep-sequencing revealed citrus bark cracking viroid (CBCVd) as a highly aggressive pathogen on hop. Plant Pathol. 64 (4), 831–842.

Jiang, D.M., Hou, W.Y., Sano, T., Kang, N., Qin, L., Wu, Z.J., Li, S.F., Xie, L.H., 2013. Rapid detection and identification of viroids in the genus *Coleviroid* using a universal probe. J. Virol. Methods 187 (2), 321–326.

Jiao, J., Kong, K.K., Han, J.M., Song, S.W., Bai, T.H., Song, C.H., Wang, M.M., Yan, Z.L., Zhang, H.T., Zhang, R.P., Feng, J.C., Zheng, X.B., 2021. Field detection of multiple RNA viruses/viroids in apple using a CRISPR/Cas12a-based visual assay. Plant Biotechnol. J. 19 (2), 394–405.

Kaminski, M.M., Abudayyeh, O.O., Gootenberg, J.S., Zhang, F., Collins, J.J., 2021. CRISPR-based diagnostics. Nat. Biomed. Eng. 5 (7), 643–656.

Kersting, S., Rausch, V., Bier, F.F., von Nickisch-Rosenegk, M., 2014. Rapid detection of *Plasmodium falciparum* with isothermal recombinase polymerase amplification and lateral flow analysis. Malaria J. 13, 99.

Kovalskaya, N., Hammond, R.W., 2022. Rapid diagnostic detection of tomato apical stunt viroid based on isothermal reverse transcription-recombinase polymerase amplification. J. Virol. Methods 300, 114353.

Kutnjak, D., Tamisier, L., Adams, I., Boonham, N., Candresse, T., Chiumenti, M., De Jonghe, K., Kreuze, J.F., Lefebvre, M., Silva, G., Malapi-Wight, M., Margaria, P., Plesko, I.M., McGreig, S., Miozzi, L., Remenant, B., Reynard, J.S., Rollin, J., Rott, M., Schumpp, O., Massart, S., Haegeman, A., 2021. A primer on the analysis of high-throughput sequencing data for detection of plant viruses. Microorganisms 9 (4), 841.

Lee, H.J., Cho, I.S., Ju, H.J., Jeong, R.D., 2021. Development of a reverse transcription droplet digital PCR assay for sensitive detection of peach latent mosaic viroid. Mol. Cell. Probes 58, 101746.

Lee, B.D., Neri, U., Oh, C.J., Simmonds, P., Koonin, E.V., 2022. ViroidDB: a database of viroids and viroid-like circular RNAs. Nucleic Acids Res. 50 (D1), D432–D438.

Lenarcic, R., Morisset, D., Mehle, N., Ravnikar, M., 2013. Fast real-time detection of potato spindle tuber viroid by RT-LAMP. Plant Pathol. 62 (5), 1147–1156.

Li, S.F., Onodera, S., Sano, T., Yoshida, K., Wang, G.P., Shikata, E., 1995. Gene diagnosis of viroids: comparisons of return-PAGE and hybridization using DIG-labeled DNA and RNA probes for practical diagnosis of hop stunt, citrus exocortis an apple scar skin viroids in their natural host plants. Ann. Phytopathol. Soc. Jpn. 61, 381–390.

Li, R.G., Gao, S., Hernandez, A.G., Wechter, W.P., Fei, Z.J., Ling, K.S., 2012. Deep sequencing of small RNAs in tomato for virus and viroid identification and strain differentiation. PLoS One 7 (5), e37127.

Li, J., Macdonald, J., von Stetten, F., 2019. Review: a comprehensive summary of a decade development of the recombinase polymerase amplification. Analyst 144 (1), 31–67.

Lobato, I.M., O'Sullivan, C.K., 2018. Recombinase polymerase amplification: basics, applications and recent advances. Trend Anal. Chem. 98, 19–35.

Lopez, M.M., Llop, P., Olmos, A., Marco-Noales, E., Cambra, M., Bertolini, E., 2009. Are molecular tools solving the challenges posed by detection of plant pathogenic bacteria and viruses? Curr. Issues Mol. Biol. 11, 13–45.

Matousek, J., Steinbachova, L., Drabkova, L.Z., Kocabek, T., Potesil, D., Mishra, A.K., Honys, D., Steger, G., 2020. Elimination of viroids from tobacco pollen involves a decrease in propagation rate and an increase of the degradation processes. Int. J. Mol. Sci. 21 (8), 3029.

Matsushita, Y., Tsuda, S., 2016. Seed transmission of potato spindle tuber viroid, tomato chlorotic dwarf viroid, tomato apical stunt viroid, and columnea latent viroid in horticultural plants. Eur. J. Plant Pathol. 145 (4), 1007–1011.

Metzker, M.L., 2010. Sequencing technologies – the next generation. Nat. Rev. Genet. 11 (1), 31–46.

Minoia, S., Navarro, B., Delgado, S., Di Serio, F., Flores, R., 2015. Viroid RNA turnover: characterization of the subgenomic RNAs of potato spindle tuber viroid accumulating in infected tissues provides insights into decay pathways operating in vivo. Nucleic Acids Res. 43 (4), 2313–2325.

Mori, Y., Nagamine, K., Tomita, N., Notomi, T., 2001. Detection of loop-mediated isothermal amplification reaction by turbidity derived from magnesium pyrophosphate formation. Biochem. Biophys. Res. Commun. 289 (1), 150–154.

Nagamine, K., Hase, T., Notomi, T., 2002. Accelerated reaction by loop-mediated isothermal amplification using loop primers. Mol. Cell. Probes 16 (3), 223–229.

Nie, X., Singh, R.P., 2001. A novel usage of random primers for multiplex RT-PCR detection of virus and viroid in aphids, leaves, and tubers. J. Virol. Methods 91 (1), 37–49.

Nie, X.Z., Singh, R.P., 2017. Viroid detection and identification by bioassay. In: Hadidi, A., Flores, R., Randles, J.W., Palukaitis, P. (Eds.), Viroids and Satellites. Academic Press, Boston, pp. 347–356.

Notomi, T., Okayama, H., Masubuchi, H., Yonekawa, T., Watanabe, K., Amino, N., Hase, T., 2000. Loop-mediated isothermal amplification of DNA. Nucleic Acids Res. 28 (12), e63.

Owens, R.A., Diener, T.O., 1981. Sensitive and rapid diagnosis of potato spindle tuber viroid disease by nucleic acid hybridization. Science 213 (4508), 670–672.

Owens, R.A., Diener, T.O., 1982. RNA intermediates in potato spindle tuber viroid replication. Proc. Natl. Acad. Sci. U. S. A. 79 (1), 113–117.

C. Viroid pathogenesis and viroid-host interaction

Owens, R.A., Flores, R., Di Serio, F., Li, S.F., Pallas, V., Randles, J.W., Sano, T., Vidalakis, G., 2012a. Viroids. In: King, A.M.Q., Adams, M.J., Carstens, E.B., Lefkowitz, E.J. (Eds.), Virus Taxonomy: Ninth Report of the International Committee on Taxonomy of Viruses. Elsevier/Academic Press, London, UK, pp. 1221–1234.

Owens, R.A., Sano, T., Duran-Vila, N., 2012b. Plant viroids: isolation, characterization/detection, and analysis. Methods Mol. Biol. 894, 253–271.

Pallas, V., Sanchez-Navarro, J.A., Kinard, G.R., Di Serio, F., 2017. Molecular hybridization techniques for detecting and studying viroids. In: Hadidi, A., Flores, R., Randles, J.W., Palukaitis, P. (Eds.), Viroids and Satellites. Academic Press, Boston, pp. 369–379.

Pallas, V., Sanchez-Navarro, J.A., James, D., 2018. Recent advances on the multiplex molecular detection of plant viruses and viroids. Front. Microbiol. 9, 2087.

Papaefthimiou, I., Hamilton, A.J., Denti, M.A., Baulcombe, D.C., Tsagris, M., Tabler, M., 2001. Replicating potato spindle tuber viroid RNA is accompanied by short RNA fragments that are characteristic of post-transcriptional gene silencing. Nucleic Acids Res. 29 (11), 2395–2400.

Piepenburg, O., Williams, C.H., Stemple, D.L., Armes, N.A., 2006. DNA detection using recombination proteins. PLoS Biol. 4 (7), e204.

Pinheiro, L.B., Coleman, V.A., Hindson, C.M., Herrmann, J., Hindson, B.J., Bhat, S., Emslie, K.R., 2012. Evaluation of a droplet digital polymerase chain reaction format for DNA copy number quantification. Anal. Biochem. 84 (2), 1003–1011.

Podleckis, E.V., Hammond, R.W., Hurtt, S.S., Hadidi, A., 1993. Chemiluminescent detection of potato and pome fruit viroids by digoxigenin-labeled dot blot and tissue blot hybridization. J. Virol. Methods 43 (2), 147–158.

Raymer, W.B., O'Brien, M.J., 1962. Transmission of potato spindle tuber virus to tomato. Am. Potato J. 39, 401–408.

Rizza, S., Nobile, G., Tessitori, M., Catara, A., Conte, E., 2009. Real time RT-PCR assay for quantitative detection of citrus viroid III in plant tissues. Plant Pathol. 58 (1), 181–185.

Saiki, R.K., Gelfand, D.H., Stoffel, S., Scharf, S.J., Higuchi, R., Horn, G.T., Mullis, K.B., Erlich, H.A., 1988. Primer-directed enzymatic amplification of DNA with a thermostable DNA polymerase. Science 239 (4839), 487–491.

Sano, T., Uyeda, I., Shikata, E., Meshi, T., Ohno, T., Okada, Y., 1985. A viroid-like RNA isolated from grapevine has high sequence homology with hop stunt viroid. J. Gen. Virol. 66 (2), 333–338.

Schumacher, J., Randles, J.W., Riesner, D., 1983. A two-dimensional electrophoretic technique for the detection of circular viroids and virusoids. Anal. Biochem. 135 (2), 288–295.

Schumacher, J., Meyer, N., Riesner, D., Weidemann, H.L., 1986. Diagnostic procedure for detection of viroids and viruses with circular RNAs by "return"–gel electrophoresis. J. Phytopathol. 115 (4), 332–343.

Shendure, J., Ji, H., 2008. Next-generation DNA sequencing. Nat. Biotechnol. 26 (10), 1135–1145.

Sykes, P.J., Neoh, S.H., Brisco, M.J., Hughes, E., Condon, J., Morley, A.A., 1992. Quantitation of targets for PCR by use of limiting dilution. BioTechniques 13 (3), 444–449.

Tomita, N., Mori, Y., Kanda, H., Notomi, T., 2008. Loop-mediated isothermal amplification (LAMP) of gene sequences and simple visual detection of products. Nat. Protoc. 3 (5), 877–882.

Tseng, Y.W., Wu, C.F., Lee, C.H., Chang, C.J., Chen, Y.K., Jan, F.J., 2021. Universal primers for rapid detection of six pospiviroids in Solanaceae plants using one-step reverse-transcription PCR and reverse-transcription loop-mediated isothermal amplification. Plant Dis. 105 (10), 2859–2864.

Tsutsumi, N., Yanagisawa, H., Fujiwara, Y., Ohara, T., 2010. Detection of potato spindle tuber viroid by reverse transcription loop-mediated isothermal amplification. Res. Bull. Plant Protect. Japan 46, 61–67.

Untergasser, A., Cutcutache, I., Koressaar, T., Ye, J., Faircloth, B.C., Remm, M., Rozen, S.G., 2012. Primer3-new capabilities and interfaces. Nucleic Acids Res. 40 (15), e115.

Verhoeven, J.T.J., Jansen, C.C.C., Willemen, T.M., Kox, L.F.F., Owens, R.A., Roenhorst, J.W., 2004. Natural infections of tomato by citrus exocortis viroid, columnea latent viroid, potato spindle tuber viroid and tomato chlorotic dwarf viroid. Eur. J. Plant Pathol. 110 (8), 823–831.

Verma, G., Raigond, B., Pathania, S., Kochhar, T., Naga, K., 2020. Development and comparison of reverse transcription-loop-mediated isothermal amplification assay (RT-LAMP), RT-PCR and real time PCR for detection of potato spindle tuber viroid in potato. Eur. J. Plant Pathol. 158 (4), 951–964.

Vogelstein, B., Kinzler, K.W., 1999. Digital PCR. Proc. Natl. Acad. Sci. U. S. A. 96 (16), 9236–9241.

Wu, Q.F., Wang, Y., Cao, M.J., Pantaleo, V., Burgyan, J., Li, W.X., Ding, S.W., 2012. Homology-independent discovery of replicating pathogenic circular RNAs by deep sequencing and a new computational algorithm. Proc. Natl. Acad. Sci. U. S. A. 109 (10), 3938–3943.

Wu, Q.F., Ding, S.W., Zhang, Y.J., Zhu, S.F., 2015. Identification of viruses and viroids by next-generation sequencing and homology-dependent and homology-independent algorithms. Annu. Rev. Phytopathol. 53, 425–444.

Xu, W.X., Hong, N., Jin, Q.T., Farooq, A.B.U., Wang, Z.Q., Song, Y.S., Wu, C.C., Wang, L.P., Wang, G.P., 2009. Probe binding to host proteins: a cause for false positive signals in viroid detection by tissue hybridization. Virus Res. 145 (1), 26–30.

Yanagisawa, H., Shiki, Y., Matsushita, Y., Ooishi, M., Takaue, N., Tsuda, S., 2017. Development of a comprehensive detection and identification molecular based system for eight pospiviroids. Eur. J. Plant Pathol. 149 (1), 11–23.

Zhang, Z.X., Qi, S.S., Tang, N., Zhang, X.X., Chen, S.S., Zhu, P.F., Ma, L., Cheng, J.P., Xu, Y., Lu, M.G., Wang, H.Q., Ding, S.W., Li, S.F., Wu, Q.F., 2014. Discovery of replicating circular RNAs by RNA-seq and computational algorithms. PLoS Pathog. 10 (12), e1004553.

Zhao, Y.X., Chen, F., Li, Q., Wang, L.H., Fan, C.H., 2015. Isothermal amplification of nucleic acids. Chem. Rev. 115 (22), 12491–12545.

Zheng, Y., Gao, S., Padmanabhan, C., Li, R.G., Galvez, M., Gutierrez, D., Fuentes, S., Lin, K.S., Kreuze, J., Fei, Z.J., 2017. VirusDetect: an automated pipeline for efficient virus discovery using deep sequencing of small RNAs. Virology 500, 130–138.

Zhu, Y.L., Green, L., Woo, Y.M., Owens, R., Ding, B., 2001. Cellular basis of potato spindle tuber viroid systemic movement. Virology 279 (1), 69–77.

C. Viroid pathogenesis and viroid-host interaction

17

Viroid disease control and strategies

Rosemarie W. Hammond

USDA ARS Molecular Plant Pathology Laboratory, Beltsville, MD, United States

Graphical representation

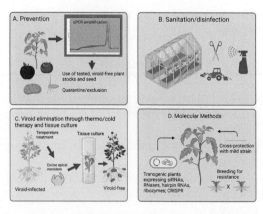

Schematic diagram of methods used to control diseases caused by viroids. (A) Use of planting stocks and seeds that are certified to be viroid-free exploiting a variety of testing methods, including sensitive and specific polymerase chain reactions (positive reactions shown as amplification curves in the inset box). These methods are utilized to exclude (quarantine) and eradicate viroids. (B) Sanitation and disinfection using a number of effective chemicals to prevent mechanical transmission of viroids. (C) A combination of thermotherapy or cold therapy and meristem-tip culture was efficient in the elimination of viroids from potato and chrysanthemum. (D) Molecular methods for viroid control include cross-protection, whereby prior inoculation with a mild strain protects against a severe strain, breeding for resistance, and transgenic plants engineered to stably express proteins or nucleic acids designed to suppress viroid replication and target viroid RNAs for destruction.

Fundamentals of Viroid Biology
https://doi.org/10.1016/B978-0-323-99688-4.00020-1

Abbreviations

dsRNA	double-stranded RNA
EPPO	European and Mediterranean Plant Protection Organization
hpRNA	hairpin RNA
miRNA	microRNA
NAPPO	North American Plant Protection Organization
PTGS	posttranscriptional gene silencing
SAM	shoot apical meristem
scFv	single-chain variable antibody
siRNA	small interfering RNA
ssRNA	single-stranded RNA
vd-sRNA	viroid-derived small RNA

Viroid species

ASSVd	apple scar skin viroid
ASBVd	avocado sunblotch viroid
CCCVd	coconut cadang cadang viroid
CChMVd	chrysanthemum chlorotic mottle viroid
CEVd	citrus exocortis viroid
CSVd	chrysanthemum stunt viroid
HSVd	hop stunt viroid
PLMVd	peach latent mosaic viroid
PSTVd	potato spindle tuber viroid

Definitions

Chemotherapy: The control of plant disease by compounds that, through their effect upon the host or pathogen reduce or nullify the effect of the pathogen after it has entered the plant (Dimond et al., 1952).

CRISPR: Clustered regularly interspaced short palindromic repeats; the basis for CRISPR-Cas 9 genome editing technology.

Cross-protection: The use of a mild strain of a virus/viroid to infect plants that are subsequently protected against economic damage caused by a severe strain of the same virus/viroid.

Cryotherapy: A novel application of plant cryopreservation techniques that allows efficient pathogen eradication.

Disease resistance: Resistance is the reduction in pathogen replication in a plant.

Disease tolerance: Plants that exhibit little disease despite substantial pathogen levels.

Hammerhead ribozyme: An RNA motif that catalyzes reversible cleavage and ligation reactions at a specific site within an RNA molecule.

Polymerase chain reaction: A method used to amplify DNA sequences.

Ribonuclease: An enzyme that catalyzes the hydrolysis of RNA.

Ribozyme: An RNA molecule capable of acting as an enzyme.

RNAi: RNA interference (RNAi) or posttranscriptional gene silencing (PTGS) is a conserved biological response to double-stranded RNA that mediates resistance to both endogenous and exogenous nucleic acids.

Thermotherapy: Consists of heat treatment of plant parts at temperature/time regimes that kill the pathogen and that are only slightly injurious to the host.

Transgenic: Denoting an organism that contains genetic material into which DNA from another organism has been stably introduced by artificial means.

Chapter outline

While many plant species are nonhosts, no naturally occurring durable resistance has been observed in most viroid host species. Current effective control methods for viroid diseases include phytosanitary certification programs, exclusion, quarantine, early detection followed by eradication, and implementation of agricultural practices, such as decontamination measures to prevent mechanical transmission. In addition, viroid elimination by heat or cold therapy combined with meristem tip culture has been successful in eliminating the viroid in some viroid-host combinations. The introduction of resistance to viroids in host species includes strategies such as cross-protection and biotechnological approaches by engineering transgenic plants developed to mediate the degradation of the single and double-stranded viroid RNAs. This chapter summarizes the current perspectives on the control of viroid disease and provides an outlook for enhancing resistance using additional approaches.

Learning objectives

- Objective 1: To learn about viroid infection/replication and implications for conventional and nonconventional control strategies
- Objective 2: To learn examples of naturally occurring resistance
- Objective 3: To learn the limits of induced resistance/cross-protection/therapies
- Objective 4: To learn how to design engineered resistance by expression of proteins and antisense RNAs in host plant species
- Objective 5: To learn how CRISPR can be employed to develop resistance in host species

Fundamental introduction

Many diseases of considerable economic importance are caused globally by viroids in both monocot and dicot plant species, and in herbaceous and woody plants. Field and greenhouse studies have demonstrated that viroids can be easily transmitted mechanically through contact with contaminated tools and farming implements, by human hands, and by contact between plants. They can also be spread through propagation, seed, pollen, and mechanically by insects. There is no known naturally occurring resistance to viroids in most host species, therefore, current control relies on detection and eradication of viroid-infected plants. Chemotherapy, thermotherapy, and cross-protection methods have been employed with success in some viroid-host combinations to eliminate viroids once the host plant is infected. Novel molecular methods for control, including engineered transgenic plants, have been developed but not yet employed commercially. These methods are illustrated in Fig. 1.

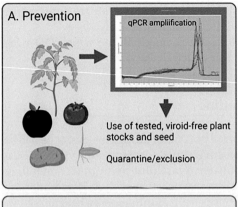

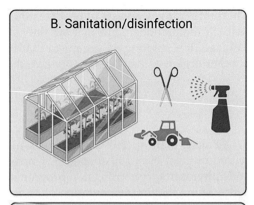

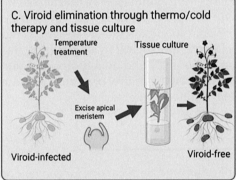

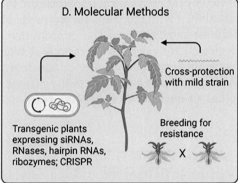

FIG. 1 Schematic diagram of methods used to control diseases caused by viroids. (A) Use of planting stocks and seeds that are certified to be viroid-free exploiting a variety of testing methods, including sensitive and specific polymerase chain reactions (positive reactions shown as amplification curves in the inset box). These methods are utilized to exclude (quarantine) and eradicate viroids. (B) Sanitation and disinfection using a number of effective chemicals to prevent mechanical transmission of viroids. (C) A combination of thermotherapy or cold therapy and meristem-tip culture was efficient in the elimination of viroids from potato and chrysanthemum. (D) Molecular methods for viroid control include cross-protection, whereby prior inoculation with a mild strain protects against a severe strain, breeding for resistance, and transgenic plants engineered to stably express proteins or nucleic acids designed to suppress viroid replication and target viroid RNAs for destruction.

Viroid replication and spread—Control of viroid diseases

As discussed in Chapters 3 and 4, viroids are small (246–401-nt), single-stranded (ss), covalently closed circular RNAs that have a partially double-stranded (ds), thermodynamically stable stem-loop structure. Viroids replicate autonomously in either the nucleus (pospiviroids) or chloroplast (avsunviroids) via a rolling circle mechanism using an asymmetric (pospiviroids) or symmetric (avsunviroids) mode of replication. Replication involves the generation of multimeric plus and minus ss- and ds-RNA intermediates, the existence of which have been exploited for the development of control strategies based on genetic engineering and the expression of RNAs or proteins (Chapter 15).

Key steps in the colonization of a host plant by viroids include their ability to move, once introduced, within the cell to the site of replication, cell-to-cell through the plasmodesmata to adjacent cells, and finally systemically in the plant through the vascular system. Viroids are also easily mechanically transmissible and spread through infected pollen and seed. Each of these critical steps presents an opportunity to design control measures, which are summarized below.

Conventional methods of control

The primary method to control viroid diseases is to prevent the introduction into the crop using clean tested propagation material (Fig. 1A) and by disinfection of equipment in greenhouse and field operations (Fig. 1B). Identification of natural resistance and breeding of resistance has been reported for selected viroids (Fig. 1D).

Naturally occurring resistance and breeding for resistant cultivars

The genetic make-up of the plant has a profound influence on the outcome of infection following inoculation by a pathogen. Nonhost resistance describes the resistance to a pathogen that is observed when all members of a plant species exhibit resistance to all members of a given pathogen species. A susceptible host response to viroid infection can be asymptomatic or result in mild to severe symptoms depending on the host cultivar and viroid strains. No known naturally occurring resistance to PSTVd has been found in potato (*Solanum tuberosum* L.) or tomato (*Solanum lycopersicum* L.) cultivars, although clones of *Solanum berthaultii* and *S. acaule* OCH 11603 were resistant to mechanical inoculation with PSTVd, but not to graft or agroinoculation, respectively (Kovalskaya and Hammond, 2014). To date, none of the tolerance has been transferred to potato cultivars. A recent report in tomato, however, demonstrated that tolerance to PSTVd infection found in a wild tomato species (*Solanum lycopersicum* var. *cerasiforme*) could be introduced into a cultivated tomato by crossing, resulting in low viroid accumulation early in infection and asymptomatic F1 hybrids upon viroid challenge, suggesting usefulness in tomato breeding programs (Naoi and Hataya, 2021).

Chrysanthemum (*Chrysanthemum morifolium*) is an important ornamental crop that is vegetatively propagated and suffers damage from chrysanthemum stunt viroid (CSVd) infection globally. Because it can be easily transmitted mechanically from plant to plant during propagation, resistant cultivars that could be used for breeding purposes would be a great advantage for disease control. Promising reports to control the disease by the breeding of CSVd-resistant chrysanthemum plants are in the literature (Nabeshima et al., 2012, 2018). Interspecific hybrids between susceptible chrysanthemum cultivars and a resistant cultivar, "Okayama Heiwa," expressed resistance in the first hybrid generation, although the mechanisms of resistance remained unclear. A later study in Japan of the moderately resistant cultivar "Utage" revealed that, although not resistant to infection, CSVd titers were very low in the shoot apical meristems in inbred seedlings a few months after infection. Recently, several screening programs in Japan have demonstrated the presence of resistance to CSVd in commercial cultivars of chrysanthemum (see review by Nabeshima et al., 2018).

Cross-protection

Infection of a host plant with a viroid that causes only mild symptoms (known as the protecting or mild strain) may protect the plant from infection by severe strains of the same or closely related viroid (Fig. 1D). This protection measure is known as cross-protection or homologous interference. Fernow (1967) described this phenomenon for mild and severe strains of PSTVd in tomato and Niblett et al. (1978) demonstrated that the protecting and challenge strains could be different viroid species in tomatoes, although they members of the same viroid family, for example, pospiviroids or avsunviroids. For example, CSVd and mild and severe strains of PSTVd protected chrysanthemum against CEVd, but not CChMVd, and a mild strain of PSTVd protected plants from the severe symptoms that are induced upon infection with the severe strain of PSTVd. In cross-protection, both the protecting and challenge strains replicate in the infected plants although symptom development by a severe strain is delayed. In potato, Singh et al. (1990) showed that in the highly susceptible Russet Burbank cultivar preinfected with a mild strain of PSTVd and challenged with a severe strain of PSTVd, the latter was not detected in any of the inoculated plants. However, in the tolerant potato, BelRus cultivar preinfected with the mild strain, the severe strain was detected in two of ten plants, and cross-protection was not completely achieved. In the same study, the second generation potato plants grown from tubers derived from the cross-protected plants were completely protected from challenge infection with mild or severe strains of PSTVd. The mechanism of cross-protection remains unclear; however, it appears to be sequence specific and may be mediated by various defense mechanisms in the host plant that may be protein or RNA-mediated.

Cross-protection is generally not recommended as a control measure for viroids in a field situation due to environmental risks. The infected plants serve as a reservoir for cross-contamination and the mild strain may mutate to a more severe strain in some plants, resulting in more severe disease that may result from the mixed infection.

Elimination through tissue culture

The application of heat and cold therapy for the elimination of viroids from infected plant material has had mixed results, although the combination of a variety of in vitro tissue culture methods has been applied successfully to several viroid/host combinations (Fig. 1C). Although the methods can take several months to complete, these strategies are especially applicable for pathogen eradication in valuable germplasm (Barba et al., 2017; Kovalskaya and Hammond, 2014).

The combination of cryotherapy and meristem-tip culture was efficient in the elimination of viroids from potato and chrysanthemum. For example, cold treatment (5–8°C) and low light, combined with meristem tip culture were successful in eliminating PSTVd from an infected potato clone (Lizárraga et al., 1980) and prolonged cold treatment (5°C for 3–6 months) of chrysanthemum infected with CSVd, CChMVd or CPFVd yielded viroid-free meristems. In peach and pear plants, cold therapy at 4°C for 3 weeks led to HSVd elimination in 18% of plants (El-Dougdoug et al., 2010), where the storage of HLVd-infected hops at low temperatures (2–4°C) in the dark for several months followed by meristem culture led to 36% recovery of viroid-free plants (Adams et al., 1996).

The application of thermotherapy for viroid elimination has had mixed results in certain viroid/host combinations (Barba et al., 2017). For example, thermotherapy (30–37°C) was not effective in eliminating PSTVd from potato, CSVd from chrysanthemum, or HSVd from pears and peaches (El-Dougdoug et al., 2010); however, Postman and Hadidi (1995) reported elimination of ASSVd from pears with a combination of thermotherapy and culture of apical meristems. Other examples of the use of a combination of thermotherapy and tissue culture to eliminate viroids from grapes and apples have been reported (Desvignes et al., 1999; Gambino et al., 2011; Howell et al., 1998).

In summary, excision of shoot apical meristems (SAMs) after thermo/cryotherapy treatment combined with meristem tip culture seems to be more effective in eliminating viroids than thermo/cryotherapy treatment alone. Each viroid/host combination requires development of specific methods for effective control (Barba et al., 2017).

Chemotherapy with the broad-spectrum antiviral agent, ribavirin, although resulting in reduced viroid titer and remission of viroid symptoms in CEVd or PSTVd-infected plants, did not eliminate viroids from infected material (see Kovalskaya and Hammond, 2014).

Horticultural practices

Most viroids are mechanically transmissible and stable in the plant sap, and as the mechanical transmission is the major route of viroid spread, several studies have identified disinfectants that can be used in greenhouse and field conditions to reduce the infectivity of contaminated surfaces (Ling, 2017) (Fig. 1B). Several chemical substances have been found to inactivate viroids and disinfect surfaces exposed to infected plant sap (examples are shown in Table 1). Mackie et al. (2015) found that PSTVd remained infectious in tomato sap on common surfaces, including wood, rubber tires, metal, and plastic, for up to 24h. The

TABLE 1 Disinfectants used to inactivate viroids.

Disinfectant	Concentration	Viroid	References
Sodium hypochlorite	0.25% (2% available chlorine)	CEVd	Garnsey and Whidden (1972)
Virkon S	0.5%–2%	PSTVd	Li et al. (2015), Mackie et al. (2015)
Hydrogen peroxide	2%–6%	ASBVd	Desjardins et al. (1987)
1% or 2% Formaldehyde + 2% sodium hydroxide		ASBVd	Desjardins et al. (1987)
		CEVd	Garnsey and Whidden (1972)
Nonfat dried skim milk	20% wt/vol	PSTVd	Mackie et al. (2015)
Pepper plant leaves	10% wt/vol	TCDVd	Matsuura et al. (2010)
Chlorox regular bleach (household)	10%–20% dilution (0.5%–1.0% sodium hypochlorite) (fresh dilutions)	CEVd	Roistacher et al. (1969)
		PSTVd	Singh et al. (1989)

C. Viroid pathogenesis and viroid-host interaction

effectiveness of disinfectants at inactivating the viroid revealed that the most effective disinfectant was a dilution of household bleach (1:4) or 20% nonfat dried milk, while 0.5% and 1% Virkon S were ineffective in inactivating PSTVd in infective sap (Mackie et al., 2015). In addition, the commercial product MENNO® clean, the only approved product for PSTVd decontamination in several countries, was not found to be effective at the recommended contact time for the recommended minimal concentration, and Virkon S and other commercial products were not found to be effective against dried sap droplets (Olivier et al., 2015). Interestingly, Matsuura et al. (2010) found that, although 0.5% sodium hypochlorite completely degraded TCDVd, crude extracts of the bell and sweet pepper leaves ground in 0.2 M phosphate buffer reduced the infection rate of TCDVd to 25%. Establishing stringent cultural practices that minimize the potential of viroid spread include decontamination of cutting tools, equipment, greenhouse benches, removal of dried plant debris, and although using chemical methods is an important component of disease control, the development and practice of stringent cultural activities that minimize the possibility of transfer of viroids in managed growing conditions is critical.

Detection, prevention, and exclusion

The international movement of infected germplasm has contributed to the worldwide distribution of viroids. In addition to strict hygiene measures to prevent transmission, the most effective means of viroid control is the prevention of the introduction of infected plant material by use of seed and germplasm certified to be viroid-free (Fig. 1A). Identification of infected material relies on rapid, sensitive, and specific detection methods (described in Chapter 16). Early methods of viroid detection involved biological assays on susceptible indicator plants, for example, tomato for pospiviroids. After the nature of the viroid physical and chemical structure was elucidated, gel electrophoresis methods were developed for the identification, purification, and molecular characterization of viroids. Nucleic acid hybridization using viroid-specific reagents allowed for relatively rapid and low cost large scale indexing of plant samples on solid supports. Finally, the development of several reverse-transcription polymerase chain reaction (RT-PCR) methods and high throughput sequencing technologies has allowed the detection of viroid molecules with increasing specificity and sensitivity. Although the latter methods require sophisticated equipment, trained personnel, and costly reagents, they have been incorporated into certification schemes worldwide.

Quarantine regulations in most countries require an assay for apple scar skin viroid (ASSVd) in imported pome fruit germplasm to prevent its introduction. Propagation from clean nursery stock (ASSVd-indexed) mother trees and removal of infected trees from orchards are effective to control measures to reduce viroid spread.

With the increase in international trade of vegetatively propagated germplasm and seeds, plant health regulatory agencies have implemented phytosanitary requirements for several viroid species (Chapter 18). NAPPO for North America and EPPO for European Union-other countries have their own regulatory agencies. Implementation of strict quarantine measures resulted in the successful eradication of PSTVd in seed potato production in Canada and the United States (Singh, 2014; Sun et al., 2004). Quarantine is a set of legislative and regulatory measures that are designed to exclude a pathogen by minimizing the introduction, transport, and spread of an organism. Certification is one means to ensure that plants are maintained under accepted conditions and that phytosanitary requirements required for movement, that is, they are free of specified pests or pathogens, are met. The certification schemes outline all the steps that are necessary to obtain healthy plants and list the pathogens that must be excluded from that plant species (Chapter 18).

Transgenic methods of control

The RNA secondary structure and replication intermediates of viroids expose them to the host RNA silencing system where they are vulnerable to targeted degradation (Chapters 11 and 15). Examination of these features led to the development of viroid resistance strategies to attenuate or prevent viroid infection using molecular transformation of the plant (Fig. 1D). Such strategies include the expression of ribonucleases that target double-stranded RNAs and catalytic antibodies endowed with intrinsic ribonuclease activity, antisense and sense RNAs, catalytic antisense RNAs derived from hammerhead ribozymes, hairpin RNAs and artificial small RNAs for RNA interference (Flores et al., 2017). Potential applications of CRISPR to target viroid RNAs for inactivation have also been proposed. These strategies are summarized below.

Protein-mediated

Ribonucleases

Engineered expression of ribonucleases (RNases) targeting dsRNA viroid replication intermediates was explored as a means of introducing resistance to PSTVd and CSVd in host plants. A naturally occurring, yeast derived dsRNA-specific RNase, Pac1, which digests long dsRNAs into short oligonucleotides, was introduced into potato plants by transgenic methods and the resulting transgenic lines were subsequently challenged with PSTVd. Five of the lines, and their progeny, were shown to be free of viroid infection (Sano et al., 1997). In chrysanthemum, transgenic expression of the Pac1 protein did not result in any abnormal phenotypes and, when challenged with CSVd, lines expressing low levels of the protein became infected and developed stunting symptoms. In the lines expressing Pac1 at the highest level, viroid RNA was only detected in 20% of the lines and the plants were asymptomatic. These combined results suggest the potential application of the technology to the development of viroid resistant herbaceous hosts.

An extension of the use of ribonucleases is a novel approach based on the transgenic expression of catalytic single-chain variable antibodies (scFv's) that possess RNase and DNase activity (3D8 scFv), thereby catalyzing the hydrolysis of nucleic acids in a sequence-specific manner. When the 3D8 scFv gene was expressed in transgenic chrysanthemum plants under the control of the constitutive Cauliflower mosaic virus (CaMV) 35S transcriptional promoter (Tran et al., 2016), several of the lines resulted in increased resistance to CSVd infection. The mechanisms of resistance are unknown and this method has not been deployed in commercial cultivars.

Nucleic acid-mediated

Most transgenic control strategies for viroids have been developed to interfere with viroid replication or result in viroid RNA degradation through nucleic acid-based technology.

Antisense and sense RNAs

Matoušek et al., 1994 first demonstrated that an antisense RNA directed against either the plus or minus sense strand of PSTVd RNA formed complexes with the target RNA in vitro. When integrated into transgenic potato plants, a significant reduction of PSTVd accumulation in plants challenged with the viroid was observed in all transgenic lines at 4 weeks postinoculation, however, at 6–8 weeks postchallenge severely infected plants were observed in all plant lines Matoušek et al. (1994). In 1995, transgenic tomato seedlings expressing antisense constructs targeting the minus sense strand of CEVd resulted in a moderate reduction in the accumulation of the viroid after challenge inoculation with CEVd, whereas targeting the plus strand resulted an increase in viroid accumulation (Atkins et al., 1995). Incorporation of ribozyme motifs in the constructs did not enhance activity. On a positive note, strong resistance to CSVd was reported in 9 of 16 lines by Jo et al. (2015) who developed genetically modified chrysanthemum plants (commercial cultivar "Vivid Scarlet") containing sense or antisense constructs of CSVd.

RNAi

RNA interference (RNAi) or posttranscriptional gene silencing (PTGS) is a conserved biological response to double-stranded RNA that mediates resistance to both endogenous and exogenous nucleic acids. Viroids are strong inducers of RNA silencing due to their mode of replication. The finding of viroid-derived small RNAs (vd-sRNAs) in plants infected with viroids suggests that they trigger the RNA silencing mechanism in the plant host. RNAi-mediated resistance was investigated by expressing a hairpin RNA (hpRNA) construct derived from PSTVd in tomato plants (Schwind et al., 2009). The plants were resistant to infection by PSTVd and resistance was correlated with a high level accumulation of hpRNA-derived silencing RNA (siRNAs) targeting the PSTVd genome. In a separate study, seven partial or truncated versions of the PSTVd genome were designed as specific hpRNAs based on functional domains of PSTVd and regions known to trigger RNA silencing in infected plants, the so-called "hot-spots" (Adkar-Purushothama et al., 2015). Of the 21 transgenic *Nicotiana benthamiana* lines generated, five displayed reduced viroid accumulation upon challenge with PSTVd and that correlated with high levels of hpRNA derived sRNAs and demonstrated that a hpRNA as small as 26–49-nt results in inhibition of PSTVd infection.

Catalytic RNAs

Ribozymes are catalytic RNAs that can cleave complementary RNAs at specific sites. Hammerhead ribozymes are unique ribozymes that form branched RNA structures and can catalyze self-cleavage through specific intermolecular interactions. A hammerhead ribozyme targeting the minus-sense strand of PSTVd in potato plants resulted in 68% of the transgenic lines possessing high levels of resistance to challenge inoculation with PSTVd (Yang et al., 1997). The resistance was stably inherited to the vegetative progenies. A trans-cleaving hammerhead ribozyme derived from PLMVd was investigated for its ability control of PSTVd infection in *N. benthamiana* when they were co-expressed transiently in tobacco leaves. The results suggested that the ribozyme may target the primary PSTVd transcript and oligomeric RNA replication intermediates and that constitutive expression in transgenic plants may efficiently control viroid infection (Carbonell et al., 2011).

Genome editing

Genome editing using CRISPR-Cas systems employs RNA-guided nucleases that can be programmed to target specific nucleotide motifs on the viroid RNA molecule that may be involved in replication and movement (Chapter 4). For example, potential guide RNA targets in the pospiviroid PSTVd include RNA motifs in the terminal left (TL) domain that contains the proposed initiation site for (−) strand RNA synthesis, secondary hairpins I and II, and the RY motif in the terminal right (TR) domain that interacts with a host viroid-binding protein 1 (ViRP1) that is proposed to play a role in viroid infection by transferring the viroid molecule to the host nucleus, the site of replication (Hadidi, 2019). PSTVd mutants defective in binding to ViRP1 do not move systemically. In addition, suppression of the host ViRP1 in protoplasts resulted in defective viroid replication. In the avsunviroids, there are predicted transcription initiation and RNA processing sequence motifs in PLMVd and in ASBVd that could be targeted (Hadidi, 2019; Hadidi and Flores, 2017). PLMVd interacts with the eukaryotic transcription factor eEF1A and downregulation of this protein may result in reduced accumulation of viroid RNA. Therefore, either the viroid molecule itself or a gene encoding a protein that is critical for viroid replication or movement may be a target for CRISPR editing to develop resistance to these pathogens. To date, there are no reports in the literature of CRISPR being used experimentally to target viroids or their interacting proteins.

Perspective/future implications

Currently, the most effective methods to control viroid disease include the use of viroid-free planting materials, which rely on rapid and sensitive detection methods, and the implementation of stringent disinfection protocols to inactivate viroids on surfaces and prevent mechanical transmission. If valuable germplasm is infected, there are several lengthy thermo/cryotherapy and tissue culture methods that can be used to obtain viroid-free planting materials. In addition, naturally occurring resistance has been found in some crops (e.g., chrysanthemum) but is limited in main cultivated crops. Transgenic resistance employing nucleases or modified RNA molecules for viroid control has been reported in crop species with promising results and CRISPR technology is on the horizon. Although the transgenic technology has been developed, it has not yet been deployed commercially, and it is still unknown if engineered, durable, multivalent resistance to viroids will be achievable and acceptable.

Questions for the reader

1. How does understanding the replication strategy and infection process of viroids inform the development of control measures?
2. What is the difference between host resistance versus nonhost resistance?

3. What methods can be used to prevent transmission of viroids from viroid-infected material?
4. What methods have been used to eliminate viroids from already infected host plants?
5. Describe two strategies that have been used to introduce viroid resistance by transgenic methods and their efficacy?

References

Adams, A.N., Barbara, G.J., Morton, A., Darby, P., 1996. The experimental transmission of hop latent viroid and its elimination by low temperature treatment and meristem culture. Ann. Appl. Biol. 128, 37–44.

Adkar-Purushothama, C.R., Kasai, A., Sugawara, K., Yamamoto, H., Yamazaki, Y., He, Y.-H., Takada, N., Goto, H., Shindo, S., Harada, T., Sano, T., 2015. RNAi mediated inhibition of viroid infection in transgenic plants expressing viroid-specific small RNAs derived from various functional domains. Sci. Rep. 5, 17949.

Atkins, D., Young, M., Uzzell, S., Kelly, L., Fillati, J., Gerlach, W.L., 1995. The expression of antisense and ribozyme genes targeting citrus exocortis viroid in transgenic plants. J. Gen. Virol. 76, 1781–1790.

Barba, M., Hosakawa, M., Wang, Q.-C., Taglienti, A., Zhang, Z., 2017. Viroid elimination by thermotherapy, cold therapy, tissue culture, in vitro micrografting, or cryotherapy. In: Hadidi, A., Flores, R., Randles, J.W., Palukaitis, P. (Eds.), Viroids and Satellites. Academic Press, London, UK, pp. 425–435.

Carbonell, A., Flores, R., Gago, S., 2011. Trans-cleaving hammerhead ribozymes with tertiary stabilizing motifs: in vitro and in vivo activity against a structured viroid RNA. Nucleic Acids Res. 39, 2432–2444.

Desjardins, P., Saski, P., Drake, R., 1987. Chemical inactivation of avocado sunblotch viroid on pruning and propagation tools. Calif. Avocado Soc. Yearb. 71, 259–262.

Desvignes, J.C., Grasseau, N., Boye, R., Cornaggia, D., Aparicio, F., Di Serio, F., Flores, R., 1999. Biological properties of apple scar skin viroid: isolates, host range, different sensitivity of apple cultivars, elimination, and natural transmission. Plant Dis. 83, 768–772.

Dimond, A.E., Davis, D., Chapman, R.A., Stoddard, E.M., 1952. Plant chemotherapy as evaluated by the Fusarium wilt assay on tomatoes. In: Conn. Agr. Expt. Sta. Bull., No. 557. 82 pp.

El-Dougdoug, K.A., Osma, M.E., Abdelkader, H.S., Dawoud, R.A., Elbaz, R.M., 2010. Elimination of Hop stunt viroid (HSVd) from infected peach and pear plants using cold therapy and chemotherapy. Aust. J. Basic Appl. Sci. 4, 54–60.

Fernow, K.H., 1967. Tomato as a test plant for detecting mild strains of potato spindle tuber virus. Phytopathology 57, 1347–1352.

Flores, R., Navarro, B., Kovalskaya, N., Hammond, R.W., Di Serio, F., 2017. Engineering resistance against viroids. Curr. Opin. Virol. 26, 1–17.

Gambino, G., Navarro, B., Vallania, R., Gribaudo, I., Di Serio, F., 2011. Somatic embryogenesis efficiently eliminates viroid infections from grapevines. Eur. J. Plant Pathol. 130, 511–519.

Garnsey, S.M., Whidden, R., 1972. Decontamination treatments to reduce the spread of citrus exocortis (CEV) by contaminated tools. Proc. Fla. Stn. Hortic. Soc. 84, 63–65.

Hadidi, A., 2019. Next-generation sequencing and CRISPR/Cas13 editing in viroid research and molecular diagnostics. Viruses 11, 120.

Hadidi, A., Flores, R., 2017. Genome editing by CRISPR-based technology: potential applications for viroids. In: Hadidi, A., Flores, R., Randles, J.W., Palukaitis, P. (Eds.), Viroids and Satellites. Academic Press, London, UK, pp. 531–540.

Howell, W.E., Burgess, J., Mink, G.I., Zhang, Y.P., 1998. Elimination of apple fruit and bark deforming agents by heat therapy. Acta Hortic. 472, 641–646.

Jo, K.M., Jo, Y., Choi, H., Chu, H., Lian, S., Yoon, J.Y., Choi, S.K., Kim, K.H., 2015. Development of genetically modified chrysanthemums resistant to chrysanthemum stunt viroid using sense and antisense RNAs. Sci. Hortic. 195, 17–24.

Kovalskaya, N., Hammond, R.W., 2014. Molecular biology of viroid-host interactions and disease control strategies. Plant Sci. 228, 48–60.

Li, R., Baysal-Gurel, F., Abdo, Z., Miller, S.A., Ling, K.-S., 2015. Evaluation of disinfectants to prevent mechanical transmission of viruses and a viroid in greenhouse tomato production. Virol. J. 12, 5.

Ling, K.-S., 2017. Decontamination measures to prevent mechanical transmission of viroids. In: Hadidi, A., Flores, R., Randles, J.W., Palukaitis, P. (Eds.), Viroids and Satellites. Academic Press, Oxford, UK, pp. 437–445.

Lizárraga, R.E., Salazar, S.F., Roca, W.M., Schilde-Rentschler, L., 1980. Elimination of potato spindle tuber viroid by low temperature and meristem culture. Phytopathology 70, 754–755.

Mackie, A.E., Coutts, B.A., Barbetti, M.J., Rodoni, B.C., McKirdy, S.J., Jones, R.A.C., 2015. *Potato spindle tuber viroid*: stability on common surfaces and inactivation with disinfectants. Plant Dis. 99, 770–775.

Matoušek, J., Shröder, A.R., Trněná, L., Reimers, M., Baumstark, T., Dědic, P., Vlasák, J., Becker, I., Kreuzaler, F., Fladung, M., 1994. Inhibition of viroid infection by antisense RNA expression in transgenic plants. Biol. Chem. Hoppe Seyler 375, 765–777.

Matsuura, S., Matsushita, Y., Usugi, T., Tsuda, S., 2010. Disinfection of *Tomato chlorotic dwarf viroid* by chemical and biological agents. Crop Prot. 29, 1157–1161.

Nabeshima, T., Hosokawa, M., Yano, S., Ohishi, K., Doi, M., 2012. Screening of Chrysanthemum cultivars with resistance to Chrysanthemum stunt viroid. J. Jpn. Soc. Hortic. Sci. 81, 285–294.

Nabeshima, T., Matsushita, Y., Hosokawa, M., 2018. Chrysanthemum stunt viroid resistance in *Chrysanthemum*. Viruses 10, 719.

Naoi, T., Hataya, W., 2021. Tolerance even to lethal strain of potato spindle tuber viroid found in wild tomato species can be introduced by crossing. Plants 10, 575.

Niblett, C.L., Dickson, E., Fernow, K.H., Horst, R.K., Zaitlin, M., 1978. Cross protection among four viroids. Virology 91, 198–203.

Olivier, T., Sveikauskas, V., Grausgruber-Gröger, S., Virscek Marn, M., Faggioli, F., Luigi, M., Pitchugina, E., Planchon, V., 2015. Efficacy of five disinfectants against *Potato spindle tuber viroid*. Crop Prot. 67, 257–260.

Postman, J., Hadidi, A., 1995. Elimination of apple scar skin viroid from pears by in vitro thermotherapy and apical meristem culture. Acta Hortic. 386, 536–543.

Roistacher, C.N., Calavan, E.C., Blue, E.L., 1969. Citrus exocortis virus—chemical inactivation on tools, tolerance to heat and separation of isolates. Plant Dis. Rep. 53, 333–336.

Sano, T., Nagayama, A., Ogawa, T., Ishida, I., Okada, Y., 1997. Transgenic potato expressing a double-stranded RNA-specific ribonuclease is resistant to potato spindle tuber viroid. Nat. Biotechnol. 15, 1290–1294.

Schwind, N., Zwiebel, M., Itaya, A., Ding, B., Wang, M.B., Krczal, G., Wassenegger, M., 2009. RNAi-mediated resistance to potato spindle tuber viroid in transgenic tomato expressing a viroid hairpin RNA construct. Mol. Plant Pathol. 10, 459–469.

Singh, R.P., 2014. The discovery and eradication of potato spindle tuber viroid in Canada. Virus Dis. 25, 415–424.

Singh, R.P., Boucher, A., Sommerville, T.H., 1989. Evaluation of chemicals for disinfection of laboratory equipment exposed to potato spindle tuber viroid. Am. Potato J. 66, 239–246.

Singh, R.P., Boucher, A., Sommerville, T.H., 1990. Cross-protection with strains of *Potato spindle tuber viroid* in the potato plant and other *Solanaceous* hosts. Phytopathology 80, 246–250.

Sun, M., Siemsen, S., Campbell, W., Guzman, P., Davidson, R., Whitworth, J.L., Bourgoin, T., Axford, J., Schrage, W., Leever, G., Westra, A., et al., 2004. Survey of potato spindle tuber viroid in seed potato growing areas of the United States. Am. J. Potato Res. 81, 227–231.

Tran, D.T., Cho, S., Hoang, P.M., Kim, J., Kil, E.-J., Lee, T.-K., et al., 2016. A codon-optimized nucleic acid hydrolyzing single-chain antibody confers resistance to chrysanthemum stunt viroid infection. Plant Mol. Biol. Report. 34, 221–232.

Yang, X., Yie, Y., Zhu, F., Liu, Y., Kang, L., Wang, X., Tien, P., 1997. Ribozyme-mediated high resistance against potato spindle tuber viroid in transgenic potatoes. Proc. Natl. Acad. Sci. U. S. A. 94, 4861–4865.

C. Viroid pathogenesis and viroid-host interaction

Policies, regulations, and production of viroid-free propagative plant materials for sustainable agriculture

Irene Lavagi-Craddock[a], Scott Harper[b], Robert Krueger[c], Paulina Quijia-Lamiña[a], and Georgios Vidalakis[a]

[a]Citrus Clonal Protection Program and the University of California, Riverside, CA, United States
[b]Clean Plant Center Northwest, Washington State University, Prosser, WA, United States
[c]USDA-ARS National Clonal Germplasm Repository for Citrus and Dates, Riverside, CA, United States

Graphical representation

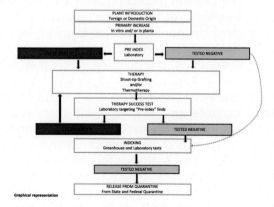

Pathways to produce viroid-free clean plant material.

Definitions

Budwood: Short pieces of young stems, free of thorns and leaves, with buds used for grafting onto rootstocks for the vegetative production of plants.

Bud: A lateral or apical undeveloped shoot, leaf, or flower occurring on the stems of plants.

Clean plant program: A program developed by a country, state, university, or research center to ensure the propagative plant materials distributed to nurseries and growers is free of regulated graft-transmissible pathogens and the fruit is true-to-type. The source plants or trees are usually registered with a state or other regulatory program, and propagative plant materials obtained from these plants can be used to produce additional registered source plants.

CFR: The Code of Federal Regulations (CFR) is an annual codification of the general and permanent rules published in the Federal Register by the executive departments and agencies of the United States (US) federal government.

Foundation block: Depending on the crop, a field planting or plantings in a protective structure (e.g., greenhouses, screenhouses, or other) with restricted access to authorized personnel, where targeted-pathogen-tested negative source plants, propagative material from which can be used for clonal propagation by nurseries, are maintained.

Germplasm: Plant or animal tissues or organs (seeds, budwood, pollen, sperm, ova) that may be used to propagate new organisms and pass to those organisms the genetic information contained in the propagative material.

Meristem: A plant tissue consisting of actively dividing, undifferentiated cells that can divide, multiply, and differentiate to form different tissues and organs of a plant. Meristems occurring at the tips of stems or roots are called apical meristems, while those occurring at other locations are called axillary meristems.

Marketing order: Agricultural policy that allows producers to promote the marketing of a particular commodity (managing supply, demand, or price) and, in some cases, support research with collective funds.

PCIP: A Plant Controlled Import Permit (PCIP) is issued by United States Department of Agriculture, Animal and Plant Health Inspection Service (USDA-APHIS) allowing the importation of propagative materials of "prohibited" plants (i.e., Not Authorized Pending Pest Risk Analysis) with specific requirements as to facilities, testing, therapeutics, etc. A PCIP is issued with the objective of eventually making the variety imported available for commercial propagation, distribution, and planting, rather than for specific research purposes.

Plant quarantine: Plant quarantines, intended as a set of legislative and regulatory measures and associated activities, are designed to safeguard against harmful pests/pathogens exotic to a country or a region. Quarantines are a method of exclusion to prevent or minimize the introduction, transport, and spread of harmful organisms, including viroids, by means of human activities. A process for ensuring disease- and pest-free plants by isolating them while performing tests to detect the presence of a problem.

The Plant Quarantine Act: A US federal law that regulated the importation and movement of nursery stock and other plants and plant products into the United States to control the dissemination of harmful plant pests and diseases. Pursuant to the Act, the Animal and Plant Health Inspection Service was given the authority to regulate the importation and interstate movement of nursery stock and other plants that may carry pests and diseases harmful to agriculture. The Plant Quarantine Act prohibited any person from importing or bringing nursery stock into the United States or accepting delivery of nursery stock moving from a foreign country into or through the United States unless the movement complied with regulations enacted by the Secretary of Agriculture to prevent dissemination of plant pests and diseases. This Act has been superseded by the Plant Protection Act of 2000, the Animal and Plant Health Inspection Service (APHIS) statute that consolidated related responsibilities previously spread over various legislative statutes (i.e., the Plant Quarantine Act, the Federal Plant Pest Act, and the Federal Noxious Weed Act of 1974).

Permit: A document typically issued by a state or federal government allowing specific actions involving biological materials (specifically in this chapter plant parts or nucleic acids) to be performed by the permit holder or the facility it refers to. Examples include permits to import plants and permits to move infected and noninfected plant materials or nucleic acids extracted from these plants. Failure to obtain and present the correct documents showing compliance with US law will lead to a Customs Agricultural Inspection Hold by US Customs and Border Protection ("Customs"). The imported article will fail a United States Department of Agriculture (USDA) inspection at the US port of entry.

Plant Protection and Quarantine Permit: The Plant Protection and Quarantine (PPQ) Permit, issued by the Plant Protection and Quarantine branch of USDA-APHIS, specifies the conditions under which plants, plant parts, plant products, or products with plant-origin ingredients may be brought into the United States.

Select plant agents: These include some fungal and bacterial strains that have been determined to have the potential to pose a severe threat to plant products.

Sustainable agriculture: An agricultural approach that uses state-of-the-art, science-based practices to maximize productivity, profit, and environmental and health benefits while minimizing negative effects on the environment and natural resources that could compromise the ability of current or future generations to meet their agricultural needs.

Therapy: The process of removing a pathogenic element from an infected plant or plant propagative unit to produce a "clean plant," that is a plant without the pathogenic element. Different therapy methods exist including thermotherapy, which is the treatment of infected plant material by heat, and shoot-tip grafting, where a shoot tip or meristem tip is grafted aseptically onto a decapitated seedling rootstock germinated in vitro.

USDA-APHIS: United States Department of Agriculture, Animal and Plant Health Inspection Service. It retains the plant quarantine function.

Viroid indexing: Greenhouse tests (bio-indexing), and molecular tests (lab-indexing) that confirm the presence or absence of a viroid.

Chapter outline

This chapter will present an overview of the existing policies, regulations, and production of viroid-free propagative plant material for sustainable agriculture. A general picture of how the spread of viroids can be prevented is depicted through specific examples of regulations and policies governing the introduction and distribution of perennial plant species germplasm, including citrus, apple, hop, and date palm trees. This chapter also provides general information on issues to consider when conducting research using plant materials and provides an example of how to apply for permits and phytosanitary certificates.

Learning objectives

On completion of this chapter, you should be able to:

- Objective 1: Understand the scope of plant quarantine programs and the services offered by plant protection agencies.
- Objective 2: Discuss the differences in the process of producing viroid-free clean plant propagative materials differs depending on the crop.
- Objective 3: Apply for a permit to move plant materials and a phytosanitary certificate with USDA-APHIS.

Fundamental introduction

Introducing viroid diseases into a country, state, or local geographic area can have potentially catastrophic consequences. Viroid plant hosts include herbaceous (e.g., potatoes, chrysanthemum) and woody species (e.g., citrus, pome and stone fruit trees, palms, avocado), and comprise agronomic and horticultural as well as ornamental species. Several viroid-induced diseases are of considerable economic importance due to the significant yield losses they can cause. A few such examples include potatoes infected with potato spindle tuber viroid (PSTVd), chrysanthemum infected with chrysanthemum stunt viroid (CSVd), citrus infected with citrus exocortis viroid (CEVd), coconut palms infected with coconut cadang-cadang viroid (CCCVd), hops and stone fruits infected with hop stunt viroid (HSVd), apples infected with apple scar skin viroid (ASSVd) or apple fruit crinkle viroid (AFCVd), peaches infected with peach latent mosaic viroid (PLMVd), and avocado infected with avocado sunblotch viroid (ASBVd) (Hadidi et al., 2017a; Rodriguez et al., 2017; Verhoeven et al., 2017). Pathogenic viroids can be easily distributed through infected propagative materials, or by cross-contamination following mechanical damage during horticultural practices, and depending on the plant species, also by insect vectors, seed, or pollen, thus becoming a potential hazard to future plantings. The principal means of viroid spread is through the propagation of infected stock, and the current rapid movement of people and plants by air transportation poses high risks for the introduction of infected plant materials by uninformed individuals. Moreover, clear diagnostic symptoms are not always obvious on viroid-infected trees in the field. Trees infected with viroids can be symptomless in various hosts, thus viroids can unknowingly be spread throughout a country by clonal propagation or by mechanical transmission. In some cases, viroids present in asymptomatic hosts can cause symptoms and disease if spread to a different host. At present, no spray or chemical can be used to cure viroid infections. Therefore, comprehensive programs to produce, maintain, and distribute propagative materials testing negative for targeted graft-transmissible viroid pathogens were initiated and currently operate in different parts of the world for various crops. Indeed, as with other

plant diseases, the most effective means of controlling viroid diseases is by exclusion or eradication of infected materials.

The availability of clean, viroid negative-tested propagative plant materials, is recognized
as essential to the establishment and maintenance of viable, sustainable, and competitive industries of various commodities, including citrus, fruit trees, and palm trees. Plants produced
using materials from a clean plant program are healthy, more uniform, higher-yielding with
better-colored fruit, and thus much more profitable to the grower. Well-established plant
indexing programs clearly show the economic benefits of such programs to the grower.
For these reasons, the services offered by plant protection agencies, that is, phytosanitary regulations and policies established by individual countries or organizations, are designed to
ensure effective clean plant programs can operate and prevent the entry and spread of viroids. Effective clean plant programs depend on reliable diagnostics; as new plant stock varieties are imported into a country or state, detecting viroids, among other systemic
pathogens, can ensure that infected plant materials are not propagated for distribution to
nurseries and growers. If infected plant materials are intercepted, viroids can be eliminated
by different therapy methods (e.g., tissue culture or thermotherapy), and the resulting viroid-
free plant stock can then be propagated. Therefore, clean plant programs prevent viroid
spread by intercepting infected plant materials, removing pathogens from them through different therapy methods if necessary, propagating, and distributing viroid-free tested clean
propagative plant materials.

It is important to appreciate that clean plant programs must comply with a series of quarantine requirements, including facility requirements (e.g., monthly inspections by USDA-
APHIS in the United States) for the location where the source plants can be established,
for example (e.g., screenhouses), to avoid infection by other graft-transmissible pathogens.
Other requirements include record keeping, training, and reporting to the local, regional,
state, national, and international agencies and organizations dealing with plant protection
services. Moreover, establishing a comprehensive clean plant program with a foundation
block, a laboratory, and greenhouse operations must be supported financially as a long-term
project. Other aspects critical to the success of a germplasm program include the training of
personnel and an education and outreach component to inform plant propagators (nurseries
or others), growers, and the public of the dangers of introduced pathogens and the benefits of
stock free of characterized pathogens.

The interplay between science and the phytosanitary regulatory framework ensures the
development of thriving economies through the production and distribution of plant stock
tested for targeted pathogens and the advancement of science and technology. Scientific experiments and peer-reviewed publications largely determine the benchmark for regulators,
who offer services based on the knowledge provided by science; for example, geographical
distribution, host range, transmissibility, availability of detection methods, and impact on
crop yield are all parameters that help plant protection agencies formulate their quarantine
programs. In addition to providing healthy plant stock, clean plant programs assist researchers by providing invaluable positive controls in the form of known infected plant material from a disease bank or readily extracted nucleic acids. Researchers performing
experiments requiring access to these samples must be well educated in the regulatory framework under which they must operate. What permits do they require to receive that type of

C. Viroid pathogenesis and viroid-host interaction

sample? Typically, this information will be needed in publications when describing how the samples were acquired.

Viroid-free planting stock is key to the cost-effective production of crops such as citrus, hops, date, and fruit trees such as apples, pears, and peaches. For these reasons, it is crucial to appreciate that propagative plant materials are handled under the auspices of plant protection agencies. This chapter focuses on perennial species of agronomic relevance, where they can be most deleterious to tree survival and fruit production.

Chapter

Citrus: The Citrus Clonal Protection Program in California, United States

In the 1930s, the United States and most citrus-growing countries initiated quarantine programs to prevent introducing exotic pests and diseases. In the period 1930–60s, importation of citrus in the United States was restricted to the seed, but the discovery that viral diseases could, in addition to being graft-transmissible, be transmitted via the seed, required a change in the importation and propagation protocols (Ferguson and Grafton-Cardwell, 2014). Citrus is a prohibited species in the United States. Citrus and its relatives are prohibited from entering the United States "in order to prevent the introduction into the United States of … citrus diseases…" per CFR § 319.19. However, citrus "… may be imported into the United States for experimental, therapeutic, or developmental purposes under the conditions specified in a Controlled Import Permit issued per CFR § 319.6." USDA-APHIS issues PCIP which "allow the importation of restricted or not authorized plant materials into the United States."

In California, the 1937 Psorosis Freedom Program, renamed the Citrus Variety Improvement Program in 1957 and finally rebranded as the Citrus Clonal Protection Program (CCPP) in 1977, has allowed diagnostic and therapeutic services against all known graft transmissible pathogens of citrus to be performed, thus playing a pivotal role in maintaining a healthy California citrus industry. In 2009, the CCPP became part of the National Clean Plant Network (NCPN), comprised of clean plant centers, scientists, educators, state and federal regulators, nurseries, and growers of specialty crops that cooperate to ensure that clean plant propagative material is available nationwide.

How does a quarantine program for citrus ensure the production of viroid-free propagative plant materials for sustainable agriculture? In the whole United States, specific permit holders can import propagative citrus plant materials; for example, there is one in Florida, one in California, one in Maryland, and one with the National Clonal Germplasm Repository of Citrus and Dates (NCGRCD). The California holder of these Controlled Import Permits is the Director of the CCPP at the University of California, Riverside, California, United States. He holds a USDA-APHIS-PPQ Plant Controlled Import Permit (PCIP) that allows the introduction of citrus into California and the United States. The Director is the holder of additional permits for inter- and intrastate movement of citrus propagative materials from domestic sources in the United States. presumed to be free of pathogens or potentially infected with graft-transmissible diseases. We will look at the citrus variety introduction program at the

CCPP as an example of a successful germplasm program supporting the sustainable production of citrus in California, valued today at $3.63 billion (Babcock, 2022).

The CCPP was developed to provide a safe mechanism for the introduction of citrus varieties into California for research, variety improvement, or commercial production, avoiding or restricting the spread of pathogens in citrus while providing the California citrus industry and researchers with primary budwood sources of commercially valuable varieties to support a thriving citrus industry, which occupies a prominent position within the sustainable agriculture framework. To achieve its goals, the CCPP incorporates a series of provisions that result in a comprehensive citrus germplasm program. The CCPP introductory pipeline includes the following requirements: 1. Introduction of Citrus Propagative Material; 2. Pathogen Detection or Disease Diagnosis; 3. Pathogen Elimination; 4. Maintenance (retesting & evaluation) of Primary Budwood Sources; 5. Distribution of Citrus Propagative Material; 6. Extension-Outreach and Research.

For California, the interstate movement of citrus germplasm is highly regulated by the federal (USDA-APHIS Plants for Planting Manual) and the state of California (Cal. Admin. Code tit. 3, § 3250 Citrus pests exterior quarantine) regulations. The CCPP introduces citrus varieties at the CCPP Rubidoux Quarantine Facility in Riverside, California, at the location of the original 1907 University of California Citrus Experiment Station, under the appropriate state and federal permits. Domestic varieties are shipped directly to the CCPP. In contrast, varieties from overseas are shipped to the USDA-APHIS, National Plant Germplasm Inspection Station, in Beltsville, Maryland, for an initial visual inspection and then forwarded to the CCPP. Upon arrival at the CCPP, citrus varieties are maintained under quarantine. Sources of introductions are established in vitro to initiate the therapy procedure. The in vitro therapy procedure used at the CCPP is called shoot-tip grafting (STG), which is essentially a propagation where a scion is reduced to a single meristem tip or shoot tip consisting of the meristematic dome and 1–3 leaf primordia, performed at a microscopic scale (Fig. 1).

STG requires apical meristems from newly developing axillary shoots and young succulent rootstock seedlings cultured on growth media under sterile conditions. Meristems, which by definition do not have any differentiated vascular tissues, cannot be easily invaded by pathogens and are therefore the ideal tissue from which to produce a pathogen-free citrus plant. Each microscopic apical meristem tip (<0.2 mm) or shoot tip explant (0.2–0.8 mm) excised from a new flushing shoot is grafted onto a 2- to 3-week-old rootstock seedling germinated in vitro after surface sterilization of a citrus seed without the outer seed coat. A few weeks after STG, the plantlets growing in vitro are removed from the growth medium (Fig. 1). After the removal of their roots, they are grafted onto a fast-growing citrus rootstock seedling to produce the first generation of source trees of the introduced varieties. These source plants must test negative for all known graft-transmissible pathogens of citrus following the variety index (VI) program to become eligible for release from quarantine and use by the industry. The VI protocol currently includes 17 biological and laboratory tests (28 pathogens including seven viroids) and is currently implementing hybrid in vitro and in silico diagnostic tests (e.g., high throughput sequencing) (Roistacher, 1991; Dang et al., 2022; Krueger and Vidalakis, 2022; Vidalakis et al., 2022) performed at the CCPP Rubidoux greenhouses and the University of California, Riverside (UCR) laboratories, respectively. Upon testing negative for all known graft-transmissible pathogens in the VI testing protocol, each variety is assigned a unique number to indicate its status. After the VI is completed with

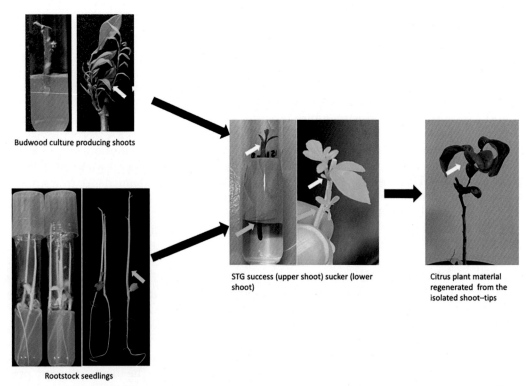

Budwood culture producing shoots

STG success (upper shoot) sucker (lower shoot)

Citrus plant material regenerated from the isolated shoot–tips

Rootstock seedlings

FIG. 1 Key shoot-tip-grafting steps in the therapy process of citrus varieties.

negative results, the Director of the CCPP writes a letter to the state and federal regulatory agencies requesting the release of the VI-tested varieties from quarantine, assigning a VI number to each accession that permanently accompanies all tree propagation records linking nursery stock and field trees to pathogen-tested budwood source at the CCPP (Roistacher, 1991; Vidalakis et al., 2014).

The public domain citrus varieties released from quarantine are deposited in the CCPP Foundation facility at the University of California Agriculture and Natural Resources, Lindcove Research and Extension Center (LREC) for distribution to the industry and citrus hobbyists (protected foundation block) and trueness-to-type evaluation (evaluation foundation block). All CCPP trees at LREC-protected blocks for budwood distribution are registered with the California Department of Food & Agriculture (CDFA) and tested annually for various graft-transmissible pathogens, including viroids, as required by the CDFA regulations (Section 3701, Citrus Nursery Stock Pest Cleanliness Program). In 2009, Senate Bill 140 was passed; it requires the CDFA to establish a mandatory Citrus Nursery Stock Pest Cleanliness Program to protect citrus nursery source propagative trees from harmful diseases, pests, and other risks and threats. The establishment of the Citrus Nursery Stock Pest Cleanliness Program clearly illustrates the more recent developments of the industry-research-regulatory continuum, which in California has a long-standing tradition of promoting a thriving and sustainable citrus industry.

It is paramount to appreciate that the California citrus industry's support from the program's early days was instrumental to its development. The dramatic CCPP progress results from the considerable investments made by the University of California, state and federal agencies, and the generous California citrus industry. In California, the Citrus Research Board (CRB), included in its 1968 Marketing Order, Article 4: "The Board is hereby authorized to carry on or support a program of variety improvement to assure the continued freedom of citrus nursery stock from pathologically harmful viruses and economically undesirable viruses and mutations. The Board may assist or otherwise support citrus registration and certification programs." It is worthy of note to clarify that the term "virus," which means poison in Latin, has been used historically in citrus pathology to describe all harmful graft-transmissible pathogens of citrus including viroids, which were not discovered until the late 1960s. CRB's Article 4 has been indeed crucial for the success of the CCPP, which today stands as a cooperative program involving the University of California, including the UCR and UC Agriculture and Natural Resources (UC-ANR); the CDFA; the US Department of Agriculture-Animal and Plant Health Inspection Service (USDA-APHIS), and the citrus industry of California, which is represented by the California Citrus Nursery Board (CCNB) and the CRB. This comprehensive cooperative program has allowed diagnostic and therapeutic services against all graft-transmissible pathogens of citrus to continue uninterrupted for decades at the CCPP, thus playing a key role in maintaining a healthy California citrus industry.

The CCPP provides a safe mechanism for the introduction of citrus germplasm into California and distributes pathogen-tested and true-to-type citrus propagative material to the California citrus industry, and via the NCPN, to the entire United States.

Budwood orders can be placed via the online budwood ordering system (https://ccpp.ucr.edu/onlineOrdersV2/), which is open to any interested party to promote the use of clean pathogen-tested propagative plant materials across the board, from the individual citrus hobbyist propagating a few trees in their home backyard to large commercial citrus nurseries. Outreach and extension events and the network of UC ANR Farm Advisors present in California provide robust tools for a far-reaching education of the public directed at enhancing the adoption of clean plant practices. The CCPP is dedicated to helping maintain California at the forefront of high-quality citrus nursery and fruit production. Since the original discovery of Psorosis, the first known graft-transmissible disease of citrus, by Dr. Fawcett at the Riverside Experiment station in 1933, the CCP has continuously performed research and implemented novel diagnostic and therapeutic technologies to produce pathogen-tested propagative materials. This continued availability of pathogen-tested propagative materials from the CCPP remains essential for the sustainability and profitability of a thriving California's citrus industry.

Fruit trees—From IR-2 to NCPN

During the late 1930s and early 1940s, significant advances were made in describing and identifying many fruit tree virus, viroid, and phytoplasma-induced diseases, including apple scar skin disease, later found to be caused by ASSVd (Hadidi et al., 2017a,b) and peach calico and peach blotch, later found to be caused by PLMVd (Flores et al., 2017). At this time, there was a recognized need in the United States and other countries, for virus- or disease-free fruit

C. Viroid pathogenesis and viroid-host interaction

tree planting stock for research, as infected trees had caused unexpected outcomes in experimental trials, and to improve the quality of nursery stock being sold to commercial growers (Fridlund, 1980). Within the United States, national consultation between scientists and state and federal regulators led to the formation of the Interregional Research Project IR-2 in July 1955, based at Washington State University's Irrigated Agriculture Research and Extension Center in Prosser, Washington. The original goal of this program was to collect, index, and hold virus-free germplasm, and to distribute this material for research or as nursery stock.

The IR-2 program began with *Prunus* material, as at the time, biological indicators used for the detection of diseases of stone fruit were better understood, though work by IR-2 scientists led to the development of indicators for *Malus*, *Pyrus*, and *Cydonia* and the program began cultivars from these genera in the 1970s (Fridlund, 1980). In the early 1990s, the IR-2 program evolved into National Research Support Project NRSP-5, which had the goal of minimizing the adverse effects of viruses in orchards of the United States by providing virus-free propagation material of important temperate fruit tree varieties from domestic and foreign sources. These were times of major change in the program, with the introduction of enzyme-linked immunosorbent assay (ELISA) and PCR, and an increasing number of plants imported via the program into the United States, including such success stories as Fuji from Japan and Cripps Pink from Australia and was credited with significant reduction in disease incidence in the supported crops across the United States and was estimated in 2003 to provide benefits to nurseries, growers, and consumers on the order of $227 million per year. In 2009, the NRSP-5 program at Prosser, renamed the Clean Plant Center Northwest (CPCNW), became one of the foundational members of the NCPN.

In California, the cherry disease management program merged with the Grapevine program to become Foundation Plant Materials Service (FPMS) in 1958. This program received much material from the Prosser IR-2 center and other sources (Rosenberg and Aichele, 1989), planting foundation orchards in the 1960s and 1970s that were registered with the California Department of Agriculture. This material was made available to propagators and nurseries on a fee-for-service basis. In 2003, FPMS became Foundation Plant Services (FPS), later joining the CPCNW as a founding member of NCPN in 2009.

Meanwhile, in South Carolina, Clemson University has maintained the Southeastern Budwood program since 2001. This was formed as a collaborative effort between growers, nurseries, and other propagators in the Southern US states to test for and reduce viral incidence in budwood blocks used for peach and other stone fruit propagation. This program joined NCPN in 2010, establishing a foundation collection and in 2021, became the Clemson Clean Plant Center (CCPC).

Led by the CPCNW, these three centers coordinate the clean plant process of diagnostics, therapeutics, and maintaining foundation collections, ensuring that there is virus and viroid-tested material available for US propagators and nurseries, preventing disease introduction and spread. Each of the three centers tests for the same economically important and harmful pathogens, using primarily molecular-based methods such as PCR and high throughput sequencing (HTS) to detect pathogens; biological indexing on indicator species such as GF305 for prunus pathogens has largely been discontinued as they have been found to show poor sensitivity, producing a symptomatic reaction of only 25% to 33% of cases when the indicator was PCR positive for the pathogen(s) in question (Harper, unpublished).

At present, a total of 16 pome fruit pathogens and 24 stone fruit pathogens are tested for by PCR, with new and emerging pathogens, as well as rare or less economically important viruses captured by HTS. It is critical to note that the current testing protocols target all pathogens regardless of pathogenicity, as state and federal regulatory agencies wish to protect the fruit tree industry from pathogens that might be a problem given the right combinator of virus (or viroid), host species or cultivar, and environmental conditions. All fruit tree-infecting viroids are regulated at the federal (USDA-APHIS) level, including fruit marking and bark cracking viroids, with importation and interstate movement if known infected material prohibited. Preventing the introduction of harmful viroids not present in the United States, such as AFCVd, is a priority.

The three clean plant centers are important hubs for the introduction and movement of plant germplasm, with new cultivars being introduced from state and federal breeding programs, private breeders, and collectors in the United States. The CPCNW and FPS also import material on behalf of US stakeholders, paralleling the USDA-APHIS-PPQ program in Beltsville, MD., because importation of fruit tree material from Canada and select EU countries such as the Netherlands is restricted by Not Authorized Pending Pest Risk Analysis (NAPPRA), and requires a Controlled Import Permit (CIP).

Irrespective of origin, the diagnostic process takes a minimum of 2 years, with testing split across a winter dormancy, and pathogen tested for at optimal times of the growing season. If pathogens are found, virus elimination techniques such as thermotherapy or meristem tissue culture are applied to remove these pathogens. This can extend the process by several years depending on how recalcitrant the pathogens are to removal.

Germplasm, which has cleared the diagnostic and virus elimination processes at one of the centers, enters that center's foundation collection. The foundations are maintained under different conditions, from smaller potted plants held in screenhouses at the CPCNW, to larger trees maintained in open fields at FPS; the CCPC has a hybrid model of both indoor and outdoor plants. Each of these foundations is tested on a 3-year rotation for a core list of economically important pathogens, with additional targeted pathogens added based on local pathogen epidemiology or specific concerns. While there are regional biases toward their neighboring fruit tree industries as a result of local financial support, all three centers serve the entire US fruit tree industry, making propagative material available to propagators across the country on a first-come-first-served basis, for a nominal fee, and what is not available at one center is normally available at another.

The most important outlet for clean plant material from these three centers is state-run fruit tree certification programs. At present, the US states of California, New York, Pennsylvania, Oregon, and Washington have active tree fruit certification programs. While there are between-state differences in approach, each of these state programs tests for a limited number of pathogens in nursery mother trees, supplemented with wide-scale visual scouting programs to detect infection in nursery increase and production stock. If pathogen infection at the nursery, or in subsequent grower plantings is detected, a traceback process can occur, all the way back to the original source at one of the three clean plant centers, if necessary, to determine where the infection occurred.

This is particularly important for viroids in fruit trees as symptoms often appear years after planting, on the saleable product, the fruit, in the case of the pome fruit infecting apscaviroids ASSVd, apple dimple fruit viroid (ADFVd), AFCVd, and PLMVd (Di Serio et al., 2017; Flores

C. Viroid pathogenesis and viroid-host interaction

et al., 2017; Hadidi et al., 2017a,b), or the growth and productivity of the tree itself in the case of pear blister canker viroid (PBCVd) or apple hammerhead viroid (AHVd) (Di Serio et al., 2017; Messmer et al., 2017). All plant material in the US clean plant centers, and in the state certification programs are supposed to be free of these pathogens. We have found, however, that legacy material in the state programs, and material introduced from breeding programs often have these viroids (Harper, unpublished data); therefore, it is a goal of both NCPN and the respective state departments of agriculture to remove these potentially harmful pathogens.

Hop

Hop cultivation has a long history in the United States, with the first recorded planting in Massachusetts in 1648, but it is a history shaped in part by disease. Production was focused in the Northeastern United States until the early 1900s, when outbreaks of downy mildew (*Pseudoperonospora humuli*), among other factors, led to a decline in hop yard acreage in the eastern and midwestern United States, and a shift to the Pacific Northwest, which possesses a climate suitable for hop growth, but not for downy mildew. At present 96% of all hops produced in the United States are grown in Idaho, Oregon, and Washington state.

Viruses, viroids, and other systemic pathogens have long been an issue for hop producers, as many of these pathogens reduce hop bine growth and vigor, as well as the alpha- and beta-acid content of hop cones, compounds that are required for aroma and bittering in brewing (Pethybridge et al., 2008). Therefore, management of systemic pathogens has long been an issue for hop growers, and the need for pathogen-free germplasm has been recognized for over 70 years, with hop certification programs in the United Kingdom and Australia dating back to the 1940s and 1950s (Pethybridge et al., 2008). In the United States, several states have historically had hop certification programs, although these are largely inactive at the time of writing.

Instead, the development, storage, and distribution of virus-tested hop germplasm fell on pathologists affiliated with hop breeding programs, such as at Washington State University, and the USDA-ARS National Clonal Germplasm Repository in Corvallis, OR. These programs, with industry support, had historically screened hop germplasm for classical disease symptoms produced by apple mosaic virus and hop mosaic virus, as well as through the use of biological indicators. With the introduction of ELISA and PCR, additional viruses were screened for, including the latent hop viruses that while not producing visual symptoms, can affect alpha and beta-acid content (Pethybridge et al., 2008), as well as viroids.

Over the past few decades, viroids have been responsible for the most significant shift in grower attitudes toward the need for virus-tested germplasm, with the discovery and outbreaks of HSVd and AFCVd in hops in Japan (Hataya et al., 2017; Sano et al., 2004), to citrus bark cracking viroid (CBCVd) in Slovenia (Jakse et al., 2015). In the United States, the discovery of HSVd in 2004 (Eastwell and Nelson, 2007) and the severe damage it caused to hop yards across the Pacific Northwest (Kappagantu et al., 2017) caused renewed interest in expanding and supporting a clean plant program, which led to hops joining the National Clean Plant Network (NCPN) in 2010.

C. Viroid pathogenesis and viroid-host interaction

At present, the Clean Plant Center Northwest (CPCNW), at Washington State University's Irrigated Agriculture Research and Extension Center in Prosser, WA, is the nation's only public clean hop germplasm program. Hop germplasm is introduced to this center from both public breeding programs such as the USDA-ARS, and private hop breeders. The CPCNW also works closely with the USDA-ARS NCGR in Corvallis, OR to test and perform virus elimination on public, off-patent, or historical lines of interest to growers across the United States irrespective of origin, all hop lines go through the same diagnostic and virus elimination process.

New lines, which arrive as dormant rhizomes, tissue-cultured plantlets, or bine cuttings are propagated in sterile media in a contained greenhouse and allowed to grow out for approximately 4 to 6 months to permit pathogen accumulation. Plants are then tested for seven viruses and two viroids (HSVd and CBCVd) by RT-PCR on a seasonal basis to account for temperature effects, with the viruses tested for in spring, and the viroids in summer. A general screen for these and other hop-infecting viruses, viroids, and phytoplasmas by high-throughput sequencing (HTS) is also performed, with nucleic acids collected during the seasonal time points and pooled to give a composite of what is in each plant. If a targeted or novel pathogen is detected, the plant undergoes virus elimination using meristem tissue culture, a process that takes anywhere between 3 to 6 months to regenerate plantlets that can be transferred to potting media for growth and retesting. Regenerated plants are tested for the pathogen(s) originally identified after a minimum growth period in a contained greenhouse of at least 3 months. If they pass this stage, a full final testing round is performed at least 6 months later by RT-PCR for the seven viruses and two viroids of economic importance.

Upon successful completion of the diagnostic process, hop lines are planted in duplicate in large (20 gals) growing boxes contained within a screenhouse with a double-door vestibule to prevent potential vectors from entering the structure. These plants are retested every year for four viruses and one viroid (HSVd) to ensure continued confidence in the material. Material is distributed as five-node cuttings, dormant rhizomes, and sterilized shoot tips in tissue culture vials to hop propagators across the United States annually. As there are currently no state hop certification programs, the CPCNW provides best management practices to end users and works closely with grower organizations to monitor for new and emerging pathogens and alert producers to prevent pathogen spread.

In 2021 USDA-APHIS-PPQ listed HSVd on the Not Authorized Pending Pest Risk Analysis (NAPPRA) list, preventing the unrestricted importation of hop germplasm, other than seeds, into the United States without a Controlled Import Permit. While a hindrance to the industry, this has provided increased protection to hop producers. The CPCNW is the only institution to be approved as a postentry quarantine site, which provides an APHIS-approved pathway for importation and will help prevent the inadvertent introduction of the two viroids not currently present in the US hops, CBCVd and AFCVd, as well as the further introduction of HSVd from Asia or Europe.

Finally, with the rise of industrial hemp production in the United States and other countries, a viroid long thought to be relatively harmless, hop latent viroid (HLVd), has become of greater interest. While HLVd does not produce obvious macroscopic symptoms in hops, it is pathogenic to hemp, causing stunting and chlorosis (Warren et al., 2019). This led growers to ask the CPCNW to develop methods to eliminate HLVd in hop germplasm to prevent further spread to hemp, and to allow them to compare whether HLVd, which is present in nearly all commercial hop lines, does potentially harm hop yield and growth.

Palms

Palms are monocotyledonous and mostly arborescent plants in the family *Arecaceae*. There are three main economic palms cultivated around the world: the date palm (*Phoenix dactylifera* L.), the coconut palm (*Cocos nucifera* L.), and the oil palms (*Elaeis guineensis* Jacq. and *E. oleifera* (Kunth) Cortés). The date palm is a subtropical crop adapted to high temperatures with abundant groundwater, whereas the coconut palm and the oil palms are tropical species. All three are cultivated for their fruit, although the uses of the fruit differ.

There are few reported viroid diseases of palms. These are CCCVd, which has killed millions of coconut trees in the Philippines and a variant of which also causes orange leaf-spotting of oil palms in Malaysia, and the coconut tinangaja viroid, which has killed many coconut trees in Guam (Rodriguez et al., 2017). The spread of these diseases is not well characterized but thought to be via seed and pollen transmission. Viroid diseases have not been reported in date palm (El Bouhssini and Faleiro de Socorro, 2018). However, other fungal and bacterial pathogens and insects attack palms, so it is beneficial to have a certification or clean stock program in place.

As monocots, the palms differ from most other cultivated trees in that they have only a single apical meristem to sustain their growth, with the lateral meristems giving rise to fruiting organs. In the case of the date palm, the lateral meristems produce offshoots as a juvenile characteristic. Offshoots are similar to tillers in grasses and can be visualized as immature trees that can be removed from the mother tree and planted as offspring trees (Fig. 2). Planting offshoots is the traditional method of propagation, as seedlings are not true to type. Offshoot production is limited to the earliest years of the palm's life and limited numbers of offshoots are therefore available for planting (Krueger, 2021). Due to the limitations of offshoot numbers and difficulties in establishing offshoots, tissue culture propagation of mass numbers of date palms has become established as an alternative method (Al-Khayri and Naik, 2017).

Because there is only a single meristem providing growth, technologies such as shoot-tip grafting are not possible with palms. Likewise, due to the bulk of the offshoots (optimum planting size is approximately 10–20 kg), thermotherapy is not feasible. It should be noted that summer temperatures in date-producing areas approximate the conditions used in the thermotherapy of other tree crops such as citrus. A potential method for pathogen elimination in date palms involves tissue culture, typically performed as meristem culture, which can eliminate pathogens if performed and monitored properly (Laimer and Barba, 2011). As with all technologies for pathogen elimination, it is vital to assay for the presence of pathogens after the pathogen elimination technology has been applied.

Although tissue culture production of date palms can potentially produce pathogen-free palms, it does not appear that there are currently any certification or clean plant programs for palm propagation. This includes any testing requirements for propagation within a country. Pest exclusion via quarantines is currently the main method for maintaining high health, low-pest-pressure palm cultivation zones.

SCHEMATIC DIAGRAM OF THE DATE PALM

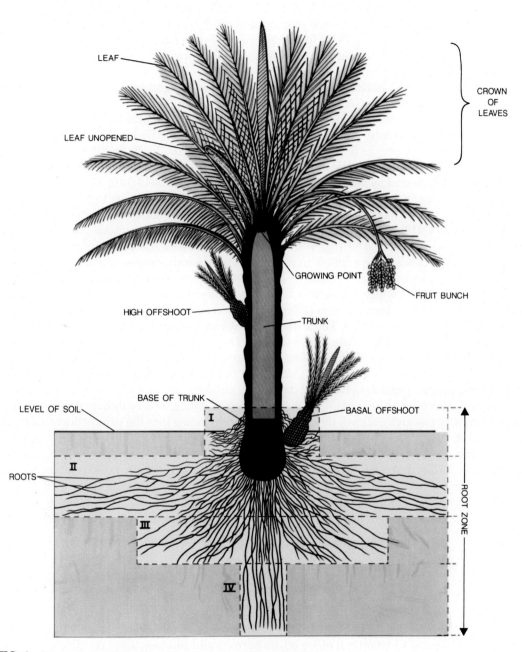

FIG. 2 Schematic representation of the date palm (*Phoenix dactylifera* L.).

C. Viroid pathogenesis and viroid-host interaction

In the United States, all three economic palms are "prohibited" (i.e., Not Authorized Pending Pest Risk Analysis), as are all ornamental palms, except for seed introductions. Therefore, they must be imported under a PCIP. A PCIP is also required for introductions as tissue-cultured plantlets. Date palms are the only current economic palm cultivated in the United States. Current conditions of a PCIP for tissue-cultured date palm plantlets require a 2-year postentry quarantine and testing free of pathogens including the viroid pathogen CCCVd, and various fungal and bacterial pathogens.

In the United States, dates are produced in a limited geographic area in Southeastern California and Southwestern Arizona having sufficient heat to mature a commercial crop (Krueger, 2015). This area corresponds to the low-elevation desert areas of Eastern Riverside County and Imperial County in California and Yuma County in Arizona. Most date palm diseases are fungal in nature (El Bouhssini and Faleiro de Socorro, 2018), and the arid nature of this region is not conducive to the development and spread of fungal pathogens. In California, this area is protected by State Interior Quarantines 3401 and 3419 (http://pi.cdfa.ca.gov/pqm/manual/htm/pqm_index.htm#interior). The former (3401) applies to all commodities that might transport *Phymatotrichum [Ozonium] omnivorum*, a root-rotting fungus, into this area. The latter (3419) applies specifically to plants, propagative materials, and tools used on *Phoenix* species. The coastal region of California commonly has palms of many species grown as ornamentals or specimen trees, including several species of *Phoenix*. *Fusarium oxysporum* f. sp. *canariensis* is present in the coastal region. This particular f. sp. can kill *P. canariensis*, which is widely planted in the urban Southern California coastal area. It is also lethal to date palms. There are also other fungal pathogens in the coastal region that are potentially lethal to date palms. Thus, the State Interior Quarantine is vital in maintaining the plant health of the California date industry. There are also State Exterior Quarantines (http://pi.cdfa.ca.gov/pqm/manual/htm/pqm_index.htm#exterior) against Ozonium (3261) and the Lethal Yellows Phytoplasma (3282) that prohibit the introduction of palms from out of state, further protecting both the date palm industry and ornamental palms growing outside the date production area. CDFA has a permits office which, under certain circumstances (mostly research), allows the movement of date palms into the production area from outside of it.

Exclusion seems to be the main method of preventing palm disease dispersal in general. For instance, in Saudi Arabia, date palms moving from one province to another must be first shown to be free of the Red Palm Weevil (*Rhynchophorus ferrugineus* (Olivier, 1790) (J.M. Al-Khayri, personal communication, 2022). Quarantine and exclusion, combined with surveillance, are currently the standard for preventing the dispersal of viroid and other pathogens of palms.

Protocols/procedures/methods

In the United States, the USDA-APHIS-PPQ (Animal Plant Health Inspection Services Plant Protection and Quarantine) office issues permits and phytosanitary certificates that allow the movement of otherwise prohibited plant materials. For plant scientists, it is of paramount importance to be able to obtain the biological materials of interest correctly and reference the appropriate permits under the existing regulations. Therefore, a researcher needs to know how to acquire the proper documentation. There is a special type of permit used to import materials for therapy and distribution (PCIP) that programs such as the CCPP, the CPCNW, and the National Clonal Germplasm Repository for Citrus and Dates (NCGRCD) use. Most plant pathologists will be interested in knowing how to apply for: (1) a Permit to Move Live Plant Pests, Biocontrol Organisms, Federal Noxious Weeds, Soil and Bees (PPQ-526); and (2) Phytosanitary certificates. Here, we provide a step-by-step guide on applying for these documents with USDA-APHIS.

How to apply for a permit

Nowadays, the online cFile application system allows one to apply conveniently for a P526 permit. A YouTube training video on the APHIS eFile Account Registration and Management is available at: https://youtu.be/zAGIbF4ylYs

While online systems may be updated in the future, the key elements of a permit application will most likely remain the same. These include the applicant's information about the receiving facility, the point of origin, and the proposed measure for escape prevention and final disposition.

For the current process, to get started, visit: https://www.aphis.usda.gov/aphis/resources/permits. After selecting eFile, the authorized user (who is eAuthenticated, a process that allows individuals in the United States to establish their electronic identity and be granted access to government websites) will land on the APHIS eFile initial page. By clicking on "Select an option" from the first dropdown menu, under "Ready to Apply?", you will select the permit option you need. Here, we show you how to apply for PPQ-526 Permit to Move Live Plant Pests, Biocontrol Organisms, Federal Noxious Weeds, Soil, and Bees. After a few screens covering select agent special specifications, whether the permit contains confidential business information, the intended use (e.g., Educational Use, Research-Laboratory, etc.), movement type, the applicant provides information regarding the articles of interest, where the article was collected and equipment access for containment. An example is shown in Fig. 3A and B. After specifying the place of origin and the destination of the article, a section to upload files in support of the application can be completed. The application is then submitted electronically so the regulatory officials can review the application, assess the risk of the proposed work with a specific pathogen, and issue a permit with the appropriate terms and conditions to minimize the risk of introduction or escape of a pathogen, such as a viroid, into the environment.

It is worthy of mention that in addition to the permit issued by the federal government, permits issued by the State Department of Agriculture of the state receiving the plant materials are sometimes required in the United States. State regulations for different commodities are available from the National Plant Board (https://nationalplantboard.org). The same concepts apply to other countries or regions. Rather than creating a detailed list of permit requirements in the world, which would be beyond the scope of this chapter, it is sufficient to appreciate that different permit requirements and different mechanisms to obtain these permits may exist in the world, but it is the responsibility of conscientious scientists to contact the appropriate authorities to establish the requirements and processes applicable to each

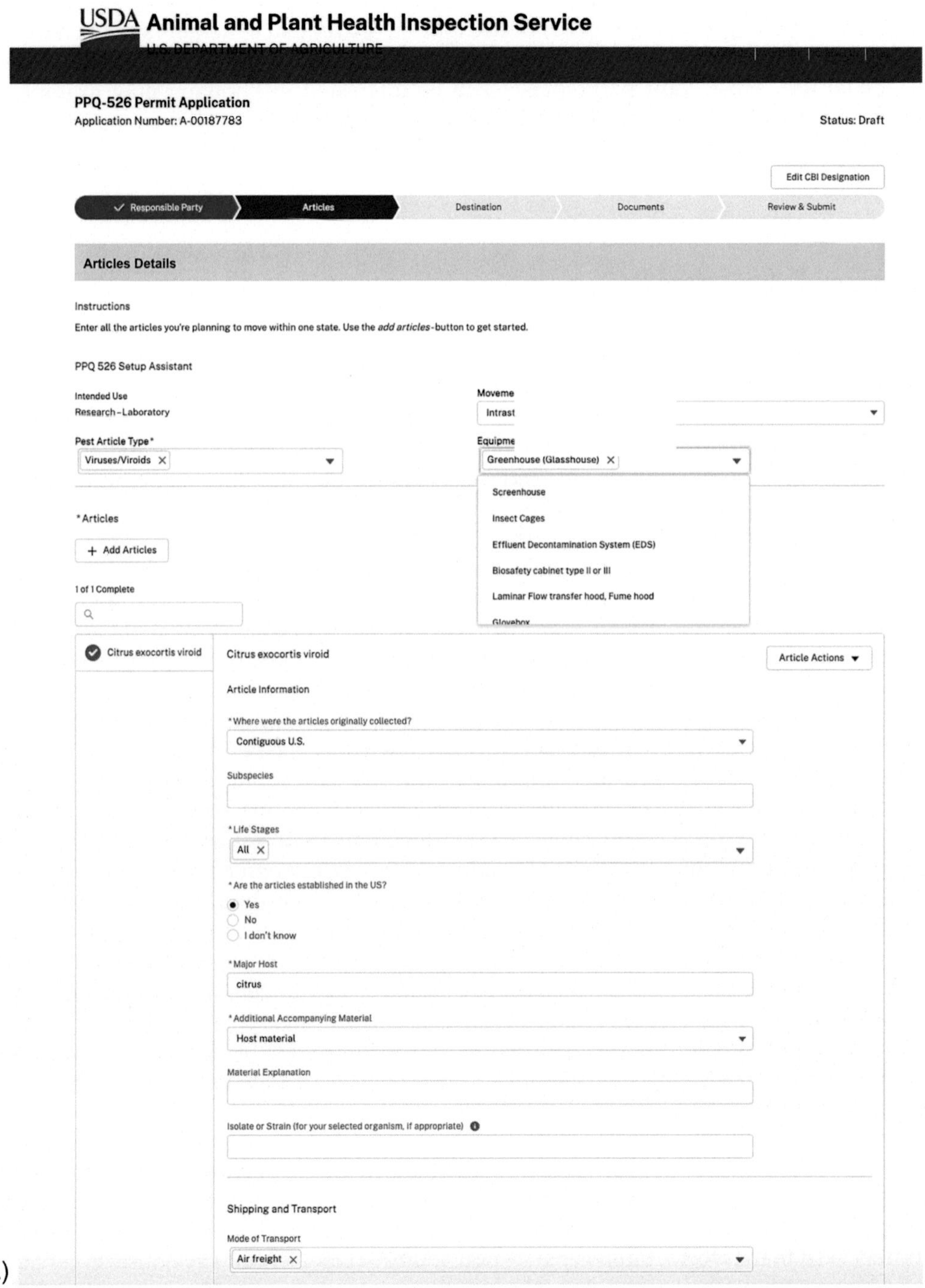

USDA Animal and Plant Health Inspection Service
U.S. DEPARTMENT OF AGRICULTURE

PPQ-526 Permit Application
Application Number: A-00187783

Status: Draft

Edit CBI Designation

✓ Responsible Party | Articles | Destination | Documents | Review & Submit

Articles Details

Instructions

Enter all the articles you're planning to move within one state. Use the *add articles*-button to get started.

PPQ 526 Setup Assistant

Intended Use
Research – Laboratory

Moveme
Intrast

Pest Article Type*
Viruses/Viroids ✕

Equipme
Greenhouse (Glasshouse) ✕

Screenhouse

Insect Cages

Effluent Decontamination System (EDS)

Biosafety cabinet type II or III

Laminar Flow transfer hood, Fume hood

Glovebox

*Articles

+ Add Articles

1 of 1 Complete

🔍

✓ Citrus exocortis viroid

Citrus exocortis viroid

Article Actions ▼

Article Information

*Where were the articles originally collected?
Contiguous U.S.

Subspecies

*Life Stages
All ✕

*Are the articles established in the US?
● Yes
○ No
○ I don't know

*Major Host
citrus

*Additional Accompanying Material
Host material

Material Explanation

Isolate or Strain (for your selected organism, if appropriate) ⓘ

Shipping and Transport

Mode of Transport
Air freight ✕

(A)

FIG. 3 Sections of an application for a permit to move live plant pests, biocontrol organisms, federal noxious weeds, soil, and bees (A and B); sections of a phytosanitary certificate application (C and D).

(Continued)

Add Articles

Instructions

Enter the following information about your articles.

* Search for and select articles

| Search | 🔍 | Add |

☐ I can't find my article

📖 **Articles**

---

1 article selected   Clear All

Citrus exocortis viroid ✕

Articles Details

* Where were the articles originally collected?

| Contiguous U.S. | ▼ |

Subspecies

| |

* Life Stages

| All ✕ | ▼ |

* Are the articles established in the US?

● Yes
○ No
○ I don't know

* Major Host

| citrus |

* Additional Accompanying Material

| Host material | ▼ |

Material Explanation

| |

Mode of Transport

(B) | Air freight ✕ | ▼ |

FIG.3, CONT'D

C. Viroid pathogenesis and viroid-host interaction

(C)

FIG. 3, CONT'D

C. Viroid pathogenesis and viroid-host interaction

(D)

FIG.3, CONT'D

situation. In the United States, the requirements of other countries for the importation of specific commodities are available in the Phytosanitary Export Database, which is accessed via the Phytosanitary Certificate Issuance and Tracking System (PCIT) (see next paragraph).

How to apply for a phytosanitary certificate

In addition to permits issued by the federal and state governments, shipments of samples or specimens are often accompanied by a phytosanitary certificate. As a scientist, you may have collaborators in other countries, and there may be instances in which you need to ship your plant materials to them. Such shipments need to be accompanied by documentation: the importation permit of your collaborator that the receiving party is responsible for and the phytosanitary certificate that you, the sender, need to obtain to fulfill the requirements of the importation permit.

In the United States, the online Phytosanitary Certificate Issuance and Tracking System (PCIT) tracks the inspection of agricultural products and certifies compliance with importing countries' plant health standards. Similarly to the ePermit service explained above, it is an online system requiring eAuthentication and can easily be accessed at https://pcit.aphis.usda.gov/pcit/faces/signIn.jsf.

After selecting the certificate type that is appropriate for the requested kind of movement (e.g., interstate or international movement), the general details of the sender and consignee, and the export and shipping details (e.g., air freight, number of packages, Fig. 3C) are entered. The commodities section is completed, as shown in Fig. 3D.

Next, the Attachments section allows one to upload the permits of the receiving state or country, which the United States needs to process the phytosanitary certificate application. The application is then submitted and can be tracked on the user side. Typically, the local government office (e.g., County Agricultural Commissioner) responsible for issuing phytosanitary certificates sends an inspector who inspects the plant materials being shipped and verifies that the information on the phytosanitary documentation matches the specific shipment.

Prospective

Disease outbreaks resulting from the introduction of pests or pathogens, including some viroids, into a country, can cause significant disruption and enormous costs for governments, growers, and consumers. The consequences of viroid infections may have different implications than those of infections caused by select plant agents, which include some fungal and bacterial strains that have been determined to have the potential to pose a severe threat to plant products. However, viroids have caused economic damages (for example, in citrus, hops, and tomato) serious enough to require permits and a regulatory pipeline. It is imperative to understand and operate within the regulatory framework with the understanding that permits are not intended to restrict research and development by the government but rather create a record of activities that the scientist can use in peer-reviewed publications or by the regulatory agencies and other programs in case of future disease outbreaks and pathogen detections. In other words, in the unfortunate event of a disease outbreak, there would be a record indicating that the researcher took all necessary measures, as described in the terms and conditions of the permit, to minimize the risk of an outbreak of a given exotic pathogen, and the chances are that the outbreak had a different origin. Before initiating any project, the scientist must identify the need for permits. It is beyond doubt that the investment in time and resources to identify any permit needs and apply and acquire these permits in advance, greatly benefits the project, the researcher, the publication of research outcomes, and the crop health of the country.

Furthermore, with increased international travel and trade and the recognition of the threat it poses, three international standard-setting bodies of the World Trade Organization's (WTO's) Agreement on Sanitary and Phytosanitary Measures (SPS Agreement) came into force. These include the International Plant Protection Convention (IPPC), which promotes fair and safe trade and protects the health of cultivated and wild plants by preventing the introduction and spread of pests. The IPPC periodically publishes a list of quarantine pests present in different regions of the world, and viroids are typically found on these lists due to: (1) the severity and degree of damage a specific viroid could cause; (2) the economic importance of the host crop that could be affected by a single viroid or a group of viroids; and (3) the wide global distribution and cultivation of crops that are known host varieties.

Future implications

Despite the current stringent national and international requirements regulating the movement of plant materials, outbreaks of viroid diseases are still observed, suggesting that either the existing control measures are insufficient or that growers may be unaware of the risks of viroid infection to their crops. Moreover, viroids have now been detected in crop species where they were not previously known to occur, in some cases as latent infections but in others, they follow a severe disease progression. These observations have several interesting implications for the future of viroid research and the control measures necessary for sustainable agriculture. Indeed, should viroid infections pose a serious threat to modern agriculture, the creation and use of viroid-resistant plants in the field could become a reality (Adkar-Purushothama et al., 2015). Regardless of the biotechnology advancements in developing plant resistance to viroids, clean plant programs need to become very efficient. They need to meet the rapid pace of current global economic competition in agriculture and accommodate the intellectual property regulations of the protected patented varieties. Typically, germplasm programs of perennial plant species require 2–3 years or longer to process an accession, after which a two 5-year period of evaluation for horticultural properties follows. A few additional years are also usually required for an accession to be propagated in large numbers in nurseries, reach maturity in field plantings, and produce fruit, which means a protected variety has already lost more than half its typical 20-year patent protection time.

Chapter summary

Plant quarantine programs prevent the accidental introduction of viroids into a country or region. Maintaining and distributing clean plants free of viroids in a germplasm program is essential for sustainable agriculture and could help prevent trade conflict. At times, pathogen-exclusion programs, which are preventative in nature, are perceived as costly. However, one must not underestimate the reality of the much more expensive management or eradication measures that must be implemented if viroids causing severe damage to crops are introduced and spread to regions where they do not exist. Typically, eradication or management costs far exceed those of prevention.

This chapter presented an overview of the existing policies, regulations, and the production of viroid-free propagative plant material for sustainable agriculture through specific examples of perennial plant species. In addition, examples of how to apply for permits and phytosanitary certificates were described.

Study questions

1. How do quarantine programs contribute to sustainable agriculture?
2. Why do researchers need permits?
3. How would you find out if the movement of the plant materials of your interest is restricted?
4. If you need to acquire plant materials from a different geographical area, how would you proceed to obtain them?
5. What plant protection programs and services are available in your country?

References

Adkar-Purushothama, C.R., et al., 2015. RNAi mediated inhibition of viroid infection in transgenic plants expressing viroid-specific small RNAs derived from various functional domains. Sci. Rep. https://doi.org/10.1038/srep17949.

Al-Khayri, J.M., Naik, P.M., 2017. Date palm micropropagation: advances and applications. Cienc. Agrotecnologia 41 (4). https://doi.org/10.1590/1413-70542017414000217.

Babcock, B.A., 2022. Economic impact of California's citrus industry in 2020. J. Citrus Pathol. 9. https://doi.org/10.5070/C49156433.

Dang, T., et al., 2022. High-throughput RNA extraction from citrus tissues for the detection of viroids. Methods Mol. Biol. 2316, 57–64.

Di Serio, F., Torchetti, E.M., Flores, R., Sano, T., 2017. Other apscaviroids infecting pome fruit trees. In: Viroids and Satellites. Academic Press, pp. 229–241.

Eastwell, K.C., Nelson, M.E., 2007. Occurrence of viroids in commercial hop (*Humulus lupulus* L.) production areas of Washington State. Plant Health Prog. 8 (1), 1. https://doi.org/10.1094/PHP-2007-1127-01-RS.

El Bouhssini, M., Faleiro de Socorro, J.R.H., 2018. Date Palm Pests and Diseases: Integrated Management Guide. ICARDA, Lebanon.

Ferguson, L., Grafton-Cardwell, E.E. (Eds.), 2014. Citrus Production Manual. UCANR Publications.

Flores, R., Navarro, B., Delgado, S., Hernández, C., Xu, W.X., Barba, M., Hadidi, A., Di Serio, F., 2017. Peach latent mosaic viroid in infected peach. In: Viroids and Satellites. Academic Press, pp. 307–316.

Fridlund, P.R., 1980. The IR-2 program for obtaining virus-free fruit trees. Plant Dis. 64 (9), 826–830.

Hadidi, A., Vidalakis, G., Sano, T., 2017a. Economic significance of fruit tree and grapevine viroids. In: Viroids and Satellites. Academic Press, San Diego, CA, pp. 15–25, https://doi.org/10.1016/b978-0-12-801498-1.00002-4.

Hadidi, A., Barba, M., Hong, N., Hallan, V., 2017b. Apple scar skin viroid. In: Viroids and Satellites. Academic Press, pp. 217–228.

Hataya, T., Tsushima, T., Sano, T., 2017. Hop stunt viroid. In: Viroids and Satellites. Academic Press, pp. 199–210.

Jakse, J., Radisek, S., Pokorn, T., Matousek, J., Javornik, B., 2015. Deep-sequencing revealed citrus bark cracking viroid (CBCV d) as a highly aggressive pathogen on hop. Plant Pathol. 64 (4), 831–842.

Kappagantu, M., Bullock, J.M., Nelson, M.E., Eastwell, K.C., 2017. Hop stunt viroid: effect on host (*Humulus lupulus*) transcriptome and its interactions with hop powdery mildew (*Podospheara macularis*). Mol. Plant-Microbe Interact. 30 (10), 842–851.

Krueger, R.R., 2015. Date palm status and perspective in the United States. In: Al-Khayri, J.M., Jain, S.M., Johnson, D.V. (Eds.), Date Palm Genetic Resources and Utilization. Springer, Dordrecht, pp. 447–485, https://doi.org/10.1007/978-94-017-9694-1_14.

Krueger, R.R., 2021. Date palm (Phoenix dactylifera L) biology and utilization. In: Al-Khayri, J., Johnson, D., Jain, S.M. (Eds.), The Date Palm Genome. Compendium of Plant Genomes, vol. 1. Springer, Cham, pp. 3–28, https://doi.org/10.1007/978-3-030-73746-7_1.

Krueger, R.R., Vidalakis, G., 2022. Study and detection of citrus viroids in woody hosts. Methods Mol. Biol. 2316, 3–21.

Laimer, M., Barba, M., 2011. Elimination of systemic pathogens by thermotherapy, tissue culture, or in vitro micrografting. In: Virus and Virus-Like Diseases of Pome and Stone Fruits. American Phytopathological Society, St. Paul, MN, pp. 389–393.

Messmer, A., Sanderson, D., Braun, G., Serra, P., Flores, R., James, D., 2017. Molecular and phylogenetic identification of unique isolates of hammerhead viroid-like RNA from 'Pacific Gala' apple (Malus domestica) in Canada. Can. J. Plant Pathol. 39 (3), 342–353.

Pethybridge, S.J., Hay, F.S., Barbara, D.J., Eastwell, K.C., Wilson, C.R., 2008. Viruses and viroids infecting hop: significance, epidemiology, and management. Plant Dis. 92 (3), 324–338.

Rodriguez, M.J.B., Vadamalai, G., Randles, J.W., 2017. Economic significance of palm tree viroids. In: Hadidi, A., Flores, R., Randles, J.W., Palukaitis, P. (Eds.), Viroids and Satellites. Academic Press, San Diego, CA, pp. 39–49. https://doi.org/10.1016/B978-0-12-801498-1.00004-8.

Roistacher, C.N., Food and Agriculture Organization of the United Nations, 1991. Graft-Transmissible Diseases of Citrus: Handbook for Detection and Diagnosis. Food & Agriculture Organization.

Rosenberg, D.Y., Aichele, M.D., 1989. Virus-free certification programs in the United States and Canada. In: Virus and Virus-Like Diseases of Pome Fruits and Simulating Noninfectious Disorders. Washington State University, pp. 299–307.

Sano, T., Yoshida, H., Goshono, M., Monma, T., Kawasaki, H., Ishizaki, K., 2004. Characterization of a new viroid strain from hops: evidence for viroid speciation by isolation in different host species. J. Gen. Plant Pathol. 70, 181–187.

Verhoeven, J.T.J., Hammond, R.W., Stancanelli, G., 2017. Economic significance of viroids in ornamental crops. In: Hadidi, A., et al. (Eds.), Viroids and Satellites. Academic Press, Boston, pp. 27–38 (Chapter 3).

Vidalakis, G., et al., 2014. The California citrus clonal protection program. In: Ferguson, L., Grafton-Cardwell, B. (Eds.), Citrus Production Manual. University of California Agricultural and Natural Resources (UC ANR) Publication 3539, pp. 117–130 (Chapter 7).

Vidalakis, G., et al., 2022. SYBR® green RT-qPCR for the universal detection of citrus viroids. Methods Mol. Biol., 211–217. https://doi.org/10.1007/978-1-0716-1464-8_18.

Warren, J.G., Mercado, J., Grace, D., 2019. Occurrence of hop latent viroid causing disease in Cannabis sativa in California. Plant Dis. 103 (10), 2699.

C. Viroid pathogenesis and viroid-host interaction

19

Bioinformatic approaches for the identification and discovery of viroid-like genomes

Maria José López-Galiano[a,b] *and Marcos de la Peña*[a]

[a]Institute of Molecular and Cellular Biology of Plants, Politechnic University of Valencia-CSIC, Valencia, Spain [b]Department of Genetics, University of Valencia, Burjassot, Valencia, Spain

Graphical representation

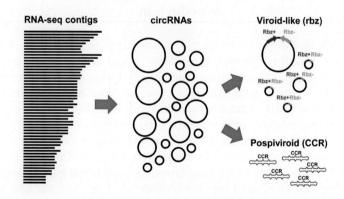

Definitions

Deep sequencing or next-generation sequencing (NGS): A massively parallel sequencing technology of nucleic acids that offers ultra-high-throughput genetic sequencing information.

Metagenomic/metatranscriptomic analyses: DNA/RNA sequencing studies from material recovered directly from environmental or clinical samples without further purification or cultivation-based methods.

Fundamentals of Viroid Biology
https://doi.org/10.1016/B978-0-323-99688-4.00016-X

Ribozymes: A group of RNA molecules capable of catalyzing specific biochemical reactions similar to the action of protein enzymes.

circRNAs: Covalently closed molecules of circular RNA.

Chapter outline

The progress in the last decades of high-throughput sequencing techniques has revolutionized many fields of biological research. The unprecedented amount of genetic information generated by these approaches requires the use and development of computational tools to efficiently analyze all these data. One of the many advances derived from this revolution is the discovery of novel viral and subviral agents such as animal deltaviruses, plant viroid-like satellites, and viroids. These unique infectious elements share exceptional genomic features compared with any other virus or microorganism, which makes their identification a challenging task. This chapter summarizes the tools and methods to computationally identify and discover viroid-like and similar minimal genomes of circular RNA.

Learning objectives

- Sequence databases are an ever-growing source of genetic information.
- Computational biology is a powerful approach for viroid discovery.
- circRNA and ribozyme sequences can be computationally identified.

Introduction

Scientific research in the many fields of biology and medicine has followed a deep revolution in the last decades. The technological advances in the acquisition of genomic information through next-generation sequencing (NGS) approaches have resulted in an unprecedented amount of genetic data, which is growing exponentially every year (Cochrane et al., 2011; Leinonen et al., 2011). This dramatic increase in high-throughput sequencing data has come in parallel to a similar development of computational tools and resources, giving rise to a young but extremely powerful field of research known as bioinformatics or computational biology. A more recent step in this revolution has come from the metagenomic and metatranscriptomic analyses, where NGS experiments explore the genetic content of multiple organisms, including the associated microorganisms such as bacteria (microbiome), fungi (mycobiome), and viral as well as subviral agents (virome). This field has dramatically increased our knowledge about the diversity, evolution, and role of microorganisms in disparate ecological and biomedical contexts, and has opened up new avenues for the development of biotechnological approaches. In biological research, metagenomic and metatranscriptomic approaches have been used to study the diversity and vital function of microbial communities in the earth's biogeochemical cycles, from animal and plant microbiomes to soil, ocean, or freshwater ecosystems. This information is critical for understanding the roles that microorganisms play in the environment, as well as for detecting and identifying novel species (Edgar et al., 2022; Rinke et al., 2013). In the last years, DNA and RNA data obtained from well-defined organisms and environmental samples allow us to realize about the extremely large and complex biodiversity which exists in every ecosystem. This is more evident when we talk about the numerous microorganisms ranging from simple and unicellular eukaryotes to prokaryotes (bacteria or archaea) and DNA and RNA viral elements. It is believed that our planet hosts as many as 1 trillion of mostly unknown prokaryotic and eukaryotic species (Locey and Lennon, 2016; Mora et al., 2011), and at least 10 times more of viruses and subviral agents (Dominguez-Huerta et al., 2022). Thus, the use of NGS technologies is continuously generating vast amounts of genetic data, which can be only processed and analyzed using computational approaches. These techniques are expanding our understanding of the microbial world, and definitely have the potential to continue to uncover not only more classical microbial and viral species but also totally novel subviral agents in the future. Recent computational data mining has advanced the existence of a huge world of subviral agents sharing similar genomic synteny and "morphology" (i.e., small and highly base-paired circular RNA genomes, usually encoding paired self-cleaving RNA motifs or ribozymes) (Edgar et al., 2022; Weinberg et al., 2021). We do not know yet even the host for most of these viroid-like agents, but they open a new field of research where computational tools will be required.

The discovery of viroid-like RNAs

As described in this book, viroid-like genomes are regarded as a small and strange group of infectious agents sharing a few common features and very limited (if any) sequence homology. They all possess a covalently closed circular RNA genome (circRNA) of small size (~200–2000 nucleotides (nt)), which shows extensive base pairing. This circular topology allows their RNA/RNA replication through a rolling-circle mechanism in eukaryotic cells of animals, plants, and, as recently shown, also fungi (Dong et al., 2022). In contrast to RNA viruses, viroid-like agents are extreme pathogens that do not encode the polymerase factor required for their own replication, parasitizing the transcriptional machinery (DNA-dependent RNA polymerases or DdRp) of the eukaryotic host in the case of viroids and deltaviruses, or the RNA-dependent RNA polymerase (RdRp) of any accompanying virus (viroid-like circRNA satellites). In the vast majority, they do not even code for any protein factor at all, a feature that together with their extreme evolutionary rates (Gago et al., 2009) makes their computational identification a really challenging task.

The first described example of a viroid agent was the potato spindle tuber viroid (PSTVd), the type member of the family *Pospiviroidae* (Diener, 1971). Since then, a few dozen of similar plant viroids affecting diverse angiosperms have been reported. In most cases, these viroids are also members of the family *Pospiviroidae*, although slightly different viroid genomes classified in a second family (*Avsunviroidae*, or self-cleaving motifs-containing viroids) have also been described. Apart from viroids, it is known that plants can be affected by another group of small infectious circRNAs, the viroid-like RNA satellites, which require the presence of specific plant viruses to complete their infectious cycle (Buzayan et al., 1995). Animals can be also infected by viroid-like agents of slightly bigger size (~1500 nt) and are generally known as deltaviruses (Flores et al., 2016). The first agent of this class was the human hepatitis deltavirus (HDV), discovered in the mid-1970s (Rizzetto, 2015). In the last years, however, we have been aware of the existence of similar HDV-like agents in diverse animals, such as mammals (Bergner et al., 2021; Iwamoto et al., 2021; Paraskevopoulou et al., 2020), sauropsids (Hetzel et al., 2019; Wille et al., 2018), or invertebrates (Chang et al., 2019), but also widespread in environmental samples with yet unknown hosts (see below) (Edgar et al., 2022). More recently, examples of small viroid-like agents have been reported in a filamentous fungus (Dong et al., 2022), indicating that these infectious agents with small genomes of circular RNA are much more common in the biosphere than previously thought.

Initial approaches to identify viroid-like RNAs through bioinformatics

Viroid-like genomes are generally regarded as noncoding RNAs (ncRNAs) with the exception of the family deltaviruses and at least one example of a plant viroid-like RNA satellite (AbouHaidar et al., 2014). The absence of protein-coding capabilities hinders their informatic search and identification in addition to allowing much faster evolutionary rates for these RNAs, complicating, even more, their detection. Nevertheless, we can exploit the unique features of these RNA agents, such as their genomic circularity or the presence of diverse conserved RNA structures (self-cleaving motifs, hairpin I/II structures, rod-like secondary structures, etc.), which combined with more classical sequence homology approaches, allow their computational hunting.

An interesting common feature in plant viroid-like satellites, avsunviroids, and metazoan deltaviruses is the presence of self-catalytic RNA domains in either one or both genomic polarities. These are small RNA motifs that allow the self-cleavage of the RNA required for processing of the genome during the rolling-circle replication. These tiny (less than 100 nt) and relatively complex motifs are conserved at structural but not at the sequence level and can be used as a way to identify the presence of a viroid-like genome. To search for these self-cleaving RNA motifs, we can use either specific software to detect secondary RNA structures, such as RNAmotif (Macke et al., 2001) or InfeRNAl (Nawrocki and Eddy, 2013), but also more simple sequence homology searches through BLAST/BLAT (Kent, 2002) followed by a filtering by structural requirements. Using these approaches, initial studies done more than a decade ago confirmed that self-cleaving motifs were not exclusive of infectious viroid-like RNA agents, but they are also widespread in the genomic DNA of most organisms along the tree of life (De la Peña and Garcia-Robles, 2010; De la Peña and García-Robles, 2010; Hammann et al., 2012; Perreault et al., 2011; Webb et al., 2009). In some cases, these genomic self-cleaving RNA motifs are related to novel small circular RNAs similar to viroid-like agents (Cervera et al., 2016; Cervera and de la Peña, 2020), which are, however, a totally different form of mobile genetic elements called retrozymes (for retroelements with hammerhead ribozymes) (De la Peña et al., 2020; De la Peña and Cervera, 2017).

De novo discovery of viroid-like genomes

Among the first trials conducted for the specific identification of typical viroid-like RNAs, we can highlight the approach called progressive filtering of overlapping small RNAs (PFOR) (Wu et al., 2012). In this work, the authors identified viroid-specific small interfering RNAs (siRNAs) for viroid genome assembly by progressively eliminating nonoverlapping small RNAs and those that overlap but cannot be assembled into a direct repeat RNA, which is synthesized from circular or multimeric repeated-sequence templates during viroid replication. That way, they readily assembled known viroids using silencing RNAs (siRNAs) sequenced from infected plants, but also identified a novel viroid-like circular RNA with hammerhead self-cleaving RNA motifs in grapevine (Wu et al., 2012). Among the drawbacks of PFOR, the software was computationally very slow and unable to discover those subviral pathogens that do not trigger in vivo accumulation of extensively overlapping small RNAs. An improved version of this approach called PFOR2 (Zhang et al., 2014) allowed the homology-independent discovery of replicating circular RNAs by analyzing longer RNAs instead of small RNAs. Analyses of libraries from grapevine and apple plants led to the tentative discovery of grapevine latent viroid (GLVd) and apple hammerhead viroid-like RNA (AHVd-like RNA) (Zhang et al., 2014).

In the last decade, the development of diverse techniques in NGS has allowed an unprecedented acquisition of genetic information. This way, genomic databases have followed an exponential increase which in some cases offer petabases (10^{15} bases) of sequencing data, such as the Sequence Read Archive (SRA) (Leinonen et al., 2011). These databases may contain viroid and viral-like nucleic acids captured incidental or not to the goals of the original studies (Moore et al., 2011) and can be systematically used for their detection and identification. In this regard, a cloud computing platform called Serratus was recently developed for ultra-high-throughput sequence alignment, screening millions of ecologically diverse sequencing libraries which represented more than 10 petabases of data (Edgar et al., 2022). This approach was used to search for the hallmark gene of most RNA viruses, the RNA-dependent RNA polymerase (RdRp), allowing the discovery of more than 130 thousand novel RNA viruses. Moreover, the search of the delta antigen sequence characteristic of the HDV allowed the identification of more than 50 novel deltavirus genomes in datasets obtained either from vertebrate and invertebrate animals or from environmental samples. A more detailed bioinformatic analysis of these and similar datasets allowed the discovery of hundreds of viroid-like agents of smaller size ($\sim$300 to 700 nt), featured by a circRNA full genome with a rod-like structure and ambisense self-cleaving RNA motifs, mostly of the HHR class. To confirm whether these genomes are putatively coming from circular molecules, we can take into account that most of the software programs used to assemble the sequencing reads into contigs assume that the molecules are linear. However, when reads are obtained from either a circular molecule or a multimeric repeat, a random nucleotide is selected as position 1 and reads overlapping this first position span the junction of the assembled linear sequence (Fig. 1). In assemblers such as SPAdes (Prjibelski et al., 2020), this spanning allows the contig to be 3′ extended with a sequence matching the 5′ end. The length of this matching region depends on the data and the assembler used, which in the case of SPAdes correspond to the last k-mer length used during the assembly (Fig. 1). This way, all the contigs showing these 5′/3′ repeats

FIG. 1   After deep sequencing of a circular RNA molecule *(circle in black)*, the obtained DNA reads *(shorter lines in grey)* map to random sections of the original genome. Assembling of the reads results in a DNA contig containing a 5′/3′ repeat of the size of the k-mer. Trimming of the k-mer repeat allows us to get the final full genome of the viroid-like agent. *(Figure inspired from Weinberg, C.E., Olzog, V.J., Eckert, I., Weinberg, Z. 2021. Identification of over 200-fold more hairpin ribozymes than previously known in diverse circular RNAs. Nucleic Acids Res. 49 (11), 6375–6388. https://doi.org/10.1093/NAR/GKAB454).*

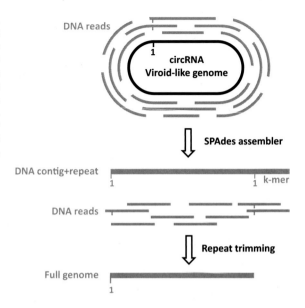

are resulting from either a full circular or a multimeric repeated-sequence. Then, the presence of characteristic viroid-like RNA motifs in these circular genomes, such as self-cleaving RNA motifs or hairpin I/II structures, can be detected using InfeRNAl software (Nawrocki and Eddy, 2013) and the corresponding covariance models available either in the literature of in the RFAM database (Kalvari et al., 2021) (Fig. 1). Further inspection of the obtained hits through secondary RNA structure prediction software, such as the ViennaRNA package (Lorenz et al., 2011), will allow us to confirm whether the obtained genomes fulfill the classical features of viroid-like genomes. Following similar approaches, Weinberg and collaborators (Weinberg et al., 2021) already described in environmental datasets up to hundreds of putative circular RNA genomes carrying self-cleaving domains in either one or both polarities. Altogether, these results indicate that viroid-like genomes similar to plant viral satellites, viroids from the family *Avsunviroidae*, and deltaviruses are widespread agents, and bioinformatic approaches followed by classical molecular characterization will definitely help us to be understood.

Perspective

The use of computational biology and bioinformatics in the analysis of high-throughput sequencing data has deeply revolutionized the discovery and understanding of novel infectious RNAs such as viruses and subviral agents. These tools enable the analysis of the vast amounts of genomic data obtained every day, providing a comprehensive view of the genetic diversity and evolution, of the heterogeneous group of subviral agents. Additionally, they allow for the rapid identification and characterization of novel RNA forms, which may allow us to better understand their origins and the origin of life itself. The integration of computational biology and bioinformatics in virology has greatly accelerated our understanding of these pathogens, leading to improved diagnostic capabilities and the creation of more effective interventions. In the future, the continued advancement of these fields will definitely play a critical role in our basic knowledge about these agents but also in controlling their spread.

Study questions

1. Why is difficult to detect most viroid-like genomes?
2. Give the name of the first methods used for the bioinformatic detection of viroids.
3. Which software would you use to detect a particular self-cleaving ribozyme?
4. How would you detect a putative circular RNA?
5. How metagenomic and metatranscriptomic analyses are changing our knowledge about infectious RNAs?

Further reading

Müller, S., Masquida, B., Winkler, W. (Eds.), 2021. Ribozymes: Principles, Methods, Applications. Wiley-VCH, Weinheim, Germany. ISBN: 978-3-527-34454-3.

Weinberg, C.E., Olzog, V.J., Eckert, I., Weinberg, Z., 2021. Identification of over 200-fold more hairpin ribozymes than previously known in diverse circular RNAs. Nucleic Acids Res. 49 (11), 6375–6388. https://doi.org/10.1093/NAR/GKAB454.

Forgia, M., Navarro, B., Daghino, S., Cervera, A., Gisel, A., Perotto, S., Aghayeva, D.N., Akinyuwa, M.F., Gobbi, E., Zheludev, I.N., Edgar, R.C., Chikhi, R., Turina, M., Babaian, A., Di Serio, F., de la Peña, M., 2023. Hybrids of RNA viruses and viroid-like elements replicate in fungi. Nat. Commun. 14 (1), 2591. https://doi.org/10.1038/s41467-023-38301-2.

References

AbouHaidar, M.G., Venkataraman, S., Golshani, A., Liu, B., Ahmad, T., 2014. Novel coding, translation, and gene expression of a replicating covalently closed circular RNA of 220 nt. Proc. Natl. Acad. Sci. U. S. A. https://doi.org/10.1073/pnas.1402814111.

Bergner, L.M., Orton, R.J., Broos, A., Tello, C., Becker, D.J., Carrera, J.E., Patel, A.H., Biek, R., Streicker, D.G., 2021. Diversification of mammalian deltaviruses by host shifting. Proc. Natl. Acad. Sci. U. S. A. 118 (3), 2019907118. https://doi.org/10.1073/PNAS.2019907118.

Buzayan, J.M., van Tol, H., Zalloua, P.A., Bruening, G., 1995. Increase of satellite tobacco ringspot virus RNA initiated by inoculating circular RNA. Virology 208 (2), 832–837. https://doi.org/10.1006/viro.1995.1221.

Cervera, A., de la Peña, M., 2020. Small circRNAs with self-cleaving ribozymes are highly expressed in diverse metazoan transcriptomes. Nucleic Acids Res. 48 (9), 5054–5064. https://doi.org/10.1093/nar/gkaa187.

Cervera, A., Urbina, D., de la Peña, M., 2016. Retrozymes are a unique family of non-autonomous retrotransposons with hammerhead ribozymes that propagate in plants through circular RNAs. Genome Biol. 17 (1), 135. https://doi.org/10.1186/s13059-016-1002-4.

Chang, W.-S., Pettersson, J.H.-O., Le Lay, C., Shi, M., Lo, N., Wille, M., Eden, J.-S., Holmes, E.C., Pettersson, J.H.-O., le Lay, C., Shi, M., Lo, N., Wille, M., Eden, J.-S., Holmes, E.C., 2019. Novel hepatitis D-like agents in vertebrates and invertebrates. Virus Evol. 5 (2), vez021. https://doi.org/10.1093/ve/vez021.

Cochrane, G., Karsch-Mizrachi, I., Nakamura, Y., 2011. The international nucleotide sequence database collaboration. Nucleic Acids Res. 39 (Database issue), D15. https://doi.org/10.1093/NAR/GKQ1150.

De la Peña, M., Ceprián, R., Cervera, A., 2020. A singular and widespread group of mobile genetic elements: RNA circles with autocatalytic ribozymes. Cells 9 (12). https://doi.org/10.3390/cells9122555.

De la Peña, M., Cervera, A., 2017. Circular RNAs with hammerhead ribozymes encoded in eukaryotic genomes: the enemy at home. RNA Biol. 14 (8), 985–991. https://doi.org/10.1080/15476286.2017.1321730.

De la Peña, M., Garcia-Robles, I., 2010. Intronic hammerhead ribozymes are ultraconserved in the human genome. EMBO Rep. 11 (9), 711–716. https://doi.org/10.1038/embor.2010.100.

De la Peña, M., García-Robles, I., 2010. Ubiquitous presence of the hammerhead ribozyme motif along the tree of life. RNA 16 (10), 1943–1950. https://doi.org/10.1261/rna.2130310.

Diener, T.O., 1971. Potato spindle tuber "virus". IV. A replicating, low molecular weight RNA. Virology 45 (2), 411–428. https://www.ncbi.nlm.nih.gov/pubmed/5095900.

Dominguez-Huerta, G., Wainaina, J.M., Zayed, A.A., Culley, A.I., Kuhn, J.H., Sullivan, M.B., Matthew Sullivan, C.B., 2022. The RNA virosphere: how big and diverse is it? Environ. Microbiol. 25 (1), 209–215. https://doi.org/10.1111/1462-2920.16312.

Dong, K., Xu, C., Kotta-Loizou, I., Jiang, J., Lv, R., Kong, L., Li, S., Hong, N., Wang, G., Coutts, R.H.A., Xu, W., 2022. Novel viroid-like RNAs naturally infect a filamentous fungus. Adv. Sci., 2204308. https://doi.org/10.1002/ADVS.202204308.

Edgar, R.C., Taylor, J., Lin, V., Altman, T., Barbera, P., Meleshko, D., Lohr, D., Novakovsky, G., Buchfink, B., Al-Shayeb, B., Banfield, J.F., de la Peña, M., Korobeynikov, A., Chikhi, R., Babaian, A., 2022. Petabase-scale sequence alignment catalyses viral discovery. Nature 602 (7895), 142–147. https://doi.org/10.1038/S41586-021-04332-2.

Flores, R., Owens, R.A., Taylor, J., 2016. Pathogenesis by subviral agents: viroids and hepatitis delta virus. In: Current Opinion in Virology. 17, pp. 87–94, https://doi.org/10.1016/j.coviro.2016.01.022.

Gago, S., Elena, S.F., Flores, R., Sanjuán, R., 2009. Extremely high mutation rate of a hammerhead viroid. Science 323 (5919), 1308. https://doi.org/10.1126/science.1169202.

Hammann, C., Luptak, A., Perreault, J., de la Peña, M., 2012. The ubiquitous hammerhead ribozyme. RNA 18 (5), 871–885. https://doi.org/10.1261/rna.031401.111.

Hetzel, U., Szirovicza, L., Smura, T., Prähauser, B., Vapalahti, O., Kipar, A., Hepojoki, J., 2019. Identification of a novel deltavirus in boa constrictors. MBio 10 (2), 1–8. https://doi.org/10.1128/MBIO.00014-19.

Iwamoto, M., Shibata, Y., Kawasaki, J., Kojima, S., Li, Y.T., Iwami, S., Muramatsu, M., Wu, H.L., Wada, K., Tomonaga, K., Watashi, K., Horie, M., 2021. Identification of novel avian and mammalian deltaviruses provides new insights into deltavirus evolution. Virus Evol. 7 (1), 3. https://doi.org/10.1093/VE/VEAB003.

Kalvari, I., Nawrocki, E.P., Ontiveros-Palacios, N., Argasinska, J., Lamkiewicz, K., Marz, M., Griffiths-Jones, S., Toffano-Nioche, C., Gautheret, D., Weinberg, Z., Rivas, E., Eddy, S.R., Finn, R.D., Bateman, A., Petrov, A.I., 2021. Rfam 14: expanded coverage of metagenomic, viral and microRNA families. Nucleic Acids Res. 49 (D1), D192–D200. https://doi.org/10.1093/NAR/GKAA1047.

Kent, W.J., 2002. BLAT- -the BLAST-like alignment tool. Genome Res. 12 (4), 656–664. https://doi.org/10.1101/gr.229202.

Leinonen, R., Sugawara, H., Shumway, M., 2011. The sequence read archive. Nucleic Acids Res. 39 (Database issue), D19. https://doi.org/10.1093/NAR/GKQ1019.

Locey, K.J., Lennon, J.T., 2016. Scaling laws predict global microbial diversity. Proc. Natl. Acad. Sci. U. S. A. 113 (21), 5970–5975. https://doi.org/10.1073/PNAS.1521291113.

Lorenz, R., Bernhart, S.H., Siederdissen, H.Z., Tafer, H., Flamm, C., Stadler, P.F., Hofacker, I.L., 2011. Vienna RNA package 2.0. Algorithms for Mol. Biol. 6, 26. https://doi.org/10.1186/1748-7188-6-26.

Macke, T.J., Ecker, D.J., Gutell, R.R., Gautheret, D., Case, D.A., Sampath, R., 2001. RNAMotif, an RNA secondary structure definition and search algorithm. Nucleic Acids Res. 29 (22), 4724–4735. http://www.ncbi.nlm.nih.gov/pubmed/11713323.

Moore, R.A., Warren, R.L., Freeman, J.D., Gustavsen, J.A., Chénard, C., Friedman, J.M., Suttle, C.A., Zhao, Y., Holt, R.A., 2011. The sensitivity of massively parallel sequencing for detecting candidate infectious agents associated with human tissue. PLoS One 6 (5). https://doi.org/10.1371/JOURNAL.PONE.0019838.

Mora, C., Tittensor, D.P., Adl, S., Simpson, A.G.B., Worm, B., 2011. How many species are there on earth and in the ocean? PLoS Biol. 9 (8), e1001127. https://doi.org/10.1371/JOURNAL.PBIO.1001127.

Nawrocki, E.P., Eddy, S.R., 2013. Infernal 1.1: 100-fold faster RNA homology searches. Bioinformatics 29 (22), 2933–2935. https://doi.org/10.1093/BIOINFORMATICS/BTT509.

Paraskevopoulou, S., Pirzer, F., Goldmann, N., Schmid, J., Corman, V.M., Gottula, L.T., Schroeder, S., Rasche, A., Muth, D., Drexler, J.F., Heni, A.C., Eibner, G.J., Page, R.A., Jones, T.C., Müller, M.A., Sommer, S., Glebe, D., Drosten, C., 2020. Mammalian deltavirus without hepadnavirus coinfection in the neotropical rodent *Proechimys semispinosus*. Proc. Natl. Acad. Sci. U. S. A. 117 (30), 17977–17983. https://doi.org/10.1073/pnas.2006750117.

Perreault, J., Weinberg, Z., Roth, A., Popescu, O., Chartrand, P., Ferbeyre, G., Breaker, R.R., 2011. Identification of hammerhead ribozymes in all domains of life reveals novel structural variations. PLoS Comput. Biol. 7 (5), e1002031. https://doi.org/10.1371/journal.pcbi.1002031.

Prjibelski, A., Antipov, D., Meleshko, D., Lapidus, A., Korobeynikov, A., 2020. Using SPAdes de novo assembler. Curr. Protoc. Bioinformatics 70 (1), e102. https://doi.org/10.1002/CPBI.102.

Rinke, C., Schwientek, P., Sczyrba, A., Ivanova, N.N., Anderson, I.J., Cheng, J.F., Darling, A., Malfatti, S., Swan, B.K., Gies, E.A., Dodsworth, J.A., Hedlund, B.P., Tsiamis, G., Sievert, S.M., Liu, W.T., Eisen, J.A., Hallam, S.J., Kyrpides, N.C., Stepanauskas, R., et al., 2013. Insights into the phylogeny and coding potential of microbial dark matter. Nature 499 (7459), 431–437. https://doi.org/10.1038/nature12352.

Rizzetto, M., 2015. Hepatitis D virus: introduction and epidemiology. Cold Spring Harbor Perspect. Med. 5 (7), 1–10. https://doi.org/10.1101/cshperspect.a021576.

Webb, C.H.T., Riccitelli, N.J., Ruminski, D.J., Lupták, A., 2009. Widespread occurrence of self-cleaving ribozymes. Science 326 (5955), 953. https://doi.org/10.1126/science.1178084.

Weinberg, C.E., Olzog, V.J., Eckert, I., Weinberg, Z., 2021. Identification of over 200-fold more hairpin ribozymes than previously known in diverse circular RNAs. Nucleic Acids Res. 49 (11), 6375–6388. https://doi.org/10.1093/NAR/GKAB454.

Wille, M., Netter, H.J., Littlejohn, M., Yuen, L., Shi, M., Eden, J.S., Klaassen, M., Holmes, E.C., Hurt, A.C., 2018. A divergent hepatitis D-like agent in birds. Viruses 10 (12). https://doi.org/10.3390/v10120720.

Wu, Q., Wang, Y., Cao, M., Pantaleo, V., Burgyan, J., Li, W.X., Ding, S.W., 2012. Homology-independent discovery of replicating pathogenic circular RNAs by deep sequencing and a new computational algorithm. Proc. Natl. Acad. Sci. U. S. A. 109 (10), 3938–3943. https://doi.org/10.1073/PNAS.1117815109.

Zhang, Z.Z., Qi, S., Tang, N., Zhang, X., Chen, S., Zhu, P., Ma, L., Cheng, J., Xu, Y., Lu, M., Wang, H., Ding, S.W., Li, S., Wu, Q., 2014. Discovery of replicating circular RNAs by RNA-seq and computational algorithms. PLoS Pathog. 10 (12), e1004553. https://doi.org/10.1371/journal.ppat.1004553.

Contributions of viroid research to methods for RNA purification, diagnostics, and secondary structure prediction

Gerhard Steger and Detlev Riesner

Institut für Physikalische Biologie, Heinrich Heine University Düsseldorf, Universitätsstraße 1, Düsseldorf, Germany

Definitions

Secondary structure: Draw the sequence of a single-stranded RNA along a circle and connect all nucleotides involved in Watson-Crick and wobble base pairs (see the following for definition) of the structure with lines inside of the circle. In the case of a secondary structure, all lines are noncrossing.

A sequence R of length N consists of nucleotides $r_1, ..., r_N$ and has Watson-Crick and wobble base pairs $r_i{:}r_j$ with positions $1 \leq i \leq j \leq N$. A secondary structure may only contain base pairs $r_i{:}r_j$ and $r_k{:}r_l$ with $j - i > 4$, $i < j < k < l$, and $i < k < l < j$. The first condition enforces at least three nonpaired nucleotides as the minimum size of a hairpin loop, the second condition allows for base pairs enclosing other base pairs as in helices or stem-loops, and the third condition allows for neighboring base pairs. A tertiary structure contains base pairs with $i < k < j < l$ or $i = k$ or $j = l$. The first condition results in pseudoknots, and the second and third conditions allow for a nucleotide to pair with more than one nucleotide as in triple pairs.

Secondary structure depiction: In this chapter, two formats to depict RNA structures are used.

    Dot-bracket format: Three different symbols are sufficient to depict any secondary structure. Usually, a dot is used for a nonpaired nucleotide, an opening bracket for a nucleotide basepaired to a 3′ located nucleotide, and a closing bracket for a nucleotide basepaired to a 5′ located nucleotide, thus the name dot-bracket format for this representation.

    Mountain plot: A mountain plot is a representation of a secondary structure. The sequence is shown on the x-axis. For a mfe structure, draw a vertical line upward by 1 if the nucleotide at that position is 3′ basepaired, draw a vertical line downward by −1 if the nucleotide at that position is 5′ basepaired, and draw a horizontal line of length 1 if the nucleotide at that position is nonpaired.

$\boldsymbol{\Delta G_T^0}$: Standard free energy or Gibbs energy is a function of temperature T (in Kelvin) and given for 1 M concentration. For a given RNA structure, ΔG_T^0 is the energy gain by structure formation (helical state H) from the ground state, which is the random-coil state C. This is calculated as the sum of the free energy contributions of each structural element i, which are helices stabilized by base pair stacking and all types of loops:

$$C \rightleftharpoons H$$

$$\Delta G_T^0 = -RT \ln K_T; \quad K_T = \frac{[H]}{[C]}\bigg|_T$$

$$= \sum \Delta G_{T,\text{stack}}^0 + \sum \Delta G_{T,\text{loop}}^0$$

$$= \sum (\Delta H_{\text{stack}}^0 - T \cdot \Delta S_{\text{stack}}^0) + \sum (\Delta H_{\text{loop}}^0 - T \cdot \Delta S_{\text{loop}}^0)$$

From the free energy, the gas constant R, and the total RNA concentration c_{total}, the equilibrium constant K, and the concentrations of the helical state [H] and the random coil state $[C] = c_{\text{total}} - [H]$ can be calculated for a given temperature. From experimental measurements of concentration dependence on temperature, the values of free energy ΔG_T^0, enthalpy ΔH^0, and entropy ΔS^0 can be determined (Schroeder and Turner, 2009).

    The most accurate ΔG_T^0 values for RNA structure prediction are valid for 1 M ionic strength at 37°C (Mathews et al., 2004). The high ionic strength is not far from in vivo conditions due to the presence of multivalent ions in cells; 37°C was the temperature of lowest error.

$\boldsymbol{\Delta G^0(\text{linear})}$ **versus** $\boldsymbol{\Delta G^0(\text{circular})}$: For a given secondary structure, the free energy differs between a linear and a circular sequence due to the open ends of the linear sequence. That is, the circular sequence has either a closed loop or an additional stacking interaction instead of the dangling end(s) of the linear sequence. This difference, however, often leads to different mfe structures and structure ensembles between linear and circular sequences.

$\boldsymbol{\Delta G_{T,\text{mfe}}^0}$: The number of possible secondary structures grows exponentially with the sequence length N ($\sim 1.8^N$; Waterman, 1995; Hofacker et al., 1998). Accordingly, all possible structures S_i of a single sequence coexist in solution with concentrations dependent on their free energies $\Delta G_T^0(S_i)$; that is, each structure is present as a fraction

$$f_{S_i} = \exp\left(-\frac{\Delta G_T^0(S_i)}{RT}\right)\Big/Q$$

based on a **partition function** Q for the ensemble of all possible structures

$$Q = \sum_{\text{all structures } S_i} \exp\left(-\frac{\Delta G_T^0(S_i)}{RT}\right).$$

The structure S_0 of lowest free energy $\Delta G_T^0(S_0) = \Delta G_{T,\text{mfe}}^0$ is called the optimal structure or **structure of minimum free energy** (mfe). But quite different structures with identical energies might exist for a single sequence. Thus, one should not assume that an RNA folds into a single, static structure.

The relative concentration of the mfe structure can be very low when there are many structural alternatives. Furthermore, the **free energy of the ensemble** is lower (more negative) than that of the mfe structure due to the entropy caused by the many different structures:

$$
\begin{aligned}
\Delta G_T^0(\text{ensemble}) \quad &= -RT \ln Q \\
&= \Delta G_T^0(S_0) - RT \ln\left[1 + \sum_{i=1}^{N} \exp\left(-\frac{\Delta G_T^0(S_i) - \Delta G_T^0(S_0)}{RT}\right)\right] \\
&\leq \Delta G_T^0(S_0)
\end{aligned}
$$

The **centroid secondary structure** only shows base pairs that have a probability of at least 0.5.

Loop types: Nonpaired nucleotides in RNA structures are called loops. Internal loops and bulge loops connect two helices; bulge loops have nonpaired nucleotides only on one side, while internal loops have nonpaired nucleotides on both sides. Hairpin loops are connected to a single helix. Multiloops or junctions connect more than two helices. Do not expect that all loop nucleotides are devoid of stacking interactions and/or hydrogen bonds; quite often they are involved in non-Watson-Crick interactions (see the following for definition).

Metastable structure: A sequence will try to reach the ensemble of structures according to the partition function; this is the favored equilibrium state. A structure is in a metastable state if its concentration is far above or below the concentration in the equilibrium state. Such a metastable structure is formed under particular conditions where the transition into the most stable structure is either prevented or very slow.

Units: Ideal gas constant $R = 8.31446\,\text{J}/(\text{Kmol}) = 1.9872\,\text{cal}/(\text{Kmol})$; factor for conversion of cal to J is 4.1868. Temperature is always in K; for example, $37°C = (273.15 + 37)\,K$.

Watson-Crick and non-Watson-Crick base pairs: Standard Watson-Crick (WC) base pairs are well known for double-stranded DNA as well as RNA. In addition to these, wobble base pairs (Crick, 1966) are found frequently in RNA structures. However, each nucleotide can pair with any other nucleotide by hydrogen bonds (Leontis and Westhof, 2001; Leontis et al., 2002); in the following such pairs are named non-WC (nWC) pairs. WC pairs are isosteric to each other; that is, any WC pair in a helix can be replaced by any other WC pair without distorting the backbone geometry; with some restrictions, this is also true for wobble pairs. Note that the hydrogen bonds of WC pairs are important for isostericity, but the thermodynamic stability of WC and wobble pairs is dominated by π-stacking of the base pairs onto each other (Yakovchuk et al., 2006).

Chapter outline

Preparation of viroid-containing crude extracts from up to 500 g of plant tissue is described in detail. For diagnostic purposes, two gel electrophoretic methods are used, two-dimensional (2D) and return gel electrophoresis. Both methods use silver staining and do not require radioactive or fluorescent labeling. For biophysical studies, especially NMR and X-ray structure determination, large amounts of viroid RNA (up to milligram amounts) with purity better than 95% are needed and can be produced by HPLC on a specially developed ion-exchange carrier.

From biophysical measurements with and without knowledge of viroid sequence(s), the peculiar structure of viroids was obvious. The native structure of *Pospiviroidae* members is rod-like, but the sequence also allows them to form very stable structural elements, which are not part of the native structure. This observation led to the concept of **metastable structures**, structures that are formed under particular conditions where the transition into the most stable, native structure is either prevented or very slow. The importance of folding kinetics is critical not only for viroids but also for further classes of noncoding RNAs as well as mRNAs.

Learning objectives

- How to produce a crude RNA extract from plant material?
- How to show the presence of a circular RNA in a crude RNA extract?
- How to purify viroid RNA?
- How to predict an RNA structure of minimum free energy?
- How to predict the RNA structure ensemble?

Fundamental introduction

Viroids exhibit structural and dynamic properties which were studied with biophysical methods in vitro; these methods are essential to understand viroid function in the cell. Studies in vitro require purified viroids in large amounts and high purity. Novel chromatographic and gel electrophoretic methods were developed for the purification and characterization of viroids; these methods were later used in molecular biology, gene technology, and prion research. Theoretical and experimental studies of RNA folding demonstrated the general biological importance of metastable structures.

Chapter

Viroid purification: The basis of plasmid-purification kits

The optimal method for the purification of viroids depends on the amount and purity of viroids required for the intended study. For example, infectivity studies and reverse transcription for sequencing need only minute amounts, and the purity need not be very high. Biochemical studies on the replication mechanism take advantage of very sensitive and sequence-specific hybridization with complementary labeled sequences. For electron microscopic studies, minute amounts of highly purified viroid are needed to differentiate the viroid RNA from other nucleic acids in the micrograph. High purity is also essential for biophysical studies such as spectroscopic, hydrodynamic, or light scattering methods, and highest purity would be needed if crystallization for X-ray scattering might be planned.

For small amounts and high purity, PAGE followed by elution of the viroid band from the gel is the method of choice, particularly because no special equipment is required. The viroid band can be identified clearly if mobility under native and denaturing conditions are compared where the viroid RNA is seen to move much more slowly under denaturing conditions because of its circularity. In order to elute specifically the viroid RNA and in particular to diagnose a viroid infection, two gel electrophoretic procedures have been developed.

In order to prepare large amounts of pure viroids, as mentioned for biophysical studies and even more than 10 mg for NMR studies, a crude RNA extract from infected plant material (typically 200–500 g) has to be prepared first and then chromatographic procedures that were developed particularly for the purification of viroids have to be applied.

Purification of the viroid RNA by HPLC

HPLC is available in nearly every laboratory. The chromatographic matrix combines size exclusion and anion exchange to purify and separate the different nucleic acids according to size. The silica carrier of an optimal pore size is modified with the anionic exchange group as depicted in Fig. 1A. The fractionation of a crude plant RNA extract by elution with a salt gradient is shown in Fig. 1B. Viroids, circular as well linearized, are eluted as purified fractions as shown by PAGE under denaturing conditions (Fig. 1C). The method described can also be used to purify supercoiled plasmids from crude bacterial extracts, DNA restriction fragments, and oligonucleotides. On the basis of these results, the company Qiagen (Hilden; Germany) developed nucleic acid purification kits that are used worldwide in molecular biology and diagnosis.

The yields of the whole procedure including the chromatographic purification have been determined as follows: from 500 g leaves, 550 µg purified viroid; from 500 g stems, 400 µg; and from 500 g roots, 200 µg purified viroid.

Purification of plant RNA by RNeasy kits

Qiagen (Hilden, Germany) currently supplies kits specially adapted for plant tissue. These are RNeasy Minikits for 500 mg, RNeasy Midikits for 500 mg, and RNeasy Maxikits for 1000 mg plant tissue. These kits are suited for the isolation of total RNA larger than 200 nucleotides (nts). If highly purified viroid RNA is required, further purification procedures have to be added. Similar to the large-scale procedures described earlier the RNeasy-kit technology is based on the homogenization of frozen material, ethanol precipitation of the RNA, binding to a

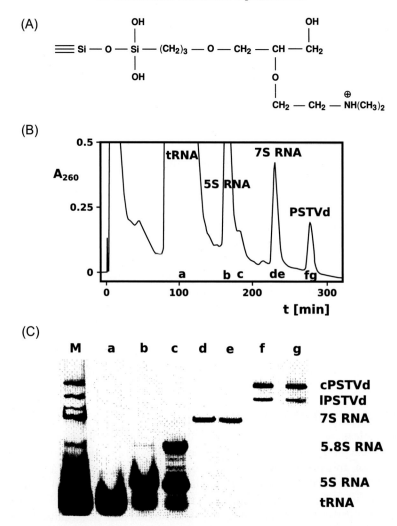

FIG. 1 Chromatography of crude RNA extract from PSTVd-infected tomato plants. (A) Unit structure of silica modified with an anionic exchange group. A silica gel (10 μm, 500 Å pore size) was reacted with (1,2-epoxy-3-propylpropoxy)-trimethoxysilane and condensed with *N,N*-dimethylaminoethanol to yield DMA-500. (B) Preparative chromatographic fractionation of 30 mg crude RNA extract from viroid-infected plants on a DMA-500 column. (C) Gel electrophoretic analysis (5% PAGE, denaturing conditions) of fractions as in B; the unfractionated sample is shown in slot M. *Modified from Colpan, M., Riesner, D., 1984. High-performance liquid chromatography of high-molecular-weight nucleic acids on the macroporous ion exchanger, nucleogen. J. Chromatogr. A 296, 339–353. https://doi.org/10.1016/S0021-9673(01)96428-3.*

membrane-bound silica gel anion exchanger, and elution from the gel matrix. Advantageously, all steps are carried out in one spin column, and centrifugal force should not be higher than 3000 g. The whole procedure is described in great detail in the Qiagen handbook. Particular precepts are given for the use of the samples in molecular biology studies including RT-PCR. The handbook contains a description of all buffers, solutions, and corresponding suppliers.

Two-dimensional and return gel electrophoresis for the diagnostics of circular RNA

PAGE is a well-established analytical and preparative method. Circular and linear viroids migrate as narrow, well-defined bands under native and fully denaturing conditions. In particular, the circularity of viroid RNA leads to extraordinarily low electrophoretic mobility under denaturing conditions, whereas the rod-like native structure behaves similarly to a dsRNA of comparable length. In 2D gel electrophoresis, native conditions are established in the first and fully denaturing conditions in the second dimension (Fig. 2A and B). Denaturing conditions can be established by high concentrations of urea or other denaturants or by high temperature. All DNAs and RNAs from a crude extract run on or faster than a diagonal-like front (Fig. 2B). If a crude extract is used, separate bands are not visible. Only the circular viroid RNA is well separated as a clear band behind the front. In a modification of the same

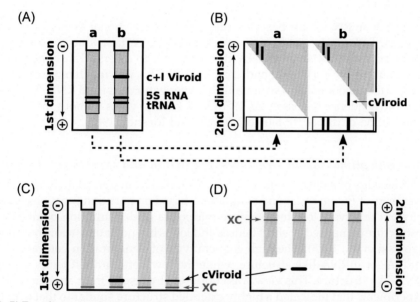

FIG. 2 (A, B) Two-dimensional gel electrophoresis. (A) The first electrophoresis is a conventional PAGE with native conditions. Circular (c) and linear (l) viroid RNAs cannot be visualized by standard staining techniques due to the high background of other nucleic acids (*gray*). After electrophoresis the lanes a and b are cut from the gel; these are polymerized at the bottom of a new gel matrix. (B) Due to the denaturing conditions used for the second dimension, the circular viroid, which migrates very slowly, is well separated from other nucleic acids. *Modified from Schumacher, J., Randles, J.W., Riesner, D., 1983. A two-dimensional electrophoretic technique for the detection of circular viroids and virusoids. Anal. Biochem. 135, 288–295. https://doi.org/10.1016/0003-2697(83)90685-1.* (C, D) Return gel electrophoresis. (C) The first electrophoresis is a conventional PAGE with high-salt conditions (89 mM Tris, 89 mM boric acid, pH 8.3) and low temperature (20°C). The viroid RNA cannot be visualized by standard staining techniques due to the high background of other nucleic acids (*gray*). If the dye xylene cyanol (XC), present in the samples, is nearly running out of the gel matrix, the buffer is changed to low-salt conditions (10 mM Tris, 10 mM boric acid, pH 8.3, 8 M urea), the temperature is raised to 60°C, and the direction of the electric field is inverted. (D) During the second electrophoresis step, all nucleic acids are denatured; then the circular viroid migrates more slowly than all other, larger nucleic acids. *Modified from Schumacher, J., Meyer, N., Riesner, D., Weidemann, H.L., 1986. Diagnostic procedure for detection of viroids and viruses with circular RNAs by "return"-gel electrophoresis. J. Phytopath. 115, 332–343. https://doi.org/10.1111/j.1439-0434. 1986.tb04346.x.*

principle (Fig. 2C and D), all nucleic acids running faster than the viroid RNA and in front of the xylene cyanol marker are allowed to exit the gel in the first dimension. The denaturing conditions are then established by changing the running buffer, raising the temperature, and reversing the direction of gel electrophoretic migration. Circular RNAs appear well separated behind all other nucleic acids.

An important advantage of both these methods is that they do not require the use of radioactively labeled or fluorescent oligonucleotides for hybridization. Viroid-like circular satellite RNAs, so-called "virusoids," can be diagnosed as well. Return gel electrophoresis can analyze more samples in a single gel and does not require repolymerization of a second gel.

Optimal, suboptimal, and metastable structures: A general problem of RNA structure research

The sequence of an RNA is the basis to understand the functional as well as structural aspects of the RNA. Today, when one becomes interested in a certain noncoding RNA (ncRNA) or untranslated region (UTR) of an mRNA, it is quite easy to reverse transcribe the RNA and sequence the resulting DNA (Walker and Lorsch, 2013). This was much more difficult during the beginning of the functional and structural characterization of viroid RNA. The elucidation of the first viroid sequence (Gross et al., 1978) took several years without the availability of reverse transcription or direct ^{32}P labeling (Gross and Riesner, 1980).

Optimal and suboptimal viroid structures: Prediction and experimental verification

Prior to knowledge of a viroid sequence, we concluded from optical melting curves of five different members of *Pospiviroidae* and simulations based on partition functions that the high cooperativity but relatively low-temperature midpoints of denaturation point to a rod-like secondary structure composed of short helices and loops but no or only a low number of junctions (Langowski et al., 1978). The high cooperativity of denaturation contradicts a highly branched tRNA- or mRNA-like structure, while the low denaturation temperature contradicts a dsRNA-like structure. This view was supported by results from analytical ultracentrifugation experiments that indicated a highly extended structure similar to that of dsDNA but contradicted a more globular shape like that of tRNA (Sänger et al., 1976).

In the following years, we used mainly "Tinoco" plots (Tinoco et al., 1971), a 2D graphic representation of all possible base pairs and their neighbors to identify thermodynamically optimal structures at temperatures relevant for interpretation of our equilibrium or kinetic denaturation curves. Fig. 3A shows an example of PSTVd with a minimum helix length of five base pairs. Most critical was the identification of two (or three) helices of the highest stability (Hairpin I and II; HPI and HPII), which were not part of the most stable rod-like structure.

In 1981, Zuker (Ottawa, Canada) published the first version of MFOLD, which was able to predict the mfe structure of a linear RNA sequence (Zuker and Stiegler, 1981). With the great help from Hans Hofmann (Computer Center, MPI Martinsried), we enhanced this program to allow us to calculate thermodynamically optimal as well as suboptimal secondary structures for circular RNAs (Steger et al., 1984; Zuker, 1989).

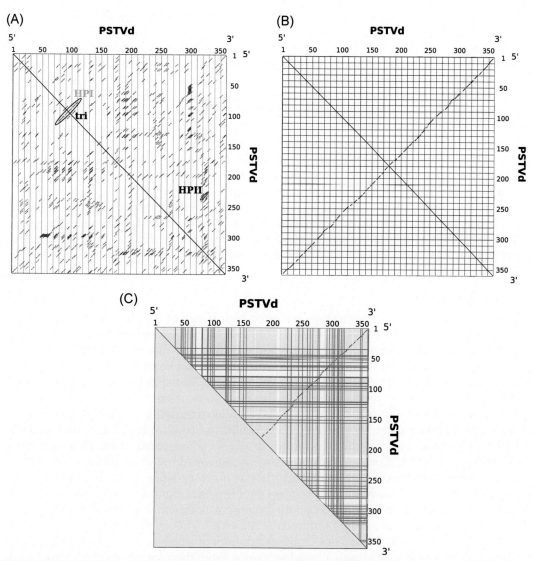

FIG. 3 Basepairing dotplots of PSTVd (GenBank Locus NC_002030). (A) Dotplot showing possible helices longer than four base pairs. The *lower left triangle* has identical content identical to the *upper right triangle*. A:U, *blue circles*; G: C, *red squares*; G:U, *purple triangles*. The tri-helical stem-loop (tri), helix of hairpin I (HPI), and helix of hairpin II (HPII) are marked; these are present only in metastable structures. Note that the tri-helical stem-loop can form only in an oligomeric PSTVd sequence. (B) Dotplot showing the base pair ensemble (*top right triangle*) and base pairs of the mfe structure (*lower left triangle*). The dot size of the ensemble base pairs is proportional to the probability that this base pair is present in any structure. The dotplot was produced by RNAFOLD with options -p -d3 -c. (C) Dotplot showing consensus base pair probabilities for an alignment of 274 full-length, nonidentical PSTVd sequences. The alignment was produced by MAFFT-X-INS-I (Katoh and Toh, 2008) with options -maxiterate 1000 -retree 100. Size (0–1) and color (*yellow to red*) of dots are proportional to the mean consensus pairing probability of the 274 structure ensembles; *white to light blue bars* denote gaps present in the alignment. The dotplot was produced by CONSTRUCT (Wilm et al., 2008).

C. Viroid pathogenesis and viroid-host interaction

Nowadays, the most recent version of MFOLD uses better and an increased number of thermodynamic parameters (Andronescu et al., 2014) and thus is still in use (Mathews et al., 1999; Zuker et al., 1999; Zuker, 2003). MFOLD has been superseded, however, by tools like UNAFOLD, RNAFOLD, or RNASTRUCTURE with advanced algorithms that make possible the prediction of partition functions (McCaskill, 1990; Steffen et al., 2006; Gruber et al., 2008; Markham and Zuker, 2008; Reuter and Mathews, 2010; Lorenz et al., 2011, 2016) or even of all suboptimal structures (Wuchty et al., 1999; Stone et al., 2015). Fig. 3B shows a prediction by RNAFOLD for PSTVd showing the base pairs of the mfe structure (lower left triangle) and all base pairs present in the structure ensemble (top right triangle). Note that the computing effort and memory usage of these tools grows with N^3 and N^2, respectively, where N is the sequence length; that is, these tools are quite fast with the sequence lengths of viroids.

On the experimental side, structural mapping with "selective 2′-hydroxyl acylation analyzed by primer extension" (SHAPE) has superseded prior techniques (Giguère et al., 2014; Steger and Perreault, 2016; López-Carrasco and Flores, 2017; Giguère and Perreault, 2017) while combinations of mutational studies and structural predictions have provided insight into the 3D structure and function of many viroid loops (Zhong et al., 2006; Wang et al., 2018; Ma et al., 2022).

Consensus structures

The error rates of viroid replication are quite high (Gago et al., 2009; Brass et al., 2016; López-Carrasco et al., 2017), which might be due to the replication by cellular DNA-dependent RNA polymerases. The high error rate fits well into the theory of the viral quasispecies (Eigen, 1971) and the theory of the inverse correlation between genome size and replication fidelity (Drake, 1991; Sanjuán et al., 2010). To maintain their biological functional structure, however, progeny sequences are only viable if single mutations are compensated for their structurally detrimental result by additional mutations. This feature was used to elucidate the structure of, for example, ribosomal RNA (Gutell et al., 1992; Pace et al., 1999; Gutell, 2014) and to analyze functional motifs in PSTVd (Zhong et al., 2006; Ding and Itaya, 2007; Wang et al., 2018). Computational tools were developed to exploit this ncRNA feature.

Performing prediction of alignment and structure is computationally highly intensive; the computing effort grows with $N^{3 \cdot k}$ for k sequences of length N (Sankoff, 1985). One alternative is to first align the sequences and then predict a consensus structure for the alignment (Wilm et al., 2008; Bernhart et al., 2008; Lindgreen et al., 2006). This method and alternatives are reviewed by Reeder et al. (2006). Fig. 3C shows a dotplot in which size of each dot is the mean of the base pair probability of each aligned sequence. From this plot, a consensus structure of the alignment members can be extracted, which avoids pitfalls caused by rare or wrong sequences as well as due to insufficient precision of thermodynamic parameters (Wilm et al., 2008).

Metastable structures in viroids: A common principle in RNA biology

From optical denaturation curves, temperature-gradient gel electrophoresis (TGGE), temperature jump experiments, and structure calculations, it was concluded that the most stable, rod-like structure of PSTVd denatures with high cooperativity to a structure containing two stable hairpins (HPI, pos. 79–87/102–110 and HPII, pos. 227–236/319–328 in PSTVd), which are not part of the rod-like structure (Henco et al., 1977, 1979; Riesner et al., 1979; Steger et al.,

(A)

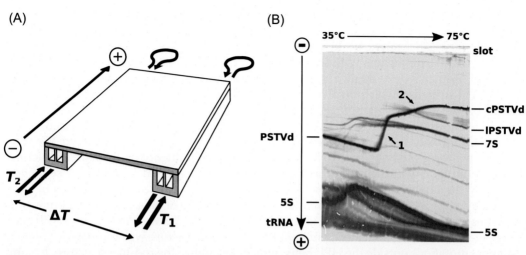

(B)

FIG. 4   Temperature-gradient gel electrophoresis (TGGE). (A) Sketch of a homemade TGGE plate constructed from aluminum. The two water circuits, the temperature gradient, and the direction of electrophoresis are shown by *arrows*. The top part of the plate is sealed by teflon film. For further details, see Rosenbaum and Riesner (1987). A similar plate was sold by Qiagen; a plate with cooling and heating based on Peltier elements was sold by Biometra. A substitute for TGGE with some disadvantages is denaturing-gradient gel electrophoresis (DGGE; Fischer and Lerman, 1983). (B) TGGE analysis of a crude RNA extract from tomato plants infected with PSTVd. A linear temperature gradient is applied perpendicular to the electric field. The sample is loaded into the broad central slot (*top*). RNAs were visualized by silver staining (Whitton et al., 1983). For further conditions, see Riesner and Steger (2014). The rod-like structure of circular PSTVd (cPSTVd; 359 nt) migrates relatively fast at low temperatures; it denatures in a highly cooperative transition (see arrow 1) into a mostly single-stranded structure with at least two extrastable hairpins I and II; these hairpins denature above the main transition (see arrow 2). Linear PSTVd (lPSTVd) molecules migrate proportionally to their length after full denaturation and faster than fully denatured cPSTVd. The temperature of the main transition of lPSTVd depends on the site of linearization; lPSTVd is not visible at low temperatures due to the low concentration of individual lPSTVd molecules with different 5'- and 3'-ends. The different 7 S RNAs (~305 nt) are separated at low temperatures due to their different thermodynamic stability but comigrate at high temperatures due to their nearly identical length. 5S: 5 S ribosomal RNA (120 nt). Note that all bands show an increase in mobility with increasing temperature; this general effect is not based on structural rearrangements, but is due to the decreasing buffer viscosity toward higher temperatures.

1984). An example of TGGE analysis is shown in Fig. 4, which clearly shows the highly cooperative main transition followed by denaturation of the hairpins at still higher temperatures. The advantage of TGGE over other biophysical methods is the gel electrophoretic separation according to RNA length and structure, while other biophysical methods need highly purified samples. Thus, the denaturation of PSTVd—and other members of *Pospiviroidae*—follows the reaction scheme (in 1 M NaCl):

$$\text{Rod-like structure} \underset{}{\overset{T \approx 77°C}{\rightleftharpoons}} \text{Structure with HPI and HPII} \underset{}{\overset{T > 80°C}{\rightleftharpoons}} \text{Denatured circle}$$

The hairpins with given length and base composition are statistically highly significant; that is, they would not be present in random sequences of length and base composition of *Pospiviroidae* members, and can be formed at similar positions in most members of *Pospiviroidae*. The condition of high temperature, however, is irrelevant for any in vivo function in plants. Consequently, the stable hairpins might have only a function at low temperatures as thermodynamically metastable structures (Riesner et al., 1992). For example, metastable structures might be kinetically preferential: RNA is synthesized at a rate of ~10 nt/s, and hairpin helices with small loops are formed in the μs range in newly synthesized RNA segments of 20–40 nt, but their rearrangements into the thermodynamically preferred global structure, which involves pairing of bases quite distant in the full-length RNA strand, is much slower.

We introduced several single- (Loss et al., 1991) and double-site mutations (Qu et al., 1993) into the HPII helix to verify the theoretical concept of metastable structures in vivo. We carefully selected these mutation sites to have only a marginal influence on the native structure. Single-site mutations in the central part of the HPII helix reverted to wild-type sequence, pointing to the importance of HPII, while single mutations, either shortening or elongating the helix by a base pair, were stably maintained after infection of tomato plants. Spontaneous reversions were generated during (−)-strand synthesis, G:U pairs were tolerated in HPII helix of the (−)-strand, and diminished stability of the hairpin structure was compensated for by additional mutations outside the HPII helix, which destabilize the rod-like structure. In summary, the appearance and rate of HPII-favoring mutations are in accordance with HPII-containing structures during replication. In addition, we demonstrated the presence of HPII-containing structures in vivo in multimeric replication intermediates of PSTVd by structure-specific hybridization with a carefully designed oligonucleotide (Schröder and Riesner, 2002).

To obtain quantitative insights into the appearance of or preference for metastable structures during PSTVd replication, we analyzed its in vitro transcription by T7-polymerase from a plasmid containing PSTVd sequence in dependence on the rate and duration of the synthesis. We varied the rate of PSTVd synthesis by altering the concentration of nucleoside triphosphates and analyzed the effect of the rate and duration of synthesis on the resulting structure distributions of the PSTVd transcripts by TGGE (Repsilber et al., 1999). Indeed, at low transcription rates—similar to in vivo rates—metastable structures were preferentially generated.

We simulated this process of RNA structure formation and structural rearrangement using a special Monte-Carlo procedure known as "simulated annealing" based on kinetic data (and not thermodynamic parameters as in MFOLD or RNAFOLD). This tool allowed us to describe the kinetically controlled folding process of any RNA (Schmitz and Steger, 1996). Together with RNA elongation rates, even the process of "sequential folding" during transcription was described. From the kinetically controlled folding of circular as well as of linear PSTVd the importance of HPI and HPII for the folding pathway was obvious and in accordance with experimental data from kinetics (Repsilber et al., 1999) and from the mutagenesis studies on HPII (see earlier; Henco et al., 1979; Loss et al., 1991; Qu et al., 1993).

In the meantime, many further examples of biological function of the mfe structure and metastable structures as well as of cotranscriptional or sequential folding have been described. Riboswitches switch between two mutually exclusive, functional structures due to the presence/absence of a ligand; thermosensors regulate gene expression in dependence on environmental temperature (Breaker, 2012; Schumann, 2012; Serganov and Nudler, 2013; Krajewski and Narberhaus, 2014; Quereda and Cossart, 2017; Gong et al., 2017). Correspondingly, computational methods based on different algorithms were developed for the prediction of the kinetic behavior of RNA folding and the design of functional RNA molecules (Schmitz and Steger, 1996; Isambert and Siggia, 2000; Shcherbakova et al., 2008; Cao et al., 2009; Zhu et al., 2013; Senter et al., 2014; Mann et al., 2014; Hofacker, 2014; Kuchařík et al., 2015; Sun et al., 2017; Meyer, 2017).

Protocols/procedures/methods

Preparation of viroid-containing RNA extract

Solutions:
  50% (w/v) PEG_{6000}; must be stored in a light-tight bottle
  5 M NaCl
  500 mM EDTA, pH 8.0
Buffer B: 10 mM Tris/HCl, 0.5 mM EDTA, pH 7.5
Buffer C: 100 mM Tris/HCl, pH 7.5, 10 mM NaCl, 10 mM EDTA, 2% SDS, 2% β-mercaptoethanol

Plant material is ground in liquid nitrogen, followed by the addition of 2 volumes (v/w) each of acidic phenol (crystalline phenol dissolved in deionized water) and buffer C. After vigorous mixing and centrifugation, the aqueous phase is transferred to a new tube and subsequently extracted with phenol/chloroform (1/1) and chloroform. The final volume of the aqueous phase (V_e) is determined, $V_e/6.7$ of 50% PEG_{6000} and $V_e/9.4$ of 5 M NaCl are added, and the mixture is inverted many times and incubated at room temperature for 30 min. Subsequent centrifugation sediments $\geq 80\%$ of plant nucleic acid, including DNA and high molecular weight RNA, but leaving tRNA and viroid RNA in solution. Addition of $V_e/2.9$ of 50% PEG_{6000} and $V_e/26.3$ of 5 M NaCl to the supernatant of the first precipitation, followed by mixing and incubation at room temperature for 30 min precipitates RNAs of intermediate size including the viroid RNA, which are collected by centrifugation. The pellet is washed with 70% ethanol and dissolved in $V_e/125$ of buffer B after the removal of ethanol.

There are two common pitfalls when using PEG_{6000}, however, which may lead to failure of the precipitation. Both relate to the very high viscosity of the 50% stock solution. First, it is important to pipet this solution very carefully and slowly to avoid adding less PEG than necessary. Second, mixing of the volume to which 50% PEG_{6000} has been added must be done very thoroughly, with flicking or inverting the tube often being more efficient than vortexing. Additionally, it must be stressed that removal of the supernatant following centrifugation must be performed promptly and completely to avoid loss of the pellet and to achieve optimal results. For larger volumes, the bulk of the supernatant is taken off quickly and the tubes are recentrifuged for 2 min to collect the rest of the supernatant at the bottom of the tube, which can then be completely removed using a micropipette. Washing the pellet with 70% ethanol removes traces of salt and PEG_{6000}, the latter being soluble in this solvent to at least 25% (w/v). For sensitive applications, washing with 70% EtOH should be performed twice.

Structural predictions

To carry out analyses similar to those illustrated in Fig. 3A–C, we recommend the local installation of the necessary tools (see Table 1).

TABLE 1 Web addresses.

Package or program	Address	Reference
Download		
RNASTRUCTURE	`https://rna.urmc.rochester.edu/RNAstructure.html`	Mathews (2014)
ViennaRNA package	`https://www.tbi.univie.ac.at/RNA/`	Lorenz et al. (2011)
RNASHAPES	`https://bibiserv.cebitec.uni-bielefeld.de/rnashapes`	Janssen and Giegerich (2014)
MAFFT	`https://mafft.cbrc.jp/alignment/software/`	Katoh et al. (2017) and Katoh and Toh (2008)
CONSTRUCT	`http://www.biophys.uni-duesseldorf.de/html/local/construct3/index.html`	Wilm et al. (2008)
EMBOSS	`http://emboss.sourceforge.net/`	Chojnacki et al. (2017)
Web services		
MFOLD	`http://unafold.rna.albany.edu/?q=mfold`	Zuker (2003)
UNAFOLD	`http://unafold.rna.albany.edu/?q=DINAMelt`	Markham and Zuker (2008)
RNASTRUCTURE	`https://rna.urmc.rochester.edu/RNAstructureWeb/`	Mathews et al. (2016)
ViennaRNA Web Services	`http://rna.tbi.univie.ac.at/`	Gruber et al. (2008)
RNAFOLD	`http://rna.tbi.univie.ac.at/cgi-bin/RNAWebSuite/RNAfold.cgi`	
RNAALIFOLD	`http://rna.tbi.univie.ac.at/cgi-bin/RNAWebSuite/RNAalifold.cgi`	
RNAEVAL	`http://rna.tbi.univie.ac.at/cgi-bin/RNAWebSuite/RNAeval.cgi`	
RNASHAPES	`https://bibiserv.cebitec.uni-bielefeld.de/rnashapes?id=rnashapes_view_download`	Janssen and Giegerich (2014)
MAFFT	`https://mafft.cbrc.jp/alignment/server/`	Katoh et al. (2017)
EMBOSS	`http://emboss.sourceforge.net/`	Buso et al. (2019)
DOTMATCHER	`https://www.bioinformatics.nl/cgi-bin/emboss/dotmatcher`	
BLASTn	`https://blast.ncbi.nlm.nih.gov/Blast.cgi`	

Base pair dotplot of a viroid sequence

Many web tools allow a user to produce dotplots showing similarities between two different sequences, but we want a dotplot showing base pairs within a single sequence (Fig. 3A). The shortest helix consists of two base pairs. If base pair i:j is part of the helix then either $(i + 1)$:$(j - 1)$ or $(i - 1)$:$(j + 1)$ is the neighboring base pair. That is, we have to feed to the standard dotplot the sequence and the reverse (but not complement) sequence, and a matrix that defines base pairs instead of base/base similarities. DOTMATCHER (Table 1) uses a matrix like the following:

```
  A T G C U
A 0 1 0 0 1
T 1 0 1 0 0
G 0 1 0 1 1
C 0 0 1 0 0
U 1 0 1 0 0
```

Store this matrix as a text file. In case one wants to use the web interface to DOTMATCHER (Table 1), select "DNA" as sequence type, upload your sequence, the reverse of the sequence, and the matrix file. Note that the matrix contains U as well as T; that is, it does not matter if your sequence(s) are formatted as RNA or as DNA. Specify the desired helix length and threshold, which should be identical to the helix length, and the desired output format.

In the case of a local EMBOSS installation, the command is

```
dotmatcher -asequence PSTVd_reverse.fa    \
   -bsequence PSTVd.fa -graph ps -matrixfile RNABP \
   -windowsize 5 -threshold 5       \
   -goutfile PSTVd_PSTVdrev_BP.ps
```

with sequence files named PSTVd.fa and PSTVd_reverse.fa, both in FASTA format, the matrix file RNABP, output format PostScript, and output file name PSTVd_PSTVdrev_BP.ps. For further options, execute tfm -program dotmatcher, which requires installation of the EMBOSS documentation.

Ensemble and mfe dotplot of a viroid sequence

To create a dotplot as shown in Fig. 3B, go to the RNAFOLD web server (Table 1) and upload your sequence. The following example uses the sequence of PSTVd variant Intermediate (GenBank AC NC_002030). At the "Advanced folding options" you have to select "assume RNA molecule to be circular." In addition, you might select "allow coaxial stacking of adjacent helices in multiloops"; this option is valid only for the mfe prediction and enhances the probability to predict multiloops. On the result page the following is given:

- $\Delta G^0_{37°C,mfe} = -160.00$kcal/mol.
- Input sequence and mfe structure in dot-bracket notation.
- $\Delta G^0_{37°C,ensemble} = -164.28$kcal/mol.
- The frequency of the mfe structure in the ensemble is 0.10%.
- The ensemble diversity is 29.42.

C. Viroid pathogenesis and viroid-host interaction

- The dotplot as shown in Fig. 3B shows the mfe structure in the lower left triangle and the structure ensemble in the top right triangle with dot size ∝ probability of base pair.
- The centroid secondary structure in dot-bracket notation with $\Delta G^0_{37°C,centroid} = -159.00$kcal/mol.
- Sketch of mfe and centroid structure with different coloring options.
- Mountain plot of mfe structure, centroid structure, and structure ensemble.

In the case of a local Vienna package installation, the command is

```
RNAfold -partfunc -dangles=3 -noLP -circ \
  < PSTVd.fa > PSTVd.log
```

or shorter

```
RNAfold -p -d3 -noLP -c < PSTVd.fa > PSTVd.log
```

with a request of partition function calculation (-p), coaxial stacking of adjacent helices in multiloops (-d3), without lonely pairs (helices of length 1; -noLP), calculation for a circular molecule (-c), reading the sequence from file PSTVd.fa in FASTA format, and writing information to PSTVd.log. If the first line of the FASTA file is NC_002030, further output files are NC_002030_dp.ps, which contains the dotplot in encapsulated PostScript format, and NC_002030_ss.ps, which contains the secondary structure sketch in encapsulated PostScript format. Note that the secondary structure is drawn anticlockwise. To rotate the sketch by 90 degrees and mirror it, use

```
rotate_ss.pl -angle 90 -m \
  NC_002030_ss.ps > NC_002030_ss_standard.ps
```

To create a mountain plot, use

```
mountain.pl NC_002030_dp.ps | xmgrace -pipe \
  -hdevice PostScript -hardcopy \
  -printfile NC_002030_mountain.ps
```

To colorize the secondary structure sketch with pair probability (option -p), with accessibility (e.g., a primer; option -a), or with positional entropy (give no option), use

```
relplot.pl NC_002030_ss.ps \
  NC_002030_dp.ps > NC_002030_ss_entropy.ps
```

Base pairs with low positional entropy have a high probability to be paired and thus no or very few structural alternatives; nonpaired nucleotides with low positional entropy have a high probability to be nonpaired; high entropy implies many structural alternatives. The perl programs ROTATE_SS.PL, MOUNTAIN.PL, and RELPLOT.PL are part of the Vienna package.

Consensus base pair dotplot of a viroid species

In this section, we outline a procedure to obtain information on the consensus sequence and structure of PSTVd sequences available from GenBank.

1. Do a BLASTN (Table 1) search with the first sequence of your choice in "Standard databases," "nucleotide collection"; here we used NC_002030. "Select the maximum number of aligned sequences to display" at "Algorithmic parameters"; the value should be sufficient to find all relevant sequences.
2. "Filter" the detected sequences according to identity and species.
3. "Download" selected sequences as "GenBank (complete sequence)" and "FASTA (complete sequence)." The file in GenBank format is nice to look up information on specific sequence variants; the file in FASTA format is useful for further processing.
4. Upload the FASTA file to the MAFFT web server (Table 1) and align the sequences with a fast method.
5. Refine the sequence set according to the preliminary alignment from the previous step:
   * Remove sequences with ambiguous nucleotides.
   * Remove incomplete sequences.
   * If the sequence set contains (−)-stranded sequences, substitute these with their reverse complement to get (+)-stranded sequences.
   * Adjust sequences that do not start with nucleotide position 1.
     Go back to step 4 and use the FASTA file with adjusted sequences.
6. Do a final alignment with an advanced MAFFT method. MAFFT-Q-INS-I is available at the web server, but is restricted to less than 200 sequences; MAFFT-X-INS-I is preferable but needs a local installation and a lot of computing time (Katoh and Toh, 2008).
7. Download the final alignment from the MAFFT web server in CLUSTAL format or do the conversion locally using, for example, CLUSTALX (Thompson et al., 1997).
8. Upload the final alignment in Clustal format to the RNAalifold web server (Table 1). At "advanced options" select "circular sequences." Submit the job. Note that the alignment has to contain less than 300 sequences of at most 3000 nt.

On the result page the following is given; we will omit the concrete values because these will depend on the set of aligned sequences:

* $\Delta G^0_{37°C}$. The first value is the mfe free energy and the second is the covariance contribution.
* Consensus sequence and mfe structure in dot-bracket notation.
* Detailed information about each basepaired position ($i{:}j$, percentage of pairs, number, and type of base pairs and nonpairs).
* A structure-annotated alignment: Red marks pairs with no sequence variation; ochre, green, cyan, blue, and violet marks pair with two to six different types of pairs. Pairs with one or two inconsistent mutations are shown in (two types of) pale colors; if there are more than two noncompatible pairs, the pair is not shown.
* $\Delta G^0_{37°C,\text{ensemble}}$.
* The frequency of the mfe structure in the ensemble.

- The dotplot similar to that shown in Fig. 3C shows the mfe structure in the lower left triangle and the structure ensemble in the top right triangle with dot size $\propto$ probability of base pair.
- Sketch of mfe structure with different coloring options.
- Mountain plot of mfe structure and structure ensemble.

In the case of a local Vienna package installation, the command is

```
RNAalifold -p -r -d2 -noLP -circ -color -aln \
  < PSTVd_xinsi.aln > PSTVd_xinsi.log
```

This results in a plot of the consensus mfe structure (alirna.ps), the annotated alignment (aln.ps), the detailed information about each basepaired position (alifold.out), the dotplot (alidot.ps), and the log file (PSTVd_xinsi.log) with mfe and ensemble structure in bracket-dot notation and $\Delta G^0_{37°C}$ values. The filenames, except that of the log file, are fixed and will be overwritten with the next run of RNAalifold.

To obtain the further plots available from the website, use the tools mentioned in the previous protocol.

Future implications

Both return and 2D gel electrophoresis are simple, inexpensive, and efficient methods to discover new viroids and viroid-like RNAs, based only on their circularity; examples of their successful use include the detection of grapevine yellow speckle viroid 1 (Koltunow and Rezaian, 1988), dahlia latent viroid (Verhoeven et al., 2012), and portulaca latent viroid (Verhoeven et al., 2015). A long-standing alternative to electrophoresis, the use of "universal," degenerate primers, for viroid discovery is being replaced by next-generation sequencing (for review, see Hadidi, 2019).

Study questions

1. The preparation of the RNA crude extract starts with grinding the plant material in liquid nitrogen. What problem(s) does this prevent?
2. What biomolecules are found for which reason in the aqueous and the organic phase produced during phenol/chloroform extraction? What biomolecules are found in the interphase, which separates aqueous and organic phases?
3. What is the probability to find a helix with eight WC as well as GU base pairs in a sequence with random base composition?
4. In a base pair dotplot, one usually marks helices of length ≥ 2. Such a plot guarantees to contain the base pairs of all possible secondary structures. Why?
5. What is the difference in $\Delta G^0_{T=37°C}$ of a linear and a circular PSTVd? Does this difference depend on the point of linearization? To answer this question, you might need the help of RNAFOLD and RNAEVAL (Table 1).

Further reading

For an introduction to the background of structure prediction and a comparison of MFOLD and RNAFOLD we refer to Steger (2014); a similar presentation including an introduction to consensus structure prediction and structure drawing is given by Steger (2022). For an alternative to the structure prediction tools mentioned in this chapter, we refer to RNASHAPES (Giegerich et al., 2004) and derived tools (Reeder and Giegerich, 2005; Steffen et al., 2006; Janssen and Giegerich, 2014). Pseudoknots do not play a role in the structure of *Pospiviroidae* members in contrast to *Avsunviroidae* members; see Dubé et al. (2011), Steger and Perreault (2016), and Sato and Kato (2022).

References

Andronescu, M., Condon, A., Turner, D.H., Mathews, D.H., 2014. The determination of RNA folding nearest neighbor parameters. Methods Mol. Biol. 1097, 45–70. https://doi.org/10.1007/978-1-62703-709-9_3.

Bernhart, S.H., Hofacker, I.L., Will, S., Gruber, A.R., Stadler, P.F., 2008. RNAalifold: improved consensus structure prediction for RNA alignments. BMC Bioinf. 9, 474. https://doi.org/10.1186/1471-2105-9-474.

Brass, J.R., Owens, R.A., Matoušek, J., Steger, G., 2016. Viroid quasispecies revealed by deep sequencing. RNA Biol. 14, 317–325. https://doi.org/10.1080/15476286.2016.1272745.

Breaker, R.R., 2012. Riboswitches and the RNA world. Cold Spring Harb. Perspect. Biol. 4, a003566. https://doi.org/10.1101/cshperspect.a003566.

Buso, N., Gur, T., Madhusoodanan, N., Basutkar, P., Tivey, A.R.N., Potter, S.C., Finn, R.D., Lopez, R., 2019. The EMBL-EBI search and sequence analysis tools APIs in 2019. Nucleic Acids Res. 47, W636 W641. https://doi.org/10.1093/nar/gkz268.

Cao, H., Xie, H.Z., Zhang, W., Wang, K., Li, W., Liu, C.Q., 2009. Dynamic extended folding: modeling the RNA secondary structures during co-transcriptional folding. J. Theor. Biol. 261, 93–99. https://doi.org/10.1016/j.jtbi.2009.07.027.

Chojnacki, S., Cowley, A., Lee, J., Foix, A., Lopez, R., 2017. Programmatic access to bioinformatics tools from EMBL-EBI update: 2017. Nucleic Acids Res. 45, W550–W553. https://doi.org/10.1093/nar/gkx273.

Crick, F.H., 1966. Codon-anticodon pairing: the wobble hypothesis. J. Mol. Biol. 19, 548–555. https://doi.org/10.1016/s0022-2836(66)80022-0.

Ding, B., Itaya, A., 2007. Viroid: a useful model for studying the basic principles of infection and RNA biology. Mol. Plant Micr. Int. 20, 7–20. https://doi.org/10.1094/MPMI-20-0007.

Drake, J.W., 1991. A constant rate of spontaneous mutation in DNA-based microbes. Proc. Natl. Acad. Sci. USA 88, 7160–7164. https://doi.org/10.1073/pnas.88.16.7160.

Dubé, A., Bolduc, F., Bisaillon, M., Perreault, J.P., 2011. Mapping studies of the Peach latent mosaic viroid reveal novel structural features. Mol. Plant Pathol. 12, 688–701. https://doi.org/10.1111/j.1364-3703.2010.00703.x.

Eigen, M., 1971. Selforganization of matter and the evolution of biological macromolecules. Naturwissenschaften 58, 465–523. https://doi.org/10.1007/BF00623322.

Fischer, S.G., Lerman, L.S., 1983. DNA fragments differing by single base-pair substitutions are separated in denaturing gradient gels: correspondence with melting theory. Proc. Natl. Acad. Sci. USA 80, 1579–1583. https://doi.org/10.1073/pnas.80.6.1579.

Gago, S., Elena, S.F., Flores, R., Sanjuán, R., 2009. Extremely high mutation rate of a hammerhead viroid. Science 323, 1308. https://doi.org/10.1126/science.1169202.

Giegerich, R., Voss, B., Rehmsmeier, M., 2004. Abstract shapes of RNA. Nucleic Acids Res. 32, 4843–4851. https://doi.org/10.1093/nar/gkh779.

Giguère, T., Perreault, J.P., 2017. Classification of the *Pospiviroidae* based on their structural hallmarks. PLoS ONE 12, e0182536. https://doi.org/10.1371/journal.pone.0182536.

Giguère, T., Adkar-Purushothama, C.R., Bolduc, F., Perreault, J.P., 2014. Elucidation of the structures of all members of the *Avsunviroidae* family. Mol. Plant Pathol. 15, 767–779. https://doi.org/10.1111/mpp.12130.

Gong, S., Wang, Y., Wang, Z., Zhang, W., 2017. Co-transcriptional folding and regulation mechanisms of riboswitches. Molecules 22, 1169. https://doi.org/10.3390/molecules22071169.

Gross, H.J., Riesner, D., 1980. Viroids: a class of subviral pathogens. Angew. Chem. Int. Ed. Engl. 19, 231–243. https://doi.org/10.1002/anie.198002313.

Gross, H.J., Domdey, H., Lossow, C., Jank, P., Raba, M., Alberty, H., Sänger, H.L., 1978. Nucleotide sequence and secondary structure of potato spindle tuber viroid. Nature 273, 203–208. https://doi.org/10.1038/273203a0.

Gruber, A.R., Lorenz, R., Bernhart, S.H., Neuböck, R., Hofacker, I.L., 2008. The Vienna RNA websuite. Nucleic Acids Res. 36, W70–W74. https://doi.org/10.1093/nar/gkn188.

Gutell, R.R., 2014. Ten lessons with Carl Woese about RNA and comparative analysis. RNA Biol. 11, 254–272. https://doi.org/10.4161/rna.28718.

Gutell, R.R., Power, A., Hertz, G.Z., Putz, E.J., Stormo, G.D., 1992. Identifying constraints on the higher-order structure of RNA: continued development and application of comparative sequence analysis methods. Nucleic Acids Res. 20, 5785–5795. https://doi.org/10.1093/nar/20.21.5785.

Hadidi, A., 2019. Next-generation sequencing and CRISPR/Cas13 editing in viroid research and molecular diagnostics. Viruses 11, 120. https://doi.org/10.3390/v11020120.

Henco, K., Riesner, D., Sänger, H.L., 1977. Conformation of viroids. Nucleic Acids Res. 4, 177–194. https://doi.org/10.1093/nar/4.1.177.

Henco, K., Sänger, H.L., Riesner, D., 1979. Fine structure melting of viroids as studied by kinetic methods. Nucleic Acids Res. 6, 3041–3059. https://doi.org/10.1093/nar/6.9.3041.

Hofacker, I.L., 2014. Energy-directed RNA structure prediction. Methods Mol. Biol. 1097, 71–84. https://doi.org/10.1007/978-1-62703-709-9_4.

Hofacker, I., Schuster, P., Stadler, P., 1998. Combinatorics of RNA secondary structures. Discrete Appl. Math. 88, 207237. https://doi.org/10.1016/S0166-218X(98)00073-0.

Isambert, H., Siggia, E.D., 2000. Modeling RNA folding paths with pseudoknots: application to hepatitis delta virus ribozyme. Proc. Natl. Acad. Sci. USA 97, 6515–6520. https://doi.org/10.1073/pnas.110533697.

Janssen, S., Giegerich, R., 2014. The RNA shapes studio. Bioinformatics 31, 423–425. https://doi.org/10.1093/bioinformatics/btu649.

Katoh, K., Toh, H., 2008. Improved accuracy of multiple ncRNA alignment by incorporating structural information into a MAFFT-based framework. BMC Bioinf. 9, 212. https://doi.org/10.1186/1471-2105-9-212.

Katoh, K., Rozewicki, J., Yamada, K.D., 2017. MAFFT online service: multiple sequence alignment, interactive sequence choice and visualization. Brief. Bioinform. https://doi.org/10.1093/bib/bbx108.

Koltunow, A.M., Rezaian, M.A., 1988. Grapevine yellow speckle viroid: structural features of a new viroid group. Nucleic Acids Res. 16, 849–864. https://doi.org/10.1093/nar/16.3.849.

Krajewski, S.S., Narberhaus, F., 2014. Temperature-driven differential gene expression by RNA thermosensors. Biochim. Biophys. Acta 1839, 978–988. https://doi.org/10.1016/j.bbagrm.2014.03.006.

Kuchařík, M., Hofacker, I.L., Stadler, P.F., Qin, J., 2015. Pseudoknots in RNA folding landscapes. Bioinformatics 32, 187–194. https://doi.org/10.1093/bioinformatics/btv572.

Langowski, J., Henco, K., Riesner, D., Sänger, H.L., 1978. Common structural features of different viroids: serial arrangement of double helical sections and internal loops. Nucleic Acids Res. 5, 1589–1610. https://doi.org/10.1093/nar/5.5.1589.

Leontis, N.B., Westhof, E., 2001. Geometric nomenclature and classification of RNA base pairs. RNA 7, 499–512. https://doi.org/10.1017/s1355838201002515.

Leontis, N.B., Stombaugh, J., Westhof, E., 2002. The non-Watson-Crick base pairs and their associated isostericity matrices. Nucleic Acids Res. 30, 3497–3531. https://doi.org/10.1093/nar/gkf481.

Lindgreen, S., Gardner, P.P., Krogh, A., 2006. Measuring covariation in RNA alignments: physical realism improves information measures. Bioinformatics 22, 2988–2995. https://doi.org/10.1093/bioinformatics/btl514.

López-Carrasco, A., Flores, R., 2017. Dissecting the secondary structure of the circular RNA of a nuclear viroid *in vivo*: a "naked" rod-like conformation similar but not identical to that observed *in vitro*. RNA Biol. 14, 1046–1054. https://doi.org/10.1080/15476286.2016.1223005.

López-Carrasco, A., Ballesteros, C., Sentandreu, V., Delgado, S., Gago-Zachert, S., Flores, R., Sanjuán, R., 2017. Different rates of spontaneous mutation of chloroplastic and nuclear viroids as determined by high-fidelity ultra-deep sequencing. PLoS Pathog. 13, e1006547. https://doi.org/10.1371/journal.ppat.1006547.

Lorenz, R., Bernhart, S.H., Höner Zu Siederdissen, C., Tafer, H., Flamm, C., Stadler, P.F., Hofacker, I.L., 2011. ViennaRNA Package 2.0. Algorithms Mol. Biol. 6, 26. https://doi.org/10.1186/1748-7188-6-26.

Lorenz, R., Hofacker, I.L., Stadler, P.F., 2016. RNA folding with hard and soft constraints. Algorithms Mol. Biol. 11, 8. https://doi.org/10.1186/s13015-016-0070-z.

Loss, P., Schmitz, M., Steger, G., Riesner, D., 1991. Formation of a thermodynamically metastable structure containing hairpin II is critical for infectivity of potato spindle tuber viroid RNA. EMBO J. 10, 719–727. https://doi.org/10.1002/j.1460-2075.1991.tb08002.x.

Ma, J., Mudiyanselage, S.D.D., Park, W.J., Wang, M., Takeda, R., Liu, B., Wang, Y., 2022. A nuclear import pathway exploited by pathogenic noncoding RNAs. Plant Cell. 34, 3543–3556. https://doi.org/10.1093/plcell/koac210.

Mann, M., Kuchařík, M., Flamm, C., Wolfinger, M.T., 2014. Memory-efficient RNA energy landscape exploration. Bioinformatics 30, 2584–2591. https://doi.org/10.1093/bioinformatics/btu337.

Markham, N.R., Zuker, M., 2008. UNAFold: software for nucleic acid folding and hybridization. Methods Mol. Biol. 453, 3–31. https://doi.org/10.1007/978-1-60327-429-6_1.

Mathews, D.H., 2014. RNA secondary structure analysis using RNAstructure. Curr. Protoc. Bioinformatics 46, 12.6.1–12.6.25. https://doi.org/10.1002/0471250953.bi1206s46.

Mathews, D.H., Sabina, J., Zuker, M., Turner, D.H., 1999. Expanded sequence dependence of thermodynamic parameters improves prediction of RNA secondary structure. J. Mol. Biol. 288, 911–940. https://doi.org/10.1006/jmbi.1999.2700.

Mathews, D.H., Disney, M.D., Childs, J.L., Schroeder, S.J., Zuker, M., Turner, D.H., 2004. Incorporating chemical modification constraints into a dynamic programming algorithm for prediction of RNA secondary structure. Proc. Natl. Acad. Sci. USA 101, 7287–7292. https://doi.org/10.1073/pnas.0401799101.

Mathews, D.H., Turner, D.H., Watson, R.M., 2016. RNA secondary structure prediction. Curr. Protoc. Nucleic Acid Chem. 67, 11.2.1–11.2.19. https://doi.org/10.1002/cpnc.19.

McCaskill, J.S.M., 1990. The equilibrium partition function and base pair binding probabilities for RNA secondary structure. Biopolymers 29, 1105–1119. https://doi.org/10.1002/bip.360290621.

Meyer, I.M., 2017. In silico methods for co-transcriptional RNA secondary structure prediction and for investigating alternative RNA structure expression. Methods 120, 3–16. https://doi.org/10.1016/j.ymeth.2017.04.009.

Pace, N.R., Thomas, B.C., Woese, C.R., 1999. Probing RNA structure, function, and history by comparative analysis. In: Gesteland, R.F., Cech, T.R., Atkins, J.F. (Eds.), The RNA World. Cold Spring Harbor Laboratory Press, New York, pp. 113–141.

Qu, F., Heinrich, C., Loss, P., Steger, G., Tien, P., Riesner, D., 1993. Multiple pathways of reversion in viroids for conservation of structural elements. EMBO J. 12, 2129–2139. https://doi.org/10.1002/j.1460-2075.1993.tb05861.x.

Quereda, J.J., Cossart, P., 2017. Regulating bacterial virulence with RNA. Annu. Rev. Microbiol. 71, 263–280. https://doi.org/10.1146/annurev-micro-030117-020335.

Reeder, J., Giegerich, R., 2005. Consensus shapes: an alternative to the Sankoff algorithm for RNA consensus structure prediction. Bioinformatics 21, 3516–3523. https://doi.org/10.1093/bioinformatics/bti577.

Reeder, J., Höchsmann, M., Rehmsmeier, M., Voss, B., Giegerich, R., 2006. Beyond Mfold: recent advances in RNA bioinformatics. J. Biotechnol. 124, 41–55. https://doi.org/10.1016/j.jbiotec.2006.01.034.

Repsilber, D., Wiese, U., Rachen, M., Schröder, A.R., Riesner, D., Steger, G., 1999. Formation of metastable RNA structures by sequential folding during transcription: time-resolved structural analysis of potato spindle tuber viroid (−)-stranded RNA by temperature-gradient gel electrophoresis. RNA 5, 574–584. https://doi.org/10.1017/S1355838299982018.

Reuter, J.S., Mathews, D.H., 2010. RNAstructure: software for RNA secondary structure prediction and analysis. BMC Bioinf. 11, 129. https://doi.org/10.1186/1471-2105-11-129.

Riesner, D., Steger, G., 2014. Temperature-gradient gel-electrophoresis. In: Hartmann, R., Bindereif, A., Schön, A., Westhof, E. (Eds.), Handbook of RNA Biochemistry. Wiley-VCH, pp. 427–444.

Riesner, D., Henco, K., Rokohl, U., Klotz, G., Kleinschmidt, A.K., Domdey, H., Jank, P., Gross, H.J., Sänger, H.L., 1979. Structure and structure formation of viroids. J. Mol. Biol. 133, 85–115. https://doi.org/10.1016/0022-2836(79)90252-3.

Riesner, D., Baumstark, T., Qu, F., Klahn, T., Loss, P., Rosenbaum, V., Schmitz, M., Steger, G., 1992. Physical basis and biological examples of metastable RNA structures. In: Lilley, D., Heumann, H., Suck, D. (Eds.), Structural Tools for the Analysis of Protein-Nucleic Acids Complexes. Advances in Life Sciences, Birkhäuser Verlag AG, Basel, pp. 401–435.

Rosenbaum, V., Riesner, D., 1987. Temperature-gradient gel electrophoresis. Thermodynamic analysis of nucleic acids and proteins in purified form and in cellular extracts. Biophys. Chem. 26, 235–246. https://doi.org/10.1016/0301-4622(87)80026-1.

Sänger, H.L., Klotz, G., Riesner, D., Gross, H.J., Kleinschmidt, A.K., 1976. Viroids are single-stranded covalently closed circular RNA molecules existing as highly base-paired rod-like structures. Proc. Natl. Acad. Sci. USA 73, 3852–3856. https://doi.org/10.1073/pnas.73.11.3852.

Sanjuán, R., Nebot, M.R., Chirico, N., Mansky, L.M., Belshaw, R., 2010. Viral mutation rates. J. Virol. 84, 9733–9748. https://doi.org/10.1128/JVI.00694-10.

Sankoff, D., 1985. Simultaneous solution of the RNA folding, alignment and protosequence problems. SIAM J. Appl. Math. 45, 810–825. https://doi.org/10.1137/0145048.

Sato, K., Kato, Y., 2022. Prediction of RNA secondary structure including pseudoknots for long sequences. Brief. Bioinform. 23, bbab395. https://doi.org/10.1093/bib/bbab395.

Schmitz, M., Steger, G., 1996. Description of RNA folding by "simulated annealing". J. Mol. Biol. 255, 254–266. https://doi.org/10.1006/jmbi.1996.0021.

Schröder, A.R., Riesner, D., 2002. Detection and analysis of hairpin II, an essential metastable structural element in viroid replication intermediates. Nucleic Acids Res. 30, 3349–3359. https://doi.org/10.1093/nar/gkf454.

Schroeder, S.J., Turner, D.H., 2009. Optical melting measurements of nucleic acid thermodynamics. Methods Enzymol. 468, 371–387. https://doi.org/10.1016/S0076-6879(09)68017-4.

Schumann, W., 2012. Thermosensor systems in eubacteria. Adv. Exp. Med. Biol. 739, 1–16. https://doi.org/10.1007/978-1-4614-1704-0_1.

Senter, E., Dotu, I., Clote, P., 2014. RNA folding pathways and kinetics using 2D energy landscapes. J. Math. Biol. 70, 173–196. https://doi.org/10.1007/s00285-014-0760-4.

Serganov, A., Nudler, E., 2013. A decade of riboswitches. Cell 152, 17–24. https://doi.org/10.1016/j.cell.2012.12.024.

Shcherbakova, I., Mitra, S., Laederach, A., Brenowitz, M., 2008. Energy barriers, pathways, and dynamics during folding of large, multidomain RNAs. Curr. Opin. Chem. Biol. 12, 655–666. https://doi.org/10.1016/j.cbpa.2008.09.017.

Steffen, P., Voss, B., Rehmsmeier, M., Reeder, J., Giegerich, R., 2006. RNAshapes: an integrated RNA analysis package based on abstract shapes. Bioinformatics 22, 500–503.

Steger, G., 2014. Secondary structure prediction. In: Hartmann, R., Bindereif, A., Schön, A., Westhof, E. (Eds.), Handbook of RNA Biochemistry. Wiley-VCH, pp. 549–577.

Steger, G., 2022. Predicting the structure of a viroid: structure, structure distribution, consensus structure, and structure drawing. Methods Mol. Biol. 2316, 331–371. https://doi.org/10.1007/978-1-0716-1464-8_26.

Steger, G., Perreault, J., 2016. Structure and associated biological functions of viroids. Adv. Virus Res. 94, 141–172. https://doi.org/10.1016/bs.aivir.2015.11.002.

Steger, G., Hofmann, H., Förtsch, J., Gross, H.J., Randles, J.W., Sänger, H.L., Riesner, D., 1984. Conformational transitions in viroids and virusoids: comparison of results from energy minimization algorithm and from experimental data. J. Biomol. Struct. Dyn. 2, 543–571. https://doi.org/10.1080/07391102.1984.10507591.

Stone, J.W., Bleckley, S., Lavelle, S., Schroeder, S.J., 2015. A parallel implementation of the Wuchty algorithm with additional experimental filters to more thoroughly explore RNA conformational space. PLoS ONE 10, e0117217. https://doi.org/10.1371/journal.pone.0117217.

Sun, L.Z., Zhang, D., Chen, S.J., 2017. Theory and modeling of RNA structure and interactions with metal ions and small molecules. Annu. Rev. Biophys. 46, 227–246. https://doi.org/10.1146/annurev-biophys-070816-033920.

Thompson, J.D., Gibson, T.J., Plewniak, F., Jeanmougin, F., Higgins, D.G., 1997. The CLUSTAL_X windows interface: flexible strategies for multiple sequence alignment aided by quality analysis tools. Nucleic Acids Res. 25 (24), 4876–4882. https://doi.org/10.1093/nar/25.24.4876.

Tinoco Jr., I., Uhlenbeck, O.C., Levine, M.D., 1971. Estimation of secondary structure in ribonucleic acids. Nature 230, 362–367. https://doi.org/10.1038/230362a0.

Verhoeven, J.T., Meekes, E.T., Roenhorst, J.W., Flores, R., Serra, P., 2012. Dahlia latent viroid: a recombinant new species of the family *Pospiviroidae* posing intriguing questions about its origin and classification. J. Gen. Virol. 94, 711–719. https://doi.org/10.1099/vir.0.048751-0.

Verhoeven, J.T., Roenhorst, J.W., Hooftman, M., Meekes, E.T., Flores, R., Serra, P., 2015. A pospiviroid from symptomless portulaca plants closely related to iresine viroid 1. Virus. Res. 205, 22–26. https://doi.org/10.1016/j.virusres.2015.05.005.

Walker, S.E., Lorsch, J., 2013. Reverse transcriptase dideoxy sequencing of RNA. Methods Enzymol. 530, 347–359. https://doi.org/10.1016/B978-0-12-420037-1.00020-8.

Wang, Y., Zirbel, C.L., Leontis, N.B., Ding, B., 2018. RNA 3-dimensional structural motifs as a critical constraint of viroid RNA evolution. PLoS Pathog. 14, e1006801. https://doi.org/10.1371/journal.ppat.1006801.

C. Viroid pathogenesis and viroid-host interaction

Waterman, M.S., 1995. Introduction to Computational Biology. Maps, Sequences and Genomes. Chapman & Hall, London.

Whitton, J.L., Hundley, F., O'Donnell, B., Desselberger, U., 1983. Silver staining of nucleic acids. Applications in virus research and in diagnostic virology. J. Virol. Methods 7, 185–198. https://doi.org/10.1016/0166-0934(83)90008-3.

Wilm, A., Linnenbrink, K., Steger, G., 2008. ConStruct: improved construction of RNA consensus structures. BMC Bioinf. 9, 219. https://doi.org/10.1186/1471-2105-9-219.

Wuchty, S., Fontana, W., Hofacker, I.L., Schuster, P., 1999. Complete suboptimal folding of RNA and the stability of secondary structures. Biopolymers 49, 145–165. https://doi.org/10.1002/(SICI)1097-0282(199902)49:2<145::AID-BIP4>3.0.CO;2-G.

Yakovchuk, P., Protozanova, E., Frank-Kamenetskii, M.D., 2006. Base-stacking and base-pairing contributions into thermal stability of the DNA double helix. Nucleic Acids Res. 34 (2), 564–574. https://doi.org/10.1093/nar/gkj454.

Zhong, X., Leontis, N., Qian, S., Itaya, A., Qi, Y., Boris-Lawrie, K., Ding, B., 2006. Tertiary structural and functional analyses of a viroid RNA motif by isostericity matrix and mutagenesis reveal its essential role in replication. J. Virol. 80, 8566–8581. https://doi.org/10.1128/JVI.00837-06.

Zhu, J.Y., Steif, A., Proctor, J.R., Meyer, I.M., 2013. Transient RNA structure features are evolutionarily conserved and can be computationally predicted. Nucleic Acids Res. 41, 6273–6285. https://doi.org/10.1093/nar/gkt319.

Zuker, M., 1989. On finding all suboptimal foldings of an RNA molecule. Science 244, 48–52. https://doi.org/10.1126/science.2468181.

Zuker, M., 2003. Mfold web server for nucleic acid folding and hybridization prediction. Nucleic Acids Res. 31, 3406–3415. https://doi.org/10.1093/nar/gkg595.

Zuker, M., Stiegler, P., 1981. Optimal computer folding of large RNA sequences using thermodynamics and auxiliary information. Nucleic Acids Res. 9, 133–148. https://doi.org/10.1093/nar/9.1.133.

Zuker, M., Mathews, D.H., Turner, D.H., 1999. Algorithms and thermodynamics for RNA secondary structure prediction: a practical guide. In: Barciszewski, J., Clark, B.F.C. (Eds.), RNA Biochemistry and Biotechnology. NATO ASI Series, Kluwer.

Future perspectives in viroid research

José-Antonio Daròs

Instituto de Biología Molecular y Celular de Plantas (Consejo Superior de Investigaciones
Científicas-Universitat Politècnica de València), Valencia, Spain

Graphical representation

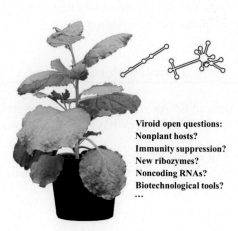

Viroid open questions:
Nonplant hosts?
Immunity suppression?
New ribozymes?
Noncoding RNAs?
Biotechnological tools?
...

Abbreviations

AGO	Argonaute
Cas	CRISPR associated
CRISPR	clustered regularly interspaced palindromic repeats
DCL	Dicer-like
dsRNA	double-stranded RNA
HDV	hepatitis deltavirus
RDR	RNA-dependent RNA polymerase
siRNA	small interfering RNA

Viroid species

ASBVd	avocado sunblotch viroid
CCCVd	coconut cadang-cadang viroid
ELVd	eggplant latent viroid
HSVd	hop stunt viroid
PSTVd	potato spindle tuber viroid

Chapter outline

A series of remarkable pieces of experimental research during the late 60s and early 70s of last century brought to the discovery and basic understanding of the unique viroids, relatively small, highly-base paired, nonprotein-coding circular RNAs that replicate through an RNA-based rolling-circle mechanism and can autonomously infect higher plants, frequently causing disease. The following five decades of viroid research served to learn many details about viroid structure and functional motifs, including self-cleaving ribozymes, viroid replication and host factors involved, mechanisms of pathogenicity, and viroid phylogeny. However, some intriguing questions are still ahead of researchers. Do viroids infect hosts other than higher plants? How do viroids evade host immune systems? Can we derive biotechnological tools out from viroid molecules, the same way we do with viruses? Are definitively viroids noncoding RNAs or viroid-encoded polypeptides hidden in infected cells? In addition to the self-cleaving hammerheads, do viroids contain other motifs with ribozyme activity? These are some personally selected questions that, among many others, researchers will most probably answer in future viroid research.

Learning objectives

- Objective 1: Viroids are unique RNAs with remarkable properties.
- Objective 2: Many aspects of viroid biology have been understood based on up-to-date research.
- Objective 3: Some intriguing questions about viroid biology still require answers.
- Objective 4: New technologies may bring findings that challenge current viroid knowledge.

Fundamental introduction

Viroids are relatively small (approximately 250–430 nt) circular RNAs able to replicate and move systemically, that is, to infect, when manage to enter into the cells of the right host plant. Viroids were considered an oddity from the very same moment when they were discovered. The causal agent of potato spindle tuber disease, later known as potato spindle tuber viroid (PSTVd), was a single naked RNA, whose size was far smaller than the genome of the smallest known virus (Diener, 1971, 1972; Diener and Raymer, 1967). Pioneering research indicated that PSTVd RNA did not code for any protein (Davies et al., 1974), an astonishing feature later reinforced by sequence analysis (Gross et al., 1978). PSTVd RNA sequencing also revealed circularity and a high-degree of self-complementarity (Gross et al., 1978), which matched what was previously observed under the electron microscope (Sogo et al., 1973). Many different viroid species have been discovered in the last decades (Di Serio et al., 2018, 2021), all of them sharing these foundational features. However, viroids may not be considered such a curiosity anymore, since an ever increasing number of small, medium, and large, some of them even circular, noncoding RNAs are recognized nowadays as playing crucial roles in all the realms of life (Fan, 2022; Zhao et al., 2022).

The currently known viroid species are classified into two families (*Pospiviroidae* and *Avsunviroidae*) according to several demarcation criteria (Chiumenti et al., 2021; Di Serio et al., 2014). All of them replicate through a rolling-circle mechanism, although with two different, asymmetric and symmetric, pathways (Branch and Robertson, 1984). Viroid replication is mediated by host enzymes, such as polymerases, transcription factors, RNases, RNA chaperones, or ligases (Wang, 2021), and is likely associated with particular subcellular structures. Interestingly, viroids in the family *Avsunviroidae*, such as avocado sunblotch viroid (ASBVd), were shown to contain hammerhead self-cleaving motifs in the strands of both polarities (Hutchins et al., 1986), later confirmed to be involved in viroid replication (Daròs et al., 1994). Viroid infections range from devastating (Randles et al., 1988) to symptom-less or latent (Daròs, 2016), and although different mechanisms likely contribute to viroid-induced symptoms, silencing of host genes induced by viroid-derived small RNAs is being revealed as one of the most prominent (Adkar-Purushothama et al., 2018; Bao et al., 2019; Navarro et al., 2012).

All these viroid properties have been thoroughly reviewed in the different chapters of this book, but what are the future challenges in viroid research?

Do viroids infect hosts other than flowering plants?

Viroids have been only so far found infecting flowering plants (clade Angiospermae). However, nothing in viroid biology converts them into obligate parasites of higher plants. This is particularly true for the viroids in the family *Pospiviroidae* that replicate in the nuclei of infected cells. Viroids in the family *Avsunviroidae* replicate in the chloroplasts of infected cells, which may definitively be a stronger restriction in terms of host permittivity. Nonetheless, nonflowering plants, algae, and photosynthetic protists also contain chloroplasts. A chloroplast-like environment appropriate for viroid replication may also occur in different lineages of photosynthetic bacteria, which in fact are the chloroplast antecessors. Efficient transcription, cleavage and circularization, although not full replication, was shown for some members of the family *Avsunviroidae* in the unicellular alga *Chlamydomonas reinhardtii* (Molina-Serrano et al., 2007).

Certainly, no viroid has been found outside the kingdom Plantae. However, the human hepatitis delta virus (HDV), although not a viroid, may be considered a viroid-like RNA (Flores et al., 2016). HDV is a defective virus that depends on a helper virus for encapsidation, and whose genome consists of a highly base-paired circular RNA molecule that contains particular self-cleaving motifs, not of the hammerhead type, in the strands of both polarities (Taylor, 2020). HDV also replicates through an RNA-to-RNA rolling-circle mechanism in the nucleus of infected cells, and transcription of HDV strands is mediated by the host

RNA polymerase II. However, HDV genome contains a coding gene, the delta antigen, in the antigenomic strand that produces two different versions of a protein. Astonishingly, HDV is not so unique as previously considered, since many HDV-like viruses, the so-called deltaviruses, have been found in vertebrate and invertebrate animals (De La Peña et al., 2021; Pérez-Vargas et al., 2021). HDV and deltaviruses are not the only case of viroid-like RNAs (Lee et al., 2022a), since different RNA satellites of plant viruses share many properties with viroids, such as being constituted of a noncoding circular RNA, or an RNA that undergoes through a circular phase at same point in the infectious cycle, RNA-to-RNA rolling-circle replication, and the presence of self-cleaving motifs (i.e., hammerhead and hairpin) in one or the two polarity strands (Palukaitis, 2016). Again, plant viroid-like RNA satellites are not viroids in restricted sense, because they require a helper virus to complete their infectious cycles. Fascinatingly, one of these viroid-like RNA satellites, that associated to rice yellow mottle virus, in striking similarity to HDV, encodes a 16-kDa protein (AbouHaidar et al., 2014). In addition, a cherry small circular viroid-like RNA that contains hammerhead self-cleaving motifs in the strands of both polarities has been suggested to be an RNA satellite, not of a plant virus, but of one of the mycoviruses that induce the leaf scorch disease in cherries (Minoia et al., 2014). A retroviriod-like element found in carnations contributed to enlarge the range of viroid-like RNAs. The carnation small viroid-like RNA contains hammerhead self-cleaving motifs in the strands of both polarities and was found associated to tandem-repeats of a DNA counterpart fused to DNA sequences of the pararetrovirus carnation etched ring virus (Daròs and Flores, 1995). A novel family of nonautonomous retrotransposons called retrozymes also produce noncoding circular RNAs in plants and animals, possibly consisting of transposition intermediates (de la Peña et al., 2020). Genomes of many metazoan have been shown to encode small circular RNAs with self-cleaving motifs that are actively transcribed (Cervera and De la Peña, 2020).

All these observations indicate a continuity among viroids and other viroid-like RNAs, not only in plants, but in all other realms of life, and suggest that we may be a step ahead of discovering a bona fide viroid species outside the clade Angiospermae. A recent preprint in bioRxiv announces a vast world of viroid-like circular RNAs not limited to plants (Lee et al., 2022b). Obviously, autonomous replication will need to be demonstrated before these viroid-like RNAs are accepted as bona fide viroids. In fact, a recent publication shows viroid-like RNAs, potential *mycoviroids*, naturally infecting a filamentous fungus (Dong et al., 2022). Finally, a preprint in bioRxiv builds on the previously unnoticed properties of fungus-infecting ambiviruses, which further connect viroid-like RNAs and deltaviruses (Forgia et al., 2022). If all these results are finally confirmed, we may be about to cross some classic lines regarding viroid hosts and virus-viroid evolutionary relationship.

How do viroid RNAs survive in the hostile environment of an infected cell?

When viroids were discovered, knowledge about plant defense mechanisms against pathogens was very limited. Back in those days, the simply mention of a plant immune system would have been read with perplexity. However, nowadays, it has been well established that plants display a large number of cell-surface and intracellular immune receptors that perceive pathogens and trigger a variety of defensive mechanisms (Han and Tsuda, 2022; Zhou and

Zhang, 2020). Chief among them to perceive viruses and viroids, but also other types of pathogens, is RNA silencing (Carr et al., 2019; Lopez-Gomollon and Baulcombe, 2022). Virus and viroid double-stranded RNAs (dsRNAs), also highly structured fragments of single-stranded RNAs, are recognized by plant Dicer-like (DCL) RNases to produce small interfering RNAs (siRNAs) that complex with Argonaute (AGO) proteins to form an RNA-induced silencing complex (RISC) and target the pathogens. In plants, RNA-dependent RNA polymerases (RDRs) contribute to production dsRNAs from the invading pathogen to feed the cycle. The defensive effectiveness of RNA silencing is not questionable, because all plant viruses, RNA and DNA, require to dedicate an important portion of their reduced genomes to encode one or more viral suppressors of RNA silencing that dismantle the host defense (Csorba et al., 2015). So, how viroids survive RNA silencing? Particularly when we know that viroid-derived siRNAs are ubiquitous in viroid infections (Itaya et al., 2001; Martínez de Alba et al., 2002). Also, when experimental evidence suggests that viroid RNAs accumulate in the infected cells as naked RNAs rather than tightly associated with protecting host proteins (López-Carrasco and Flores, 2017). The answer may be in the very particular viroid structure, highly base-paired, but with relatively short dsRNA stretches interrupted by loops and bulges (Giguère et al., 2014a,b). Compartmentalization in specific locations of nuclei (*Pospiviroidae*) and chloroplasts (*Avsunviroidae*) may also contribute to protection; RNA silencing has not been described yet in chloroplasts. However, viroids still need to cross the cytosol for cell-to-cell and long distance movement. Research is definitively required to understand how viroids evade plant immune systems, a knowledge that may be most useful to design durable resistance strategies to viroid infections.

Do viroids contain auto-catalytic motifs other than hammerhead?

Soon after RNA catalytic activity was discovered in some self-splicing introns and ribonuclease P (Cech et al., 1981; Guerrier-Takada et al., 1983), it was also found in plant virus satellite RNAs (Prody et al., 1986), particularly in viroid-like satellite RNAs (Forster and Symons, 1987), and ASBVd (Hutchins et al., 1986). Later, the presence of hammerhead self-cleaving motifs in the strands of both polarities of the viroids belonging to the *Avsunviroidae* has become a landmark of the family (Di Serio et al., 2018; Flores et al., 2001). Efficient auto-catalytic self-ligation activity was shown in peach latent mosaic viroid (PLMVd, family *Avsunviroidae*) RNAs both in vitro and in vivo (Côté et al., 2001; Côté and Perreault, 1997; Lafontaine et al., 1995), although producing a $2',5'$-phosphodiester linkage, whose physiological relevance is still uncertain. Since viroids are compact noncoding RNAs, the concept of displaying auto-catalytic activity fits appropriately. In fact, the question is whether only the viroids in the family *Avsunviroidae* contain hammerhead ribozyme to mediate self-cleavage. In other words, one could be expecting auto-catalytic activities to be more spread in viroid RNAs. Viroid RNA motifs and domains must interact with many host factors and structures to mediate the different steps of infection cycle. Past reports suggested that transcripts of the coconut cadang-cadang viroid (CCCVd; *Pospiviroidae*) may also self-cleave (Liu and Symons, 1998) and that the molecules of hop stunt viroid (HSVd; *Pospiviroidae*) variants contained a conserved hammerhead-like domain (Amari et al., 2001). However, none of these reports was followed with a clear demonstration of auto-catalytic activity outside the hammerhead domain of the *Avsunviroidae*. Despite these discouraging results, we cannot discard that novel viroid ribozymes with unexpected roles may remain unnoticed in viroid RNA molecules.

May viroids still encode proteins?

Viroids are defined as noncoding RNAs. Early analyses using several in vitro translation systems failed to produce substantial amounts of proteins (Davies et al., 1974). Viroid RNAs did not interfere either with in vitro translatability of bona fide mRNAs (Hall et al., 1974). Pioneering analysis of PSTVd full sequence also precluded speculation about potential protein-coding activity (Gross et al., 1978). Nowadays, with the sequence of more than 40 viroid species at hand, and the availability of many sequence variants for most of them (Lee et al., 2022a; Rocheleau and Pelchat, 2006), no clear pattern for protein translation has been reported. Relatively long open reading frames are missing or conservation is not supported in different species and sequence variants. Elements that may mediate protein translation, such as internal ribosome entry sites are also missing. No viroid-encoded proteins have ever been identified in infected tissues. Nonetheless, the absence of evidence does not mean evidence of absence, and small viroid-encoded peptides could still be produced in infected cells. High sensitivity mass spectrometry analyses may help in this search. However, one would expect that these viroid-encoded peptides play a relevant role in the infectious cycle. Spurious translation of viroid RNA fragments, particularly in the context of diseased tissue, may not be either discarded, although with reduced physiological relevance.

A recent published point of view hypothesizes, based on some computational and experimental observations, that viroids can act, under certain cellular conditions, as noncanonical translatable transcripts (Marquez-Molins et al., 2021). Although another recent research re-visited the question of viroid translation in the case of pospiviroids and concluded that, despite PSTVd association to ribosomes in infected tissue, no viroid-derived peptides could be detected experimentally (Katsarou et al., 2022). Whether viroids remain as prototypical noncoding RNAs or we change or mind about this foundational feature is a challenge of future research.

Are viroids useful for biotechnology?

Viroids are endowed with mighty features, small noncoding RNAs able to manipulate plant cells in such a way that they manage to replicate and traffic into specific organelles, cell-to-cell and long distance through the plant. In addition, viroids, accidentally or as a consequence of evolutionary selection, alter expression of host genes by RNA silencing mechanisms based on specific viroid derived siRNAs (Ramesh et al., 2021). The question is whether these features can be incorporated into biotechnological developments. Earlier work showed viroid usefulness for cross-protection against more severe viroid strains (Khoury et al., 1988; Niblett et al., 1978), to produce a desired phenotype, such as dwarfing, in the infected host (Tessitori et al., 2013), or to contribute to the desired fruit quality (Eiras et al., 2010). More recent work has shown that viroid RNA sequence fragments can be used to target heterologous gene products to the nucleoli (Gómez and Pallás, 2007) or chloroplasts (Gómez and Pallás, 2010), although possibly as a consequence of accidental, or not, coding activities

(Marquez-Molins et al., 2021). However, in remarkable contrast to viroids, plant viruses are being extensively used in many biotechnological applications, such as vectors to express recombinant proteins (Abrahamian et al., 2020), to induce gene silencing in the host for reverse genetic analyses (Rössner et al., 2022), to produce nanoparticles, some of them with remarkable therapeutic properties (Nkanga and Steinmetz, 2021), to reprogramming crops for agronomic performance (Torti et al., 2021), or to express CRISPR-Cas genome editing reaction components (Uranga and Daròs, 2022), just to cite some of them.

Viroid molecules are apparently more difficult to manipulate than those from their viral counterparts due to extreme condensation of many biological functions in a small RNA. Thus, insertion of relatively long heterologous sequences into viroid genomes is expected to render noninfectious constructs. Only insertion of small motifs in particular locations may be tolerated although quickly mutated or even deleted in the progeny (Martínez et al., 2009). However, two remarkable recent examples indicate that viroid application to biotechnology, although not so straight forward as that in plant viruses, is still possible. Eggplant latent viroid (ELVd), in combination with the eggplant tRNA ligase, the enzyme that circularizes the viroid RNA during replication (Nohales et al., 2012), was used to produce large amounts of recombinant RNA in *Escherichia coli* (Daròs et al., 2018), including the dsRNAs that are used in RNA interference (RNAi) applications against pests and pathogens (Ortolá et al., 2021). This same viroid has also been recently used for viroid-induced gene silencing in eggplant, by inserting RNA fragments homologous to host genes in a way that preserves viroid secondary structure (Marquez-Molins et al., 2022). These unprecedented demonstrations support that novel biotechnological developments may be achieved in the coming years, either using full viroids or specific domains and motifs present in viroid molecules.

Questions for the reader

1. Could viroid evolutionary origin be firmly stablished in the future?
2. How do viroids escape plant RNA silencing?
3. Should researchers look for additional ribozyme activities in viroid RNAs?
4. Do you think viroid molecules could still encode some functional polypeptides?
5. Can you think of further viroid biotechnological applications beyond those explained in the chapter?

Acknowledgments

This work was supported by grant PID2020-114691RB-I00 from the Spanish Ministerio de Ciencia e Innovación through the Agencia Estatal de Investigación.

References

AbouHaidar, M.G., Venkataraman, S., Golshani, A., Liu, B., Ahmad, T., 2014. Novel coding, translation, and gene expression of a replicating covalently closed circular RNA of 220 nt. Proc. Natl. Acad. Sci. USA 111, 14542–14547. https://doi.org/10.1073/pnas.1402814111.

Abrahamian, P., Hammond, R.W., Hammond, J., 2020. Plant virus-derived vectors: applications in agricultural and medical biotechnology. Annu. Rev. Virol. 7, 513–535. https://doi.org/10.1146/annurev-virology-010720-054958.

Adkar-Purushothama, C.R., Sano, T., Perreault, J.P., 2018. Viroid-derived small RNA induces early flowering in tomato plants by RNA silencing. Mol. Plant Pathol. 19, 2446–2458. https://doi.org/10.1111/mpp.12721.

Amari, K., Gomez, G., Myrta, A., Di Terlizzi, B., Pall, S.V., 2001. The molecular characterization of 16 new sequence variants of hop stunt viroid reveals the existence of invariable regions and a conserved hammerhead-like structure on the viroid molecule. J. Gen. Virol. 82, 953–962.

Bao, S., Owens, R.A., Sun, Q., Song, H., Liu, Y., Eamens, A.L., Feng, H., Tian, H., Wang, M.B., Zhang, R., 2019. Silencing of transcription factor encoding gene StTCP23 by small RNAs derived from the virulence modulating region of potato spindle tuber viroid is associated with symptom development in potato. PLoS Pathog. 15, e1008110. https://doi.org/10.1371/journal.ppat.1008110.

Branch, A.D., Robertson, H.D., 1984. A replication cycle for viroids and other small infectious RNAs. Science 223, 450–455.

Carr, J.P., Murphy, A.M., Tungadi, T., Yoon, J.Y., 2019. Plant defense signals: players and pawns in plant-virus-vector interactions. Plant Sci. 279, 87–95. https://doi.org/10.1016/j.plantsci.2018.04.011.

Cech, T.R., Zaug, A.J., Grabowski, P.J., 1981. In vitro splicing of the ribosomal RNA precursor of tetrahymena: involvement of a guanosine nucleotide in the excision of the intervening sequence. Cell 27, 487–496. https://doi.org/10.1016/0092-8674(81)90390-1.

Cervera, A., De la Peña, M., 2020. Small circRNAs with self-cleaving ribozymes are highly expressed in diverse metazoan transcriptomes. Nucleic Acids Res. 48, 5054–5064. https://doi.org/10.1093/nar/gkaa187.

Chiumenti, M., Navarro, B., Candresse, T., Flores, R., Di Serio, F., 2021. Reassessing species demarcation criteria in viroid taxonomy by pairwise identity matrices. Virus Evol. 7, veab001. https://doi.org/10.1093/ve/veab001.

Côté, F., Perreault, J.P., 1997. Peach latent mosaic viroid is locked by a 2′,5′-phosphodiester bond produced by in vitro self-ligation. J. Mol. Biol. 273, 533–543.

Côté, F., Lévesque, D., Perreault, J.P., 2001. Natural 2′,5′-phosphodiester bonds found at the ligation sites of peach latent mosaic viroid. J. Virol. 75, 19–25.

Csorba, T., Kontra, L., Burgyàn, J., 2015. Viral silencing suppressors: tools forged to fine-tune host-pathogen coexistence. Virology 479–480, 85–103. https://doi.org/10.1016/j.virol.2015.02.028.

Daròs, J.A., 2016. Eggplant latent viroid: a friendly experimental system in the family *Avsunviroidae*. Mol. Plant Pathol. 17, 1170–1177. https://doi.org/10.1111/mpp.12358.

Daròs, J.A., Flores, R., 1995. Identification of a retroviroid-like element from plants. Proc. Natl. Acad. Sci. USA 92, 6856–6860. https://doi.org/10.1073/pnas.92.15.6856.

Daròs, J.A., Marcos, J.F., Hernández, C., Flores, R., 1994. Replication of avocado sunblotch viroid: evidence for a symmetric pathway with two rolling circles and hammerhead ribozyme processing. Proc. Natl. Acad. Sci. USA 91, 12813–12817. https://doi.org/10.1073/pnas.91.26.12813.

Daròs, J.A., Aragonés, V., Cordero, T., 2018. A viroid-derived system to produce large amounts of recombinant RNA in *Escherichia coli*. Sci. Rep. 8, 1904. https://doi.org/10.1038/s41598-018-20314-3.

Davies, J.W., Kaesberg, P., Diener, T.O., 1974. Potato spindle tuber viroid XII. An investigation of viroid RNA as a messenger for protein synthesis. Virology 61, 281–286.

de la Peña, M., Ceprián, R., Cervera, A., 2020. A singular and widespread group of mobile genetic elements: RNA circles with autocatalytic ribozymes. Cell 9, 2555. https://doi.org/10.3390/cells9122555.

De La Peña, M., Ceprian, R., Casey, J.L., Cervera, A., 2021. Hepatitis delta virus-like circular RNAs from diverse metazoans encode conserved hammerhead ribozymes. Virus Evol. 7, veab016. https://doi.org/10.1093/ve/veab016.

Di Serio, F., Flores, R., Verhoeven, J.T., Li, S.F., Pallás, V., Randles, J.W., Sano, T., Vidalakis, G., Owens, R.A., 2014. Current status of viroid taxonomy. Arch. Virol. 159, 3467–3478. https://doi.org/10.1007/s00705-014-2200-6.

Di Serio, F., Li, S.F., Matousek, J., Owens, R.A., Pallás, V., Randles, J.W., Sano, T., Verhoeven, J.T.J., Vidalakis, G., Flores, R., Consortium, I.R, 2018. ICTV virus taxonomy profile: *Avsunviroidae*. J. Gen. Virol. 99, 611–612. https://doi.org/10.1099/jgv.0.001045.

Di Serio, F., Owens, R.A., Li, S.F., Matoušek, J., Pallás, V., Randles, J.W., Sano, T., Verhoeven, J.T.J., Vidalakis, G., Flores, R., 2021. ICTV virus taxonomy profile: *Pospiviroidae*. J. Gen. Virol. 102, 001543. https://doi.org/10.1099/jgv.0.001543.

Diener, T.O., 1971. Potato spindle tuber "virus" IV. A replicating, low molecular weight RNA. Virology 45, 411–428.

Diener, T.O., 1972. Potato spindle tuber viroid VIII. Correlation of infectivity with a UV-absorbing component and thermal denaturation properties of the RNA. Virology 50, 606–609.

Diener, T.O., Raymer, W.B., 1967. Potato spindle tuber virus: a plant virus with properties of a free nucleic acid. Science 158, 378–381. https://doi.org/10.1126/science.158.3799.378.

Dong, K., Xu, C., Kotta-Loizou, I., Jiang, J., Lv, R., Kong, L., Li, S., Hong, N., Wang, G., Coutts, R.H.A., Xu, W., 2022. Novel viroid-like RNAs naturally infect a filamentous fungus. Adv. Sci., e2204308. https://doi.org/10.1002/ADVS.202204308.

Eiras, M., Silva, S.R., Eduardo, S., Flores, R., Daròs, J.-A., 2010. Viroid species associated with the bark-cracking phenotype of "Tahiti" acid lime in the state of São Paulo, Brazil. Trop. Plant Pathol. 35, 303–309.

Fan, J., 2022. Literature mining of disease associated noncoding RNA in the omics era. Molecules 27, 4710. https://doi.org/10.3390/molecules27154710.

Flores, R., Hernandez, C., De la Peña, M., Vera, A., Daros, J.A., 2001. Hammerhead ribozyme structure and function in plant RNA replication. Methods Enzymol. 341, 540–552. https://doi.org/10.1016/S0076-6879(01)41175-X.

Flores, R., Owens, R.A., Taylor, J., 2016. Pathogenesis by subviral agents: viroids and hepatitis delta virus. Curr. Opin. Virol. 17, 87–94. https://doi.org/10.1016/j.coviro.2016.01.022.

Forgia, M., Navarro, B., Daghino, S., Cervera, A., Gisel, A., Perotto, S., Aghayeva, D.N., Akinyuwa, M.F., Gobbi, E., Zheludev, I.N., Edgar, R.C., Chikhi, R., Turina, M., Babaian, A., Di Serio, F., de la Peña, M., 2022. Extant hybrids of RNA viruses and viroid-like elements. bioRxiv. https://doi.org/10.1101/2022.08.21.504695.

Forster, A.C., Symons, R.H., 1987. Self-cleavage of plus and minus RNAs of a virusoid and a structural model for the active-sites. Cell 49, 211–220.

Giguère, T., Adkar-Purushothama, C.R., Bolduc, F., Perreault, J.P., 2014a. Elucidation of the structures of all members of the Avsunviroidae family. Mol. Plant Pathol. 15, 767–779.

Giguère, T., Adkar-Purushothama, C.R., Perreault, J.P., 2014b. Comprehensive secondary structure elucidation of four genera of the family Pospiviroidae. PLoS One 9, e98655. https://doi.org/10.1371/journal.pone.0098655.

Gómez, G., Pallás, V., 2007. A peptide derived from a single-modified viroid-RNA can be used as an "in vivo" nucleolar marker. J. Virol. Methods 144, 169–171. https://doi.org/10.1016/j.jviromet.2007.04.009.

Gómez, G., Pallás, V., 2010. Noncoding RNA mediated traffic of foreign mRNA into chloroplasts reveals a novel signaling mechanism in plants. PLoS One 5, e12269. https://doi.org/10.1371/journal.pone.0012269.

Gross, H.J., Domdey, H., Lossow, C., Jank, P., Raba, M., Alberty, H., Sänger, H.L., 1978. Nucleotide sequence and secondary structure of potato spindle tuber viroid. Nature 273, 203–208.

Guerrier-Takada, C., Gardiner, K., Marsh, T., Pace, N., Altman, S., 1983. The RNA moiety of ribonuclease P is the catalytic subunit of the enzyme. Cell 35, 849–857. https://doi.org/10.1016/0092-8674(83)90117-4.

Hall, T.C., Wepprich, R.K., Davies, J.W., Weathers, L.G., Semancik, J.S., 1974. Functional distinctions between the ribonucleic acids from citrus exocortis viroid and plant viruses: cell-free translation and aminoacylation reactions. Virology 61, 486–492. https://doi.org/10.1016/0042-6822(74)90284-0.

Han, X., Tsuda, K., 2022. Evolutionary footprint of plant immunity. Curr. Opin. Plant Biol. 67, 102209. https://doi.org/10.1016/j.pbi.2022.102209.

Hutchins, C.J., Rathjen, P.D., Forster, A.C., Symons, R.H., 1986. Self-cleavage of plus and minus RNA transcripts of avocado sunblotch viroid. Nucleic Acids Res. 14, 3627–3640. https://doi.org/10.1093/nar/14.9.3627.

Itaya, A., Folimonov, A., Matsuda, Y., Nelson, R.S., Ding, B., 2001. Potato spindle tuber viroid as inducer of RNA silencing in infected tomato. Mol. Plant-Microbe Interact. 14, 1332–1334. https://doi.org/10.1094/MPMI.2001.14.11.1332.

Katsarou, K., Adkar-Purushothama, C.R., Tassios, E., Samiotaki, M., Andronis, C., Lisón, P., Nikolaou, C., Perreault, J.P., Kalantidis, K., 2022. Revisiting the non-coding nature of pospiviroids. Cell 11, 265. https://doi.org/10.3390/cells11020265.

Khoury, J., Singh, R.P., Boucher, A., Coombs, D.H., 1988. Concentration and distribution of mild and sever strains of potato spindle tuber viroid in cross-protected tomato plants. Phytopathology 78, 1331–1336. https://doi.org/10.1094/Phyto-78-1331.

Lafontaine, D., Beaudry, D., Marquis, P., Perreault, J.P., 1995. Intra- and intermolecular nonenzymatic ligations occur within transcripts derived from the peach latent mosaic viroid. Virology 212, 705–709. https://doi.org/10.1006/viro.1995.1528.

Lee, B.D., Neri, U., Oh, C.J., Simmonds, P., Koonin, E.V., 2022a. ViroidDB: a database of viroids and viroid-like circular RNAs. Nucleic Acids Res. 50, D432–D438. https://doi.org/10.1093/NAR/GKAB974.

C. Viroid pathogenesis and viroid-host interaction

Lee, B.D., Neri, U., Roux, S., Wolf, Y.I., Camargo, A.P., Krupovic, M., Consortium, R.V.D., Simmonds, P., Kyrpides, N.-C., Gophna, U., Dolja, V.V., Koonin, E., 2022b. A vast world of viroid-like circular RNAs revealed by mining metatranscriptomes. SSRN Electron. J. https://doi.org/10.2139/ssrn.4174577.

Liu, Y.H., Symons, R.H., 1998. Specific RNA self-cleavage in coconut cadang cadang viroid: potential for a role in rolling circle replication. RNA 4, 418–429.

López-Carrasco, A., Flores, R., 2017. Dissecting the secondary structure of the circular RNA of a nuclear viroid in vivo: a "naked" rod-like conformation similar but not identical to that observed in vitro. RNA Biol. 14, 1046–1054. https://doi.org/10.1080/15476286.2016.1223005.

Lopez-Gomollon, S., Baulcombe, D.C., 2022. Roles of RNA silencing in viral and non-viral plant immunity and in the crosstalk between disease resistance systems. Nat. Rev. Mol. Cell Biol. 23, 645–662. https://doi.org/10.1038/s41580-022-00496-5.

Marquez-Molins, J., Navarro, J.A., Seco, L.C., Pallas, V., Gomez, G., 2021. Might exogenous circular RNAs act as protein-coding transcripts in plants? RNA Biol. 18, 98–107. https://doi.org/10.1080/15476286.2021.1962670.

Marquez-Molins, J., Hernandez-Azurdia, A.G., Urrutia-Perez, M., Pallas, V., Gomez, G., 2022. A circular RNA vector for targeted plant gene silencing based on an asymptomatic viroid. Plant J. 112, 284–293. https://doi.org/10.1111/tpj.15929.

Martínez de Alba, A.E., Flores, R., Hernández, C., 2002. Two chloroplastic viroids induce the accumulation of small RNAs associated with posttranscriptional gene silencing. J. Virol. 76, 13094–13096.

Martínez, F., Marqués, J., Salvador, M.L., Darós, J.A., 2009. Mutational analysis of eggplant latent viroid RNA processing in *Chlamydomonas reinhardtii* chloroplast. J. Gen. Virol. 90, 3057–3065. https://doi.org/10.1099/vir.0.013425-0.

Minoia, S., Navarro, B., Covelli, L., Barone, M., Garcia-Becedas, M.T., Ragozzino, A., Alioto, D., Flores, R., Di Serio, F., 2014. Viroid-like RNAs from cherry trees affected by leaf scorch disease: further data supporting their association with mycoviral double-stranded RNAs. Arch. Virol. 159, 589–593.

Molina-Serrano, D., Suay, L., Salvador, M.L., Flores, R., Daròs, J.-A., 2007. Processing of RNAs of the family Avsunviroidae in *Chlamydomonas reinhardtii* chloroplasts. J. Virol. 81, 4363–4366. https://doi.org/10.1128/jvi.02556-06.

Navarro, B., Gisel, A., Rodio, M.E., Delgado, S., Flores, R., Di Serio, F., 2012. Small RNAs containing the pathogenic determinant of a chloroplast-replicating viroid guide the degradation of a host mRNA as predicted by RNA silencing. Plant J. 70, 991–1003. https://doi.org/10.1111/j.1365-313X.2012.04940.x.

Niblett, C.L., Dickson, E., Fernow, K.H., Horst, R.K., Zaitlin, M., 1978. Cross protection among four viroids. Virology 91, 198–203.

Nkanga, C.I., Steinmetz, N.F., 2021. The pharmacology of plant virus nanoparticles. Virology 556, 39–61. https://doi.org/10.1016/j.virol.2021.01.012.

Nohales, M.-A., Molina-Serrano, D., Flores, R., Daròs, J.-A., 2012. Involvement of the chloroplastic isoform of tRNA ligase in the replication of viroids belonging to the family Avsunviroidae. J. Virol. 86, 8269–8276. https://doi.org/10.1128/jvi.00629-12.

Ortolá, B., Cordero, T., Hu, X., Darós, J.A., 2021. Intron-assisted, viroid-based production of insecticidal circular double-stranded RNA in *Escherichia coli*. RNA Biol. 18, 1846–1857. https://doi.org/10.1080/15476286.2021.1872962.

Palukaitis, P., 2016. Satellite RNAs and satellite viruses. Mol. Plant-Microbe Interact. 29, 181–186. https://doi.org/10.1094/MPMI-10-15-0232-FI.

Pérez-Vargas, J., de Oliveira, R.P., Jacquet, S., Pontier, D., Cosset, F.L., Freitas, N., 2021. HDV-like viruses. Viruses 13, 1207. https://doi.org/10.3390/v13071207.

Prody, G.A., Bakos, J.T., Buzayan, J.M., Schneider, I.R., Bruening, G., 1986. Autolytic processing of dimeric plant-virus satellite RNA. Science 231, 1577–1580.

Ramesh, S.V., Yogindran, S., Gnanasekaran, P., Chakraborty, S., Winter, S., Pappu, H.R., 2021. Virus and viroid-derived small RNAs as modulators of host gene expression: molecular insights into pathogenesis. Front. Microbiol. 11, 614231. https://doi.org/10.3389/fmicb.2020.614231.

Randles, J.W., Rodriguez, M.J., Imperial, J.S., 1988. Cadang-cadang disease of coconut palm. Microbiol. Sci. 5, 18–22.

Rocheleau, L., Pelchat, M., 2006. The subviral RNA database: a toolbox for viroids, the hepatitis delta virus and satellite RNAs research. BMC Microbiol. 6, 24. https://doi.org/10.1186/1471-2180-6-24.

Rössner, C., Lotz, D., Becker, A., 2022. VIGS goes viral: how VIGS transforms our understanding of plant science. Annu. Rev. Plant Biol. 73, 703–728. https://doi.org/10.1146/annurev-arplant-102820-020542.

Sogo, J.M., Koller, T., Diener, T.O., 1973. Potato spindle tuber viroid. X. Visualization and size determination by electron microscopy. Virology 55, 70–80.

Taylor, J.M., 2020. Infection by hepatitis delta virus. Viruses 12, 648. https://doi.org/10.3390/v12060648.

Tessitori, M., Rizza, S., Reina, A., Causarano, G., Di Serio, F., 2013. The genetic diversity of citrus dwarfing viroid populations is mainly dependent on the infected host species. J. Gen. Virol. 94, 687–693. https://doi.org/10.1099/vir.0.048025-0.

Torti, S., Schlesier, R., Thümmler, A., Bartels, D., Römer, P., Koch, B., Werner, S., Panwar, V., Kanyuka, K., von Wirén, N., Jones, J.D.G., Hause, G., Giritch, A., Gleba, Y., 2021. Transient reprogramming of crop plants for agronomic performance. Nat. Plants 7, 159–171. https://doi.org/10.1038/S41477-021-00851-Y.

Uranga, M., Daròs, J., 2022. Tools and targets: the dual role of plant viruses in CRISPR–Cas genome editing. Plant Genome, e20220. https://doi.org/10.1002/tpg2.20220.

Wang, Y., 2021. Current view and perspectives in viroid replication. Curr. Opin. Virol. 47, 32–37. https://doi.org/10.1016/j.coviro.2020.12.004.

Zhao, X., Zhong, Y., Wang, X., Shen, J., An, W., 2022. Advances in circular RNA and its applications. Int. J. Med. Sci. 19, 975–985. https://doi.org/10.7150/ijms.71840.

Zhou, J.M., Zhang, Y., 2020. Plant immunity: danger perception and signaling. Cell 181, 978–989. https://doi.org/10.1016/j.cell.2020.04.028.

Index

Note: Page numbers followed by *f* indicate figures and *t* indicate tables.